REMOTE SENSING OF SOIL AND LAND SURFACE PROCESSES

EARTH OBSERVATION

REMOTE SENSING OF SOIL AND LAND SURFACE PROCESSES

MONITORING, MAPPING, AND MODELING

Edited by

ASSEFA M. MELESSE
Florida International University, Miami, FL, United States

OMID RAHMATI
Soil Conservation and Watershed Management Research Department, Kurdistan Agricultural and Natural Resources Research and Education Center, Agricultural Research, Education and Extension Organization (AREEO), Sanandaj, Iran

KHABAT KHOSRAVI
Florida International University, Miami, FL, United States
Ferdowsi University of Mashhad, Mashhad, Iran

Series Editor

GEORGE P. PETROPOULOS

ELSEVIER

Elsevier
Radarweg 29, PO Box 211, 1000 AE Amsterdam, Netherlands
The Boulevard, Langford Lane, Kidlington, Oxford OX5 1GB, United Kingdom
50 Hampshire Street, 5th Floor, Cambridge, MA 02139, United States

ISBN: 978-0-443-15341-9

For information on all Elsevier publications visit our website at
https://www.elsevier.com/books-and-journals

Publisher: Candice Janco
Acquisitions Editor: Peter Llewellyn
Editorial Project Manager: Rupinder Heron
Production Project Manager: Kumar Anbazhagan
Cover Designer: Christian J. Bilbow

Typeset by TNQ Technologies
Transferred to Digital Printing in 2024

Contents

7. Unraveling the changes in soil properties availed by UAV-derivative data in an arid floodplain: Lessons learned and things to fathom

Sedigheh Maleki, Aiding Kornejady, and Minoo Haghani Shirazi

8. Investigating the land use changes effects on the surface temperature using Landsat satellite data

Hadi Eskandari Damaneh, Hassan Khosravi, and Hamed Eskandari Damaneh

9. The application of remote sensing on wetlands spatio-temporal change detection

Saleh Yousefi, Sayed Naeim Emami, Aleksandar Valjerević, Yaghoub Iranmanesh, Mohammad Nekoeimehr, and Nastaran Naderi

10. Machine learning modeling of the wind-erodible fraction of soils

Bahram Choubin, Nikoo Hamzehpour, Fereshteh Alizadeh, and Amir Mosavi

11. Application of remote sensing techniques for evaluating land surface vegetation

Esmaeil Heydari Alamdarloo, Azam Abolhasani, Maliheh Behrang Manesh, and Hassan Khosravi

12. A brief review of digital soil mapping in Iran

Ruhollah Taghizadeh-Mehrjardi, Mojtaba Zeraatpisheh, Alireza Amirian-Chakan, and Thomas Scholten

13. Impacts of land use and land cover changes on soil erosion

Carla S.S. Ferreira, Milica Kašanin-Grubin, Marijana Kapović Solomun, and Zahra Kalantari

14. Road-side slope erosion using MLS and remote sensing

Saleh Yousefi, Somayeh Mirzaee, and Christopher Gomez

15. Suspended sediment load prediction and tree-based algorithms

Salim Heddam, Amir Naghibi, Khabat Khosravi, and Shailesh K. Singh

16. Soil erosion and sediment change detection using UAV technology

Somayeh Mirzaee, Christopher Gomez, Mehdi Pajouhesh, and Khodayar Abdollahi

17. Monitoring and detection of land subsidence

Shamsollah Asgari, Kourosh Shirani, and Freidoon Soleimani

18. Drought mapping, modeling, and remote sensing

Davood Mafi-Gholami and Abolfazl Jaafari

19. Predictive pedometric mapping of soil texture in small catchments: Application of the integrated computer-assisted digital maps, machine learning, and limited soil data

Sedigheh Maleki and Aiding Kornejady

20. Object-based image analysis approach for gully erosion detection

Ramyar Aliramaee, Omid Rahmati, Farnoush Mohammadi, and Seyed Masoud Soleimanpour

21. Landslide detection and monitoring using remote sensing approach

Freidoon Soleimani, Kourosh Shirani, Shamsollah Asgari, and Zeinab Hazbavi

22. Classification algorithms for remotely sensed images

Neda Mohamadzadeh and Amin Ahmadisharaf

23. Spatial analysis of sediment connectivity and its applications

Raana Javidan, Omid Rahmati, Artemi Cerdà, Francisco Escriva-Saneugenio, and Saskia D. Keesstra

24. Soil properties mapping using the Google Earth Engine platform

Raana Javidan, Omid Rahmati, Seyed Masoud Soleimanpour, and Farnoush Mohammadi

25. Supportive role of remote sensing techniques for landslide susceptibility modeling

Abolfazl Jaafari and Davood Mafi-Gholami

26. An overview of remotely sensed fuel variables for the prediction of wildfires

Xingwen Quan, Binbin He, Abolfazl Jaafari, Zhanmang Liao, Xiangzhuo Liu, Chao Yue, and Rui Chen

27. Improving landslide susceptibility mapping using integration of ResU-Net technique and optimized machine learning algorithms

Fatemeh Rezaie, Mahdi Panahi, Sayed M. Bateni, Zahra Kalantari, Omid Rahmati, Saro Lee, and Arip Syaripudin Nur

Contributors

Khodayar Abdollahi Faculty of Earth Science, Shahrekord University, Shahrekord, Iran

Azam Abolhasani Department of Reclamation of Arid and Mountainous Regions, Faculty of Natural Resource, University of Tehran, Tehran, Iran

Amin Ahmadisharaf Kansas State University, Manhattan, KS, United States

Ramyar Aliramaee Department of Watershed Management Engineering, Faculty of Natural Resources, Tarbiat Modares University, Noor, Iran

Fereshteh Alizadeh Department of Soil Sciences, Faculty of Agriculture, University of Maragheh, Maragheh, Iran

Alireza Amirian-Chakan Faculty of Agriculture and Natural Resources, Lorestan University, Khorramabad, Iran

Shamsollah Asgari Soil Conservation and Watershed Management Research Institute, Ilam Agricultural and Natural Resources Research and Education Center, Agricultural Research, Education and Extension Organization (AREEO), Ilam, Iran

Sayed M. Bateni Department of Civil and Environmental Engineering and Water Resources Research Center, University of Hawaii at Manoa, Honolulu, HI, United States

Artemi Cerdà Soil Erosion and Degradation Research Group, Departament de Geografia, Universidad de Valencia - Universitat de València, Valencia, Spain

Rui Chen School of Resources and Environment, University of Electronic Science and Technology of China, Chengdu, China

Bahram Choubin Soil Conservation and Watershed Management Research Institute, West Azarbaijan Agricultural and Natural Resources Research and Education Center, AREEO, Urmia, Iran

Sayed Naeim Emami Soil Conservation and Watershed Management Research Institute, Chaharmahal and Bakhtiari Agricultural and Natural Resources Research and Education Center, AREEO, Shahrekord, Iran

Francisco Escriva-Saneugenio Soil Erosion and Degradation Research Group, Departament de Geografia, Universidad de Valencia - Universitat de València, Valencia, Spain

Hadi Eskandari Damaneh Department of Reclamation of Arid and Mountainous Regions, Faculty of Natural Resource, University of Tehran, Tehran, Iran

Hamed Eskandari Damaneh Department of Reclamation of Arid and Mountainous Regions, Faculty of Natural Resource, University of Tehran, Tehran, Iran

Carla S.S. Ferreira Department of Physical Geography and Bolin Centre for Climate Research, Stockholm University, Stockholm, Sweden; Polytechnic Institute of Coimbra, Applied Research Institute, Coimbra, Portugal; Research Centre for Natural Resources Environment and Society (CERNAS), Polytechnic Institute of Coimbra, Coimbra, Portugal

Eslam Galehban Department of Remote Sensing & Geographical Information System, Faculty of Geography, University of Tehran, Tehran, Iran

Christopher Gomez Graduate School of Oceanology, Laboratory of Sediment Hazards and Disaster Risk, Kobe University, Kobe, Japan; LofHazs Laboratory (Sabo), Graduate School of Oceaonology, Kobe University, Kobe, Japan

Nikoo Hamzehpour Department of Soil Sciences, Faculty of Agriculture, University of Maragheh, Maragheh, Iran

Zeinab Hazbavi Rangeland and Watershed Management Department, Faculty of Agriculture and Natural Resources, University of Mohaghegh Ardabili, Ardabil, Iran

Binbin He School of Resources and Environment, University of Electronic Science and Technology of China, Chengdu, China

Salim Heddam Faculty of Science, Agronomy Department, University 20 Août 1955, Skikda, Algeria

Esmaeil Heydari Alamdarloo Department of Reclamation of Arid and Mountainous Regions, Faculty of Natural Resource, University of Tehran, Tehran, Iran

Yaghoub Iranmanesh Research Division of Natural Resources, Chaharmahal and Bakhtiari Agricultural and Natural Resources Research and Education Center, AREEO, Shahrekord, Iran

Abolfazl Jaafari Research Institute of Forests and Rangelands, Agricultural Research, Education and Extension Organization (AREEO), Tehran, Iran

Raana Javidan University of Environment, Karaj, Iran

Zahra Kalantari Department of Physical Geography and Bolin Centre for Climate Research, Stockholm University, Stockholm, Sweden; Department of Sustainable Development, Environmental Science and Engineering (SEED), KTH Royal Institute of Technology, Stockholm, Sweden

Marijana Kapović Solomun Department of Forest Ecology, Faculty of Forestry, University of Banja Luka, Banja Luka, Bosnia and Herzegovina

Milica Kašanin-Grubin Institute of Chemistry, Technology, and Metallurgy, University of Belgrade, Belgrade, Serbia

Saskia D. Keesstra Climate-Kic Holding B.V., Amsterdam, the Netherlands; University of Granada, Granada, Spain

Hassan Khosravi Department of Reclamation of Arid and Mountainous Regions, Faculty of Natural Resource, University of Tehran, Tehran, Iran

Khabat Khosravi Department of Earth and Environment, Florida International University, Miami, FL, United States

Aiding Kornejady Research Department of Natural Resources, Golestan Agricultural and Natural Resources Research and Education Center, Agricultural Research, Education and Extension Organization (AREEO), Gorgan, Iran

Saro Lee Geoscience Data Center, Korea Institute of Geoscience and Mineral Resources (KIGAM), Daejeon, Republic of Korea; Department of Geophysical Exploration, University of Science and Technology, Daejeon, Republic of Korea

Zhanmang Liao School of Resources and Environment, University of Electronic Science and Technology of China, Chengdu, China

Xiangzhuo Liu INRAE, UMR1391 ISPA, Université de Bordeaux, Villenave d'Ornon, France

Davood Mafi-Gholami Department of Forest Sciences, Faculty of Natural Resources and Earth Sciences, Shahrekord University, Shahrekord, Iran

Sedigheh Maleki Department of Plant Production, Faculty of Agriculture, University of Torbat Heydarieh, Torbat Heydarieh, Iran; Saffron Institute, University of Torbat Heydarieh, Torbat Heydarieh, Iran

Maliheh Behrang Manesh Department of Reclamation of Arid and Mountainous Regions, Faculty of Natural Resource, University of Tehran, Tehran, Iran

Assefa Melesse Department of Earth and Environment, Florida International University, Miami, FL, United States

Somayeh Mirzaee Faculty of Earth Science, Shahrekord University, Shahrekord, Iran; Department of Natural Resources, Shahrekord University, Shahrekord, Iran

Neda Mohamadzadeh Department of Geography and Geospatial Sciences, Kansas State University, Manhattan, KS, United States

Farnoush Mohammadi Department of Reclamation of Arid and Mountainous Regions, Faculty of Natural Resources, University of Tehran, Karaj, Iran

Amir Mosavi Institute of Automation, Kalman Kando Faculty of Electrical Engineering, Obuda University, Budapest, Hungary; Institute of Advanced Studies Koszeg (IASK), Koszeg, Hungary

Nastaran Naderi Islamic Azad University of Tonekabon, Mazandaran, Iran

Amir Naghibi Department of Water Resources Engineering & Center for Advanced Middle Eastern Studies, Lund University, Lund, Sweden

Mohammad Nekoeimehr Soil Conservation and Watershed Management Research Institute, Chaharmahal and Bakhtiari Agricultural and Natural Resources Research and Education Center, AREEO, Shahrekord, Iran

Mehdi Pajouhesh Faculty of Earth Science, Shahrekord University, Shahrekord, Iran

Mahdi Panahi Division of Science Education, Kangwon National University, Chuncheon-si, Gangwon-do, Republic of Korea

Mehrdad Pasandi Department of Geology, University of Isfahan, Isfahan, Iran

Xingwen Quan School of Resources and Environment, University of Electronic Science and Technology of China, Chengdu, China; Yangtze Delta Region Institute (Huzhou), University of Electronic Science and Technology of China, Huzhou, China

Omid Rahmati Soil Conservation and Watershed Management Research Institute, Agricultural Research, Education and Extension Organization (AREEO), Kurdistan Agricultural and Natural Resources Research and Education Center, Sanandaj, Iran

Fatemeh Rezaie Geoscience Data Center, Korea Institute of Geoscience and Mineral Resources (KIGAM), Daejeon, Republic of Korea; Department of Geophysical Exploration, University of Science and Technology, Daejeon, Republic of Korea

Thomas Scholten Department of Geosciences, University of Tübingen, Tübingen, Germany

Kourosh Shirani Soil Conservation and Watershed Management Research Institute, Agricultural Research, Education and Extension Organization (AREEO), Tehran, Iran

Minoo Haghani Shirazi Spatial Sciences Innovators Consulting Engineering Company, Tehran, Iran

Shailesh K. Singh National Institute of Water and Atmospheric Research, Christchurch, New Zealand

Freidoon Soleimani Soil Conservation and Watershed Management Research Institute, Ilam Agricultural and Natural Resources Research and Education Center, Agricultural Research, Education and Extension Organization (AREEO), Ilam, Iran

Seyed Masoud Soleimanpour Soil Conservation and Watershed Management Research Institute, Fars Agricultural and Natural Resources Research and Education Center, Agricultural Research, Education and Extension Organization (AREEO), Shiraz, Iran

Arip Syaripudin Nur Division of Science Education, Kangwon National University, Chuncheon-si, Gangwon-do, Republic of Korea

Ruhollah Taghizadeh-Mehrjardi Department of Geosciences, University of Tübingen, Tübingen, Germany

Aleksandar Valjerević Department of Geospatial and Environmental Science, Faculty of Geography, University of Belgrade, Belgrade, Serbia

Shadman Veysi Soil and Water Research Institute (SWRI), Agricultural Research, Education and Extension Organization (AREEO), Karaj, Iran

Saleh Yousefi Soil Conservation and Watershed Management Research Institute, Chaharmahal and Bakhtiari Agricultural and Natural Resources Research and Education Center, AREEO, Shahrekord, Iran

Chao Yue State Key Laboratory of Soil Erosion and Dryland Farming on the Loess Plateau, Northwest A&F University, Yangling, China

Mojtaba Zeraatpisheh Rubenstein School of Environment and Natural Resources, University of Vermont, Burlington, VT, United States

Preface

The intention of the editors and authors of this book is to provide up-to-date knowledge on the applications of remote sensing and spatial analysis in soil-related and land surface processes. Soils and sediments are natural resources that are the result of geological and hydro-geomorphological processes over a long time. Sediments are composed of soils of varying textures, minerals, and nutrient contents. Sediments are transported mainly by water and deposited in low-lying areas or water bodies, including lakes, reservoirs, rivers, streams, wetlands, and others. Soil and sediment characteristics vary over space and time. Watershed land cover, geomorphology, topography, management, geology, hydrology, and other factors determine the development of soils and their characteristics. Understanding soil properties and sediment transport and deposition from field observations and sampling is critical for monitoring water quality and planning watershed management measures. Land surface processes including landslides, land subsidence, debris-flow, land degradation, and gully erosion influence and their interactions with vegetation dynamic are crucial in shaping the spatiotemporal characteristics of soils and sediments in a watershed system.

The role of remote sensing in mapping natural resources such as land cover, forest resources, water bodies, and other landscape features is well documented. Mapping soils and sediments from remote sensing for larger area applications is a relatively new area of focus. Land degradation and deforestation have led to the loss of soil fertility and exposed surfaces for soil erosion and sediment transport and deposition in critical sensitive areas of the landscape.

New satellites and sensors have not proven to be useful for improving the spatiotemporal mapping of landscape features, including soils and sediments. The use of artificial intelligence (AI)—based data analysis methods has also been shown to be effective in the spatial characterization of soils and sediments in areas where field data are scarce. In this book, Remote Sensing of Soil and Land Surface Processes, 27 chapters are included focusing on various areas of research. Topics covered in this book include groundwater-induced land subsidence zonation mapping, prediction of soil organic carbon, modeling of soil moisture, land surface temperature mapping, deriving soil properties from UAV-derived data, mapping wetland spatiotemporal change detection, modeling of wind-erodible fractions of soils, estimation of suspended sediment load, drought mapping, object-based image analysis (OBIA) for gully erosion identification, identifying sediment sources, modeling of soil erosion and sediment, landslide susceptibility modeling, prediction of wildfires, and more.

This book is a very valuable reference and guide for researchers, students, policy makers, and water resource managers.

Editors

CHAPTER

1

Introduction to soil and sediment

Assefa Melesse[1], *Omid Rahmati*[2] *and Khabat Khosravi*[1]

[1]Department of Earth and Environment, Florida International University, Miami, FL, United States; [2]Soil Conservation and Watershed Management Research Institute, Agricultural Research, Education and Extension Organization (AREEO), Kurdistan Agricultural and Natural Resources Research and Education Center, Sanandaj, Iran

Exchanges in energy, mass, and momentum between the atmosphere and land are governed by several important factors, such as turbulent and molecular motions in the atmosphere, surface soil hydrological processes, and the behavior of the continental biosphere. Hence, the land surface is the layer including the lower part of the atmospheric boundary layer, the top few meters of the soil, the continental biosphere, and surface water bodies (Niyogi, 2019). Soil top-surface processes have a significant impact on land surface processes and can be considered one of the main features of Earth surface processes and the development of different landforms.

Soils implement crucial functions to sustain human, animal, and plant life, control and adjust water flow, filter and buffer pollutants, and cycle nutrients. It has been reported that more than 90% of human food, livestock feed, and fuel are provided by soil, and more than 2.6 billion people depend on agricultural activities (Hurni, 1998).

Soil erosion, especially soil water erosion, is a significant global challenge and threat, leading to the degradation of agricultural lands, reduced soil fertility, reduced food production, and increased sedimentation in rivers and reservoirs and leading to a serious socioeconomic issue (Pimentel et al., 1995), while wind erosion leads to several environmental problems and human health issues, such as asthma and other health problems. According to Food and Agriculture Organization (FAO)-Global Soil Partnership was (GSP) reports at a global-scale, about 75 billion tons of soil are removed from lands and loose areas per year, and this leads to an around US$400 billion financial loss (GSP, 2017). Several criteria, such as developing cost-effective soil conservation strategies and understanding the spatial and temporal dynamics of soil erosion and land surface changes, are essential for managing soil erosion and mitigating its impact (Evans, 1997). Soil erosion can be impacted by both anthropogenic and natural component, such as soil properties, land use and vegetation cover, geology, watershed characteristics, hydroclimatic component, and generally land/watershed management. The soil erosion rate has speeded up due to population growth, pressure on

Remote Sensing of Soil and Land Surface Processes
https://doi.org/10.1016/B978-0-443-15341-9.00028-9

natural resources, land-use, and climate changes (Scott, 2001) and is extending beyond arid and semiarid regions into humid areas (Boudjemline and Semar, 2018). Anthropogenic and human activities have a significant impact on natural erosion processes and increase the soil loss rate (Poesen, 2018). In addition to soil erosion, there are numerous kinds of soil loss and land degradation, such as gully erosion, landslides, and land subsidence, which lead to vast amounts of damage not only to the environment and infrastructure but also to human life and their properties.

Accurate assessment of soil erosion rate loss and general land degradation and sediment prediction are critical for land management, suggesting soil conservation measures and reducing land degradation and soil loss. Over the years, various techniques have been employed to monitor and predict soil erosion, ranging from field-based measurements to computer simulations. To date, different kinds of models including empirical/linear, conceptual, or physically based models have been provided at different spatiotemporal scales to predict soil erosion processes (Merritt et al., 2003; Morgan and Nearing, 2011; Nearing, 2013). The Universal Soil Loss Equation (USLE) and Revised USLE (RUSLE) (Renard and Ferreira, 1993) are among the most practical and efficient empirical models for soil erosion rate prediction. Golkarian et al. (2023) compared the prediction accuracy of RUSLE with several machine learning models (convolutional neural networks optimized with gray wolf optimized), support vector machine, group method of data handling, and finally stated that RUSLE results have a high uncertainty. Implementation of process-oriented models, such as KINematic runoff and EROSion (KINEROS; Woolhiser et al., 1990) and the Watershed Erosion Prediction Project (WEPP; Nearing et al., 1989) can be restricted due to lack of information, time-consuming calibration, and poor quality of data (Nigel and Rughooputh, 2010).

Recently, geographic information systems (GISs), remote sensing (RS), and their combination with machine learning techniques have emerged as powerful tools for monitoring land surface changes and modeling soil erosion processes and have attracted the attention of researchers worldwide (Ganasri and Ramesh, 2016). This introductory chapter provides an overview of the role of remote sensing in soil erosion and land surface modeling, highlighting its advantages and recent advancements.

Remote sensing in land degradation and soil erosion processes is the process of monitoring and acquiring information about the Earth's surface using sensors mounted on satellites, aircraft, or ground-based platforms without direct contact with the object or area being observed (Lillesand et al., 2015). In the context of soil erosion and land surface modeling, remote sensing offers several advantages over traditional field–based methods, such as investigating large-scale areas, high-resolution, and repetitive observations of the Earth's surface, enabling the analysis of soil erosion processes and land surface changes over time. Furthermore, remote sensing data can be rapidly acquired and processed, offering a cost-effective and efficient means to monitor soil erosion, particularly in large or inaccessible study areas (Ganasri and Ramesh, 2016). Additionally, the integration of remote sensing data with other datasets, such as topography, climate, and land use, allows for a more comprehensive understanding of the complex processes governing soil erosion and land surface dynamics. Panahi et al. (2022) delineated land subsidence occurrence across Iran using the Interferometric Synthetic Aperture Radar (InSAR) technique and finally generated a land subsidence susceptibility map for the entire country using a machine learning model.

Satellite image acquisition can be divided into two main groups: active (radar) and passive microwave (optical) methods. Passive sensors collect and detect reflected energy emitted from the Sun, while active sensors do not rely on the Sun's energy, and they have their own source of light. The ability to collect and measure imagery at night and day and to avoid clouds and poor weather conditions are two main advantages of active remote sensing (ARS). ARS includes RADARSAT-1 and RADARSAT-2, SAR, light detection and ranging (LiDAR), and Airbus Defense and Space TerraSAR-X Radar Satellite. Landsat is the oldest and longest running Earth observation passive sensor for more than 40 years. Another example of a satellite with passive sensors is the Moderate Resolution Imaging Spectroradiometer (MODIS), which is widely used in geoscience.

This book is organized into several chapters that delve into the various aspects of remote sensing in soil erosion and land surface modeling. The first chapter provides an in-depth introduction to remote sensing, its principles, and its applications in soil erosion studies. Subsequent chapters discuss the different types of remote sensing data and their uses in land surface modeling, the integration of remote sensing data with other datasets, and the application of machine learning techniques in erosion prediction and sediment modeling. Finally, the book concludes with a comprehensive summary and future perspectives on the role of remote sensing and machine learning in advancing our understanding of soil erosion and land surface modeling.

References

Boudjemline, F., Semar, A., 2018. Assessment and mapping of desertification sensitivity with MEDALUS model and GIS – case study: basin of Hodna, Algeria. J. Water Land Dev. 36, 17–26. https://doi.org/10.2478/jwld-2018-0002.

Evans, R., 1997. Soil erosion in the UK initiated by grazing animals. Appl. Geogr. 17, 127–141. https://doi.org/10.1016/S0143-6228(97)00002-7.

Ganasri, B.P., Ramesh, H., 2016. Assessment of soil erosion by RUSLE model using remote sensing and GIS—a case study of nethravathi basin. Geosci. Front. 7, 953–961. https://doi.org/10.1016/j.gsf.2015.10.007.

Golkarian, A., Khosravi, K., Panahi, M., Clague, J., 2023. Spatial variability of soil water erosion: comparing empirical and intelligent techniques. Geosci. Front. 14 (1), 101456.

GSP, 2017. Global Soil Partnership Endorses Guidelines on Sustainable Soil Management. http://www.fao.org/global-soil partnership/resources/highlights/detail/en/c/416516/.

Hurni, H., 1998. A multi-level stakeholder approach to sustainable land management. Adv Geoecol. 31, 827–836. https://doi.org/10.7892/boris.77579. In this issue.

Merritt, W.S., Letcher, R.A., Jakeman, A.J., 2003. A review of erosion and sediment transport models. Environ. Model. Software 18, 761–799.

Morgan, R.P.C., Nearing, M.A., 2011. Handbook of Erosion Modelling. https://doi.org/10.1002/9781444328455.

Nearing, M.A., 2013. Soil erosion and conservation. In: Environmental Modelling: Finding Simplicity in Complexity, second ed., pp. 365–378. https://doi.org/10.1002/9781118351475.ch22.

Nearing, M.A., Foster, G.R., Lane, L.J., Finkner, S.C., 1989. A process-based soil erosion model for USDA-water erosion prediction project technology. Transaction of ASAE 32, 1587–1593.

Nigel, R., Rughooputh, S.D., 2010. Soil erosion risk mapping with new datasets: an improved identification and prioritisation of high erosion risk areas. Catena 82 (3), 191–205.

Niyogi, D., 2019. Land surface processes. In: Randall, D., Srinivasan, J., Nanjundiah, R., Mukhopadhyay (Eds.), Current Trends in the Representation of Physical Processes in Weather and Climate Models. Springer Atmospheric Sciences, Singapore. https://doi.org/10.1007/978-981-13-3396-5_17.

Panahi, M., Khosravi, K., Golkarian, A., Roostaei, M., et al., 2022. A country-wide assessment of Iran's land subsidence susceptibility using satellite-based InSAR and machine learning. Geocarto Int. 1–23.

Pimentel, D.C., Harvey, C., Resosudarmo, I.A., et al., 1995. Environmental and economic cost of soil erosion and conservation benefits. Science 267 (5201), 1117–1123. https://doi.org/10.1126/science.267.5201.1117.

Poesen, J., 2018. Soil erosion in the Anthropocene: research needs. Earth Surf. Process. Landform. 84, 64–84. https://doi.org/10.1002/esp.4250.

Renard, K.G., Ferreira, V.A., 1993. RUSLE model description and database sensitivity. J. Environ. Qual. 22 (3), 458–466. In this issue.

Scott, A., 2001. Water Erosion in the Murray-Darling Basin: Learning from the Past.

Woolhiser, D.A., Smith, R.E., Goodrich, D.C., 1990. KINEROS:a Kinematic Runoff and Erosion Model: Documentation and User Manual.

Further reading

Es, H.M., 2017. A new definition of soil. CSA New. https://doi.org/10.2134/csa2017.62.1016.

Irannejad, P., Shao, Y., 2002. Land surface processes. In: Peng, G., Leslie, L.M., Shao, Y. (Eds.), Environmental Modelling and Prediction. Springer, Berlin, Heidelberg. https://doi.org/10.1007/978-3-662-04868-9_5.

Lillesand, T.M., Kiefer, R.W., Chipman, J.W., 2015. Remote Sensing and Image Interpretation, seventh ed. Wiley, New York.

CHAPTER

2

DInSAR-based assessment of groundwater-induced land subsidence zonation map

Kourosh Shirani[1] *and Mehrdad Pasandi*[2]

[1]**Soil Conservation and Watershed Management Research Institute, Agricultural Research, Education and Extension Organization (AREEO), Tehran, Iran;** [2]**Department of Geology, University of Isfahan, Isfahan, Iran**

2.1 Introduction

Land subsidence refers to the gradual sinking or settling of the Earth's surface (Poland, 1984; Galloway et al., 1999; Bawden et al., 2001). It can be caused by natural or human-induced factors and can have significant consequences for human communities and the environment (Holzer and Galloway, 2005; Mazzotti et al., 2009). Land subsidence is a significant challenge facing many communities worldwide, particularly in areas where groundwater pumping is widespread. This phenomenon can have significant consequences including infrastructure damage, flooding, and ecosystem damage. By understanding the causes and effects of subsidence and implementing effective mitigation strategies, the risks can be reduced. In the following, the causes and effects of land subsidence will be discussed, and potential mitigation strategies will be explored.

Land subsidence can have a range of effects, depending on its severity and location (Holzer and Galloway, 2005). Some of the most significant effects include:

1. Infrastructure damage: Buildings, roads, and other structures can be damaged or destroyed by subsidence (Fig. 2.1). This can lead to costly repairs and disruptions to daily life. The subsidence can cause cracks in buildings, making them unstable and unsafe for use.
2. Flooding: When land subsides, it can cause drainage systems to fail, leading to increased flooding in low-lying areas (Mazzotti et al., 2009; Faunt et al., 2016). In some cases, the subsidence can cause the land to sink below sea level, leading to flooding due to sea level rise.

Remote Sensing of Soil and Land Surface Processes
https://doi.org/10.1016/B978-0-443-15341-9.00011-3

FIGURE 2.1 Examples of land subsidence in Isfahan, Iran. (a) Damage caused by ground subsidence to Si-o-se-pol bridge in Isfahan. (b) Subsidence of a section of the Kashan-Natanz highway to Tehran. (c) Subsidence of building walls in the Isfahan province due to ground subsidence. (d) Subsidence in the north of Isfahan near Mehr housing estate of Habibabad, and (e) cracks (fissures) created in Mahyar Plain in the south of Isfahan.

3. Saltwater intrusion: In coastal areas, land subsidence can cause the land to sink below sea level, allowing saltwater to intrude into freshwater aquifers (Ferguson and Gleeson, 2012). This can have significant impacts on agriculture, as crops cannot grow in salty soil.
4. Ecosystem damage: Land subsidence can also have significant impacts on natural ecosystems, including wetlands, forests, and grasslands. The loss of natural habitats due to subsidence can lead to a decline in biodiversity and ecosystem services.

2.1.1 Natural causes of land subsidence

Natural causes of land subsidence include the gradual settling of soil and rock, which can be caused by tectonic activity, erosion, or changes in the Earth's crust. These factors can cause the ground to slowly settle or sink, which can result in a range of effects depending on the severity and location of the subsidence. The gradual subsidence of the ground due to the movement of tectonic plates is a natural cause of land subsidence. Tectonic activity can cause subsidence when the Earth's crust is stretched or compressed, leading to the formation of faults or fractures that can cause the ground to sink (Bawden et al., 2001). Similarly, soil erosion caused by wind or water can lead to the subsidence of land over time. Erosion can cause subsidence when soil is gradually worn away, leaving behind voids that can cause the land above to sink. In addition, changes in the Earth's crust due to processes like volcanic activity or glacier movement can also result in land subsidence. The melting of glaciers or the movement of underground water can cause subsidence over time. Additionally, earthquakes and tectonic movements can cause land to sink or rise, leading to subsidence or uplift (Lyons and Sandwell, 2003).

2.1.2 Human-induced causes of land subsidence

Human activities can also cause land subsidence, and these causes are more varied than natural causes. One of the most significant causes of land subsidence is the excessive pumping of groundwater for human use (Galloway et al., 1999; Konikow and Kendy, 2005). This is especially prevalent in areas where the demand for water exceeds the supply, such as in arid or semi-arid regions. As groundwater is pumped out of an aquifer, the water pressure in the soil decreases, causing the soil to compress and the land surface to sink. When groundwater is extracted from an aquifer, the weight of the overlying soil and rock is no longer supported, and the land surface can sink. The rate and severity of land subsidence depend on the rate of pumping and the characteristics of the soil and aquifer.

Fine-grained layers, such as clay, silt, and other low-permeability soils, can play an important role in land subsidence.

When groundwater is pumped out of an aquifer, the water pressure in the soil decreases. This causes the fine-grained layers to compact and compress, reducing the volume of the soil and causing the land above to sink. Fine-grained layers are particularly susceptible to this process, as they have low permeability and are unable to replenish the water that is being extracted (Fig. 2.2). In addition, fine-grained layers can also act as a barrier to the flow of water within an aquifer. This can lead to the formation of localized zones of high water pressure, which can further accelerate the compaction and settling of the surrounding soil. The effects

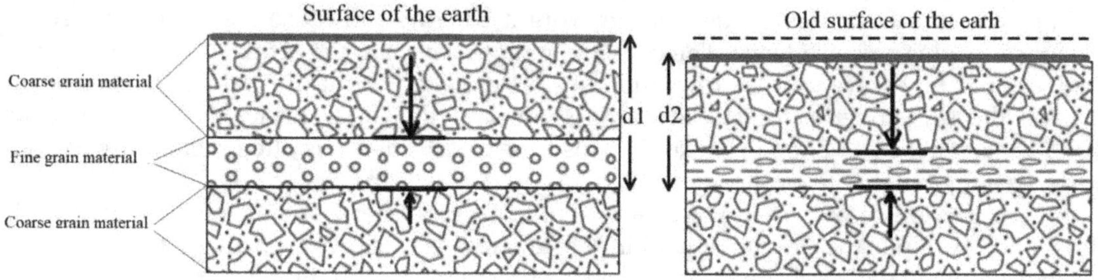

FIGURE 2.2 Reduction in volume of the fine-grained layer due to drainage.

of fine-grained layers on land subsidence can vary depending on the characteristics of the soil and the hydrogeological conditions. For example, in areas with thick layers of clay or silt, the subsidence may be more severe and occur over a larger area. In contrast, in areas with more permeable soils, the subsidence may be less pronounced.

The subsurface is composed of different layers of soil and rock, each with varying properties and characteristics. The thickness of these layers can have a significant impact on the occurrence of land subsidence. The specific relationship between sediment thickness and subsidence depends on several factors, including sediment compressibility, layering, buried substrate geometry, and groundwater levels. Sediments are deposited over time and accumulate, increasing the total thickness of overlying layers. As more sediments deposit on top, the underlying layers experience an increase in pressure and compress, resulting in a loss of layer thickness and lowering of the ground surface. This process involves the expulsion of water from the pores of the sediment, leading to a reduction in volume and, consequently, land subsidence. More compressible sediments with higher porosity, such as unconsolidated silts and clays, are more prone to compaction and subsidence. Interbedded permeable and impermeable layers can also enhance subsidence. The presence of voids and cracks in the subsurface layers can contribute to land subsidence. The shape and depth of the underlying bedrock interface impact how overlying sediment load is distributed. Higher groundwater levels lead to increased pore pressure, accelerating compaction (Poland, 1975). Quantifying the exact correlation between sediment thickness and subsidence is complex in natural systems. The link between thickness and subsidence may be nonlinear due to feedbacks between sediment loading, pore pressure, and ground failure. However, in general, greater sediment thicknesses are associated with higher long-term subsidence rates, especially where sediments are poorly consolidated and groundwater levels are high.

Mining activities, particularly underground mining, can also cause land subsidence. The extraction of minerals or fossil fuels from underground can also cause land subsidence (Holzer and Galloway, 2005; Zhao et al., 2021). When the voids left by mining are not filled or stabilized, the overlying land can collapse, leading to subsidence. The extraction of oil and gas from underground can cause subsidence in a similar way to mining. The removal of oil and gas from reservoirs can cause the overlying rock and soil to compress, leading to subsidence, particularly when it involves the removal of large amounts of material from below the surface.

Land subsidence can significantly alter the geo-hydrological conditions of a region, including changes to the direction and velocity of underground water flow, as well as the overall groundwater balance (Fig. 2.3).

The weight of buildings, roads, and other infrastructure can cause the underlying soil to compact, leading to subsidence. The construction of heavy infrastructure, such as buildings, bridges, and highways, can lead to land subsidence. The weight of the infrastructure compresses the soil underneath, causing it to settle and the land surface to sink (Fellenius, 2004; Feng et al., 2008; Zhou et al., 2020). This type of subsidence is often seen in urban areas with dense populations and heavy traffic.

A combination of human-induced and natural factors can also contribute to land subsidence. For example, the sinking of land in coastal areas is caused by a combination of natural geological processes, such as erosion and sedimentation, and human activities, such as groundwater pumping and oil and gas extraction (Dokka, 2011).

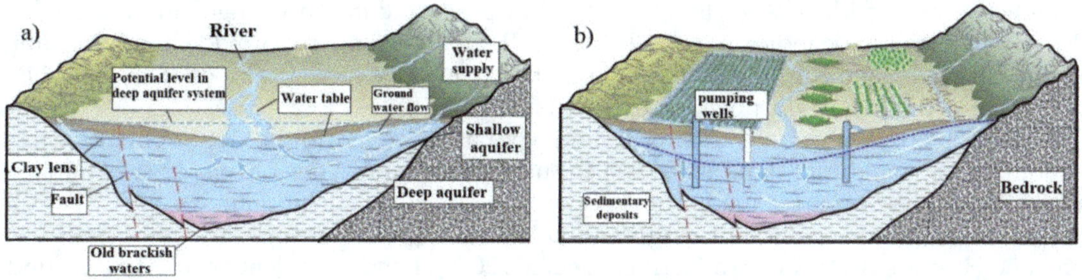

FIGURE 2.3 Impact of groundwater pumping on water flow direction in a valley (A) shows the natural direction of water flow from the mountains toward the center of the valley before groundwater pumping. (B) The photo illustrates how groundwater pumping causes the water table to lower, resulting in a change of water flow direction from the mountains toward the pumping centers.

2.1.3 Mitigation strategies

Effective mitigation strategies for land subsidence require a comprehensive understanding of the causes and effects of the phenomenon (Nyer, 1998). One of the most effective ways to mitigate land subsidence is to manage groundwater resources carefully (Cui et al., 2018; Scanlon et al., 2023). This can include reducing groundwater pumping rates, increasing recharge rates, and promoting water conservation.

Careful land-use planning can help reduce the impacts of subsidence. This can include avoiding the development of high-risk areas. In some cases, engineering solutions can be used to mitigate the effects of subsidence. For example, buildings and other structures can be designed to be more resistant to subsidence, and drainage systems can be upgraded to prevent flooding.

Regular monitoring of subsidence can help identify potential risks early, allowing for timely action to be taken to mitigate the impacts (Bawden et al., 2001).

2.1.4 Application of remote sensing in monitoring of land subsidence

Remote sensing is a technique used to gather information about the Earth's surface without direct physical contact (Merchant, 2000; Lillesand et al., 2015). Remote sensing can be particularly useful in areas where ground-based instruments cannot be deployed, such as in remote regions or over large urban areas. Remote sensing allows for the collection of large amounts of data over a wide area in a noninvasive and cost-effective manner. It involves using sensors on board satellites, airplanes, or drones to collect data that can be used to monitor changes on the Earth's surface, including land subsidence (Liu et al., 2023). Remote sensing techniques have emerged as a powerful tool for monitoring land subsidence, allowing researchers and decision-makers to detect and analyze changes in the Earth's surface over time. Remote sensing can provide valuable information for land-use planning and management, as well as for addressing environmental and societal challenges related to land subsidence. One of the main benefits of remote sensing for monitoring land subsidence is its ability to provide data over large areas and at regular intervals. This allows researchers to identify patterns and trends in subsidence, as well as to identify areas that are at high risk of subsidence. Satellite imagery can provide data on changes in land surface elevation, which can be indicative of subsidence. Remote sensing can provide highly accurate measurements of land subsidence and deformation over large areas (Kampes et al., 2003; Deilami and Hashim, 2011; Karimzadeh and Matsuoka, 2020).

Recent advances in remote sensing technology have also led to the development of new techniques for monitoring land subsidence. There are several types of remote sensing techniques that can be used for the monitoring of land subsidence including synthetic aperture radar (SAR), light detection and ranging (LiDAR), and optical sensors (Hanssen, 2001; Deilami and Hashim, 2011; Navarro-Hernández et al., 2020). Land subsidence caused by groundwater extraction can be monitored using remote sensing techniques like InSAR and GPS (Li et al., 2005; Crosetto et al., 2016). InSAR is one example of the application of remote sensing in monitoring land subsidence (Rosen et al., 2000; Liu et al., 2023). Synthetic aperture radar (SAR) and interferometric SAR (InSAR) are particularly useful for monitoring land subsidence, as radar can penetrate through clouds and vegetation to provide high-resolution images of the Earth's surface (Jiang et al., 2021). interferometric synthetic aperture radar (InSAR) is a powerful technique to monitor land subsidence due to its ability to detect and measure very small changes in surface elevation. InSAR can be used to monitor land subsidence caused by groundwater extraction, natural compaction, and other land use practices (Hanssen, 2001). Radar sensors emit microwave radiation and can penetrate through clouds, allowing them to capture images of the Earth's surface even in adverse weather conditions. They can detect small changes in the surface elevation, making them useful for monitoring land subsidence (Massonnet and Feigl, 1998). Researchers have developed algorithms that can process large amounts of SAR data to create detailed maps of subsidence patterns over time. Light detection and ranging (LiDAR) is also useful for this purpose, as it can provide highly accurate elevation measurements of the Earth's surface especially in urban areas with high building densities. LiDAR sensors emit laser pulses and can create 3D maps of the Earth's surface with high precision, making them valuable for identifying changes in elevation over time. Optical sensors, such as those used in satellite imagery, can also provide valuable information on land subsidence. Optical sensors capture visible light and can be used to create images of the Earth's surface. They can detect changes in vegetation, water bodies, and other features that can indicate land subsidence (Deilami and Hashim, 2011; Coates et al., 2015). These techniques can help identify areas experiencing land subsidence and provide data on the magnitude and rate of subsidence. Remote sensing can also help identify the causes of subsidence, such as groundwater depletion, oil and gas extraction, and geological processes.

Over the past two decades, data of a number of widely used radar sensors has been made available to researchers around the world, including the Advanced Synthetic Aperture Radar (ASAR), Phased Array L-band Synthetic Aperture Radar (PALSAR), and Sentinel-1 sensors. These sensors have provided researchers with valuable image data, which has been freely available from the European Space Agency (ESA). Data from the ASAR and PALSAR sensors are available from 2003 to 2011, while data from the Sentinel-1 sensor are available from 2015 to the present. Earth science researchers have used these images to identify, monitor, and zonate the risk of natural hazards, including land subsidence and landslides. Table 2.1 provides specifications for these images.

TABLE 2.1 Widely used and freely available spacecraft radar systems provided by the European Space Agency. The key specifications of radar systems commonly used by researchers are listed.

Name	Duration	Frequency	Polarization	Band width (MHZ)	Off-nadeir	Mode-swath (km)
Envisat-ASAR (ESA)	2003–10	C	D(HH + VV)	16M	15–45	Strip-ScanSAR: 100–400
ALOS-PALSAR (JAXA)	2006–11	L	S, D, Q	28/14	9.7–52	Strip-ScanSAR: 70–350
SENTINEL1A & B	2014-Now	C	HH, HV,VV, VH	84/42/28/14	Jul-59	Strip-ScanSAR-Spotlight: 50–350/490–25

Nowadays, there are several software options available for analyzing and processing SAR data using various methods such as DInSAR and PSInSAR. Some of these options are noncommercial or academic with open access, while others are commercial and require a license. Table 2.2 provides a summary of the key features of some of these software tools.

The temporal and spatial resolutions of remote sensing technologies have significantly improved in recent years, allowing for more accurate and frequent monitoring of land subsidence (Bai et al., 2022). Recent developments in the application of remote sensing for monitoring land subsidence include the use of advanced algorithms and machine learning techniques to process large amounts of data and extract useful information (Schmidt and Bürgmann, 2003; Leake and Galloway, 2007). For example, researchers have developed algorithms that can analyze satellite imagery to identify areas of land subsidence and track changes over time.

TABLE 2.2 General specifications of different software tools for processing SAR data using various methods, including DInSAR and PSInSAR".

Software name	Production location	Operating system	Noncommercial usage	Commercial usage
SNAP	European Spatial Agency	Windows	*	
ESA-SAR	European Spatial Agency	Linux	*	
DORIS	Delft University	Linux	*	
Orfeo toolbox	French Open Source license	Linux	*	
ROI_PAC	NASA, USA	Linux	*	
STAMPS	Stanford University/ Delft University	Linux	*	
SARScape	Sarmap SA Switzerland	Windows/Linux		*
ERDAS	ESRI/LAICA Switzerland	Windows		*
GAMMA	Switzerland	Windows/Linux		*
PALSAR	French spatial Agency	Windows/Linux		*

One of the main challenges in using remote sensing to monitor land subsidence is distinguishing between natural and human-induced subsidence. To overcome this challenge, researchers often use a combination of remote sensing techniques and ground-based measurements, such as GPS and water-level monitoring.

Another recent development is the use of multiple remote sensing techniques in combination to obtain more accurate and detailed information (Osmanoğlu et al., 2011; Qin and Perissin, 2015; Kim and Lu, 2018). For example, researchers have combined data from radar and optical sensors to create 3D maps of land subsidence in urban areas.

In addition to monitoring land subsidence, remote sensing can also be used to identify the causes of subsidence, such as groundwater extraction or natural processes (Famiglietti et al., 2011; Shirzaei and Bürgmann, 2018). Data from remote sensing can be combined with other types of information, such as geological data and groundwater extraction records, to better understand the causes of land subsidence (Orhan, 2021).

Recent research studies have demonstrated the effectiveness of remote sensing techniques in a variety of settings, and ongoing advances in technology are likely to further improve the accuracy and effectiveness of remote sensing for monitoring land subsidence in the future.

2.2 Application of DInSAR in monitoring of land subsidence

Differential Interferometric Synthetic Aperture Radar (DInSAR) is a remote sensing technique that has become increasingly popular in the monitoring of land subsidence (Osmanoğlu et al., 2011). This technique uses radar data from satellites to measure changes in the Earth's surface with high precision. DInSAR has several advantages over traditional monitoring techniques, including its ability to measure changes over large areas and at regular intervals and its ability to precisely detect small changes in elevation. This allows researchers to identify patterns and trends in subsidence, as well as to identify areas that are at high risk of subsidence. DInSAR can be used to monitor land subsidence in a variety of settings, including urban areas, agricultural regions, and natural areas.

The main principle behind DInSAR is to measure the phase difference between two SAR images that are acquired at different times over the same area. This is accomplished by comparing pairs of radar images taken at different times and using interferometric techniques to measure the changes in the radar signal between the two images (Lyons and Sandwell, 2003). This allows researchers to measure changes in the Earth's surface with millimeter-level precision, making DInSAR a valuable tool for detecting and monitoring land subsidence.

In addition to monitoring subsidence, DInSAR can also be used to monitor other types of surface deformation, including landslides, glacier movement, and volcanic activity. This makes it a versatile tool for monitoring changes in the Earth's surface and understanding the underlying geological processes that drive those changes.

However, DInSAR also has some limitations that must be considered (Hooper, 2008; Bekaert et al., 2016). For example, it can be sensitive to atmospheric conditions, which can affect the accuracy of the measurements. Additionally, DInSAR requires a high degree of technical expertise and specialized software to process the radar data and generate accurate measurements. Despite these limitations, DInSAR has become a valuable tool for monitoring land subsidence and other types of surface deformation (Hooper and Zebker, 2007). Ongoing advances in satellite technology and processing techniques are likely to further improve the accuracy and effectiveness of DInSAR for monitoring land subsidence in the future. In recent years, several advances in satellite technology and SAR data processing have enabled the development of advanced DInSAR techniques. For example, the launch of the Sentinel-1 mission by the European Space Agency (ESA) has provided unprecedented SAR data coverage with high spatial and temporal resolution. This has enabled the development of new techniques such as Time-Series InSAR (TS-InSAR), which uses a series of interferograms acquired over a period of time to detect and monitor surface deformation with high accuracy and temporal resolution (Ferretti et al., 2000).

2.3 The DInSAR theory

The first step in the DInSAR process is to acquire two SAR images over the same area and preferably with a short temporal baseline from satellite sensors (Massonnet et al., 1993). This typically involves using sensors that operate at a specific wavelength, such as C-band or L-band, which are well-suited for measuring surface deformation (Massonnet et al., 1993). The data are acquired over a specific time period, typically several months to several years, and at regular intervals (Ferretti et al., 2001).

Once the radar data have been acquired, the next step is to process the data to generate a set of interferograms from SAR images before and after the deformation event. Interferograms are phase difference maps created by taking pairs of radar images and subtracting the phase information from one image from the other image (Massonnet et al., 1993). This process results in a new image that shows the differences in the radar signal between the two images. These differences represent the changes in the Earth's surface that have occurred between the two image acquisition times (Mora et al., 2003).

After generating the interferograms, the next step is to process the data to create a displacement map. This involves using advanced processing techniques to combine multiple interferograms and generate a more accurate displacement map (Massonnet et al., 1993).

To extract surface deformation information from the interferogram, several steps are typically required. These steps include unwrapping the interferogram to remove phase ambiguities, filtering the interferogram to remove noise, and geocoding the interferogram to convert it to a geographic coordinate system. Once these steps are completed, the interferogram can be used to derive quantitative information about surface deformation (Ferretti et al., 2001).

However, one of the challenges of DInSAR is that the interferometric phase used in DInSAR measurements is affected by various sources of errors and artifacts that need to be corrected to obtain accurate deformation estimates. The accuracy of DInSAR can also be affected by factors such as spatial resolution, temporal baseline, and signal-to-noise ratio. To achieve high accuracy, it is important to optimize the processing parameters and to choose appropriate acquisition strategies (Zebker and Villasenor, 1992). To minimize these errors, several advanced processing techniques have been developed. One of these techniques is Persistent Scatterer Interferometry (PSI), which is a multitemporal approach that exploits the coherence properties of SAR data to identify stable points on the ground surface that can be used as reference points for deformation measurement (Ferretti et al., 2001). Another technique that has been developed to overcome the limitations of DInSAR is Small Baseline Subset (SBAS) analysis. SBAS uses a subset of interferograms with a small temporal baseline to minimize the effect of atmospheric artifacts and decorrelating noise. SBAS has been used successfully in many applications, including the monitoring of ground deformation in urban areas, volcanoes, and other natural hazards (Berardino et al., 2002).

Multiple correction techniques can be used in combination to improve the accuracy of DInSAR measurements (Li et al., 2005). In the following, the different types of DInSAR corrections that are commonly used to correct these errors and artifacts will be reviewed:

1. Topographic phase correction: One of the major sources of error in DInSAR is the topographic phase, which is caused by variations in the topography of the ground surface evident between two SAR images (Mora et al., 2003; Li et al., 2005). The topographic phase refers to the component of interferometric phase that is caused by terrain elevation differences between the ground imaging positions of the SAR antennas. Vertical deformation produces topographic phase changes that must be accurately accounted for by DInSAR (Ferretti et al., 2000).

This phase can be removed by subtracting a digital elevation model (DEM) from the interferogram. The DEM is generated from radar data or other sources, and it represents the height of the ground surface. The accuracy of the DEM is crucial for the quality of the topographic phase correction, and external DEMs can be supplemented with high-resolution topography mapping data to increase accuracy (Ferretti et al., 2000, 2011).

Topographic correction can be used to correct for phase delays caused by changes in elevation and terrain in DInSAR measurements, with various techniques available such as Range-Doppler Terrain-Correction (RDTC), Shuttle Radar Topography Mission (SRTM), and GEO method.

The Range-Doppler Terrain-Correction (RDTC) technique is an effective method for mitigating the errors caused by topographic distortions in DInSAR. In repeat-pass DInSAR, interferometric phase shifts induced by terrain heights lead to significant geolocation errors and make it difficult to separate ground deformations from topographic signals. The RDTC technique utilizes SAR full-resolution data to generate SAR simulated images and remove the topographic phase in the interferometric processing. The major steps of RDTC include: (1) using an SAR image and an external DEM to simulate SAR image pairs; (2) forming interferograms from simulated SAR images; (3) unwrapping the interferograms to get the absolute topographic phase; and (4) subtracting the topographic phase from the original interferogram to get the residual interferogram dominated by ground deformation and atmospheric artifacts. Compared to using only DEMs, the RDTC method further utilizes the SAR imaging geometry and characteristics to achieve better phase simulation accuracy. Since topographic artifacts are a primary error source in DInSAR measurements, the RDTC correction has significantly improved the results of various deformation monitoring studies. With advanced high-resolution SAR sensors and precise DEM data available, RDTC provides an important means of terrain correction to achieve meter-scale deformation accuracy in complex terrains (Madsen et al., 1993).

The SRTM method is a commonly used technique for topographic correction in DInSAR measurements (Ferretti et al., 2000). The Shuttle Radar Topography Mission (SRTM) produced high resolution topographic data for over 80% of the Earth's land surface, which can be leveraged for topographic correction in DInSAR processing. The SRTM mission used a dual-antenna C-band interferometric synthetic aperture radar to generate precise 3D topographic maps. The SRTM DEMs have a resolution of 1 arc s (approximately 30 m) and a vertical accuracy of ~16 m (Farr et al., 2007). The extensive spatial coverage and precision of SRTM DEMs make them well suited to serve as topographic references in DInSAR studies, especially in areas lacking other high-quality DEM data.

The GEO method is another approach that can be used for topographic correction in DInSAR measurements (Ferretti et al., 2011). In some areas, global DEMs lack the resolution or accuracy for precise DInSAR topographic correction. An alternative is using geological maps, which provide elevation data interpolated from observations of rock outcrops and terrain features. Geological maps illustrate spatial variations in land cover and rock/soil units, often with associated elevation ranges. Elevation data in geological maps are interpolated to a regular grid, producing a preliminary DEM. By incorporating knowledge of local terrain and landforms, geological DEMs can achieve higher accuracy than global models alone. However, the density of elevation data depends on availability of field observations, so resolution may vary in different map areas. While geological DEMs can improve DInSAR topographic correction, there are some limitations. The DEMs are still interpolated from sparse data, so may not fully capture abrupt elevation changes. The resolution and precision depend on the map scale and density of field observations. Geological maps only cover terrain within the map borders, limiting spatial coverage. Accessing maps and digitizing data can also be labor-intensive. However, for areas lacking detailed elevation data, geological maps offer an alternative strategy for high-quality DInSAR topographic correction.

Another correction method for removing the topographic phase is the coherent pixel technique (CPT) (Ferretti et al., 2001). The CPT utilizes a high-resolution DEM to estimate the phase contribution from topography at each coherent pixel in the interferogram (Chen and Zebker, 2002). The CPT method separates the topographic phase and the deformation signal based on their coherence properties and amplitude. The CPT method can be extended to estimate deformation rates at a high temporal resolution (Rosen et al., 2000). By knowing the baseline, wavelength, look angle, and slant range for each coherent pixel, the expected phase due to topography can be calculated using a DEM. This topographic phase estimate is then subtracted from the total phase of the interferogram to isolate the deformation signal. The CPT requires an accurate DEM that is properly co-registered to the SAR data. Any errors in the DEM or co-registration will translate directly to errors in the topographic correction (Lopez-Quiroz et al., 2009).

The most common technique for topographic phase correction is the Goldstein filter, which uses a weighted average of neighboring pixels to estimate the topographic phase (Goldstein and Werner, 1998). Unlike the CPT which utilizes an external DEM, the Goldstein filter estimates the topographic phase directly from the interferometric SAR data itself (Goldstein et al., 1988). The key assumption in the Goldstein filter is that the topographic phase varies slowly in space, while the deformation phase varies more rapidly. By applying a spatial frequency filter, the low-frequency topographic phase can be isolated and subtracted. A key parameter in the Goldstein filter is the size of the filtering window. A larger window will remove higher frequency deformation signals, improving topographic correction but reducing spatial resolution. An appropriate window size depends on the characteristics of the topography and deformation in the area of interest. The filter also depends on the topographic phase being correlated over the window, so it may perform poorly in areas of rapidly changing topography (Fattahi and Amelung, 2013).

One approach to remove the topographic phase that accounts for deformation is the Small Baseline Subset (SBAS) technique (Ferretti et al., 2000). In the SBAS method, interferograms with similar temporal and spatial baselines are combined to estimate the deformation signal. The SBAS approach uses multiple interferograms with small spatial and temporal baselines to isolate the topographic phase from other contributions like surface deformation (Berardino et al., 2002). Small baselines mean that the SAR data pairs used to generate the interferograms have small differences in imaging geometry and acquisition time. This results in interferograms that have reduced decorrelation and atmospheric artifacts, allowing for the clear extraction of the topographic phase. By combining several of these "small baseline" interferograms, more accurate topographic information can be generated through the SBAS optimization procedure. The key steps in the SBAS processing are (1) generation of small baseline interferograms, (2) phase unwrapping of the interferograms, (3) topographic phase extraction through the weighted least squares solution, and (4) removal of the topographic phase component from the original interferograms to get the corrected deformation signals (Hooper et al., 2007).

Multiple topographic correction methods can also be used in combination for improved accuracy in DInSAR measurements (Ferretti et al., 2011).

2. Atmospheric phase correction: Another source of error in DInSAR is the atmospheric phase, which is caused by variations in the propagation of radar waves through the atmosphere. InSAR phase observations are affected by the atmosphere propagation delay, which causes a spatially varying phase shift in the radar signal (Massonnet and Feigl, 1998). The atmospheric phase delay is caused by changes in the refractive index of air due to atmospheric pressure, temperature, and relative humidity variations in the time between radar image acquisitions. This can be a significant source of error, especially for long temporal baselines between images. This phase can be corrected by using atmospheric models or by measuring the atmospheric phase with a GPS receiver. Therefore, accurate meteorological data and atmospheric models are necessary for effective atmospheric correction in DInSAR measurements (Bekaert et al., 2016).

The most common atmospheric correction techniques are the empirical and model-based approaches. The empirical approach uses a set of reference pixels that are assumed to be stable over time, while the model-based approach uses atmospheric models to estimate the atmospheric phase (Ferretti et al., 2007). The Atmospheric Correction with Split Spectrum (ACOSS) model is a specific approach that can be used for atmospheric correction in DInSAR measurements (Ferretti et al., 2001). The ACOSS model is an empirical approach to estimate and reduce atmospheric phase artifacts in DInSAR data. The ACOSS method utilizes the dispersive nature of radar signals to estimate the atmospheric phase delay. As radar signals travel through the atmosphere, lower frequency components travel slightly slower than higher frequency components. By splitting the radar spectrum into multiple sub-bands and analyzing the phase delay between them, atmospheric artifacts can be estimated and removed (Wadge et al., 2002).

One method to correct for atmospheric effects is using the atmospheric phase screen (APS) technique. Another approach to correct for atmospheric delays is based on the integration of meteorological data with DInSAR processing. The APS technique requires the estimation of an additional phase term, while the meteorological method requires prior knowledge of the atmospheric properties (Hanssen, 2001). The APS technique is a statistical approach for estimating and removing atmospheric phase artifacts from DInSAR data. The APS method models the atmospheric phase screen as a combination of atmospheric parameters that can be estimated from external data sources, such as meteorological models, radiosondes, and ground-based GPS stations. By incorporating auxiliary atmospheric data, the APS technique aims to build a more accurate estimation of the atmospheric phase screen compared to methods based solely on the radar interferogram (Jolivet et al., 2011). While promising, the accuracy of the APS technique depends on the availability, resolution, and quality of the atmospheric data used. The APS method can effectively reduce atmospheric noise and improve ground deformation measurements from DInSAR data, especially when constrained by high-resolution in situ atmospheric data.

Another method to reduce atmospheric phase delay in DInSAR data is the Delaunay minimum cost flow (MCF) algorithm correction technique (Chen and Zebker, 2001). The Delaunay MCF algorithm treats the unwrapped interferometric phase as a cost surface, with higher phase values representing higher cost. It then calculates the minimum cost flow through this surface to find connections between phase discontinuities, which are likely caused by atmospheric artifacts. By identifying these connections, the atmospheric contribution can be estimated and removed.

3. Orbital phase correction: The orbital phase is caused by errors in the satellite orbit determination and positioning produced by the uncertainties in the radar satellite orbits (Massonnet and Feigl, 1998; Zebker and Villasenor, 1992; Osmanoglu et al., 2011). Orientation of the radar and changes in elevation and terrain can cause phase delays that require correction in DInSAR measurements (Kampes et al., 2003). This phase can be corrected by using precise orbit determination and positioning data. The orbital errors are corrected through the removal of the long wavelength phase and the shift of the interferogram (Hanssen, 2001). The use of external orbit data and the accuracy of precise orbit determination can help improve the effectiveness of correcting orbital errors in DInSAR measurements (Zebker and Villasenor, 1992; Zhu and Bamler, 2010). The most common technique for orbital phase correction is the orbit interpolation method, which uses precise orbit data to interpolate the satellite position and velocity at the time of image acquisition (Massonnet and Feigl, 1998; Zhu and Bamler, 2010). The orbit correction can be performed by using precise orbital data or by applying the orbital correction using permanent scatterers (PS) (Hanssen, 2001). PS are point sources on the earth's surface that exhibit a stable and persistent radar backscatter signal over time (Ferretti et al., 2000). PS are used as calibration targets, and their deformations can be reliably estimated after removing the other error sources (Ferretti et al., 2001). PS can be detected by using several techniques, such as the SqueeSAR (Small Baseline Subset) method and Persistent Scatterers Interferometry (PSI) (Hooper et al., 2004).

The Small Baseline Subset (SBAS) approach, also known as SqueeSAR, is a refined interferometric technique for measuring ground deformation and detecting PS in areas where the traditional DInSAR method is limited by spatial and temporal decorrelation (Ferretti et al., 2001). Unlike the traditional DInSAR approach that utilizes two SAR images acquired at different times, SqueeSAR leverages multiple SAR images acquired over the same area to identify PS, which are point targets with stable phase characteristics over time. The key concept of SqueeSAR is to partition the set of available SAR images into small baseline subsets, each composed of images with small temporal and spatial baselines. For each subset, a standard DInSAR process can be applied to generate differential interferograms with minimal decorrelation. The phase of each PS can then be connected across these small-baseline interferograms using a weighted least squares solution (Hooper et al., 2004). By jointly analyzing groups of interferograms with small baselines, more PS can be identified and their phase histories over long time periods can be recovered, allowing for reliable ground deformation monitoring and measurement. Compared to traditional DInSAR, SqueeSAR provides significant improvements in terms of the density and quality of measurement points, especially in nonurban areas (Ferretti, 2014).

Persistent scatterers interferometry (PSI) is an advanced DInSAR technique for detecting and monitoring PS—point targets with stable radar phase properties over time (Ferretti et al., 2001; Hooper et al., 2004). Rather than using pairs of SAR images to generate interferograms, PSI utilizes large stacks of SAR data acquired over the same area to identify PS and track their phase histories. This approach overcomes the limitations of traditional DInSAR from temporal and spatial decorrelation, allowing for long-term surface deformation analysis with dense spatial sampling. The key steps in PSI processing include: (1) identification of potential PS from amplitude dispersion and phase stability analysis; (2) phase unwrapping and modeling for each PS to recover the absolute phase; (3) estimating and removing the atmospheric phase screen in each SAR data; (4) computing deformations along the radar line of sight by differencing the phases of each PS; and (5) geocoding the deformations into a map coordinate system for spatiotemporal mapping and analysis (Ferretti et al., 2014). A major advantage of PSI is the ability to separate different deformation components, including linear deformation rates and periodic components at different timescales.

4. Temporal decorrelation correction: Temporal decorrelation is caused by changes in the surface properties between the two SAR acquisitions (Zebker and Villasenor, 1992; Zhu and Bamler, 2010). Low coherence of the radar signal can reduce the accuracy of DInSAR measurements, illustrating the importance of maintaining signal coherence for accurate results. The length of the time interval between measurements and the coherence of the radar signal are two factors that can affect the accuracy of DInSAR measurements (Kampes et al., 2003). These can result in a loss of coherence between the two images. Longer time intervals between measurements and lower coherence of the radar signal can decrease the sensitivity of DInSAR measurements (Zhu and Bamler, 2010). Temporal decorrelation can be corrected by using multitemporal SAR images or by modeling the temporal decorrelation with statistical models. Factors that affect the coherence of the radar signal in DInSAR measurements, such as vegetation cover, should be investigated to improve the accuracy of the measurements (Ketelaar, 2009). The most common technique for temporal decorrelation correction is the SBAS method, which uses a subset of images with a small temporal baseline to reduce the effect of temporal decorrelation (Berardino et al., 2002).

5. Phase unwrapping: Phase unwrapping is the process of removing the 2π phase ambiguities that occur in the interferogram. This is necessary to obtain a continuous phase map that can be used to estimate the ground deformation. The phase values in the interferogram are wrapped into the interval $(-\pi, \pi)$. To obtain the absolute phase values and correctly interpret the ground deformation, an unwrapping process is required to remove 2π phase ambiguities (Ghiglia and Pritt, 1998). Phase unwrapping can be performed using several techniques, including the minimum cost flow (MCF) method, the branch cut method, and the region growing method. These techniques use different algorithms to estimate the unwrapped phase, and the choice of technique depends on the quality of the interferogram and the characteristics of the deformation (Zhang et al., 2019).

Minimum cost flow (MCF) phase unwrapping is one of the most effective unwrapping techniques. It formulates phase unwrapping as a global optimization problem by constructing a graph with nodes representing pixels and edges representing the dependency between neighboring pixels. Each edge is assigned a cost that depends on the phase gradient between the two pixels. The unwrapped solution corresponds to the minimum cost path that flows through the entire graph (Costantini, 1998). The key advantage of the MCF method is that it exploits the connectivity of the entire image, enabling a global optimized solution and the ability to bridge regions with low coherence. The MCF technique has been shown to outperform other unwrapping methods in the presence of noise and regions of low coherence (Perissin and Wang, 2011). However, the computational complexity of the MCF algorithm is relatively high due to the global optimization. Approximation techniques are often used to reduce the computational burden while still achieving high unwrapping accuracy.

Branch cut is one of the earliest phase unwrapping algorithms. It initializes phase unwrapping by identifying phase discontinuities, called branch cuts, that divide the interferogram into distinct regions. The unwrapped phase is then determined independently within each region (Ghiglia and Pritt, 1998). Branch cuts are identified by analyzing phase gradients and unwrapping paths. Pixels with phase gradients above a threshold are marked as branch points. Branch cuts are placed between branch points to minimize unwrapping errors. The unwrapping path starts from a pixel with high coherence and low phase gradient and unwraps neighboring pixels with phase gradients below the threshold. When the path meets a branch cut, it is redirected along the cut (Ghiglia and Romero, 1994). The key advantage of the branch cut method is its low computational complexity as it performs unwrapping locally within each region. However, its performance highly depends on the threshold used to detect branch points, and it may generate unwrapping errors for complex interferograms with extensive noise and decorrelation. Approaches have been developed to refine the branch cut placement and combine it with other techniques to improve unwrapping accuracy while maintaining low computational cost (Chen and Zebker, 2001; Judge and Bryanston-Cross, 1994).

Region growing is a phase unwrapping technique that expands unwrapped regions from seed pixels with high coherence and reliability. It starts from a pixel with a reliably unwrapped phase and low phase gradient, known as a seed pixel, and unwraps its neighboring pixels if their phase differences are within a threshold. This process continues iteratively by unwrapping pixels at the border of unwrapped regions until the entire interferogram is unwrapped (Ghiglia and Pritt, 1998). The key advantage of region growing is its low computational complexity as it performs unwrapping locally. It is also less sensitive to noise since unwrapping only propagates to neighboring pixels within a certain threshold. However, region growing relies heavily on the selection of suitable seed pixels and threshold and can generate unwrapping errors for complex interferograms with extensive decorrelation if not properly optimized (Chen and Zebker, 2001). Approaches have been developed to improve seed pixel selection by considering phase gradients and coherence (Xu and Cumming, 1999; Witoszynskyj et al., 2009). Statistical models can also be used to determine adaptive thresholds for different regions (Chen and Zebker, 2000). Region growing is often used as a preliminary unwrapping method to obtain an initial solution and identify areas of low reliability. The initial unwrapping result can then be refined using other advanced techniques that can bridge unwrapped regions, such as minimum cost flow phase unwrapping. Combined with other methods, region growing provides a fast way to unwrap most coherent areas while limiting unwrapping errors, thus improving the computational efficiency and robustness.

6. Geocoding correction: Geocoding is the process of converting the interferogram from radar coordinates to geographic coordinates. This correction is necessary to relate the interferometric phase to the ground deformation. Geocoding can be performed using several techniques, including the range Doppler method and the terrain correction method. These techniques use different algorithms to convert the radar coordinates to geographic coordinates, and the choice of technique depends on the characteristics of the SAR data and the accuracy requirements of the application. In conventional geocoding, pixels are mapped onto a geographic grid based on the satellite ephemeris and terrain height. However, this approach does not account for radial satellite velocity and terrain slope effects, which can shift the pixel locations by up to 100 m (Eineder et al., 2010). Range Doppler geocoding improves upon this by using the Doppler frequency information in the radar signal to dynamically correct for radial velocity effects during SAR processing. It also integrates a high-resolution DEM to correct for terrain-induced geometric distortions (Cumming and Wong, 2005). The accuracy of DInSAR measurements can also be degraded by terrain-induced geometric distortions during geocoding. Terrain correction is an important technique to mitigate these distortions and enable precise DInSAR deformation monitoring, especially in mountainous areas. In conventional geocoding, pixels are mapped onto a geographic grid based only on satellite orbital position, with an assumption of flat terrain. However, surface topography causes additional path length variations that shift the actual pixel locations. The magnitude of these shifts depends on terrain slope and aspect relative to the radar look direction and can be up to several kilometers for areas with steep topography (Bamler and Hartl, 1998). Terrain correction improves geocoding accuracy by integrating a high-resolution DEM to estimate and correct for the topography-induced path length effects.
7. Phase gradient correction: Phase gradients are caused by variations in the terrain slope and azimuth angle. These gradients can affect the accuracy of the DInSAR (Ferretti et al., 2007). Phase gradient correction relies on the fact that deformation phases are relatively smooth in space, while noise phases tend to be rough and vary rapidly. By calculating the spatial phase gradient between pixels, noise phases can be identified and reduced. A commonly used approach is to apply a low-pass filter to the phase gradient values, which smoothing the rough noise gradients while preserving the broader deformation signal. The smoothed phase gradients are then integrated to recover an improved deformation phase measurement (Just and Bamler, 1994).

2.4 Differential radar interferometry methodology

As noted before, the simplest DInSAR technique involves using at least two images from radar sensors with artificial aperture (Massonnet and Feigl, 1998; Rosen et al., 2000). Fig. 2.4 shows the different processing stages of the DInSAR algorithm in order to prepare the land subsidence map and then its comparison with data on groundwater drawdown.

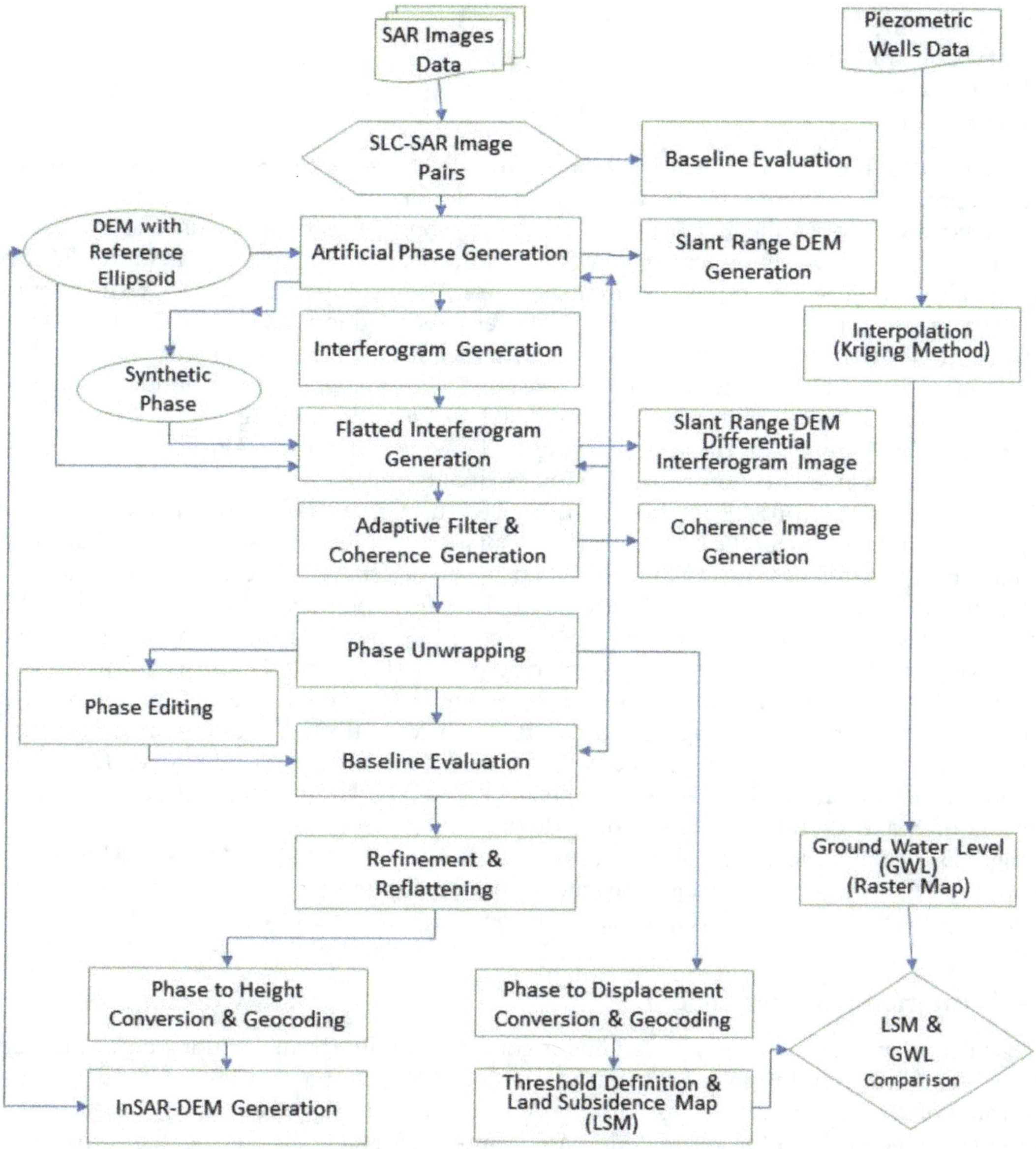

FIGURE 2.4 The various stages involved in processing radar images using the Differential Interferometric Synthetic Aperture Radar (DInSAR) method to generate a land subsidence map, which is then validated by comparing with data on groundwater drawdown.

2.4.1 Selection of pairs of SAR images for interferometry

The selection of suitable SAR images for interferometric processing is a key step and has a significant impact on the quality of the final results. The selection criteria depend on the specific application of the SAR images. Some important criteria include:

- Angle of view (ascending and descending crossing)
- Normal baseline
- Time baseline
- Data acquisition time
- Weather conditions
- Image quality

It is important to consider these criteria carefully when selecting SAR images for interferometric processing. There are several tools available for identifying and selecting SAR images. One such tool is the Earth Observation Link (EOLI) software, which is exclusively available on the European Space Agency (ESA)/European Space Research Institute (ESRIN) site. EOLI allows users to quickly search satellite missions supported by ESA and identify relevant images for a specific area. If the "Interferometry" query mode is selected, the relevant baseline information is also provided. Another tool for obtaining information about SAR images is the Display Earth remote sensing Swath Coverage for Windows (DESCW) program. This offline program can also be used to identify viewing angle, temporal and geometrical baseline, and data acquisition time. However, a complete set of information is not yet available in the section related to image quality.

In the chapter, the DInSAR method is introduced through a case study on land subsidence monitoring in the Najafabad plain, Iran. The Najafabad aquifer covers an area of 1720 square kilometers and stretches approximately 30 km west of Isfahan. The Zayandeh Rud river flows in southeast of this district. This region holds high potential for agriculture in Isfahan province. Despite the implementation of water supply projects in the Zayandeh Rud basin, excessive extraction from the alluvial aquifer by over 14,000 exploitation wells has severely depleted the aquifer. Therefore, it has become essential to monitor the resulting subsidence and determine its rate to manage groundwater resources effectively in the area. The subsidence in this region has been estimated by DInSAR processing of ASAR, PALSAR, and Sentinel-1 radar images. The results have been validated by comparison with the differential leveling data and groundwater level drawdown.

Fig. 2.5 displays the geographic area of interest and the range of ASAR, PALSAR, and Sentinel-1 radar images selected for the study. These radar images were chosen based on their suitability for the research objectives and the specific characteristics of the study area.

2.4.2 Interferogram production

Each pixel in the SAR image contains a complex number that includes amplitude and phase. While the amplitude is known, the phase is not absolute but is measured based on the value of 2π. The received phase difference from the reflections caused by two nearby orbits changes slowly, and the calculated phase difference for each pixel is the interferogram. In other words, the interferogram provides information about changes related to distance of the

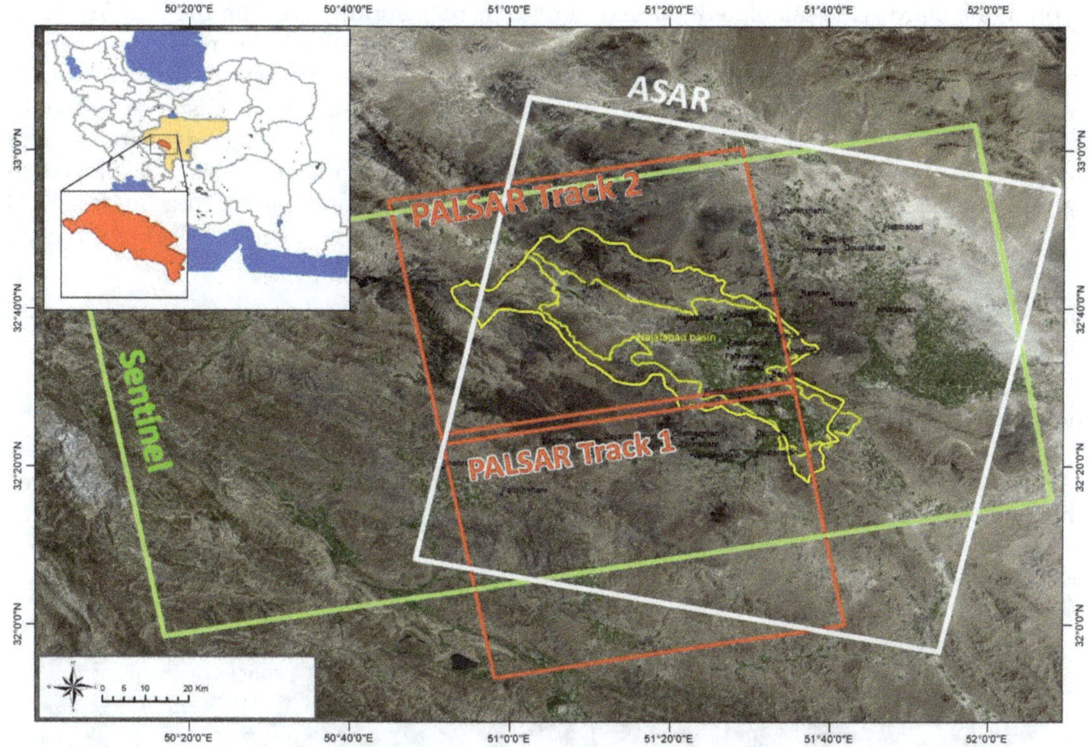

FIGURE 2.5 The geographic position and range of Advanced Synthetic Aperture Radar (ASAR), Phased Array L-band Synthetic Aperture Radar (PALSAR), and Sentinel-1 radar images, including both descending and ascending passes, superimposed on a Google Earth image of the study area. This image provides a visual representation of the coverage and spatial resolution of the radar images used in the study.

target to radar sensor. These changes are affected by satellite orbital errors ($\Delta\varphi G$), topography ($\Delta\varphi T$), noise ($\Delta\varphi$noise), atmospheric effects ($\Delta\varphi$Atm), and deformation ($\Delta\varphi D$). To extract the amount of surface displacement ($\Delta\varphi D$), all other effects must be removed from the interferogram (Eq. 2.1) (Ferretti et al., 2007):

$$\varphi_2 - \varphi_1 = \Delta\varphi G + \Delta\varphi T + \Delta\varphi \text{Atm} + \Delta\varphi D + \Delta\varphi \text{noise} \tag{2.1}$$

One simple method to address the topographical component in DInSAR is to minimize the vertical baseline of a pair of selected images. This approach helps to reduce the impact of topography in creating a phase difference. However, to further improve the accuracy of DInSAR results, aDEM of the Earth's surface can be used to remove the remaining topographical effects from the interferometric images. A DEM is a representation of the Earth's surface topography, usually in the form of a grid of elevation values. By using a DEM, it is possible to estimate the topography-induced phase difference between two images and remove it from the interferogram. To do this, the DEM is first aligned with the interferogram. Then, the topography-induced phase difference is calculated by simulating the radar signal path through the DEM. This calculated phase difference is then subtracted from the interferogram, leaving only the phase difference caused by surface deformation. The resulting image is known as a deformation map and can be used to identify areas of surface displacement.

The production of interferograms in DInSAR involves measuring the distance difference between a point on the ground and the position of the sensor in two imaging positions using the phase difference (φ) between two recovered mixed SAR images. This is achieved by multiplying one image with another image, which is made up of the interferogram (Fig. 2.6).

The interferometric phase is expressed as follows (Massonnet and Feigl, 1998; Ferretti et al., 2007):

$$\varphi = 4\pi \,/\, \lambda \times (R_1 - R_2) \tag{2.2}$$

where λ is the radar wavelength, R_1 and R_2 are the distances between the ground point and the sensor in the first and second imaging positions, respectively.

By calculating the interferometric phase for each pixel in the SAR image, it is possible to generate an interferogram that represents the phase difference between the two images. This interferogram can then be used to identify areas of surface displacement.

The interferometric phase is also expressed as follows (Guarnieri et al., 2003):

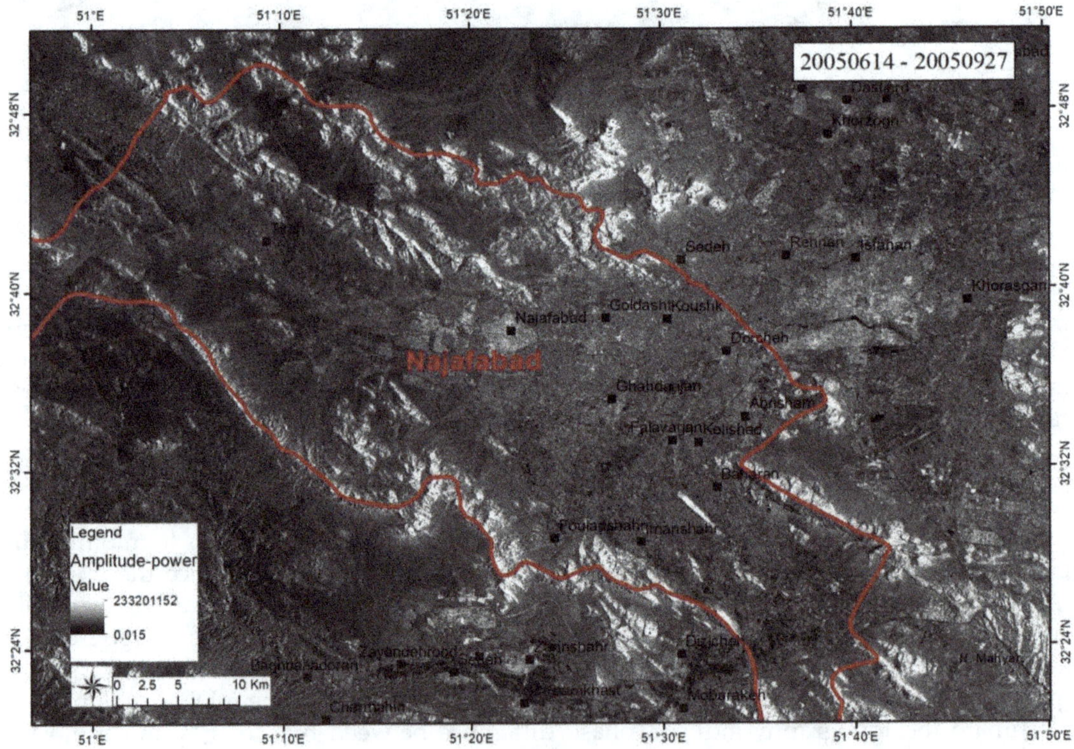

FIGURE 2.6 The primary interferogram generated from a pair of Advanced Synthetic Aperture Radar (ASAR) radar images captured at two different times.

$$\text{Phase} = A\tan[\text{Image}(I) / \text{Real}(I)] \tag{2.3}$$

Image(I) and Real(I) are the virtual and real parts of the interferogram, respectively.

During the production process of DInSAR, spectral shift and general Doppler bandwidth filtering are performed. Spectral shift is necessary to correct for the amplitude of the spectral shift caused by the variable SAR viewing angle on scattered targets. Additionally, Doppler bandwidth filtering is needed to compensate for the shifted azimuth spectrum due to different Doppler shifts. The azimuth filter applied during the interferometer production process is capable of fully recording scenes with potential coherence.

To produce an interferogram, two single-look complex (SLC) images must be fixed or co-referenced with an accuracy of about 1/100 pixel dimension. Several methods are used for this, including cross-correlation of range values for pixels, maximum value for dependence (Prati et al., 1994), maximum value of signal-to-noise ratio (SNR) in the fringe spectrum (Gabriel and Goldstein, 1988), and minimization of mean changes in fuzzy variance changes (Lin et al., 1992).

After matching the two images, the phase difference between them is obtained by differentiating the phase obtained from the two SLC images (Fig. 2.6). This phase difference is then produced by complex conjugate multiplication of the two images. The final output of this phase is a planar interferogram, with the stationary phase (according to the imaging geometry) and topographic phase (if a DEM file is provided as input) removed.

2.4.3 Apply adaptive filter and generate coherent map

After producing the planar interferogram in the previous step, filtering is applied to generate a filtered interferogram with reduced noise phase. Additionally, an interferometric coherence map that shows the phase quality and a filtered master image are produced.

Three filtering methods can be used at this stage, including the adaptive filter.

- **Adaptive filter**

In this type of filter, coherence values are used to adjust the size of the filter window. The average intensity difference among adjacent pixels is used to identify a fixed region, which defines the maximum dimensions and shape of the filter window. This process aims to preserve even the smallest patterns in the interferometric fringes. However, identifying the threshold value for the similarity average factor (which measures the similarity between the measured redistribution values in the intensity data of the master and slave images) can be time-consuming. Based on this factor, areas of the image where the signal intensity is constant are identified. The selection of pixels in each fixed region is based on the difference between the average pixels in the region (M_{all}) and the new pixel value (M_{New}) that is potentially a candidate (this difference is normalized for (M_{all}). New candidate pixels are identified using a regional growth method. The equation related to the average similarity factor, which is displayed with a digital number on a linear scale, can be written as follows (Goldstein and Werner, 1998; Baran et al., 2003):

$$\frac{(M_{\text{all}} - M_{\text{New}})}{M_{\text{all}}} \tag{2.4}$$

It should be noted that the process of applying the adaptive filter method is typically repeated several times to determine the optimal value of the average similarity factor. This iterative process helps to achieve the best possible filtering results and improve the quality of the final interferometric products.

- **Boxcar filter**

The boxcar filter method uses the local fringe frequency to optimize the intermediate filter. Its purpose, like the adaptive filter method, is to preserve even the smallest patterns in the interferometric fringes.

- **Goldstein filter**

The Goldstein filter method derives the variable bandwidth of the filter directly from the spectral power of the fringes. To optimize the filter's performance, the "alpha" parameter is used in an adaptive method based on local coherence, where a lower coherence leads to a stronger filter being applied. This filtering method significantly improves the visibility of fringes and reduces the noise resulting from time and baseline inconsistencies. The alpha parameter is dependent on the coherence, meaning that incoherent areas are filtered more than coherent areas. This minimizes the loss of signal while greatly reducing the noise level (Fig. 2.7).

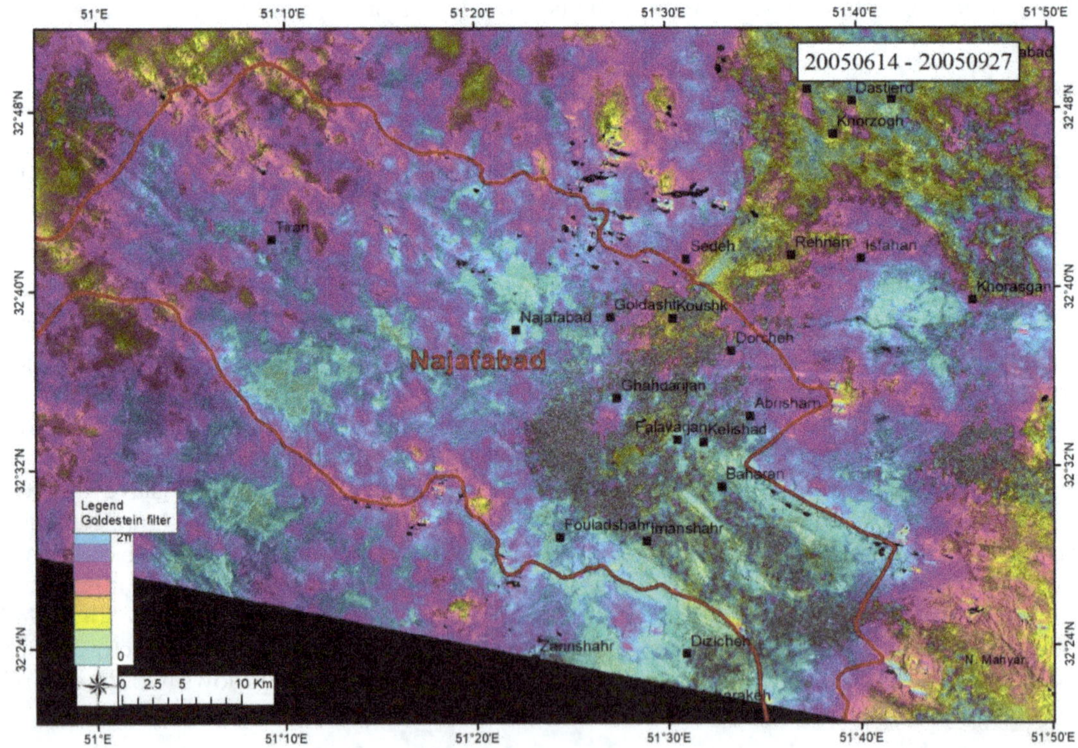

FIGURE 2.7 Application of the Goldstein adaptive filter and atmospheric correction to correct the phase in a pair of Advanced Synthetic Aperture Radar (ASAR) images captured in 2005.

However, it should be noted that the coherence obtained from the generated interferogram should be used carefully if the coherence from Fint option is selected. The phase smoothing produced by the filtering itself can lead to an overestimation of coherence, resulting in higher coherence values. If the goal is not to recover regions that appear coherent but are highly noisy, coherence can be produced from an unfiltered interferogram or an interferogram filtered using the boxcar method.

Correlation or interferometric coherence (γ) is the ratio between the sum of coherence and inconsistency (Goldstein and Werner, 1998; Baran et al., 2003; Ghulam et al., 2010):

$$\gamma = \frac{|\sum s_1(x) \cdot s_2(x)^*|}{\sqrt{\sum |s_1(x)|^2 \cdot \sum |s_2(x)|^2}} \tag{2.5}$$

The estimated coherence of the differential radar interferometry (DInSAR) data ranges between 0 and 1 and is affected by systemic spatial incoherence (such as extra noise) and time mismatch between the master and slave data.

When working with Single Look data (for example, when multilooking factors, range, and azimuth are in the same data), increasing the coherence window size (in range and azimuth) while paying attention to the number of samples and avoiding problems with high estimation of coherence can be helpful.

To correct the atmospheric phase during processing by the differential interferometric method (DInSAR), high-pass filters such as Goldstein can be applied. These filters can make land subsidence in the investigated areas more evident, as shown in Fig. 2.7. Producing a coherence map serves a dual purpose. First, it helps to determine quality of the measurement, such as phase interferometry. In general, InSAR pairs with low coherence should not be used for reliable phase correlation measurements. Second, the coherence map can be used to extract thematic information related to the characteristics of land features and their temporal changes. This information is revealed when associated with the backscattering coefficient of the master and slave intensity data.

A high temporal correlation will result in coherence with average values higher than 0.5, leading to a more accurate estimate of the displacement or subsidence of the earth, as shown in Fig. 2.8. The figure provides a visual representation of the level of coherence between the two radar images used in the DInSAR processing for monitoring land subsidence. Coherence values closer to 1 indicate a stronger correlation between the two radar images and therefore a higher level of accuracy in the DInSAR analysis.

As a general rule, the coherence decreases with the increase of the time interval between the master and slave images. Additionally, a coherent image can be obtained from either an unfiltered or filtered interferogram.

2.4.4 Phase unwrapping

As the complexity of the terrain increases, the phase difference also increases accordingly. Since periodic phase changes are a function of the 2π value, values exceeding 2π are automatically wrapped back to the range of $0-2\pi$. Consequently, the phase obtained from interferometry must be unwrapped with special techniques to solve the complexity

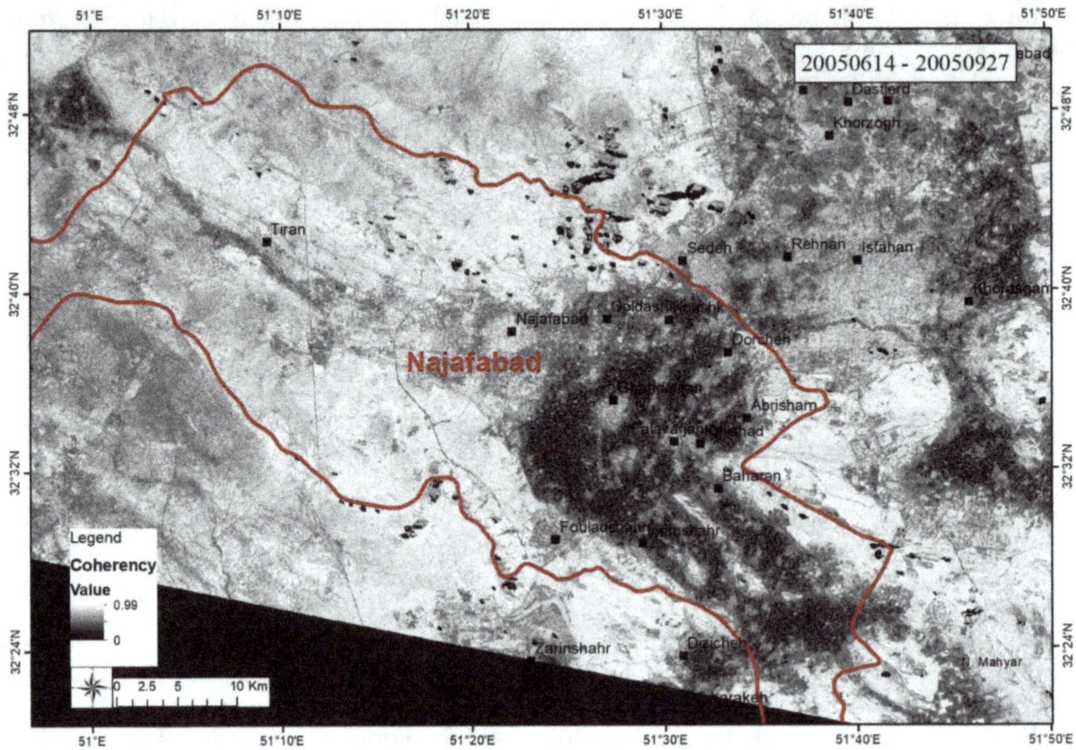

FIGURE 2.8 Coherence image generated from a pair of Advanced Synthetic Aperture Radar (ASAR) interferograms (the minimum coherence in this image is 0.004, the maximum coherence is 0.998, and the average coherence of the entire image is 0.661).

and recover the true phase values. The general process and a simple expression of this problem can be described as follows (Reigber and Moreira, 1997; Costantini, 1998; Ferretti et al., 2007):

$$\varphi_{\text{up}} = \varphi_{\text{wp}} + 2\pi \times n \tag{2.6}$$

In the above expression, φ_{up} represents the unwrapped phase, φ_{wp} is the wrapped or complex phase, and n is a positive integer value (with 0 representing a point located on the reference surface).

The phase of the interferogram is restricted to the 2π scale, meaning that whenever the phase change exceeds 2π, it starts over and repeats its cycle. Phase correction is a process that resolves this 2π ambiguity. Several algorithms have been developed for this purpose, such as Branch-Cuts, Region Growing Filter, Minimum Cost Flow, Minimum Least Squares, and several baselines. However, none of these algorithms are complete or universally applicable, and a combination of methods may be required to achieve the desired results (Fig. 2.9).

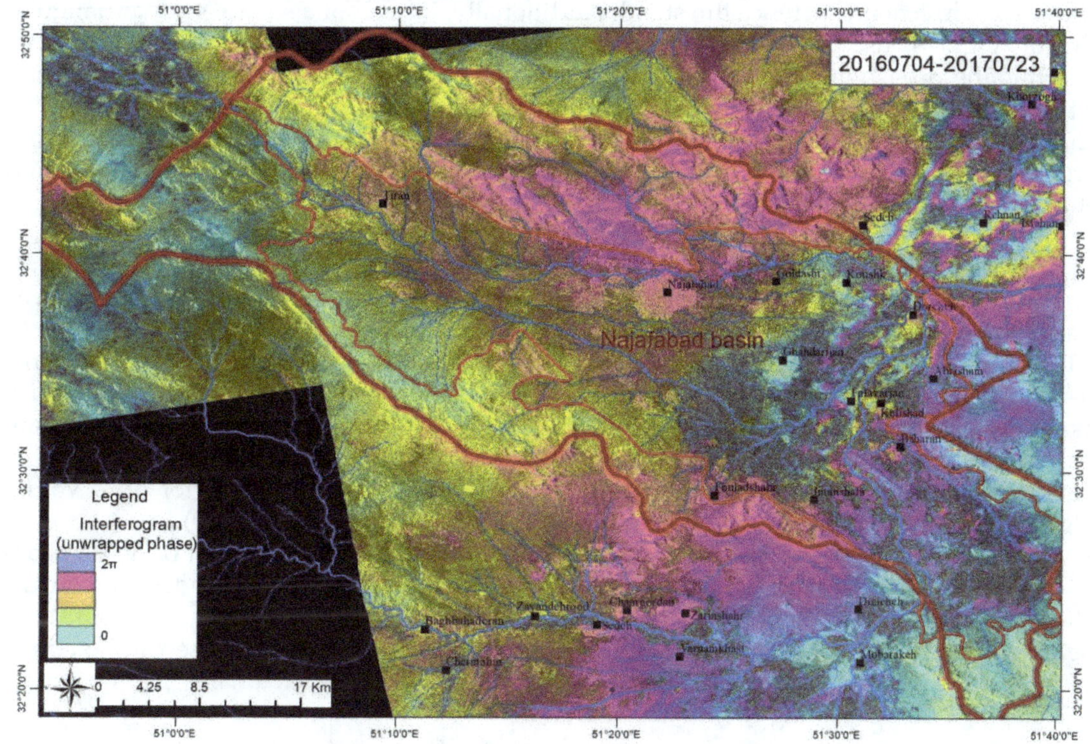

FIGURE 2.9 Phase unwrapping and correction of the interferogram from a pair of C-band Sentinel-1 radar images captured in 2016 and 2017 over Isfahan, Najafabad, and Northern Mahyar areas.

Depending on the characteristics of the data, additional phase editing may be necessary to correct errors in the modified interferogram. To correct the phase, methods such as the regional growth algorithm, the minimum cost flow (MCF) algorithm, and the Delaunay MCF algorithm can be used.

During the phase unwrapping and correction stage, areas affected by ground subsidence are well-defined. The fringes formed during interferometry, filtering, and phase unwrapping are shown in Fig. 2.9.

The regional growth algorithm is suitable for cases where the phase correction process is not complicated by large regions with low coherence or other growth-limiting factors. The MCF algorithm should be used in such cases, as it is capable of obtaining better results than the regional growth method. This method considers a square grid throughout the pixels of the image, excluding all pixels whose coherence value is lower than the unwrapping coherence threshold from the network coverage.

- **Delaunay MCF algorithm:**

This algorithm is similar to the minimum cost flow (MCF) method, with the only difference being that the square grid does not cover all pixels but only those with a coherence value above the correction coherence threshold. Additionally, instead of a square grid, a triangular Delaunay grid is used. This method corrects only the points with good coherence, without affecting the pixels with low coherence. The use of Delaunay triangles is particularly useful when there are various areas with low coherence scattered throughout the image, such as water or densely vegetated areas. In such cases, other phase correction methods can produce phase transitions, while the Delaunay method can minimize these transitions (Guarnieri et al., 2003; Ghulam et al., 2010).

- **Decomposition level:**

This process involves sampling the original pixel with the minimum decomposition level. Using the decomposition process for multispectral purposes involves an iterative cycle, where the interferogram is modified at a low resolution and then reconstructed at the original resolution. The decomposition process can help reduce errors in the correction process, such as in cases of sparse regions with low coherence, as well as reduce processing time and limitations of computer resources. The number of repetition cycles (decomposition) should be determined at this stage, with each cycle corresponding to a sampling factor of 3. Values higher than 3 should be avoided (Guarnieri et al., 2003; Ghulam et al., 2010).

In cases where there are large displacements or steep topography with fast or dense phase scattering in the fringe, the use of the decomposition process may cause undesirable visual effects. In such cases, values of -1 should be used to prevent these factors. It should also be noted that if errors occur during the phase correction process, the phase can be edited to correct 2π cycle coefficients and discontinuities in the corrected phase. The output of this step is a file containing the modified phase.

2.4.5 Refinement and phase correction

This step is critical for correctly converting the corrected phase information into height or displacement values. It enables the correction of errors and phase deviations as well as the removal of possible phase ramps. Performing this step is necessary to produce a digital height model and displacement map. A ground control points file must be created in advance for the implementation of this step (Fig. 2.10).

After correcting the phase, the number of 2π cycle coefficients should be determined and added to the resulting phase value, and the absolute phase value should be calculated. This step is also known as phase calibration. The absolute value of the phase at any point can be expressed as follows (Massonnet and Feigl, 1998; Ferretti et al., 2007):

$$\varphi_{abs} = \varphi_{up} + \varphi_{offset} \tag{2.7}$$

A simple method for determining φ_{offset} is to identify a point on the interferogram whose position and height information is known. Using this information, the integer value corresponding to the number of phase cycles for the entire interferogram can be calculated. This value can be determined directly from the data without prior knowledge of the investigated level, using a series of automatic algorithms. Various algorithms have been developed for this purpose.

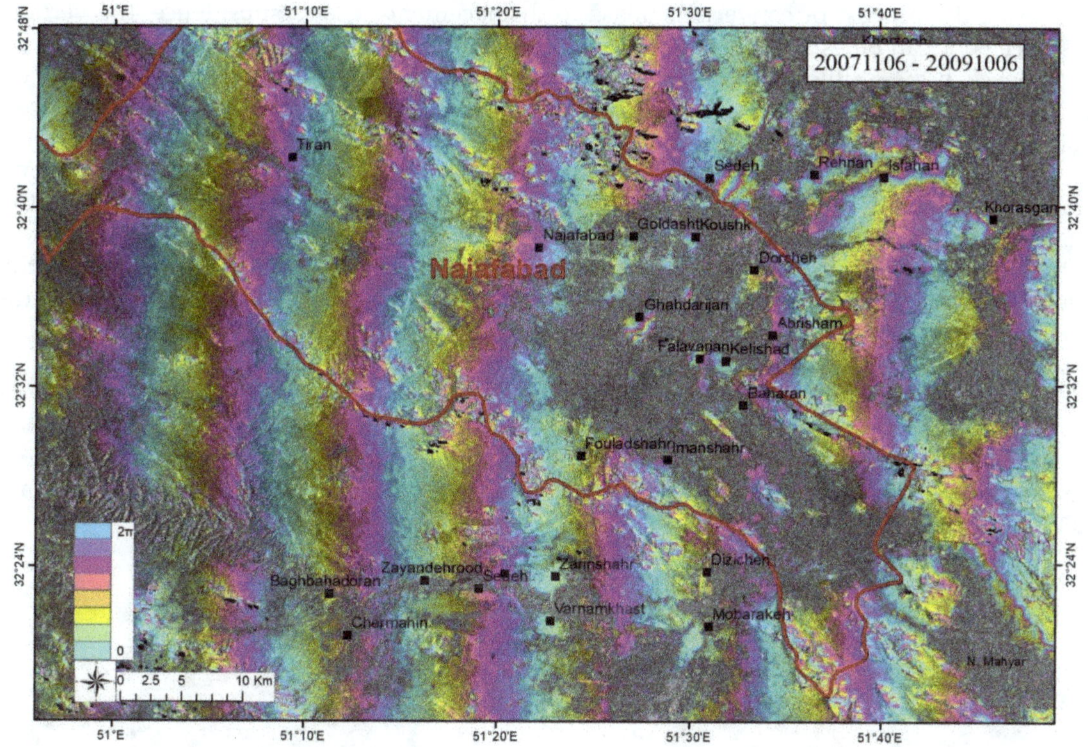

FIGURE 2.10 Interferogram generated from C-band Advanced Synthetic Aperture Radar (ASAR) radar images captured on 6/11/2007 and 6/10/2009, in the western region of Isfahan. The ramp error, which was corrected during processing, is visible in the image.

2.4.6 Phase conversion to displacement and geocoding

At this stage, the corrected and calibrated absolute phase values are directly converted into displacement and geocoded using the range Doppler method and coordinate transformation through cartographic methods. The main difference with conventional geocoding is that the amplitude-Doppler equations are applied simultaneously to both antennas, enabling the determination of not only the displacement of each pixel, but also their location (east and north coordinates) in a given reference system. The output of this step also includes a harmony map that is geocoded.

Each 2π cycle (interferometric fringe) of the differential phase corresponds to a half-wavelength displacement along the slant range direction (SAR sensor viewing direction). Any vector can be determined in which the measured displacement is in the direction of the slant range component of the deformation in the direction of the satellite image. This vector represents the component of deformation along the slanted range reimaged in one direction on the ground, specified by the user (e.g., "vertical displacement" in the case of subsidence; "slope displacement" in the case of landslides; "movement in the desired direction" in other cases).

During the process of converting phase to displacement or land subsidence, the vertical displacement can be estimated. For instance, Fig. 2.11 shows the estimated amount of land subsidence in the southeast and east regions of the Najafabad basin in Isfahan, Iran (marked in red). The urban area of Isfahan in the northeast and east regions has also experienced subsidence.

The output map displays the displacement in meters and can provide valuable information for monitoring the extent and severity of land subsidence over time (Fig. 2.12). In the case of displacement in the direction of the slanted range, a positive sign indicates movements related to the reduction of the distance in the direction of the slanted range of the sensor to the target (slave image compared to the master image (Fig. 2.12). In the desired direction mode, a positive sign corresponds to movements in the direction defined by the user (slave image compared to the master image).

Additionally, a precision file is produced as an output at this stage. This file is derived from parameters such as coherence and wavelength and provides an estimate of the accuracy of the measurements made. High values of this parameter indicate low accuracy of the measurements. The formula used to calculate accuracy is defined as follows (Guarnieri et al., 2003; Ghulam et al., 2010):

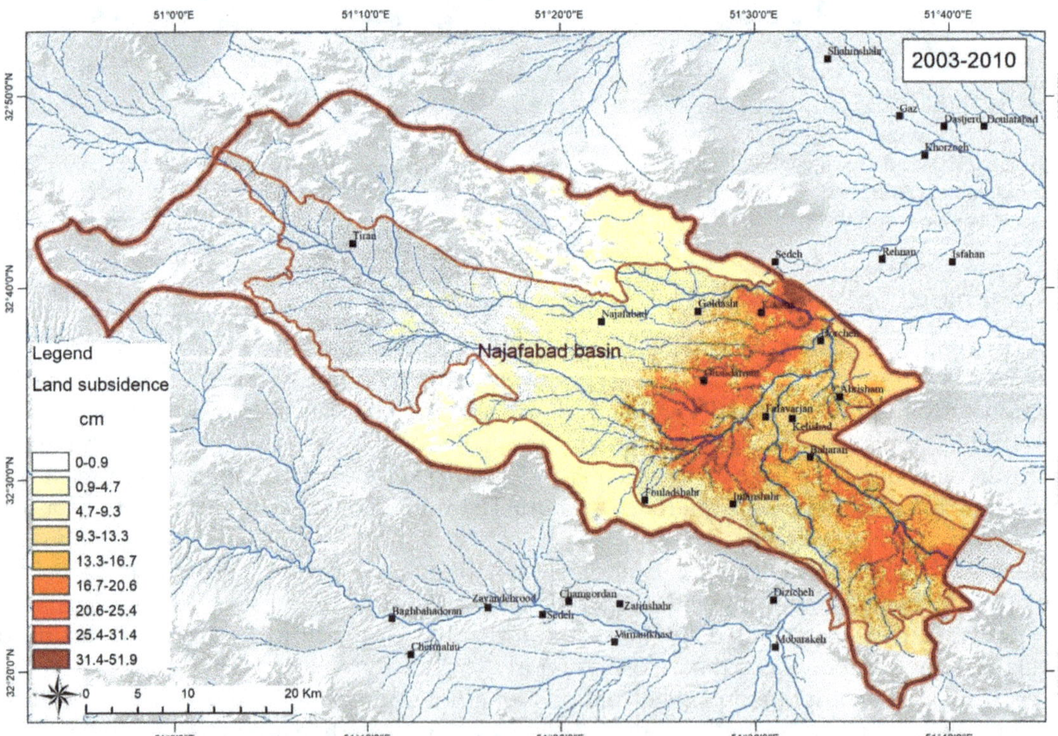

FIGURE 2.11 Phase-to-displacement conversion of a pair of Advanced Synthetic Aperture Radar (ASAR) radar images captured between 2003 and 2010 over the Najafabad plain of Isfahan, Iran.

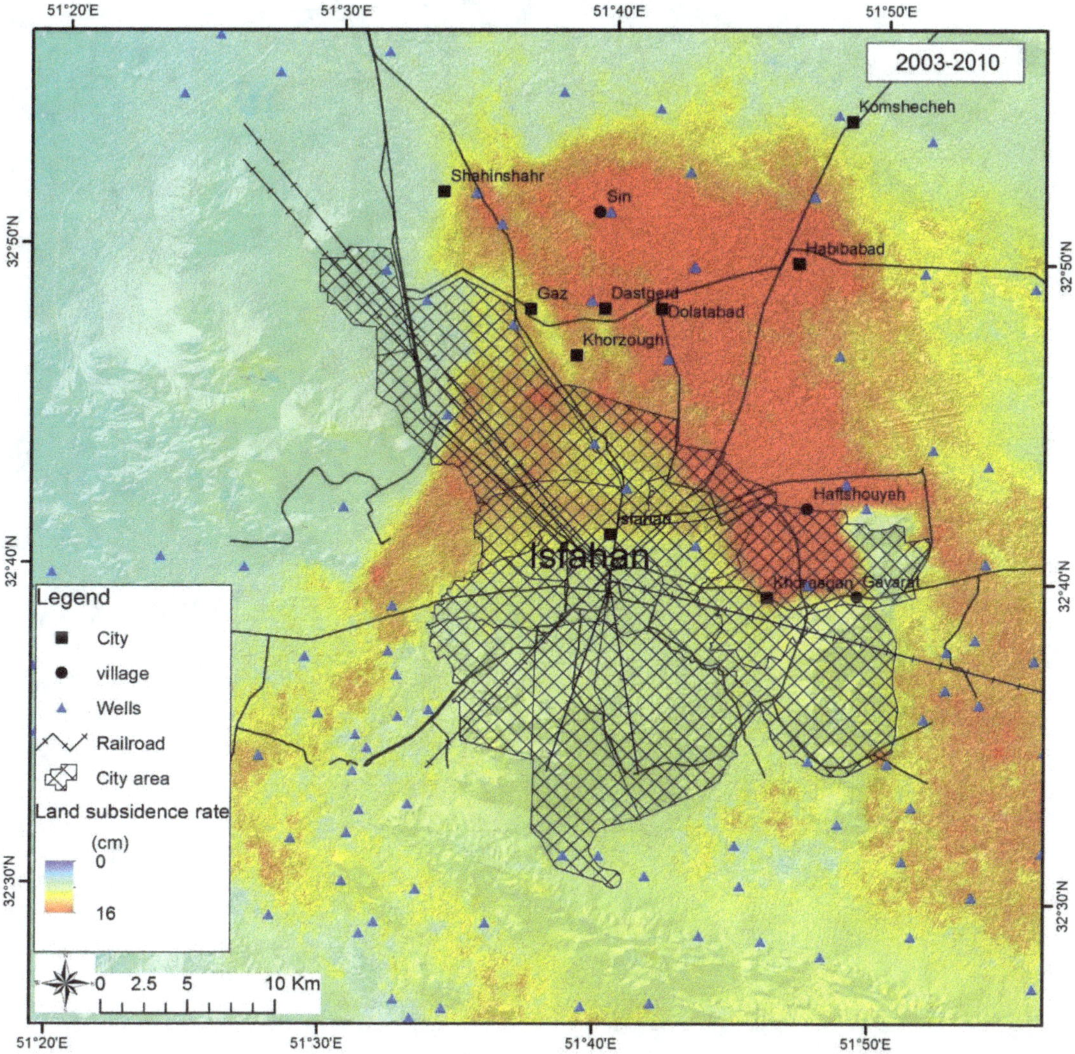

FIGURE 2.12 Phase-to-displacement conversion of a pair of Advanced Synthetic Aperture Radar (ASAR) radar images captured between 2003 and 2010 as part of a time series analysis of land subsidence in Isfahan, Iran.

$$\sqrt{\frac{1-\gamma^2}{2\gamma^2}}.\frac{\lambda}{4\pi} \tag{2.8}$$

where γ represents interferometric coherence

Fig. 2.13 illustrates matching of the differential leveling data with a land subsidence map generated using time series of ASAR radar images and the DInSAR method. The figure shows that the maximum ground subsidence observed through differential leveling

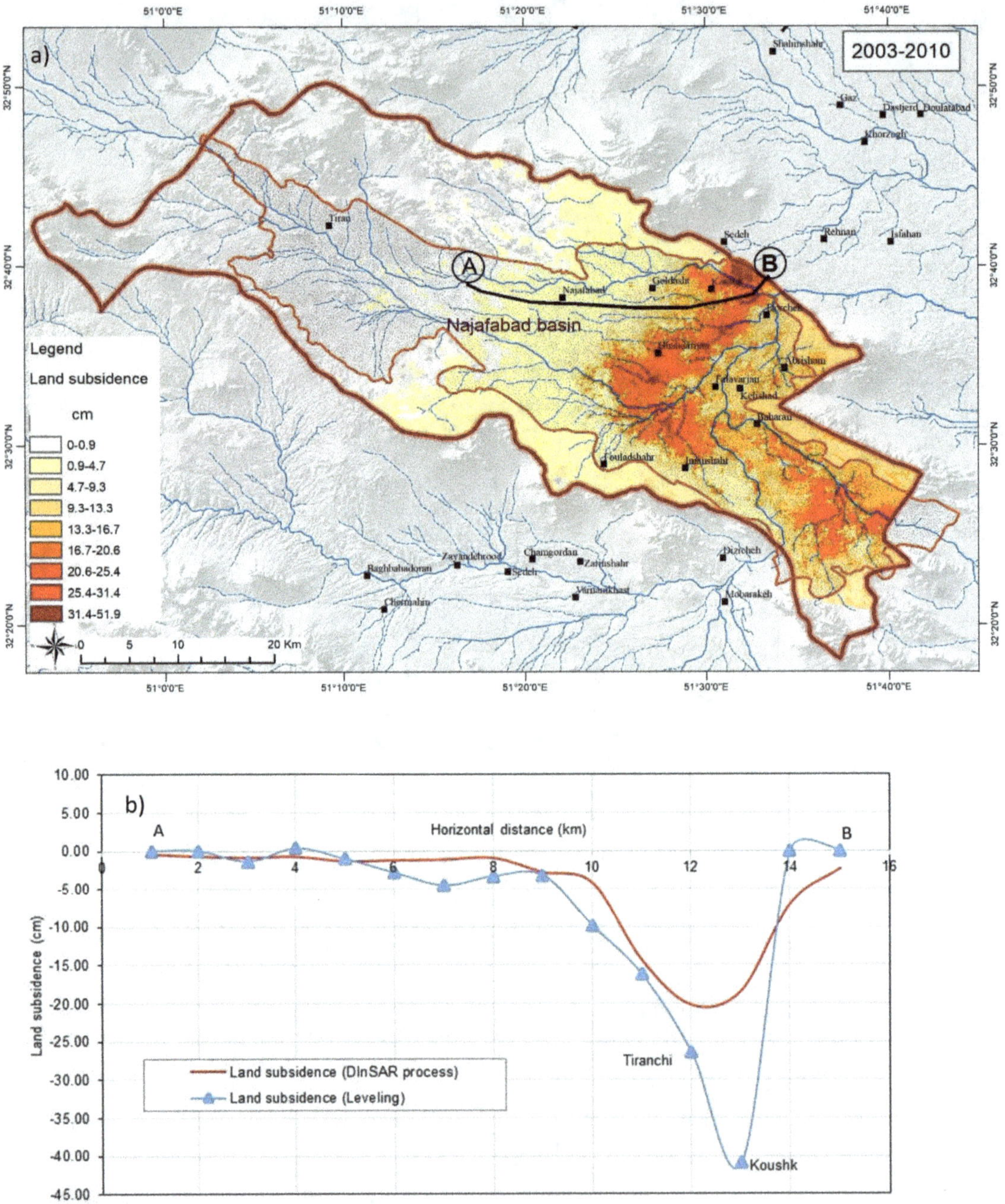

FIGURE 2.13 (A) Land subsidence monitoring using Differential Interferometric Synthetic Aperture Radar (DInSAR) and differential leveling. The leveling data with an east–west trend shown as solid black circles in figure. (B) The figure indicates that the highest level of land subsidence has occurred in the Koshk and Tiranchi areas of the Najafabad plain in Isfahan, Iran.

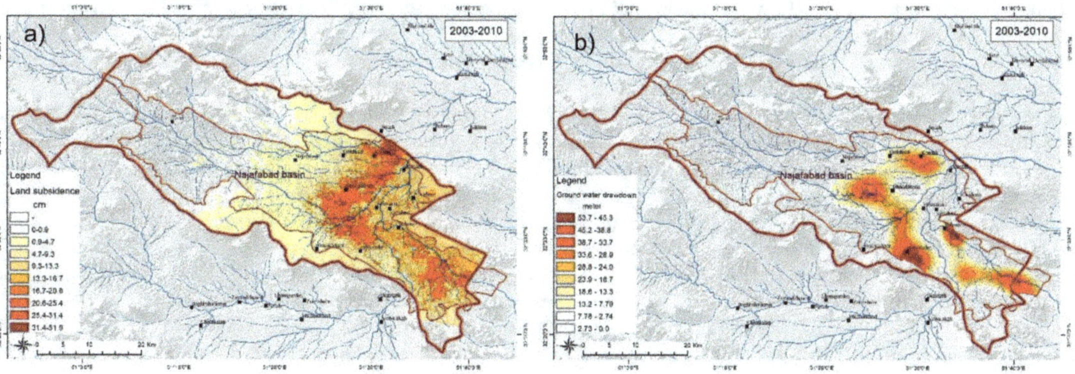

FIGURE 2.14 a) Total land subsidence generated from Differential Interferometric Synthetic Aperture Radar (DInSAR) processing and, b) the cumulative groundwater drawdown between 2003 and 2010.

corresponds to the maximum subsidence detected through the processing of radar images using the DInSAR method. The close correlation between the results of DInSAR and differential leveling highlights the effectiveness of DInSAR in accurately monitoring land subsidence over time.

To verify the accuracy of the land subsidence zones, they are compared not only with leveling data but also with zones of groundwater drawdown. Comparing the land subsidence generated by DInSAR processing with groundwater drawdown over a specific time period (Fig. 2.14) ensures the correctness of the zoning maps. The figures clearly show that the areas with the highest groundwater drawdown (Fig. 2.14B) correspond to the areas with maximum land subsidence (Fig. 2.14A).

In conclusion, the DInSAR technique has become an essential tool for monitoring land subsidence worldwide. Its accuracy, cost-effectiveness, and ability to monitor large areas make it an invaluable aid for decision-making processes related to land use and water planning and management. The continued development and application of this technique will improve understanding of land subsidence and its impacts on natural and human systems.

References

Bai, Z., Wang, Y., Balz, T., 2022. Beijing land subsidence revealed using PS-InSAR with long time series TerraSAR-X SAR data. Rem. Sens. 14 (11), 2529.

Bamler, R., Hartl, P., 1998. Synthetic aperture radar interferometry. Inverse Probl. 14 (4), R1–R54.

Baran, I., Stewart, M.P., Kampes, B.M., Perski, Z., Lilly, P., 2003. A modification to the Goldstein radar interferogram filter. IEEE Trans. Geosci. Rem. Sens. 41 (9), 2114–2118.

Bawden, G.W., Thatcher, W., Stein, R.S., Hudnut, K.W., Peltzer, G., 2001. Tectonic contraction across Los Angeles after removal of groundwater pumping effects. Nature 412, 812–815.

Bekaert, D.P.S., Hooper, A., Wright, T.J., 2016. A spatially variable power law tropospheric correction technique for InSAR data. J. Geophys. Res. Solid Earth 121, 559–574.

Berardino, P., Fornaro, G., Lanari, R., Sansosti, E., 2002. A new algorithm for surface deformation monitoring based on small baseline differential SAR interferograms. IEEE Trans. Geosci. Rem. Sens. 40 (11), 2375–2383.

Chen, C.W., Zebker, H.A., 2000. Network approaches to two-dimensional phase unwrapping: intractability and two new algorithms. JOSA A 17 (3), 401–414.

Chen, C.W., Zebker, H.A., 2001. Two-dimensional phase unwrapping with use of statistical models for cost functions in nonlinear optimization. JOSA A 18 (2), 338–351.

Chen, C.W., Zebker, H.A., 2002. Phase unwrapping for large SAR interferograms: statistical segmentation and generalized network models. IEEE Trans. Geosci. Rem. Sens. 40 (8), 1709–1719.

Coates, A.R., Dennison, P.E., Roberts, D.A., Roth, K.L., 2015. Monitoring the impacts of severe drought on southern California chaparral species using hyperspectral and thermal infrared imagery. Rem. Sens. 7 (11), 14276–14291.

Costantini, M., 1998. A novel phase unwrapping method based on network programming. IEEE Trans. Geosci. Rem. Sens. 36 (3), 813–821.

Crosetto, M., Monserrat, O., Cuevas-González, M., Devanthéry, N., Crippa, B., 2016. Persistent scatterer interferometry: a review. ISPRS J. Photogrammetry Remote Sens. 115, 78–89.

Cui, R.Y., Calvin, K., Clarke, L., Hejazi, M., Kim, S., Kyle, P., Patel, P., Turner, S., Wise, M., 2018. Regional responses to future, demand-driven water scarcity. Environ. Res. Lett. 13 (9), 094006.

Cumming, I.G., Wong, F.H., 2005. Digital processing of synthetic aperture radar data. Artech House 1 (3), 108–110.

Deilami, K., Hashim, M., 2011. Very high resolution optical satellites for DEM generation: a review. Eur. J. Sci. Res. 49 (4), 542–554.

Dokka, R.K., 2011. The role of deep processes in late 20th century subsidence of New Orleans and coastal areas of southern Louisiana and Mississippi. J. Geophys. Res. 116. B06403.

Eineder, M., Minet, C., Cong, X., Fritz, T., Steigenberger, P., 2010. Towards imaging geodesy-Achieving centimetre pixel localization accuracy with TerraSAR-X. In: 8th European Conference on Synthetic Aperture Radar. VDE, pp. 1–4.

Famiglietti, J.S., Lo, M., Ho, S.L., Bethune, J., Anderson, K.J., Syed, T.H., Swenson, S.C., de Linage, C.R., Rodell, M., 2011. Satellites measure recent rates of groundwater depletion in California's Central Valley. Geophys. Res. Lett. 38. https://doi.org/10.1029/2010GL046442.

Farr, T.G., Rosen, P.A., Caro, E., Crippen, R., Duren, R., Hensley, S., Kobrick, M., Paller, M., Rodriguez, E., Roth, L., Seal, D., 2007. The shuttle radar topography mission. Rev. Geophys. 45 (2).

Fattahi, H., Amelung, F., 2013. DEM error correction in InSAR time series. IEEE Trans. Geosci. Rem. Sens. 51 (7), 4249–4259.

Faunt, C.C., Sneed, M., Traum, J., Brandt, J.T., 2016. Water availability and land subsidence in the Central Valley, California, USA. Hydrogeol. J. 24 (3), 675.

Fellenius, B.H., 2004. Unified design of piled foundations with emphasis on settlement analysis. In: Current Practices and Future Trends in Deep Foundations, pp. 253–275.

Feng, Q.Y., Liu, G.J., Lei, M.E.N.G., Fu, E.J., Zhang, H.R., Zhang, K.F., 2008. Land subsidence induced by groundwater extraction and building damage level assessment—a case study of Datun, China. J. China Univ. Min. Technol. 18 (4), 556–560.

Ferguson, G., Gleeson, T., 2012. Vulnerability of coastal aquifers to groundwater use and climate change. Nat. Clim. Change 2, 342–345.

Ferretti, A., Prati, C., Rocca, F., 2000. Nonlinear subsidence rate estimation using permanent scatterers in differential SAR interferometry. IEEE Trans. Geosci. Rem. Sens. 38 (5), 2202–2212.

Ferretti, A., Prati, C., Rocca, F., 2001. Permanent scatterers in SAR interferometry. IEEE Trans. Geosci. Rem. Sens. 39 (1), 8–20.

Ferretti, A., Monti-Guarnieri, A., Prati, C., Rocca, F., Massonet, D., 2007. InSAR Principles-Guidelines for SAR Interferometry Processing and Interpretation, vol. 19.

Ferretti, A., Fumagalli, A., Novali, F., Prati, C., Rocca, F., Rucci, A., 2011. A new algorithm for processing interferometric data-stacks: SqueeSAR. IEEE Trans. Geosci. Rem. Sens. 49 (9), 3460–3470.

Ferretti, A., 2014. Satellite InSAR Data: Reservoir Monitoring from Space (EET 9). The European Association of Geoscientists and Engineers (EAGE) Publications.

Gabriel, A.K., Goldstein, R.M., 1988. Crossed orbit interferometry: theory and experimental results from SIR-B. Int. J. Remote Sens. 9 (15), 857–872.

Galloway, D., Jones, D.R., Ingebritsen, S.E., 1999. Land subsidence in the United States. U.S. Geol. Surv. Circular 1182, 175.

Ghiglia, D.C., Romero, L.A., 1994. Robust two-dimensional weighted and unweighted phase unwrapping that uses fast transforms and iterative methods. JOSA A 11 (1), 107–117.

Ghiglia, D.C., Pritt, M.D., 1998. Two-dimensional Phase Unwrapping: Theory, Algorithms, and Software. Wiely-Interscience.

Ghulam, A., Amer, R., Ripperdan, R., 2010. A filtering approach to improve deformation accuracy using large baseline, low coherence DInSAR phase images. In: 2010 IEEE International Geoscience and Remote Sensing Symposium. IEEE, pp. 3494–3497.

Goldstein, R.M., Werner, C.L., 1998. Radar interferogram filtering for geophysical applications. Geophys. Res. Lett. 25 (21), 4035–4038.

Goldstein, R.M., Zebker, H.A., Werner, C.L., 1988. Satellite radar interferometry: two-dimensional phase unwrapping. Radio Sci. 23 (4), 713–720.

Guarnieri, A.M., Guccione, P., Pasquali, P., Desnos, Y.L., 2003. Multi-mode ENVISAT ASAR interferometry: techniques and preliminary results. IEE Proc.–Radar, Sonar Navig. 150 (3), 193–200.

Hanssen, R.F., 2001. Radar Interferometry: Data Interpretation and Error Analysis. Springer Science & Business Media.

Holzer, T.L., Galloway, D., 2005. Impacts of land subsidence caused by withdrawal of underground fluids in the United States. In: Humans as Geologic Agents, 16, p. 87. https://doi.org/10.1130/2005.4016(08).

Hooper, A., Zebker, H.A., 2007. Phase unwrapping in three dimensions with application to InSAR time series. JOSA A 24 (9), 2737–2747.

Hooper, A., Segall, P., Zebker, H., 2007. Persistent scatterer interferometric synthetic aperture radar for crustal deformation analysis, with application to Volcán Alcedo, Galápagos. J. Geophys. Res. Solid Earth 112 (B7).

Hooper, A., 2008. A multi-temporal InSAR method incorporating both persistent scatterer and small baseline approaches. Geophys. Res. Lett. 35 (16).

Hooper, A., Zebker, H., Segall, P., Kampes, B., 2004. A new method for measuring deformation on volcanoes and other natural terrains using InSAR persistent scatterers. Geophys. Res. Lett. 31 (23).

Jiang, H., Balz, T., Cigna, F., Tapete, D., 2021. Land subsidence in Wuhan revealed using a non-linear PSInSAR approach with long time series of COSMO-SkyMed SAR data. Rem. Sens. 13 (7), 1256.

Jolivet, R., Grandin, R., Lasserre, C., Doin, M.P., Peltzer, G., 2011. Systematic InSAR tropospheric phase delay corrections from global meteorological reanalysis data. Geophys. Res. Lett. 38 (17).

Judge, T.R., Bryanston-Cross, P.J., 1994. A review of phase unwrapping techniques in fringe analysis. Opt Laser. Eng. 21 (4), 199–239.

Just, D., Bamler, R., 1994. Phase statistics of interferograms with applications to synthetic aperture radar. Appl. Opt. 33 (20), 4361–4368.

Kampes, B.M., Hanssen, R.F., Perski, Z., 2003. Radar interferometry with public domain tools. In: Proceedings of FRINGE, vol. 3.

Karimzadeh, S., Matsuoka, M., 2020. Remote Sensing X-Band SAR Data for land subsidence and pavement monitoring. Sensors 20 (17), 4751.

Ketelaar, V.G., 2009. Satellite Radar Interferometry: Subsidence Monitoring Techniques, vol. 14. Springer Science & Business Media.

Kim, J.W., Lu, Z., 2018. Association between localized geohazards in West Texas and human activities, recognized by Sentinel-1A/B satellite radar imagery. Sci. Rep. 8 (1), 4727.

Konikow, L.F., Kendy, E., 2005. Groundwater depletion: a global problem. Hydrogeol. J. 13, 317–320.

Leake, S.A., Galloway, D.L., 2007. MODFLOW Ground-Water Model: User Guide to the Subsidence and Aquifer-System Compaction Package (SUB-WT) for Water-Table Aquifers. US Geological Survey.

Li, Z., Muller, J.P., Cross, P., Fielding, E.J., 2005. Interferometric synthetic aperture radar (InSAR) atmospheric correction: GPS, Moderate Resolution Imaging Spectroradiometer (MODIS), and InSAR integration. J. Geophys. Res. Solid Earth 110 (B3).

Lillesand, T.M., Kiefer, R.W., Chipman, J.W., 2015. Remote Sensing and Image Interpretation, seventh ed. John Wiley & Sons.

Lin, Q., Vesecky, J.F., Zebker, H.A., 1992. New approaches in interferometric SAR data processing. IEEE Trans. Geosci. Rem. Sens. 30 (3), 560–567.

Liu, X., Zhu, W., Lian, X., Xu, X., 2023. Monitoring mining surface subsidence with multi-temporal three-dimensional unmanned aerial vehicle point cloud. Rem. Sens. 15 (2), 374.

López-Quiroz, P., Doin, M.P., Tupin, F., Briole, P., Nicolas, J.M., 2009. Time series analysis of Mexico City subsidence constrained by radar interferometry. J. Appl. Geophys. 69 (1), 1–15.

Lyons, S.N., Sandwell, D.T., 2003. Fault creep along the southern San Andreas from interferometric synthetic aperture radar, permanent scatterers, and stacking. J. Geophys. Res. Solid Earth. https://doi.org/10.1029/2002JB001831.

Madsen, S.N., Zebker, H.A., Martin, J.A.N.M., 1993. Topographic mapping using radar interferometry: processing techniques. IEEE Trans. Geosci. Rem. Sens. 31 (1), 246–256.

Massonnet, D., Feigl, K.L., 1998. Radar interferometry and its application to changes in the earth's surface. Rev. Geophys. 36 (4). https://doi.org/10.1029/97RG03139.

Massonnet, D., Rossi, M., Carmona, C., Adragna, F., Peltzer, G., Feigl, K., Rabaute, T., 1993. The displacement field of the Landers earthquake mapped by radar interferometry. Nature 364, 138–142.

Mazzotti, S., Lambert, A., Van der Kooij, M., Mainville, A., 2009. Impact of anthropogenic subsidence on relative sea-level rise in the Fraser River delta. Geology 37 (9), 771–774. https://doi.org/10.1130/G25640A.1.

Merchant, J.W., 2000. Remote sensing of the environment: an earth resource perspective. Cartogr. Geogr. Inf. Sci. 27 (4), 311.

Mora, O., Mallorqui, J.J., Broquetas, A., 2003. Linear and nonlinear terrain deformation maps from a reduced set of interferometric SAR images. IEEE Trans. Geosci. Rem. Sens. 41 (10), 2243–2253.

Navarro-Hernández, M.I., Tomás, R., Lopez-Sanchez, J.M., Cárdenas-Tristán, A., Mallorquí, J.J., 2020. Spatial analysis of land subsidence in the San Luis potosi valley induced by aquifer overexploitation using the coherent pixels technique (CPT) and sentinel-1 insar observation. Rem. Sens. 12 (22), 3822.

Nyer, E.K., 1998. Groundwater and Soil Remediation: Practical Methods and Strategies, vol. 2. CRC Press.

Orhan, O., 2021. Monitoring of land subsidence due to excessive groundwater extraction using small baseline subset technique in Konya, Turkey. Environ. Monit. Assess. 193 (4), 174.

Osmanoğlu, B., Dixon, T.H., Wdowinski, S., Cabral-Cano, E., Jiang, Y., 2011. Mexico City subsidence observed with persistent scatterer InSAR. Int. J. Appl. Earth Obs. Geoinf. 13 (1), 1–12.

Perissin, D., Wang, T., 2011. Repeat-pass SAR interferometry with partially coherent targets. IEEE Trans. Geosci. Rem. Sens. 50 (1), 271–280.

Poland, J.F., 1975. Land Subsidence, in the San Joaquin Valley, California, as of 1972: A History of Land Subsidence Caused by Water-Level Decline in the San Joaquin Valley, from the 1920's to 1972, vol. 437. US Government Printing Office.

Poland, J.F., 1984. Guidebook to Studies of Land Subsidence Due to Ground-Water Withdrawal. UNESCO, Paris, p. 305.

Prati, C., Rocca, F., Monti Guarnieri, A., Pasquali, P., 1994. ERS-1 SAR Interferometric Techniques and Applications. ESA Report No. 3- 7439/92/HGE-1.

Qin, Y., Perissin, D., 2015. Monitoring ground subsidence in Hong Kong via spaceborne Radar: experiments and validation. Rem. Sens. 7 (8), 10715–10736.

Reigber, A., Moreira, J., 1997. August. Phase unwrapping by fusion of local and global methods. In: IGARSS'97. 1997 IEEE International Geoscience and Remote Sensing Symposium Proceedings. Remote Sensing-A Scientific Vision for Sustainable Development, vol. 2. IEEE, pp. 869–871.

Rosen, P.A., Hensley, S., Joughin, I.R., Li, F.K., Madsen, S.N., Rodriguez, E., Goldstein, R.M., 2000. Synthetic aperture radar interferometry. Proc. IEEE 88 (3), 333–382.

Scanlon, B.R., Fakhreddine, S., Rateb, A., de Graaf, I., Famiglietti, J., Gleeson, T., Grafton, R.Q., Jobbagy, E., Kebede, S., Kolusu, S.R., Konikow, L.F., 2023. Global water resources and the role of groundwater in a resilient water future. Nat. Rev. Earth Environ. 1–15.

Schmidt, D.A., Bürgmann, R., 2003. Time-dependent land uplift and subsidence in the Santa Clara valley, California, from a large interferometric synthetic aperture radar data set. J. Geophys. Res. Solid Earth 108 (B9).

Shirzaei, M., Bürgmann, R., 2018. Global climate change and local land subsidence exacerbate inundation risk to the San Francisco Bay Area. Sci. Adv. 4. https://doi.org/10.1126/sciadv.aap7566 eaap7566.

Wadge, G., Webley, P.W., James, I.N., Bingley, R., Dodson, A., Waugh, S., Veneboer, T., Puglisi, G., Mattia, M., Baker, D., Edwards, S.C., 2002. Atmospheric models, GPS and InSAR measurements of the tropospheric water vapour field over Mount Etna. Geophys. Res. Lett. 29 (19), 1–11.

Witoszynskyj, S., Rauscher, A., Reichenbach, J.R., Barth, M., 2009. Phase unwrapping of MR images using ΦUN—A fast and robust region growing algorithm. Med. Image Anal. 13 (2), 257–268.

Xu, W., Cumming, I., 1999. A region-growing algorithm for InSAR phase unwrapping. IEEE Trans. Geosci. Rem. Sens. 37 (1), 124–134.

Zebker, H.A., Villasenor, J., 1992. Decorrelation in interferometric radar echoes. IEEE Trans. Geosci. Rem. Sens. 30 (5), 950–959.

Zhang, B., Li, J., Ren, H., 2019. Using phase unwrapping methods to apply D-InSAR in mining areas. Can. J. Rem. Sens. 45 (2), 225–233.

Zhao, Y., Zhao, X., Dai, J., Yu, W., 2021. Analysis of the surface subsidence induced by mining near-surface thick lead-zinc deposit based on numerical simulation. Processes 9, 717. https://doi.org/10.3390/pr9040717.

Zhou, C., Lan, H., Gong, H., Zhang, Y., Warner, T.A., Clague, J.J., Wu, Y., 2020. Reduced rate of land subsidence since 2016 in Beijing, China: evidence from Tomo-PSInSAR using RadarSAT-2 and sentinel-1 datasets. Int. J. Rem. Sens. 41 (4), 1259–1285.

Zhu, X.X., Bamler, R., 2010. Very high resolution spaceborne SAR tomography in urban environment. IEEE Trans. Geosci. Rem. Sens. 48 (12), 4296–4308. https://studycrumb.com/alphabetizer.

CHAPTER

3

Remotely sensed prediction of soil organic carbon

Kourosh Shirani[1] *and Mehrdad Pasandi*[2]

[1]Soil Conservation and Watershed Management Research Institute, Agricultural Research, Education and Extension Organization (AREEO), Tehran, Iran; [2]Department of Geology, University of Isfahan, Isfahan, Iran

3.1 Introduction

Organic carbon is an important indicator of soil quality affecting many physical, chemical, and biological properties of the soil (Basche and DeLonge, 2019) and a vital component in achieving sustainable agriculture and effective ecosystem management. In dry ecosystems, it plays a crucial role in reducing the effects of greenhouse gases, increasing crop yield and soil fertility, reducing soil erosion, and increasing the water and food storage capacity, among others. Thus, it is considered one of the factors that determine soil health, with titles such as "the heart of sustainable agriculture" and a vital part of production systems. Managing organic carbon is an unavoidable necessity, especially in soils where organic carbon concentration is less than 1%.

Studies indicate that from the 1920s to the early 1990s, soil erosion control was a priority of the U.S. Department of Agriculture and many other countries (Paustian et al., 2016). The objective was to control soil erosion to a level that maintains soil fertility in the long term. However, with extensive research, many additional functions of organic carbon in terrestrial ecosystems have been discovered. Following this discovery, the U.S. Department of Agriculture replaced soil-organic matter management with soil conservation management programs (Lal, 2004). By integrating organic matter management, in addition to controlling soil erosion, other vital functions are also achieved. These functions include providing carbon and energy needs for soil microorganisms forming soil grains and maintaining soil particles (erosion control), improving the soil air retention capacity, and improving its ventilation. It also optimizes maintenance of soil nutrients by increasing cationic and anion exchange capacity, increasing soil fertility and performance, reducing the effects of greenhouse gases through carbon storage in the soil, increasing plant resistance against root and leaf diseases (Tsiafouli et al., 2015),

https://doi.org/10.1016/B978-0-443-15341-9.00026-5

improving soil structure and increasing its permeability (Bronick and Lal, 2005), and increasing biological diversity and soil microbial biomass, etc.

In addition to the benefits mentioned above, the concentration of organic carbon in soil is also linked to the amount of water it can retain (Rawls et al., 2003). In dry environments, for example, soils with higher organic carbon concentrations can retain moisture for longer periods, making them better able to support plant growth and thrive in such environments (Bronick and Lal, 2005). Furthermore, the concentration of organic carbon in soil is also linked to the soil's ability to sequester carbon from the atmosphere, thus mitigating the effects of climate change (Lal, 2004).

Therefore, managing organic carbon in soil is essential for maintaining soil health while also achieving a variety of essential ecological, agricultural, and environmental goals (Basche and DeLonge, 2019). The management of organic carbon facilitates the achievement of multiple ecological, agricultural, and environmental goals while also ensuring soil health and the reduction of greenhouse gas emissions (Smith et al., 2016; Minasny et al., 2017; Basche and DeLonge, 2019). However, managing organic carbon is not always straightforward, and factors such as land use, management practices, and climate all play a significant role in predicting the outcomes of various strategies.

Despite the advantages attributed to soil organic carbon (SOC), there are also challenges and disagreements among researchers. The idea that the higher the organic carbon content of the soil, the better the condition of the soil and is widely debated when discussing the concept of soil quality (Andrews et al., 2002). Another challenge is the definition of SOC by different researchers. Baldock (2007), by examining various sources, classified the definitions into five groups:

1. Surface plant residue, which is a combination of plant material residues on the soil surface, including leaves, pasture materials, or crops, and takes several days to several years to decompose.
2. Buried plant residue, which includes materials larger than 2 mm found in the soil profile and requires a period of several days to several years to decompose.
3. Particulate organic matter (POC), which includes semi-degraded organic materials smaller than 2 mm and larger than 50 μm and also takes a long time to decompose, similar to the previous definitions.
4. Soil organic carbon, known as mineral-associated organic carbon or humus, refers to organic materials that are optimally decomposed, smaller than 50 μm, and attached to soil particles. Decomposition of this category of soil organic matter (SOM) can take several years or even several decades.
5. SOC is resistant organic carbon (ROC), which includes charcoal or burnt materials obtained from the combustion of organic matter that are resistant to biological decomposition. Decomposition of this category of SOM, however, can take decades to even thousands of years.

Considering the above cases, improving the quality of pastures and addressing the challenges of chaotic ecosystem management rely heavily on effective carbon reserve management in the soil (Minasny et al., 2017; Basche and DeLonge, 2019). Accurate information on the current state of organic carbon reserves in pastures is crucial for planning and decision-making (Smith et al., 2016). However, obtaining basic information about current

SOC reserves can be difficult due to the prevailing conditions of semi-steppe pastures, which are large and difficult to access (Smith et al., 2016).

Determining the organic carbon content of surface soil is important for several reasons.

Firstly, SOC is a critical component of soil health and fertility. It plays a key role in soil structure, water retention, nutrient cycling, and the overall productivity of agricultural and natural ecosystems (Lal, 2004). Understanding the level and distribution of SOC in a given area can help inform land management practices and improve soil health. Secondly, SOC is a major component of the global carbon cycle and has important implications for climate change. Soil acts as both a source and sink of carbon, with SOC sequestering carbon from the atmosphere and reducing the concentration of greenhouse gases (Lal et al., 2008). Accurately measuring the SOC content of soil is essential for understanding the role of soils in the carbon cycle and for developing effective strategies for mitigating climate change. Finally, the determination of SOC content is important for monitoring the quality and health of soil resources, which are essential for food security and sustainable development. In this regard, SOC content is often used as an indicator of soil quality, and monitoring changes in SOC over time can help identify areas of degradation or improvement in soil health.

To estimate SOC, two methods are used: direct and indirect. The direct method involves sampling and determining SOC content in different soil or plant types. Nonetheless, the main drawback of this method is that it is only capable of distinguishing differences in SOC content between different types of soils or plants and not within a single type of plant or soil. In contrast, the indirect method, a process-oriented method, is based on the carbon cycle model of an ecosystem. This method estimates the amount of organic carbon in the soil by utilizing key and more important parameters that simplify the complexity of process-oriented models such as the carbon-cycle and sustainability analysis (CASA) model (Potter et al., 1993) and CENTRY model (Parton et al., 1988).

The CASA model can be used to estimate SOC at regional and global scales. The model uses inputs of climate, soil properties, vegetation, and land cover to estimate SOC in agricultural, forest, and grassland ecosystems. The CASA model mathematically simulates carbon transfers and transformations in vegetation and soils based on known processes and rates. The model divides SOC into three pools: active (fast turnover, 0–1 year), slow (5–15 years), and passive (20–500 years). The size and decomposition rates of these pools depend on soil texture, moisture, temperature, and quality of organic inputs. Heterotrophic respiration releases CO_2 from decomposing SOM and plant inputs, with higher rates of decomposition and respiration occurring under warm, wet conditions. Photosynthesis fixes CO_2 from the atmosphere into living plant biomass, and the model tracks subsequent litterfall and mortality inputs to SOM. The model requires soil data on texture, bulk density, and moisture as well as daily weather data including minimum and maximum temperature, precipitation, and solar radiation or percent of annual sunshine hours. Mean monthly net primary productivity values are used to specify carbon inputs from vegetation. Steady-state simulations of 100–200 years are needed to produce realistic SOC results for a given location. Model performance can be tested against field measurements of SOC from long-term experiments or chronosequences.

The CENTURY model is a SOM simulation model used to estimate SOC dynamics over long timescales. CENTURY uses inputs of climate, soil properties, plant productivity, and land management to model carbon and nutrient transfers through vegetation, litter, and

SOM. The model divides SOC and litter into active (fast turnover), slow, and passive (slowest turnover) pools. Carbon inputs from plant litter and roots are balanced with outputs through decomposition and heterotrophic respiration. CENTURY requires daily weather data (precipitation, maximum/minimum temperature), soil texture, bulk density, and pH. Monthly net primary productivity values determine carbon inputs from vegetation. The model incorporates the effects of temperature, moisture, and soil properties on decomposition and SOC stabilization. CENTURY also accounts for the effects of agricultural management such as fertilizer applications, erosion, and tillage on carbon inputs and losses. CENTURY has been applied at scales ranging from individual farm fields to the entire globe. The model can be used to study historical SOC change, evaluate sustainable management practices, and project future SOC levels under different climate and policy scenarios. Some key limitations of CENTURY include its simplified simulation of soil aggregation and protection processes and omission of lateral carbon flows.

Another notable process-oriented model is the Pasture Simulation Model (PaSim) is a process-based model that simulates SOC (SOC) dynamics in grassland systems. PaSim was developed by Riedo et al. (1998) and has been applied to study SOC in relation to climate, soil properties, and grassland management. The model divides SOC and plant litter into metabolic (fast turnover), structural (medium turnover), and recalcitrant (slow turnover) pools. Carbon inputs include monthly net primary productivity of vegetation and roots, while outputs include CO_2 losses from decomposition. PaSim requires inputs of monthly weather data, soil properties (texture, bulk density, pH), plant productivity, and management (grazing, fertilization). The model uses these inputs to simulate plant growth and senescence, litter production, humification, and mineralization based on interacting factors such as temperature, moisture, substrate quality, and soil microbial activity (Riedo et al., 1998; Lardy et al., 2011). Higher decomposition rates occur under warm, moist conditions, while SOC stabilization depends on soil clay content and chemical recalcitrance of organic inputs. Some limitations of PaSim include its simplified water balance and N dynamics relative to other grassland models. Additional field testing and calibration may be needed when applying PaSim to new geographic regions or management systems.

However, the use of simplified models increases the level of uncertainty in estimating the amount of organic carbon in the soil (Zhou et al., 2008). Traditional methods for determining SOC are time-consuming and expensive and also limited in providing continuous information, which makes map production challenging (Basche and DeLonge, 2019). Traditional methods for estimating SOC are only applicable at small, local scales and are not feasible at larger regional, national, and international scales due to the high time and financial costs involved (Basche and DeLonge, 2019). Therefore, new methods of predicting the amount of organic carbon reserves in grassland soil are necessary (Minasny et al., 2017). These methods should provide accurate results at a low cost and offer a continuous map of the area of interest (Smith et al., 2016). In recent years, alternative methods that are faster and less expensive have been developed to measure SOC due to the time-consuming and expensive nature of traditional measurement methods in chemical laboratories. To address this issue, many researchers have turned to remote sensing, which provides a low-cost and efficient way of accessing information about remote areas and evaluating spatial changes in carbon over time (Minasny et al., 2017). Remote sensing offers a suitable solution to this need by eliminating the need for physical presence and extensive coverage of the studied area, and remote sensing allows for a more comprehensive view of the region of interest (Basche and DeLonge, 2019). Remote sensing techniques at large spatial levels have been identified as a viable option to reduce costs (Zucca et al., 2015; Glenn et al., 2008; Zucca et al., 2015). The cost of using this technology is low as satellite images cover a large surface area, and manpower requirements are minimal. Remote sensing is an economically viable and efficient alternative to traditional methods for predicting and monitoring organic carbon reserves in grassland soil (Smith et al., 2016; Minasny et al., 2017).

3.1.1 Factors affecting the organic carbon content of soil

There are many factors that can affect the organic carbon content of soil. Here are some of the most important ones:

1. Climate: Temperature and precipitation can influence the rate of decomposition of organic matter in soil, which in turn affects the amount of organic carbon stored in the soil (Davidson and Janssens, 2006).
2. Soil texture: The texture of soil affects its ability to hold water and nutrients, which can impact the growth and decomposition of plants, as well as the amount of organic matter that accumulates in the soil (Virto et al., 2012).
3. Vegetation: The type and amount of vegetation in an area can impact the amount of organic matter that accumulates in the soil. For example, grasslands tend to have higher levels of organic carbon than forests, which have more carbon stored in woody biomass.
4. Land use: Land use practices such as crop rotation, tillage, and grazing can impact the amount of organic matter that accumulates in soil. For example, tillage can accelerate the decomposition of organic matter, while no-till practices can help retain carbon in the soil.
5. Soil pH: Soil pH can impact the activity of microorganisms that decompose organic matter, which can in turn affect the amount of organic carbon stored in the soil (Rousk et al., 2009).
6. Soil drainage: Poor soil drainage can limit the growth of vegetation and the accumulation of organic matter in soil, while well-drained soils can support more vegetation and organic matter accumulation (Schipper et al., 2007).
7. Parent material: The type of rock or sediment that soil develops from can impact its fertility and the amount of organic matter that accumulates in the soil. For example, soils that develop from volcanic ash tend to be more fertile and have higher organic carbon content than soils that develop from sandstone (Zhou et al., 2008).
8. Soil biota: Soil organisms such as bacteria, fungi, and earthworms play a crucial role in the decomposition of organic matter and the cycling of nutrients. Changes in the composition or abundance of soil biota can impact the rate of decomposition and the amount of organic carbon stored in soil (Fontaine et al., 2003).
9. Management practices: Management practices such as fertilization, irrigation, and pesticide use can impact the growth of vegetation and the amount of organic matter that accumulates in soil (Simpson et al., 2004).

10. Fire: Wildfires can result in the loss of organic matter and carbon from soil, while prescribed burning can stimulate the growth of vegetation and increase the accumulation of organic matter (Coleman et al., 2017).
11. Geographical location: SOC content can vary widely depending on the geographical location of the soil. For example, soils in tropical rainforests tend to have higher levels of organic carbon than soils in arid regions (Eswaran et al., 1993).
12. Time: The amount of time that organic matter is present in soil can impact the amount of carbon that is stored. For example, ancient soils that have been accumulating organic matter for thousands of years tend to have high levels of organic carbon compared to newly formed soils.

Understanding the complex interactions between the factors is essential for accurately predicting and managing SOC content and for developing effective strategies for climate change mitigation and adaptation.

3.1.2 Remote sensing

Remote sensing was once exclusively a scientific and research subject, but it has since expanded and become widely utilized across various fields. Today, this technology is considered one of the most important sources for obtaining spatial and thematic data. Remote sensing is used in various practical and scientific fields, such as estimating agricultural production and the extent of agricultural land, investigating subsurface cultivation of specific products, determining product condition and characteristics, classifying forest cover and type, evaluating pastures, producing land use maps, analyzing geology, soil science, and geomorphology, supervising and regulating urban development and road networks, studying water pollution, evaluating water reserves, measuring snow, studying glaciers, analyzing oceanography, and quantifying the damage caused by earthquakes, floods, fires, wars, and droughts, and for military and security purposes.

Due to the high correlation between plant spectral indices and SOC, many studies now use these indices as the input layers to create predictive models. Vegetation spectral indices help to provide information about vegetation on the ground by measuring reflections in different parts of the electromagnetic spectrum of a sensor (Campbell and Wynne, 2011).

These indices are determined by the spectral behavior of vegetation in different bands. These bands can be composed of different values such as gray degrees (digital number or DN), radiance, or reflectance. Gray values, which represent numeric values of pixels, are based on the energy recorded in the sensor and are the simplest and most accessible level of data available to users. Reflectance, on the other hand, is the ratio of the energy reflected from the surface to the total energy reaching it.

Calculating plant indices can be challenging due to various factors such as the effect of the soil background in mixed pixels, a different effect of the atmosphere in different bands, and the signal-to-noise ratio in different bands (Jackson et al., 1992). The type of input data used is particularly important in the calculation of plant indices. Jackson et al. (1992) noted that plant indices can be obtained from various data, such as sensor output voltage, reflectance, and the calculated pixel value. To solve the problem of atmosphere effect on plant indices, reflectance data are necessary, where the effect of the atmosphere is already corrected. Parente (2013) compared vegetation indices obtained using pixel values and reflectance data and found that reflectance data improved the performance of indices in separating water and soil. Similarly, Song et al. (2001) suggested that gray values (pixels) must be converted into reflectance values first to calculate plant indices.

3.1.3 DSM and prediction map approaches

The foundations for soil prediction maps were established by researchers such as Dokuchaev in 1899 and Jenny in 1941 (Florinsky, 2012). One of the earliest examples of an SOC map is the soil humus map of a part of Europe that was prepared by Dokuchaev in 1983 on a scale of 1:4,200,000 (Hartemink et al., 2013). In the 1990s, quantitative methods in soil surveying and mapping were used to analyze the relationship between soil properties, soil constituents, and environmental variables (Moore et al., 1993a). The concept of digital soil mapping (DSM) was proposed by McBratney et al. (2003), which led to the prediction of soil layers and characteristics using quantitative methods.

In a general pattern, the DSM is used to prepare a digital soil map by identifying independent variables (covariates) that affect the response variable (training data). A statistical relationship between the independent variables and the response variable is established at the location of the sampling points to obtain the prediction model (f) for the dependent variable. The continuity of the independent (auxiliary) variables enables estimation of the value of the response variable in unsampled areas of the region using the prediction model (f) and the area map. In general, the following section explains, step by step, how to implement a DSM, with emphasis on the general pattern described above (Fig. 3.1).

The spatial distribution of soil carbon is a function of various factors such as soil characteristics, climate, living organisms (including human activities, animals and plants), land use, and land roughness, which include shape of the land surface, parent materials of soil formation, and age or duration of soil weathering processes. Minasny et al. (2013) presented an SOC prediction function. SOC changes are closely related to the soil development process and the soil formation factors proposed by Jenny (1941), Hobley et al. (2015). Several methods have been used to model and predict SOC using limited terrestrial data. Geostatistical methods are among the methods used in this field (Chen et al., 2017; Imamoglu and Sertel, 2016).

However, due to the heterogeneous distribution of soil properties and the need for a large amount of ground sampling data, geostatistical methods may not offer good efficiency and accuracy. In recent years, with the availability of satellite remote sensing data and digital elevation data (DEM) with appropriate spatial and vertical resolution, researchers have shifted toward digital mapping of soil characteristics. In the DSM method, environmental independent variables based on spectral features of satellite images and nonspectral features of DEM are prepared (Costa et al., 2018; Mahmoudabadi et al., 2017; Castaldi et al., 2016). One of the strengths of this data is the possibility of updating it with low cost, high speed, wide coverage, and high accuracy (Castaldi et al., 2016; Costa et al., 2018).

To quantify the correlation and relationship between independent variables and soil properties, various methods such as geostatistical models, vector machine learning, support vector machine, decision tree, random forest, step-by-step multivariate linear regression and partial least squares regression, geographically weighted regression, regression tree, factor analysis, principal component analysis, clustering analysis, stochastic gradient boosting, cubist,

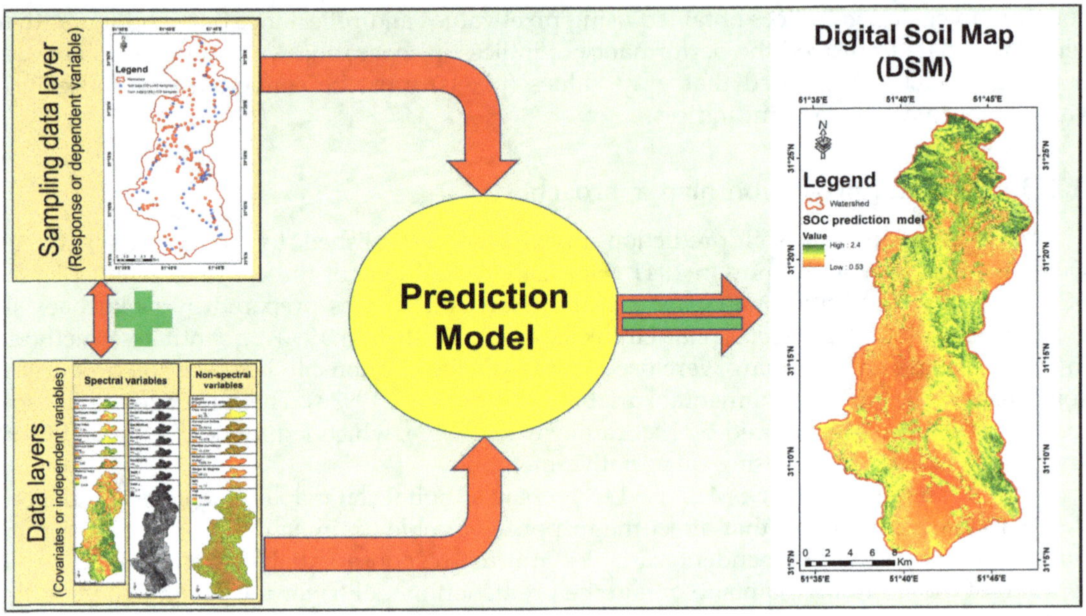

FIGURE 3.1 General scheme for preparing a digital soil map (DSM).

artificial neural network, and genetic algorithm have been used (Pahlavan-Rad et al., 2018; Forkuor et al., 2017; Wang et al., 2018, 2019; Zeraatpisheh et al., 2019).

In this chapter, a remotely sensed model is developed to predict the amount of surface SOC in a study area located in the Semirom region of Isfahan province, Iran. The model utilizes Landsat 8 satellite reflectance spectral information as well as nonspectral indicators, such as topographic indices that were extracted from the digital elevation model (DEM). Factor analysis and multiple regression models are used to summarize information layers with minimum loss of information. The spatial distribution model of SOC obtained from this research can be used as an important sub-model in designing complex models such as biomass models in dry ecosystems and models of soil sensitivity to erosion.

3.2 Materials and methods

In recent years, DSM has been used with a wide range of environmental factors and various data mining and forecasting algorithms to optimize the efficiency of SOC prediction. The digital map of SOC zoning and forecasting was prepared by considering the contents stated introduction (Section 3.1.3). For this purpose, the combined method of factor analysis and multivariate regression was used in the Semirom region of Isfahan province, Iran. After determining the homogeneous units, 218 soil samples were taken from them. The amount of organic carbon in the samples collected in the laboratory was determined as the ground reality (dependent variable). The aforementioned integrated method was performed using 22 spectral and nonspectral variables and two sets of training data (70%) and

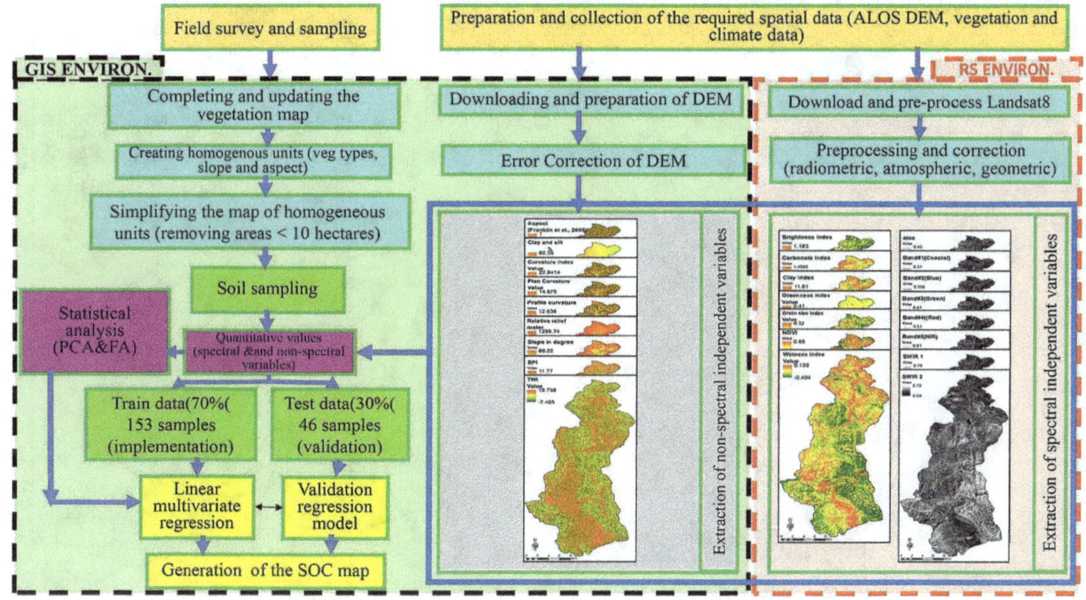

FIGURE 3.2 Flowchart illustrating the process for creating a soil organic carbon prediction zoning map using a combined method of factor analysis and multiple regression.

test (30%) of organic carbon of soil samples, in order to implement and validate the model, respectively. After prioritizing the effective factors and determining the main components, the spatial zoning map of SOC was developed. Finally, the accuracy of the model was validated and evaluated in the training and testing phase by using error measurement criteria (Fig. 3.2).

3.2.1 A case studied area

Considering the importance of carbon sequestration and the augmentation of SOC for improving soil fertility in semi-steppe to semi-arid regions, the chapter introduces remotely sensed zoning and predicting SOC content through a case study on the Semirom region in Isfahan province, situated in the center of Iran (Fig. 3.3). The area of interest lies within the Zagros region, which displays a wide array of climatic features. Employing Dumarten's climate classification, the southern portion of the basin, adjacent to Hanna in the south (Fig. 3.3), experiences a semi-arid to humid and semi-humid climate. In the central regions, extending from the north of the basin toward Garmouk (Fig. 3.3), a semi-arid to moderate climate dominates, while at the northern extremity of the basin, a semi-arid to humid climate prevails once more. The basin receives an average precipitation of 512 mm and maintains an average temperature of 12°C. The primary soil families in this basin consist of Typic Calcixerepts, Lithic Calcixerepts, Typic Xerofluvents, and Lithic Xerorthents soils. The basin hosts a variety of pasture vegetation, such as Artemisia, Astragalus, cousinia, Daphnes, and Bromus. The average elevation of the basin is 2664 m above sea level, with slopes ranging between 8

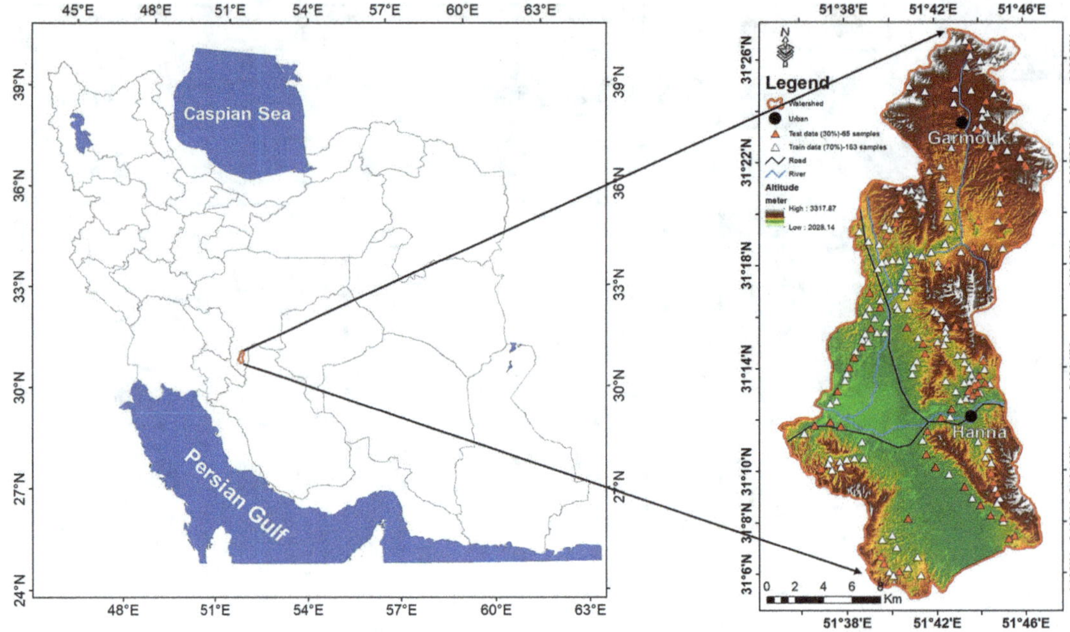

FIGURE 3.3 Geographic location of soil sampling.

and 25%, accounting for the most extensive area within the basin. Based on these characteristics, well-adapted and dense plant communities have been established in the pasture soils of this region, fostering the enhancement of SOC. Fig. 3.3 illustrates the geographical location of this area.

3.2.2 Landsat 8 satellite and its features

The Landsat 8 satellite, launched on February 11, 2013, is the eighth satellite in the Landsat series. It orbits at a height of 705 km, with a deviation angle of 98.2° and a return period of 16 days. The width of its images is 185 km. The Landsat 8 satellite has OLI and TIRS sensors. The OLI sensor has nine bands in the visible, near-infrared, and short-wavelength infrared ranges. The TIRS sensor has two bands with central wavelengths of 10.895 and 12.005 μm in the thermal infrared range. Table 3.1 lists the characteristics of the different bands of this sensor (USGS, 2016).

The satellite data used in this research are obtained from the website of the US Geological Survey (USGS). These data are Level 1 information from the Landsat 8 satellite. The metadata file accompanying the image provides information on the satellite's transit time in Greenwich Mean Time, as well as the sun's azimuth and elevation angles in degrees. The received image is in the Geographic Tagged Image File Format (GeoTIFF) with a WGS84 ellipse base.

TABLE 3.1 Characteristics of the different bands and sensors on the Landsat 8 satellite.

Sensor	Band number	Band name	Spatial resolution (m)	Wavelength (μm)
OLI	1	Dust/Coastal	30	0.0–435.451
OLI	2	Blue (visible)	30	0.0–452.512
OLI	3	Green (visible)	30	0.0–533.590
OLI	4	Red (visible)	30	0.0–636.673
OLI	5	Near infrared (NIR)	30	0.0–851.879
OLI	6	(SWIR1) short wave infrared	30	1.1–566.651
OLI	7	(SWIR2) short wave infrared	30	2.2–107.294
OLI	8	Panchromatic	15	0.0–503.676
OLI	9	Cyrus	30	1.1–363.384
TIRs	10	Thermal (TIR1)	100	10.11–60.19
TIRs	11	Thermal (TIR2)	100	11.12–50.51

3.2.3 Determination of homogeneous work units for sampling

To determine the organic carbon content of surface soil in the range of plant types, the first step was to prepare a vegetation map of the study area. This research utilized ENVI5.1 and ArcGIS10.8 software for extracting information from satellite images, as well as SPSS26.0 and Excel2019 software for statistical analysis. To prepare a map of the elevational strata model of the study area, DEM from the PALSAR2 sensor on the Japanese ALOS satellite were used in the preparation of nonspectral factors. The spatial accuracy of the DEM was 12.5 m by 12.5 m, and preprocessing operations, including noise and error removal, sink filling, and data preparation, were performed using SAGA GIS 7.7.1 and ArcGIS 10.8 software. Using the vegetation map and DEM data, slope maps, geographical directions, and height classes were produced to characterize the vegetation of the study area. From the overlaying of these three vector layers, a map of landform units (homogeneous work units) was created for sampling purposes.

3.2.4 Time of soil sampling and image acquisition

Water content in soil can affect the absorption and reflection of electromagnetic waves, with higher soil moisture leading to decreased reflectance in the visible and infrared parts of the spectrum. To mitigate the effects of soil moisture on spectral response, soil sampling should be conducted during the driest month of the year. Long-term rainfall and temperature data from the closest meteorological station were used to construct an ambrothermic curve for the study area (Fig. 3.4), which was used to determine the driest month.

The ambrothermic curve is a graph that shows the monthly temperature and rainfall patterns for a particular location. It can be used to determine the driest month of the year, which

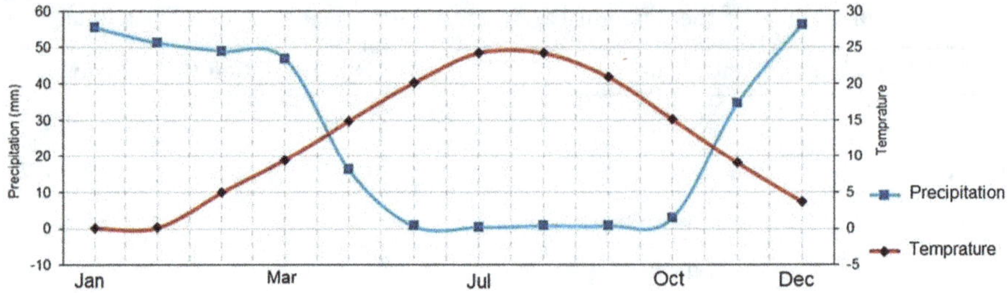

FIGURE 3.4 Ambrothermic curve of the studied area.

is important for soil sampling because soil moisture content can affect the spectral response of soil. In general, the driest month is the one with the lowest rainfall and highest temperature, as these conditions are associated with lower soil moisture content. Soil sampling can then be scheduled for this month to minimize the effects of soil moisture on spectral response.

The ambrothermic curve can also be used to determine the best time for satellite image acquisition, as the optimal time for image acquisition is when the spectral response of soil is most similar to the response at the time of soil sampling. By selecting an image acquisition date that corresponds to the driest month, the effects of soil moisture on spectral response can be minimized, improving the accuracy of the analysis.

Based on this, the optimal time for soil sampling and satellite image acquisition (using Landsat 8) is determined to minimize the time difference between the two measurements. The ambrothermic curve for each region differs based on its geographical location and climatic conditions such as precipitation and temperature. Fig. 3.4 is an ambrothermic curve related to the Semirom basin, located in the south of Isfahan province in Iran.

3.2.5 Sampling operations in homogeneous work units

Sampling in each homogeneous work unit is carried out using a stratified random sampling pattern (McCoy, 2005) (Fig. 3.5). The purpose of stratified random sampling is to improve the representativeness and accuracy of a sample by taking into account the variability within a population. By using stratified random sampling, researchers can reduce the variability in the sample and increase the precision of their estimates. This ensures that the sample is representative of the entire population and that each stratum is adequately represented in the sample.

First, the study area is divided into several homogeneous layers based on the previously described method. Then, using the Sampling Design Tool in ArcGIS 10.8 software, random points are selected in each homogeneous layer in proportion to the size of the unit. The location of each point is determined using a GPS device, topographic maps, and field surveys.

At each sampling point, a sample of the surface soil (0–20 cm) is collected using a shovel. Samples from each homogeneous unit are combined and mixed to create a 2-kg sample. This sample is labeled and transported to the soil testing laboratory for analysis of SOC content.

It should be noted that some designated sampling points may be inaccessible due to high slopes or other factors. In such cases, new random points are selected for sampling.

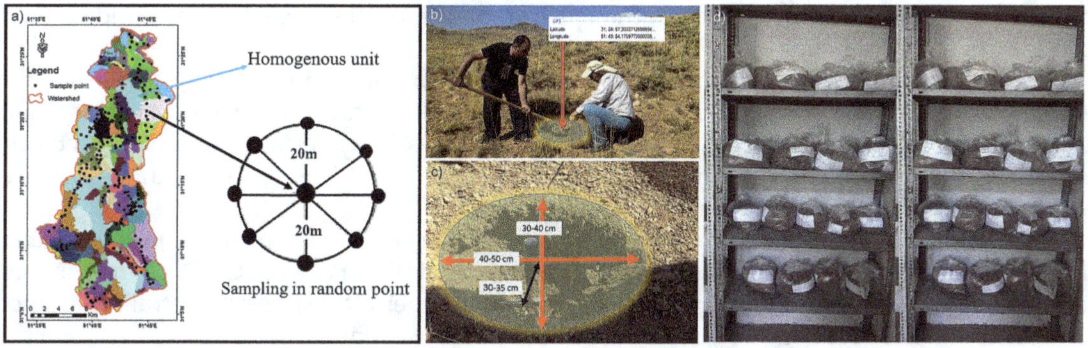

FIGURE 3.5 (A) Diagram of typical stratified random sampling, (B) field survey and sampling, (C) dimensions of sampling point excavation, and (D) samples archive.

3.2.5.1 *Soil tests*

Upon arrival at the soil science laboratory, all soil samples are dried, ground, and passed through a 2 mm sieve. The organic carbon content of each soil sample is determined using the back-alkaline titration method. This method involves adding a known amount of potassium dichromate to a soil sample in the presence of sulfuric acid. The organic carbon in the soil is oxidized by the dichromate, and the heat generated by the reaction provides the necessary energy for the oxidation process. The remaining amount of dichromate in the reaction mixture is then determined by a back titration with a solution of ferrous ammonium sulfate. This titration is carried out after adding an excess of sodium carbonate or sodium hydroxide to the reaction mixture, which converts any remaining dichromate to chromate. The excess sodium carbonate or hydroxide is then titrated with the ferrous ammonium sulfate solution to determine the amount of dichromate that was consumed by the SOC (Nelson and Sommers, 1983). The content of SOC is calculated by the following equation:

$$\%\mathrm{OC} = M * 039 * \left[\frac{v_1 - v_2}{s}\right] \tag{3.1}$$

OC represents the percent of SOC, M represents the normality of ferrous ammonium sulfate, v_1 and v_2 represent the volumes of ferrous ammonium sulfate used for control and sample, respectively, in milliliters, and s represents weight of the soil dried in open air in grams. This method is recommended for use in neutral and calcareous soils.

It should be noted that some of the soil samples were used for model calibration, while the remaining samples (a smaller amount) were used for model validation.

3.2.6 Converting gray values to high atmospheric reflectance (preprocessing)

Converting pixel values to upper atmospheric reflectance is an important step in remote sensing data processing, as it allows for the comparison of spectral values between different images and reduces the influence of atmospheric conditions on the data. The upper atmospheric reflectance is a more stable and consistent value than pixel values, which can vary depending on the viewing angle and atmospheric conditions. It is a more reliable indicator of the spectral characteristics of the surface being imaged.

The atmosphere can interact with incoming and outgoing radiation, causing changes in the spectral values of remotely sensed data. These changes can be caused by factors such as cloud cover, haze, and water vapor. By converting pixel values to upper atmospheric reflectance, the effects of atmospheric conditions can be removed, allowing for more accurate comparisons of data between different images and over time.

The preprocessing operation was performed using ENVI5.1 software in two steps, following the instructions from the United States Geological Survey (USGS, 2016) for correcting the angle of the sun's elevation and obtaining the upper atmospheric spectral reflectance of each of the Landsat 8 OLI sensor bands two through seven.

To perform radiometric corrections of bands two to seven and convert pixel values to upper atmospheric reflectance, Eq. (3.2) was used in the ENVI5.1 software environment:

$$\rho_\lambda = \frac{\rho_{\lambda'}}{\mathrm{Sin}(\theta)} \tag{3.2}$$

ρ_λ represents spectral reflectance above the atmosphere, while $\rho_{\lambda'}$ stands for reflectance without correction for the sun's elevation angle. $\rho_{\lambda'}$ is obtained using Eq. (3.3). The elevation angle of the sun, θ, is calculated from the metadata file of the satellite image.

$$\rho_{\lambda'} = (M_\rho \times Q_{\mathrm{cal}}) + A_\rho \tag{3.3}$$

In the above equation, Q_{cal} represents gray values (DN) of the pixels in the corresponding band. M_ρ and A_ρ are band-specific values obtained from metadata file of the satellite image.

3.2.7 Data used for prediction of surface soil organic carbon

To estimate SOC, two categories of auxiliary variables were used: spectral (reflectance data) and nonspectral (Table 3.2). All of these factors were considered primary independent variables in the surface soil carbon prediction model.

Spectral variables are derived from the spectral reflectance of the soil surface, which is measured using remote sensing data such as Landsat 8 images. Reflectance data from different spectral bands can be used to calculate various vegetation indices, such as the Normalized Difference Vegetation Index (NDVI), which are related to vegetation cover and can be used to estimate SOC content. Other spectral variables, such as brightness, hue, and saturation, can also be used to estimate soil properties.

Nonspectral variables are derived from other environmental and soil-related data, such as topography, climate, soil texture, and vegetation cover. These variables can provide additional information about soil properties that cannot be obtained from spectral data alone. For example, soil texture can affect the water-holding capacity of soil, which in turn can affect the decomposition of organic matter and the accumulation of SOC. Similarly, topography can affect the drainage and water availability of a site, which can impact the growth and decay of vegetation and the accumulation of organic matter.

TABLE 3.2 Characteristics of variables used in predicting soil organic carbon.

Number	Variable name	Spatial resolution (m)	Spatial data source	Spatial layer format	Figure
1	Band2(Blue)	30	Landsat 8(OLI)	Raster image	3.6A
2	Band3(Green)	30	Landsat 8(OLI)	Raster image	3.6B
3	Band4(Red)	30	Landsat 8(OLI)	Raster image	3.6C
4	Band5(IR)	30	Landsat 8(OLI)	Raster image	3.6D
5	Band6(SWIR1)	30	Landsat 8(OLI)	Raster image	3.6E
6	Band7(SWIR2)	30	Landsat 8(OLI)	Raster image	3.6F
7	Surface albedo	30	Landsat 8(OLI)	Raster image	3.6G
8	Brightness index	30	Landsat 8(OLI)	Raster image	3.6H
9	Greenness index	30	Landsat 8(OLI)	Raster image	3.6I
10	Wetness index	30	Landsat 8(OLI)	Raster image	3.6J
11	Clay index	30	Landsat 8(OLI)	Raster image	3.6K
12	Carbonate index	30	Landsat 8(OLI)	Raster image	3.6L
13	Grain size index	30	Landsat 8(OLI)	Raster image	3.6M
14	NDVI	30	Landsat 8(OLI)	Raster image	3.6N
15	Relative relief (meter)	12.5 m resample to 30 m	DEMPALSARALOS	Raster tiff	3.6O
16	Slope aspect (Franklin et al., 2000)	12.5 m resample to 30 m	DEMPALSARALOS	Raster tiff	3.6P
17	Slope gradient (percent)	12.5 m resample to 30 m	DEMPALSARALOS	Raster tiff	3.6Q
18	General curvature	12.5 m resample to 30 m	DEMPALSARALOS	Raster tiff	3.6R
19	Plan curvature	12.5 m resample to 30 m	DEMPALSARALOS	Raster tiff	3.6S
20	Profile curvature	12.5 m resample to 30 m	DEMPALSARALOS	Raster tiff	3.6T
21	TWI	12.5 m resample to 30 m	DEMPALSARALOS	Raster tiff	3.6U
22	Silt and clay	–	Field sampling-vector point	Raster tiff	3.6V

By combining both spectral and nonspectral variables in the prediction model, researchers can improve the accuracy and precision of their estimates of SOC content. The specific variables used in the model and their relative importance will depend on the study area and the specific research question being addressed.

As Landsat 8 images are provided to users in a digital number (DN) format, preprocessing is necessary before calculating spectral variables. Using the radiometric correction coefficients in the metadata file of the obtained images, upper atmospheric reflectance values of bands 1 to 7 (OLI sensor) were calculated (USGS, 2016). To ensure accuracy in modeling results, the actual position of each pixel relative to the ground reality was taken into account. Ground control points and topographic maps of the region were used to rectify the images and eliminate errors caused by pixel displacement (Fig. 3.6).

It should be noted that while Landsat 8 images are provided in a ground-based format, the accuracy of ground truthing is essential for accurate modeling results.

Table 3.2 presents the characteristics of both environmental (nonspectral) and spectral variables used in predicting organic carbon. The table includes the variable name, spatial resolution, data source of each variable, and spatial layer format (Fig. 3.7).

3.2.7.1 *Spectral variables (reflectance)*

Advancements in technology have enabled humans to access information with greater volume and accuracy in less time, expanding its use in various sciences. Soil science has been able to contribute significantly to this progress through remote sensing. SOM is influenced by factors such as vegetation, soil characteristics, and the climate of the region. If its amount in the soil is more than 2%, it can be effective as a dominant factor in determining the spectral behavior of the soil (Ray et al., 2004; Stephens et al., 2005). An increase in soil moisture and the amount of organic matter leads to a decrease in the amount of reflection from the ground. Near-infrared bands (0.76–0.9 μm) show the highest correlation with the spectral characteristics of organic matter, while soil texture, like soil moisture and SOM, has significant effects on the resulting reflective patterns. The brightness index is directly related to changes in SOM. Organic matter has the highest correlation in the visible spectrum region (Barnes et al., 2003). Early remote sensing studies for SOC modeling showed that, in general, increasing SOC leads to a decrease in the amount of reflection of the electromagnetic spectrum in the range of visible light, near-infrared, and short wavelength infrared (Croft et al., 2012; Dharumarajan et al., 2017; Gholizadeh et al., 2018; Ramcharan et al., 2018). Therefore, the reflective features of different bands of satellite images can be used as effective variables in predicting the amount of SOC, based on their correlation with soil compounds, especially SOC.

Reflective remote sensing is utilized to create the studied variables, using the spectral information of Bands two to seven of the Landsat 8 OLI sensor, which covers the wavelength range of 0.452–2.294 μm. In reflective remote sensing, the Sun is the main source of energy. The spectral (reflectance) variables used in this research consist of a total of 14 layers of information. Seven layers are related to the spectral information of bands two to seven and are modified for the reflectivity of the upper atmosphere of the OLI sensor (Fig. 3.7A to F). The other eight variables include the surface albedo spectral index, clay index, carbonate index, grain size index, normalized differential vegetation index (NDVI), and three Tasseled Cap transformation indices (brightness, greenness, and wetness).

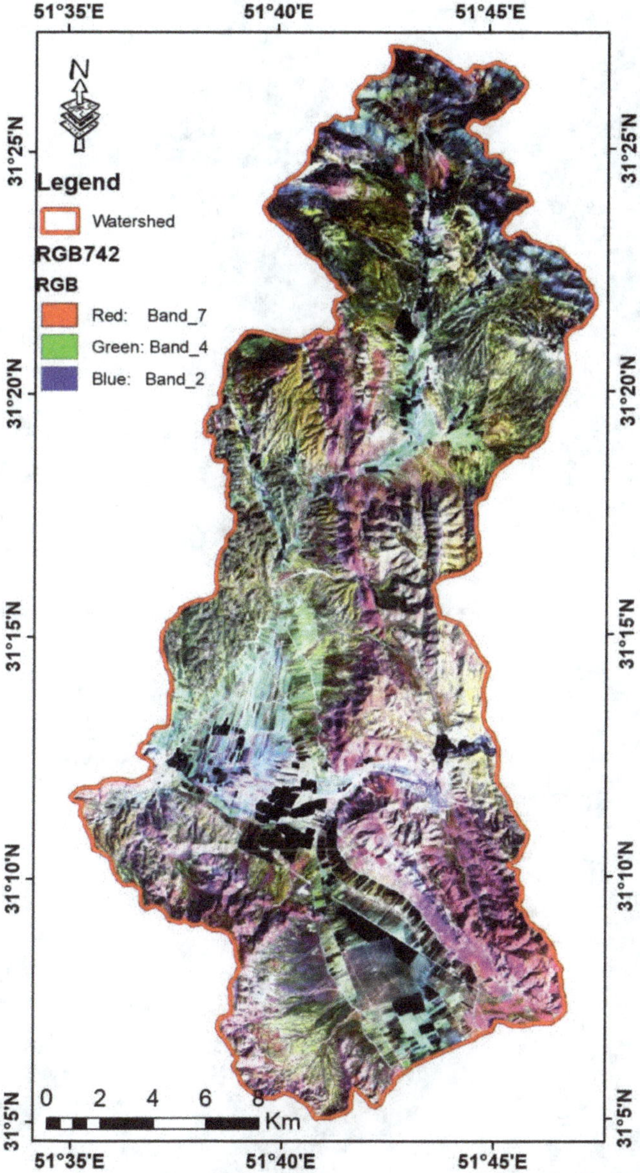

FIGURE 3.6 The false color composite in RGB #742 (Landsat 8).

3.2.7.1.1 Surface albedo

To calculate surface albedo, Eq. (3.4) is used (Liang et al., 2003) (Fig. 3.7G):

$$\alpha = \frac{\alpha_{\text{toa}} - \alpha_{\text{path radiance}}}{\tau_{sw}^2} \tag{3.4}$$

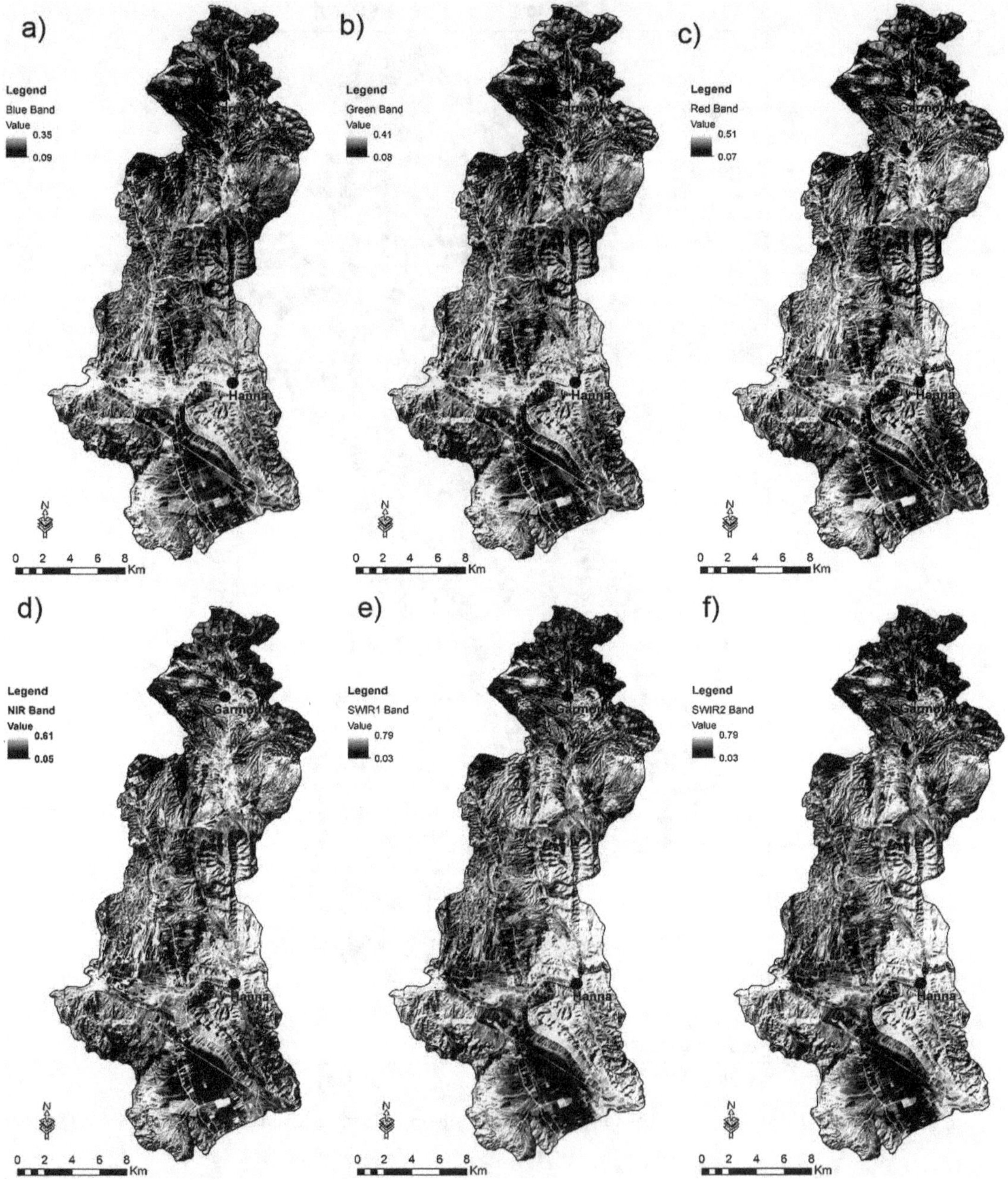

FIGURE 3.7 Maps of the independent variables (spectral and nonspectral parameters), (A) blue band, (B) green band, (C) red band, (D) NIR band, (E) SWIR1 band, (F) SWIR2 band, (G) albedo top of atmosphere, (H) brightness index, (I) greenness index, (J) wetness index, (K) clay index, (L) carbonate index, (M) grain size index, (N) NDVI, (O) relative relief, (P) slope aspect, (Q) slope gradient, (R) general curvature, (S) plan curvature, (T) profile curvature, (U) TWI, and (V) clay and silt percent.

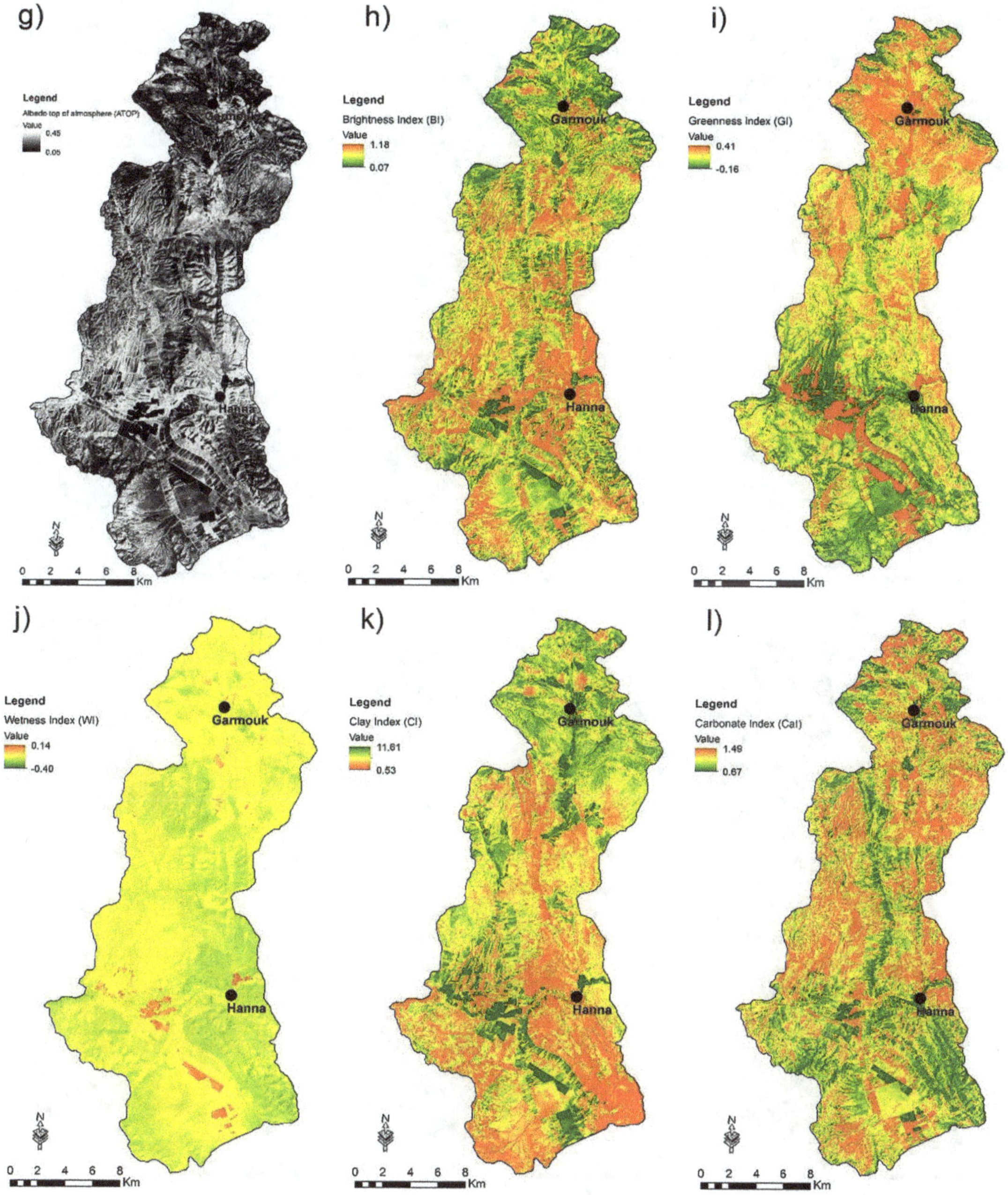

FIGURE 3.7 cont'd

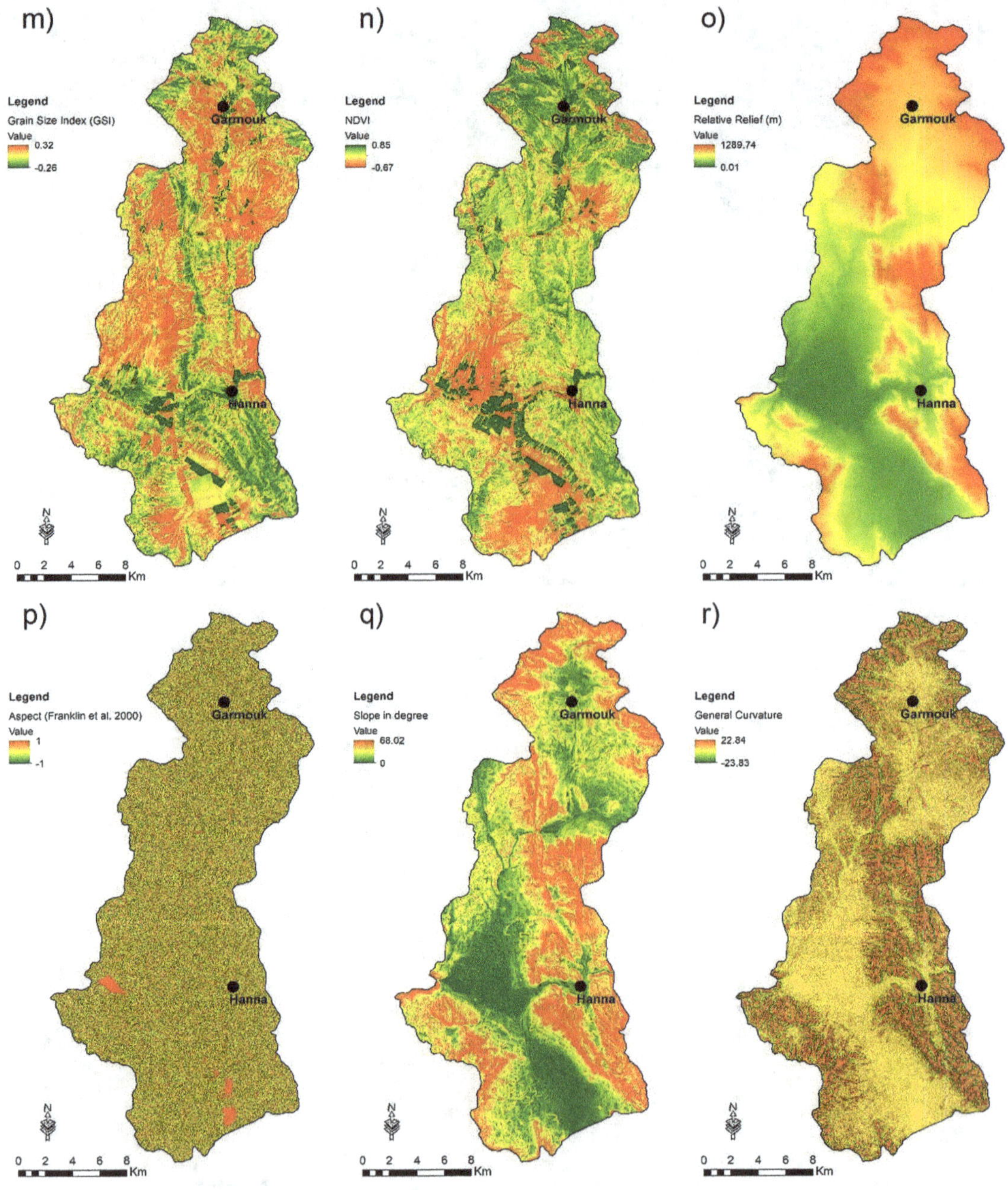

FIGURE 3.7 cont'd

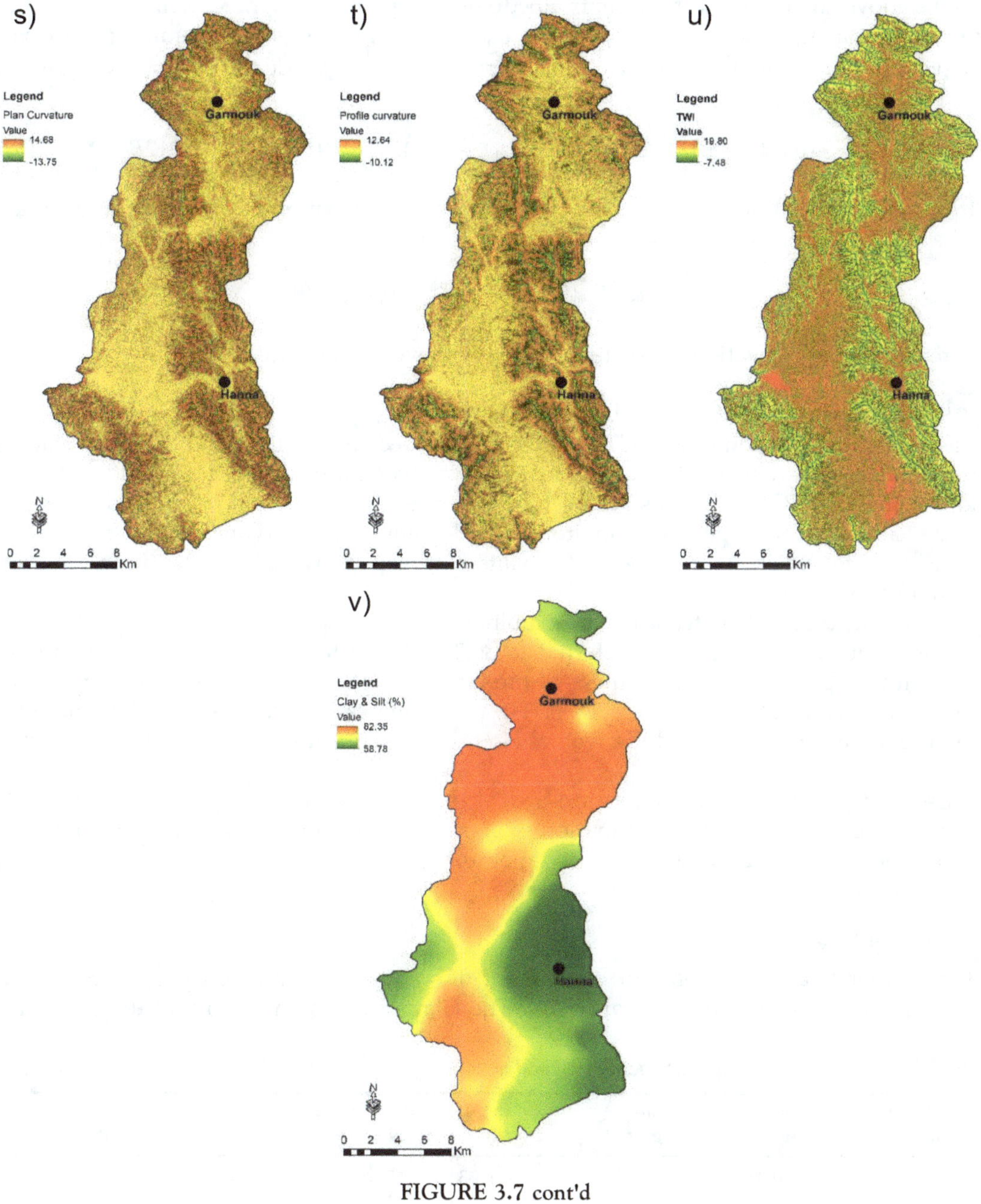

FIGURE 3.7 cont'd

In the above equation, α_{toa} represents albedo of the upper atmosphere, which is calculated for Landsat 8 images using Eq. (3.5). $\alpha_{\text{path radiance}}$ represents the albedo resulting from path radiance and is considered equal to 0.03. τ_{sw} represents atmospheric transmissivity and is obtained using Eq. (3.6).

$$\alpha_{toa} = [(0.3B_2) + (0.277B_3) + (0.233B_4) + (0.14B_5) + (0.036B_6) + (0.012B_7)] \quad (3.5)$$

In the above equation, B_n stands for the reflectance information of the upper atmosphere for the nth band.

$$\tau_{sw} = (2 \times 10^{-5} \times Z) + 0.75 \quad (3.6)$$

Z represents the elevation of the area above sea level, measured in meters.

3.2.7.1.2 Tasseled-cap transformation (TCT) indicators

Tasseled cap transformation (TCT) is a technique used to transform multispectral satellite data into a set of new bands that are more interpretable and meaningful for land cover and land use analysis. TCT indicators are useful for various applications such as land cover and land use classification, vegetation monitoring, and change detection analysis. They allow for a more accurate and efficient analysis of satellite data, especially for areas with complex land cover and topography.

The TCT method calculates a linear combination of the original spectral bands to create new composite bands, which are referred to as TCT indicators. These indicators represent the brightness, greenness, and wetness of the land surface. The brightness index measures the total reflectance of the surface (Fig. 3.7H), the greenness index measures the amount of green vegetation (Fig. 3.7I), and the wetness index measures moisture content of the surface (Fig. 3.7J).

Tasseled cap transformation indicators are calculated using Landsat 8 OLI sensor data. The upper atmosphere reflectance information in bands two to seven is linearly combined using the coefficients presented in Table 3.3 (Baig et al., 2014) to obtain each of the Tasseled Cap conversion indices.

3.2.7.1.3 Clay index (CI)

Clay minerals are an important component of soils, influencing soil properties such as texture, structure, and nutrient availability. The Clay Index is an important tool in remote

TABLE 3.3 The Tasseled cap coefficients for Landsat 8

Tasseled cap indicator	Band 2 (blue)	Band 3 (green)	Band 4 (red)	Band 5 (NIR)	Band 6 (SWIR1)	Band 7 (SWIR2)
(BI)	0.3029	0.2786	0.4733	0.5599	0.508	0.1872
(GI)	−0.2941	−0.243	−0.5424	0.7276	0.0713	−0.1608
(WI)	0.1511	0.1973	0.3283	0.3407	−0.7117	−0.4559

sensing studies of soils and land use, providing valuable information on the distribution and abundance of clay minerals in the landscape. The Clay Index (CI) is a spectral index used in remote sensing to estimate the content and distribution of clay minerals in soils. It is based on the fact that clay minerals have specific spectral features in the visible and near-infrared regions of the electromagnetic spectrum.

By using CI to estimate the distribution of clay minerals in soils, researchers and land managers can gain insights into soil fertility and productivity, as well as the potential for soil erosion and other environmental issues.

CI is calculated as the difference in reflectance between the 1.6 and 2.2 micron wavelength bands, divided by the sum of the reflectance values in these bands. The index ranges from −1 to 1, with higher values indicating a greater abundance of clay minerals. Eq. (3.7) is used to calculate the CI (Hengle, 2009):

$$CI = \frac{B_5}{B_7} \tag{3.7}$$

In the above equation, B_5 and B_7 represent the amount of high atmospheric reflectance for bands five and seven of the Landsat 8 OLI sensor, respectively (Fig. 3.7K).

3.2.7.1.4 Carbonate index (CaI)

Eq. (3.8) is used to calculate the Carbonate Index (CaI) (Boettinger et al., 2008):

$$CaI = \frac{B_3}{B_2} \tag{3.8}$$

In the above equation, B_2 and B_3 represent the amount of high atmospheric reflectance in bands two and three of the Landsat 8 OLI sensor, respectively (Fig. 3.7I).

3.2.7.1.5 Grain size index (GSI)

Fine sand particles are an important component of soils, influencing soil properties such as texture, water-holding capacity, and nutrient availability. The Grain Size Index (GSI) is a spectral index used in remote sensing to estimate the content of fine sand in soil. It is based on the fact that soil particles of different sizes have specific spectral features in the visible and near-infrared regions of the electromagnetic spectrum. By using GSI to estimate the content of fine sand in soils, researchers and land managers can gain insights into soil fertility and productivity, as well as the potential for soil erosion and other environmental issues. The accuracy of GSI in estimating fine sand content in soil may vary depending on factors such as soil type, soil moisture, and vegetation cover. Therefore, while GSI can be a useful tool in estimating the content of fine sand in soil, its applicability and accuracy may depend on the specific soil and environmental conditions in an area.

GSI is calculated as the ratio of reflectance in the red and green bands, divided by the reflectance in the blue band. The index ranges from 0 to 1, with higher values indicating a greater abundance of fine sand particles. It is calculated using Eq. (3.9) (Xiao et al.,

2006; Piccini et al., 2014). The Particle Size Index has a high correlation with fine sand content in soil.

$$\mathrm{GSI} = \frac{(R - B)}{(R + G + B)} \tag{3.9}$$

In the above equation, R, G, and B represent the amount of upper atmosphere reflections in the red, green, and blue bands, respectively (Fig. 3.7M).

3.2.7.1.6 Normalized Difference Vegetation Index (NDVI)

The Normalized Differential Vegetation Index (NDVI) is a widely used spectral index in remote sensing studies that can provide information on vegetation cover and health (Rouse et al., 1974). In this research, NDVI is one of the spectral variables used to estimate SOC content. NDVI has been shown to be strongly correlated with vegetation cover (Fig. 3.7N) and productivity, as well as SOC content. Higher NDVI values may indicate greater plant biomass and productivity, which can lead to higher rates of organic matter accumulation and increased SOC content.

NDVI is calculated as the difference between the near-infrared (NIR) and red reflectance values, divided by their sum. NDVI is obtained from the following relationship:

$$\mathrm{NDVI} = \frac{\rho_{\mathrm{NIR}} - \rho_R}{\rho_{\mathrm{NIR}} + \rho_R} \tag{3.10}$$

In the above relationship, ρ_{NIR} and ρ_R represent the amount of reflection in the infrared and red regions, respectively. The NDVI index ranges from -1 to $+1$, with higher values indicating a greater amount of photosynthetically active vegetation. Negative values indicating water bodies, values from 0 to 0.2 indicating bare soil, and values greater than 0.2 indicating vegetation. The index approaches a value of one for denser vegetation.

3.2.7.2 Nonspectral variables (morphometric indices)

Spatial environmental variables have been widely used in the preparation of digital maps of SOM. The use of environmental variables increases accuracy and reduces the need for sampling to prepare organic carbon distribution maps from three aspects. Firstly, environmental variables extracted from digital elevation maps are provided at a low cost. Secondly, environmental variables of the land are complete and spatially wide, providing data and input information with suitable content in areas without soil sampling to investigate this phenomenon. Thirdly, there is a significant correlation between environmental variables and SOC (Pei et al., 2010).

The importance and correlation of topographic indicators with the amount of SOC have been investigated and determined by several researchers (Moore et al., 1993a; Florinsky et al., 2002; Forkuor et al., 2017; Mondal et al., 2017; Schillaci et al., 2017; Pahlavan-Rad et al., 2018; Wang et al., 2018; Zeraatpisheh et al., 2019; Zhou et al., 2019; Wang et al., 2020).

A topographic landscape is a description of the shape and characteristics of landforms, and topographic information can be obtained from digital elevation models (DEM) extracted from remote sensing data (Glenn et al., 2008). Up to 43% of soil redistribution variation can be explained by topographic features extracted from DEM. Topographic effects on organic carbon distribution in extrapolating field observations to analyze soil erosion and patterns. The place of SOC is considered. Topography is an important factor in regulating and balancing soil erosion because it affects surface runoff, soil texture, and vegetation (Li et al., 2018).

There is usually a process-oriented correlation between soil characteristics and landforms (Moore et al., 1993b; Park et al., 2001). Since elevation and elevation are the main factors in the formation and development of soil (Jenny, 1941), and there are orbital process relationships between land surface characteristics and soil characteristics, especially in areas with axial hill physiography, finding a correlation through experimentation with landforms is very common. Soil-landscape modeling involves analyzing land features such as height, slope, slope direction, curvature, etc., to estimate soil properties and prepare soil spatial prediction maps, especially SOC (Moore et al., 1993a, b; Park et al., 2001; Li et al., 2018; Zeraatpisheh et al., 2019).

Morphometric indices are used to characterize the topography and drainage properties of a watershed. They provide information about the shape, size, and slope of the basin, as well as the drainage network and the flow paths of water within the basin. This information is important for understanding the hydrological behavior of the watershed, such as the patterns of runoff, erosion, and sediment transport.

Morphometric parameters are nonspectral variables that are derived from the DEM. These morphometric indices are useful for various applications such as hydrological modeling, soil erosion assessment, and land use planning. The accuracy of these parameters depends on the accuracy of the DEM.

To obtain the correct DEM, it is necessary to set the DEM image system to UTM (metric) first. Next, DEM preprocessing, including the Fill Sink and Filter operations, should be done in a geographic information system (GIS) environment, such as ArcGIS10.8. Afterward, nonspectral raster layers are generated using the modified DEM. Nine morphometric indices are then calculated using ArcGIS10.8 software, including relative relief (RR), slope, slope azimuth, landscape general curvature (concave, convex, or flat), horizontal curvature (planform curvature), vertical curvature (profile curvature), and Topographic Wetness Index (TWI).

The nonspectral information layers of the raster are prepared using the modified DEM as follows:

3.2.7.2.1 Relative relief

The elevation layer and relative elevation (RR) are obtained by calculating the height difference compared to the lowest point of the area in a raster format in the GIS environment (Fig. 3.7O).

3.2.7.2.2 Slope aspect

This index affects the maintenance of humidity due to the side or opposite sides of the slopes toward the sun's radiation, and humidity can also play a role in the amount of organic carbon storage. Therefore, it is used as an influential variable in predicting organic carbon. Aspect of the slope is calculated between 0 and 360° using the aspect function in the ArcGIS10.8 environment. Then, the relationship proposed by Franklin et al. (2000) is used to access quantitative normal values between one and -1 (Fig. 3.7P). Furthermore, to convert the slope azimuth parameter, Eq. (3.11) is utilized (Franklin et al., 2000):

$$\mathrm{Cosin}(\mathrm{Aspect} - 225^\circ) \tag{3.11}$$

3.2.7.2.3 Slope gradient

This is usually calculated as a percentage using the slope function in the ArcGIS10.8 environment. In studies of soil erosion, elevation and slope are introduced as effective factors in predicting soil redistribution and examining the distribution of SOC in pasture and agricultural lands (Theocharopoulos et al., 2003; Polyakov and Lal, 2008) (Fig. 3.7Q).

3.2.7.2.4 General curvature

The general shape of the domain curvature (landscape general curvature) is determined as concave (pixels with negative values), convex (pixels with positive values), or flat (pixels with zero values) using the curvature function in the ArcGIS10.8 environment (Fig. 3.7R).

3.2.7.2.5 Plan curvature

This index shows the intensity of changes in direction of the range and indicates whether the flow is convergent (convex) or divergent (concave) when passing through the desired surface. The values of raster map pixels for this index are determined and calculated as concave (pixels with negative values), convex (pixels with positive values), or flat (pixels with zero values) using the curvature function in the ArcGIS10.8 environment (Fig. 3.7S).

3.2.7.2.6 Profile curvature

This index shows the intensity of changes in slope and indicates whether the flow speed increases or decreases when passing through the desired surface. Pixel values of the raster map for this index are opposite to horizontal curvature values of the domain (planform curvature) and vertical curvature of the domain (profile curvature) as concave (pixels with positive values), convex (pixels with negative values), or flat (pixels with zero values) using the curvature function in ArcGIS10.8 environment (Fig. 3.7T).

3.2.7.2.7 Topographic Wetness Index (TWI)

Studies have shown that soils with high moisture content increase SOC by supporting and promoting plant growth and reducing the decomposition of organic matter (Pei et al., 2010). Since soil moisture can be quantitatively modeled using land topographical indicators, SOM (SOM) should also have a significant relationship with land topographical variables (Jenny, 1941; Moore et al., 1993a; Pei et al., 2010). The Topographic Wetness Index is used to indicate soil moisture content by calculating the amount of water that accumulates on the surface in the GIS environment (Fig. 3.7U).

The TWI describes the tendency of water to accumulate at any point in the basin (in terms of A_s) and the tendency of gravity forces to transport water downstream (in terms of $\tan\beta$ as an approximate hydraulic gradient). Eq. (3.12) is used to calculate the TWI, where A_s is characteristic level of the field and β represents the slope (Beven and Kirkby, 1979):

$$\text{TWI} = \ln\left(\frac{A_s}{\tan \beta}\right) \tag{3.12}$$

3.2.7.2.8 Clay and Silt

Size of soil particles plays an important role in absorbing and storing organic carbon. Generally, as the size of soil particles changes from coarse (sand) to fine grained (clay and silt), the amount of SOC storage increases. The percentage ratio of clay and silt is calculated for the mentioned samples in the soil laboratory. The clay and silt percent raster map is plotted in the ArcGIS10.8 (Fig. 3.7V).

3.2.8 Statistical analysis and methods used for soil organic carbon prediction modeling

Prediction of surface SOC can be performed using both spectral and nonspectral variables, utilizing factor analysis through principal component analysis (PCA) by means of SPSS software. The following briefly outlines the concepts of factor analysis and the process of obtaining main factors based on the use of spectral and nonspectral variables:

3.2.8.1 Factor analysis

Factor analysis is a statistical technique used to analyze the relationships between a large set of seemingly unrelated variables under a hypothetical model. Unlike multiple regression, the variables do not appear directly in the structure of the relational model. Instead, the technique identifies a smaller number of factors, which are linear combinations of the main variables and represent a unique feature of the relationships between the variables. The number of factors is typically far less than the number of main variables. Factor analysis helps to identify patterns and relationships in the data, making it easier to interpret and understand complex data sets.

3.2.8.2 Bartlett's sphericity test and KMO index

Bartlett's sphericity test and the KMO index are commonly used to assess the suitability of data for factor analysis. The KMO index measures the sampling adequacy of the data and ranges from zero to one. Bartlett's sphericity test assesses the assumption of the equality of correlation coefficients matrix. Failure to reject this hypothesis suggests that all the variables are independent of each other and that the commonality between the variables is not significant. The index is computed using simple and partial correlation coefficients, as shown in Eq. (3.13).

$$\text{KMO} = \frac{\sum_{i=1}^{p}\sum_{j=1}^{p} r_{ij}^2}{\sum_{i=1}^{p}\sum_{j=1}^{p} r_{ij}^2 + \sum_{i=1}^{p}\sum_{j=1}^{p} a_{ij}^2} \quad i \neq j \tag{3.13}$$

In Eq. (3.13), r_{ij} is the simple correlation coefficient between variables i and j, and a_{ij} is the partial correlation coefficient between variables i and j, with other variables held constant. The KMO index, which is computed using these correlation coefficients, ranges from zero to one and measures the sampling adequacy of the data for factor analysis. A KMO index closer to one indicates the suitability of the data for factor analysis. A value greater than 0.6 suggests that factor analysis can be performed on the input variables (Howard, 2016).

3.2.8.3 *Standardization of input variables*

In this step, the input variables were standardized to have a zero mean and a standard deviation of one, using the following formula:

$$Z = \frac{X - \mu}{\delta} \tag{3.14}$$

In the above formula, Z represents the standardized version of the observations, X represents the raw input observations, μ represents the mean of each variable, and δ represents the standard deviation for each variable.

3.2.8.4 *Extraction of factors in exploratory factor analysis*

Extraction of factors in exploratory factor analysis can be performed using various methods through SPSS software, including unweighted least squares, generalized least squares, maximum likelihood, principal components, and principal axis factoring. Among these methods, principal components and principal axis factoring are the most commonly used. While Thompson (1992) believes that there is not much difference in the results obtained from these two methods. Using the Principal Components method is suggested to extract factors when the goal is to summarize variables and obtain a limited number of factors. In factor analysis, using the principal components method, all variables are used to create the factors, thereby presenting the information of the primary variables with the least loss through the factors and preserving the information content of the main parameters (Mohan and Arumugan, 1996).

3.2.8.5 *The number of factors in exploratory factor analysis*

Determining the number of factors to extract is a critical step in exploratory factor analysis. While the goal of exploratory factor analysis is to extract the minimum number of factors that explain the most variance in the data, selecting the appropriate number of factors is crucial as it directly affects the model results.

Various rules are used to determine the number of factors in exploratory factor analysis, including the Kaiser criteria, the scree plot, and the cumulative percentage of variance explained. Although there are other criteria, such as parallel analysis, minimum least partial correlation, and Bartlett's chi-square test, they are less commonly used. The Kaiser rule is the most widely used method, whereby the number of factors extracted is determined by the eigenvalues greater than or equal to one (Kaiser, 1960). In this study, the Kaiser method is used to determine the number of factors.

3.2.8.6 Methods for factor rotation in exploratory factor analysis

To make factors easier to interpret without altering their contribution rates, statisticians have provided methods to rotate the factors, as many variables may depend on a specific component or even on multiple components, making interpretation difficult (Helena et al., 2000; Simeonov et al., 2003). Factor rotation methods can be classified into two major groups: orthogonal and oblique. Orthogonal rotation is used when the factors are independent of each other. The factors are rotated orthogonally so that they remain perpendicular to each other after the rotation. There are three methods of orthogonal rotation: Varimax, Quartimax, and Equimax, with Varimax being the most commonly used method in softwares such as SPSS. In Varimax rotation, the explanatory variables are associated with the factors in terms of columns, such that each variable is placed under the column of the factor with which it has the highest correlation. The elements in the factor matrix are called factor loadings, which indicate the degree of correlation between variables and factors.

Quartimax rotation is performed in a linear manner to establish the highest relationship between explanatory variables and factors. Equimax rotation is performed to establish the most relevant explanatory variables with the factors, both in terms of columns and rows. Since most studies employ Varimax rotation (Beavers et al., 2013; Howard, 2016), it is also known as the standard rotation method.

Diagonal rotation is used when the extracted factors are correlated. There are two types of diagonal rotation: Direct Oblimin and Promax. Both methods use parameters to indicate the degree of relationship between factors. The Direct Oblimin method uses the delta parameter, with a maximum value of 0.8 in SPSS software and a selected value of zero. The numerical value of delta parameter is selected according to the degree of relationship between factors: a larger numerical value is considered for higher relationship and a negative numerical value is considered for lower relationship. The Promax method uses the kappa parameter, with a valid numerical value from one to higher. The selected numerical value for kappa in SPSS software is four.

3.2.8.7 Interpretation and naming of factors

To interpret the results of factor analysis and classify variables in relation to factors, it is necessary to assign each variable to the factor with which it has the highest correlation after factor rotation, while considering its relationship with other factors (Courtney, 2013).

Naming factors requires paying attention to the concept that encompasses all variables under a factor and aligning it with the theoretical framework of the research. This process is both a science and an art. It involves grouping variables into those with high factor loads and those with lower factor loads and then naming the factor based on the combination of these two groups in relation to the research topic. A thorough understanding of theoretical and empirical literature related to the research topic can aid in naming factors, as variables derive their meaning from the subject of investigation. For example, if a factor contains variables such as Normalized Vegetation Index and Tasseled Cap Greenness Index, the factor could be named "Vegetation" as both variables share a common feature.

Naming factors is one of the most challenging aspects of exploratory factor analysis as factor names may not accurately reflect nature of the variables within the factor. Additionally, the interpretation of some variables may be difficult as some variables may be assigned to a factor that is not related to that variable (Yong and Pearce, 2013).

3.2.8.8 Calculation of factor scores

Factor scores are obtained by multiplying the factor weight of each variable by its standardized numerical value. In SPSS software, three methods are available to calculate factor scores: the Regression Method, Bartlett Scores, and Anderson-Rubin Method. Factor scores calculated using the regression and Bartlett methods may still be correlated even when orthogonal rotation methods are used. However, the Anderson-Rubin method, which is inspired by Bartlett's method but modified to ensure the orthogonality of factors, produces factor scores with a mean of zero, standard deviation of one, and no correlations with each other Therefore, the Anderson–Rubin method was used to calculate factor scores in this study.

3.2.9 Zonation and prediction map of soil organic carbon

The zonation of surface SOC is achieved using the multiple regression method. To accomplish this, a point map of surface SOC was created from the sampled points and used as the dependent variable (response). Independent variables are the factors resulting from factor analysis using PCA. By establishing a regression relationship between the dependent variable and independent variables, a prediction model for surface SOC is created. Eq. (3.15) shows general form of the multiple regression equation based on the independent variables (factors).

$$\mathrm{SOC} = \beta_0 + \beta_1(PC_1) + \beta_2(PC_2) + \cdots + \beta_n(PC_n) + \varepsilon \tag{3.15}$$

In the aforementioned equation, β_0 represents the intercept, β_n denotes the regression coefficient, PC_n refers to the nth factor, and ε is residual error of the regression model.

This study presents a prediction map of SOC using spectral and nonspectral effective factors and a combined method of principal components and regression, which was generated for the study area (Fig. 3.8).

3.2.10 Validation of the proposed model

When creating a model, it is important to avoid using the entire soil sample for calibration. Instead, it is recommended to use 80% of the total soil samples taken from the region for model calibration and the remaining 20% for model validation. For instance, if 218 soil samples were collected, 174 samples should be used for model calibration and the remaining 44 samples for model validation.

To validate the proposed model, three criteria are used: the correlation coefficient between the actual and predicted values of organic carbon, the root mean square error (RMSE) (Eq. 3.16), and the mean absolute relative error (MARE) (Eq. 3.17).

$$\mathrm{RMSE} = \sqrt{\frac{1}{n}\sum_{i=1}^{n}(O_i - P_i)^2} \tag{3.16}$$

$$\mathrm{MARE} = \frac{1}{n}\sum_{i=1}^{n}\frac{|O_i - P_i|}{O_i} \tag{3.17}$$

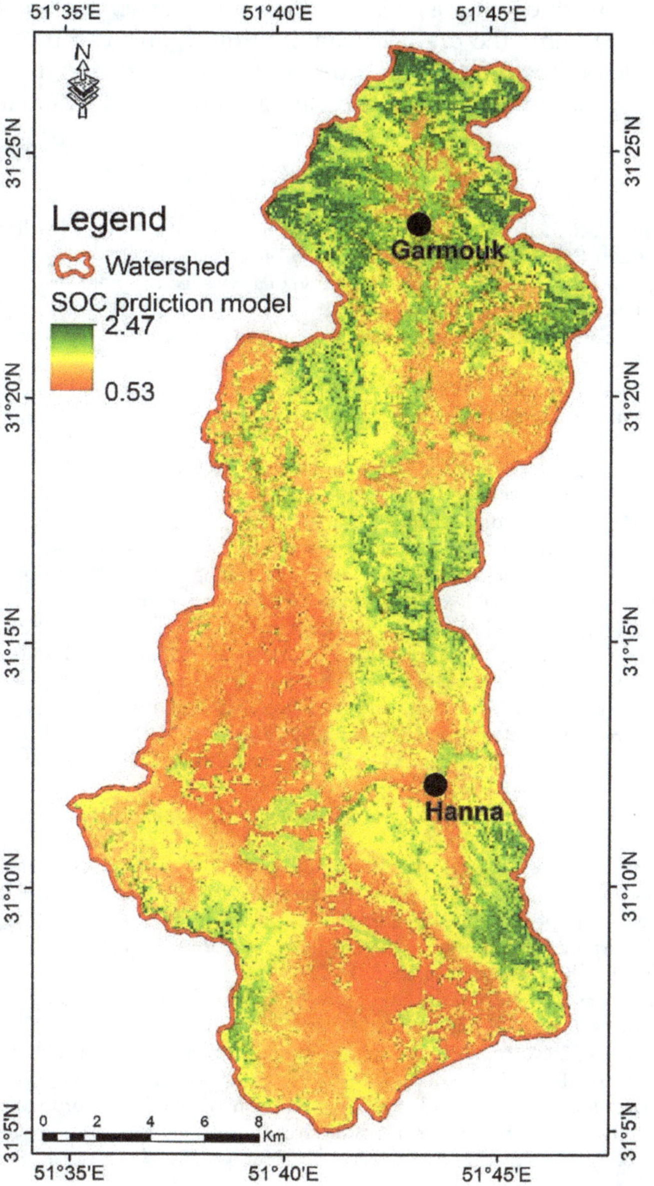

FIGURE 3.8 Predicted map of soil organic carbon in the semirom region, located in the south of isfahan province in Iran.

In the aforementioned equations, O_i represents the observed (measured) amount of organic carbon, P_i represents the predicted amount of organic carbon, and n represents the number of observations. It is worth noting that in an ideal scenario, the values of the aforementioned indicators should equal zero, indicating an unbiased estimate.

References

Andrews, S.S., Karlen, D.L., Mitchell, J.P., 2002. A comparison of soil quality indexing methods for vegetable production systems in Northern California. Agric. Ecosyst. Environ. 90 (1), 25–45.

Baig, M.H.A., Zhang, L., Shuai, T., Tong, Q., 2014. Derivation of a tasselled cap transformation based on Landsat 8 at-satellite reflectance. Remote Sens. Lett. 5 (5), 423–431.

Baldock, J.A., 2007. Composition and cycling of organic carbon in soil. Nutrient Cycling in Terrestrial Ecosystems, pp. 1–35.

Barnes, E.M., Sudduth, K.A., Hummel, J.W., Lesch, S.M., Corwin, D.L., Yang, C., Daughtry, C.S.T., Bausch, W.C., 2003. Remote and ground-Based sensore techniques to map soil properties. Photogramm. Eng. Remote Sens. 69 (6), 619–630.

Basche, A.D., DeLonge, M.S., 2019. Comparing infiltration rates in soils managed with conventional and alternative farming methods: a meta-analysis. PLoS One 14 (9), e0215702.

Beavers, A.S., Lounsbury, J.W., Richards, J.K., Huck, S.W., Skolits, G.J., Esquivel, S.L., 2013. Practical considerations for using exploratory factor analysis in educational research. Practical Assess. Res. Eval. 18 (1), 6.

Beven, K.J., Kirkby, M.J., 1979. A physically based, variable contributing area model of basin hydrology/Un modèle à base physique de zone d'appel variable de l'hydrologie du bassin versant. Hydrol. Sci. J. 24 (1), 43–69.

Boettinger, J.L., Ramsey, R.D., Bodily, J.M., Cole, N.J., Kienast-Brown, S., Nield, S.J., Saunders, A.M., Stum, A.K., 2008. Landsat spectral data for digital soil mapping. Digital Soil Mapping with Limited Data, pp. 193–202.

Bronick, C.J., Lal, R., 2005. Soil structure and management: a review. Geoderma 124 (1–2), 3–22.

Campbell, J.B., Wynne, R.H., 2011. Introduction to Remote Sensing. Guilford Press.

Castaldi, F., Palombo, A., Santini, F., Pascucci, S., Pignatti, S., Casa, R., 2016. Evaluation of the potential of the current and forthcoming multispectral and hyperspectral imagers to estimate soil texture and organic carbon. Remote Sens. Environ. 179, 54–65.

Chen, Z., Mohanty, B.P., Rodriguez-Iturbe, I., 2017. Space-time modeling of soil moisture. Adv. Water Resour. 109, 343–354.

Coleman, D.C., Callaham, M., Crossley Jr., D.A., 2017. Fundamentals of Soil Ecology. Academic press.

Costa, E.M., Samuel-Rosa, A., Anjos, L.H.C.D., 2018. Digital elevation model quality on digital soil mapping prediction accuracy. Ciência e Agrotecnol. 42, 608–622.

Courtney, M.G., 2013. Determining the number of factors to retain in EFA: using the SPSS R-menu v2 0 to make more judicious estimations. Practical Assess. Res. Eval. 18 (1), 8.

Croft, H., Kuhn, N., Anderson, K., 2012. On the use of remote sensing techniques for monitoring spatio-temporal soil organic carbon dynamics in agricultural systems. Catena 94, 64–74.

Davidson, E.A., Janssens, I.A., 2006. Temperature sensitivity of soil carbon decomposition and feedbacks to climate change. Nature 440 (7081), 165–173.

Dharumarajan, S., Hegde, R., Singh, S., 2017. Spatial prediction of major soil properties using Random Forest techniques-A case study in semiarid tropics of South India. Geoderma Reg. 10, 154–162.

Eswaran, H., Van Den Berg, E., Reich, P., 1993. Organic carbon in soils of the world. Soil Sci. Soc. Am. J. 57 (1), 192–194.

Florinsky, I.V., 2012. The Dokuchaev hypothesis as a basis for predictive digital soil mapping (on the 125th anniversary of its publication). Eurasian Soil Sci. 45 (4), 445–451.

Florinsky, I.V., Eilers, R.G., Manning, G.R., Fuller, L.G., 2002. Prediction of soil properties by digital terrain modelling. Environ. Model. Soft. 17 (3), 295–311.

Fontaine, S., Mariotti, A., Abbadie, L., 2003. The priming effect of organic matter: a question of microbial competition? Soil Biol. Biochem. 35 (6), 837–843.

Forkuor, G., Hounkpatin, O.K., Welp, G., Thiel, M., 2017. High resolution mapping of soil properties using remote sensing variables in south-western Burkina Faso: a comparison of machine learning and multiple linear regression models. PLoS One 12 (1), e0170478.

Franklin, J., McCullough, P., Gray, C., 2000. Terrain variables used for predictive mapping of vegetation communities in Southern California. In: Wilson, J.P. (Ed.), Terrain Analysis: Principles and Applications/. John C. Gallant.

Gholizadeh, A., Žižala, D., Saberioon, M., Borůvka, L., 2018. Soil organic carbon and texture retrieving and mapping using proximal, airborne and Sentinel-2 spectral imaging. Remote Sens. Environ. 218, 89–103.

Glenn, E.P., Huete, A.R., Nagler, P.L., Nelson, S.G., 2008. Relationship between remotely-sensed vegetation indices, canopy attributes and plant physiological processes: what vegetation indices can and cannot tell us about the landscape. Sensors 8 (4), 2136–2160.

Hartemink, A.E., Krasilnikov, P., Bockheim, J.G., 2013. Soil maps of the world. Geoderma 256–267.

Helena, B., Pardo, R., Vega, M., Barrado, E., Fernandez, J.M., Fernandez, L., 2000. Temporal evolution of groundwater composition in an alluvial aquifer (Pisuerga River, Spain) by principal component analysis. Water Res. 34 (3), 807–816.

Hengl, T., 2009. A Practical Guide to Geostatistical Mapping, second ed. University of Amsterdam, p. 291.

Hobley, E., Wilson, B., Wilkie, A., Gray, J., Koen, T., 2015. Drivers of soil organic carbon storage and vertical distribution in Eastern Australia. Plant Soil 390 (1–2), 111–127.

Howard, M.C., 2016. A review of exploratory factor analysis decisions and overview of current practices: what we are doing and how can we improve? Int. J. Hum. Comput. Interact. 32 (1), 51–62.

İmamoğlu, M.Z., Sertel, E., 2016. Analysis of different interpolation methods for soil moisture mapping using field measurements and remotely sensed data. Int. J. Environ. Geoinform. 3 (3), 11–25.

Jackson, R.D., Clarke, T.R., Moran, M.S., 1992. Bidirectional calibration results for 11 Spectralon and 16 BaSO4 reference reflectance panels. Remote Sens. Environ. 40 (3), 231–239.

Jenny, H., 1941. Factors of Soil Formation, A System of Quantitative Pedology. Graw-Hill, New York.

Kaiser, H.F., 1960. The application of electronic computers to factor analysis. Educ. Psychol. Meas. 20 (1), 141–151.

Lal, R., 2004. Soil carbon sequestration impacts on global climate change and food security. Science 304 (5677), 1623–1627.

Lal, R., 2008. The role of soil organic matter in the global carbon cycle. Soil and Environ. Pollut 116, 353–362.

Lardy, R., Bellocchi, G., Soussana, J.F., 2011. A new method to determine soil organic carbon equilibrium. Environ. Model. Software 26 (12), 1759–1763.

Li, X., McCarty, G.W., Karlen, D.L., Cambardella, C.A., 2018. Topographic metric predictions of soil redistribution and organic carbon in Iowa cropland fields. Catena 160, 222–232.

Liang, S., Shuey, C.J., Russ, A.L., Fang, H., Chen, M., Walthall, C.L., Daughtry, C.S., Hunt Jr., R., 2003. Narrowband to broadband conversions of land surface albedo: II. Validation. Remote Sens. Environ. 84 (1), 25–41.

Mahmoudabadi, E., Karimi, A., Haghnia, G.H., Sepehr, A., 2017. Digital soil mapping using remote sensing indices, terrain attributes, and vegetation features in the rangelands of northeastern Iran. Environ. Monit. Assess. 189, 1–20.

McBratney, A.B., Santos, M.M., Minasny, B., 2003. On digital soil mapping. Geoderma 117 (1–2), 3–52.

McCoy, R.M., 2005. Field Methods in Remote Sensing. Guilford Press.

Minasny, B., McBratney, A.B., Malone, B.P., Wheeler, I., 2013. Digital mapping of soil carbon. In: Sparks, D.L. (Ed.), Advances in Agronomy. Academic Press, pp. 1–47.

Minasny, B., Malone, B.P., McBratney, A.B., Angers, D.A., Arrouays, D., Chambers, A., Chaplot, V., Chen, Z.S., Cheng, K., Das, B.S., Field, D.J., 2017. Soil carbon 4 per mille. Geoderma 292, 59–86.

Mohan, S., Arumugam, N., 1996. Relative importance of meteorological variables in evapotranspiration: factor analysis approach. Water Resour. Manag. 10, 1–20.

Mondal, A., Khare, D., Kundu, S., Mondal, S., Mukherjee, S., Mukhopadhyay, A., 2017. Spatial soil organic carbon (SOC) prediction by regression kriging using remote sensing data. Egypt. J. Remote Sens. Space Sci. 20 (1), 61–70.

Moore, I.D., Gessler, P.E., Nielsen, G.A., Peterson, G.A., 1993a. Soil attributes prediction using terrain analysis. Soil Sci. Soc. Am. J. 57, 443–452.

Moore, I.D., Gessler, P.E., Nielsen, G.A., Peterson, G.A., 1993b, 443–452. In: Amundson, R.G., et al. (Eds.), Soil-Landscape Continuum, vol. 57, pp. 127–145. J.

Nelson, D.A., Sommers, L., 1983. Total carbon, organic carbon, and organic matter. Methods of Soil Analysis: Part 2 Chemical and Microbiological Properties, vol. 9, pp. 539–579.

Pahlavan-Rad, M.R., Dahmardeh, K., Brungard, C., 2018. Predicting soil organic carbon concentrations in a low relief landscape, eastern Iran. Geoderma Reg. 15, e00195.

Parente, C., 2013. TOA reflectance and NDVI calculation for Landsat 7 ETM+ images of Sicily. In: Proceedings in Electronic International Interdisciplinary Conference-The 2nd Electronic International Interdisciplinary Conference. EDIS-Publishing Institution of the University of Zilina, pp. 351–354.

Park, S.J., McSweeney, K., Lowery, B., 2001. Identification of the spatial distribution of soils using a processbased terrain characterization. Geoderma 103, 249–272.

Parton, W.J., Stewart, J.W., Cole, C.V., 1988. Dynamics of C, N, P and S in grassland soils: a model. Biogeochemistry 5, 109–131.

Paustian, K., Lehmann, J., Ogle, S., Reay, D., Robertson, G.P., Smith, P., 2016. Climate-smart soils. Nature 532 (7597), 49–57.

Pei, T., Qin, C.Z., Zhu, A.X., Yang, L., luo, M., Li, B., Zhou, C., 2010. Mapping soil organic matter using the topographic wetness index: a comparative study based on different flow-direction algorithms and kriging methods. Ecol. Indicat. 10, 610–619.

Piccini, C., Marchetti, A., Francaviglia, R., 2014. Estimation of soil organic matter by geostatistical methods: use of auxiliary information in agricultural and environmental assessment. Ecol. Indicat. 36, 301–314.

Polyakov, V.O., Lal, R., 2008. Soil organic matter and CO_2 emission as affected by water erosion on field runoff plots. Geoderma 143 (1–2), 216–222.

Potter, C.S., Randerson, J.T., Field, C.B., Matson, P.A., Vitousek, P.M., Mooney, H.A., Klooster, S.A., 1993. Terrestrial ecosystem production: a process model based on global satellite and surface data. Global Biogeochem. Cycles 7 (4), 811–841.

Ramcharan, A., Hengl, T., Nauman, T., Brungard, C., Waltman, S., Wills, S., Thompson, J., 2018. Soil property and class maps of the conterminous United States at 100-meter spatial resolution. Soil Sci. Soc. Am. J. 82 (1), 186–201.

Rawls, W.J., Pachepsky, Y.A., Ritchie, J.C., Sobecki, T.M., Bloodworth, H., 2003. Effect of soil organic carbon on soil water retention. Geoderma 116 (1–2), 61–76.

Ray, S.S., Singh, J.P., Dasa, G., Panigrahy, S., 2004. Use of high resolution remote sensing data for generating sitespecific soil mangement plan. In: Proceedings of the 4th International Society for Photogrammetry and Remote Sensing Congress. July 12-23, Istanbul, Turkey.

Riedo, M., Grub, A., Rosset, M., Fuhrer, J., 1998. A pasture simulation model for dry matter production, and fluxes of carbon, nitrogen, water and energy. Ecol. Model. 105 (2–3), 141–183.

Rouse Jr., J.W., Haas, R.H., Deering, D.W., Schell, J.A., Harlan, J.C., 1974. Monitoring the Vernal Advancement and Retrogradation (Green Wave Effect) of Natural Vegetation (No. E75-10354).

Rousk, J., Brookes, P.C., Bååth, E., 2009. Contrasting soil pH effects on fungal and bacterial growth suggest functional redundancy in carbon mineralization. Appl. Environ. Microbiol. 75 (6), 1589–1596.

Schillaci, C., Acutis, M., Lombardo, L., Lipani, A., Fantappiè, M., Märker, M., Saia, S., 2017. Spatio-temporal topsoil organic carbon mapping of a semi-arid Mediterranean region: the role of land use, soil texture, topographic indices and the influence of remote sensing data to modeling. Sci. Total Environ. 601–602, 821–832.

Schipper, L.A., Baisden, W.T., Parfitt, R.L., Ross, C., Claydon, J.J., Arnold, G., 2007. Large losses of soil C and N from soil profiles under pasture in New Zealand during the past 20 years. Glob. Change Biol. 13 (6), 1138–1144.

Simeonov, V., Stratis, J.A., Samara, C., Zachariadis, G., Voutsa, D., Anthemidis, A., Sofoniou, M., Kouimtzis, T., 2003. Assessment of the surface water quality in Northern Greece. Water Res. 37 (17), 4119–4124.

Simpson, R.T., Frey, S.D., Six, J., Thiet, R.K., 2004. Preferential accumulation of microbial carbon in aggregate structures of no-tillage soils. Soil Sci. Soc. Am. J. 68 (4), 1249–1255.

Smith, P., House, J.I., Bustamante, M., Sobocká, J., Harper, R., Pan, G., West, P.C., Clark, J.M., Adhya, T., Rumpel, C., Paustian, K., 2016. Global change pressures on soils from land use and management. Glob. Change Biol. 22 (3), 1008–1028.

Song, C., Woodcock, C.E., Seto, K.C., Lenney, M.P., Macomber, S.A., 2001. Classification and change detection using Landsat TM data: when and how to correct atmospheric effects? Remote Sens. Environ. 75 (2), 230–244.

Stephens, S.C., Rasmussen, V.P., Ramsey, R.D., Whitesides, R.E., Searle, G.S., Newhall, R.L., 2005. Remote sensing organic carbon in soil. USU/NASA SGEP Projects available online (15-09-05). www.extnasa.usu.edu/link_pages/downloads/remote_sensing_carbon.pdf.

Theocharopoulos, S.P., Florou, H., Walling, D.E., Kalantzakos, H., Christou, M., Tountas, P., Nikolaou, T., 2003. Soil erosion and deposition rates in a cultivated catchment area in central Greece, estimated using the 137Cs technique. Soil Till. Res. 69 (1–2), 153–162.

Thompson, B., 1992. A partial test distribution for cosines among factors across samples. Advances in Social Science Methodology, vol. 2, pp. 81–97.

Tsiafouli, M.A., Thébault, E., Sgardelis, S.P., De Ruiter, P.C., Van Der Putten, W.H., Birkhofer, K., Hemerik, L., De Vries, F.T., Bardgett, R.D., Brady, M.V., Bjornlund, L., 2015. Intensive agriculture reduces soil biodiversity across Europe. Glob. Change Biol. 21 (2), 973–985.

USGS, 2016. Landsat 8 (L8) Data Users Handbook Version 2.0. Department of the Interior, U.S. Geological Survey.

Virto, I., Barré, P., Burlot, A., Chenu, C., 2012. Carbon input differences as the main factor explaining the variability in soil organic C storage in no-tilled compared to inversion tilled agrosystems. Biogeochemistry 108, 17–26.

Wang, X., Zhang, X., Li, H., Zhang, X., Liu, H., Dou, X., Yu, Z., 2019. The minimum level for soil allocation using topsoil reflectance spectra: Genus or species? Catena 174, 36–47.

Wang, S., Zhuang, Q., Jia, S., Jin, X., Wang, Q., 2018. Spatial variations of soil organic carbon stocks in a coastal hilly area of China. Geoderma 314, 8–19.

Wang, S., Xu, L., Zhuang, Q., He, N., 2020. Investigating the Spatio-Temporal Variability of Soil Organic Carbon Stocks in Different Ecosystems of China. Science of the Total Environment. https://doi.org/10.1016/j.scitotenv.2020.143644.

Xiao, J., Shen, Y., Tateishi, R., Bayaer, W., 2006. Development of topsoil grain size index for monitoring desertification in arid land using remote sensing. Int. J. Rem. Sens. 27 (12), 2411–2422.

Yong, A.G., Pearce, S., 2013. A beginner's guide to factor analysis: focusing on exploratory factor analysis. Tutor. Quant. Methods Psychol. 9 (2), 79–94.

Zeraatpisheha, M., Ayoubia, S., Jafaric, A., Tajika, S., Finke, P., 2019. Digital mapping of soil properties using multiple machine learning in a semi-arid region, central Iran. Geoderma 338, 445–452.

Zhou, T., Shi, P., Luo, J., Shao, Z., 2008. Estimation of soil organic carbon based on remote sensing and process model. Front. For. China 3 (2), 139–147.

Zhou, Y., Webster, R., Viscarra Rossel, R.A., Shi, Z., Chen, S., 2019. Baseline map of soil organic carbon in Tibet and its uncertainty in the 1980s. Geoderma 334, 124–133.

Zucca, C., Wu, W., Dessena, L., Mulas, M., 2015. Assessing the effectiveness of land restoration interventions in dry lands by multitemporal remote sensing–a case study in ouled dlim (Marrakech, Morocco). Land Degrad. Dev. 26 (1), 80–91. Original source: https://studycrumb.com/alphabetizer.

CHAPTER

4

Conceptual of soil moisture based on remote sensing and reanalysis dataset

Shadman Veysi[1] *and Eslam Galehban*[2]

[1]Soil and Water Research Institute (SWRI), Agricultural Research, Education and Extension Organization (AREEO), Karaj, Iran; [2]Department of Remote Sensing & Geographical Information System, Faculty of Geography, University of Tehran, Tehran, Iran

4.1 Introduction

Soil moisture is defined as the amount of water, by volume, present within a given volume of soil. It is a critical factor in sustainable water resource management, particularly in agriculture, which is the largest consumer of freshwater resources (Petropoulos et al., 2015). Accurate soil moisture is essential for better understanding the water cycle and its impact on global and regional freshwater budgets (Wang et al., 2007).

The level of soil moisture has far-reaching implications for various hydrological, ecological, and meteorological processes (Chang and Islam, 2003). In agriculture, soil moisture management is crucial for increasing crop productivity, improving water utilization efficiency, and reducing water loss through evaporation. Furthermore, soil moisture data are valuable for weather forecasting, climate modeling, and estimating the water available for plant growth.

Soil moisture, despite being stored in small quantities, plays a critical role in the water-soil-plant-atmosphere relationship. According to Brusseau et al. (2019), the amount of freshwater present on Earth, soil moisture accounts for only 1 milligram for every liter. The amount of soil moisture varies according to the soil type and its water-holding capacity. Soils with a higher clay content, for example, tend to have more pores, allowing water to penetrate and remain in the soil for longer periods, resulting in higher water retention.

The dynamic nature of soil moisture and its complex relationships with precipitation, soil texture, and evapotranspiration make it challenging to accurately measure. Traditionally, point measurements were employed to estimate soil moisture over extensive regions. This method is associated with significant uncertainties. To overcome these challenges, a new approach is needed, particularly in large-scale agricultural lands.

Remote Sensing of Soil and Land Surface Processes
https://doi.org/10.1016/B978-0-443-15341-9.00020-4

Enter the growing fleet of Earth Observation (EO) satellites in orbit, with more than 18 different satellites focused on different components of energy balance and carbon cycles (Wang et al., 2011). Due to its role in the hydrological cycle, soil moisture is a vital component that significantly influences energy exchange between the soil and atmosphere. This makes remote sensing and similar tools ideal for surveying soil moisture based on energy laws. The science of remote sensing relies on the electromagnetic spectrum to describe various phenomena and determining which parts of the electromagnetic spectrum can be actively used for soil moisture estimation is crucial.

Yang et al. (2019) suggests that soil moisture variation can be detected in nearly all regions of the electromagnetic spectrum. Soil moisture is an energy-oriented phenomenon, and the law of the electromagnetic spectrum shifting from short to long wavelengths means that the amount of energy decreases, leading to different descriptions of soil moisture based on environmental conditions. The behavior of different parts of the electromagnetic spectrum in estimating soil moisture is further discussed in Fig. 4.1, taking into account these findings.

The current chapter is structured into five distinct sections to provide a comprehensive overview of soil moisture measurement and its significance. In Section 4.2, the crucial role of scale in measuring soil water content is highlighted and discussed. The following section, Section 4.3, focuses on the role of Earth Observation (EO) satellites in soil moisture estimation and its significance in the larger context of water resource management. Section 4.4 explores the various methods used to retrieve soil moisture information through remote sensing, utilizing the electromagnetic spectrum. Finally, in Section 4.5, a critical analysis of the reanalysis dataset, which extracts soil moisture information, is presented.

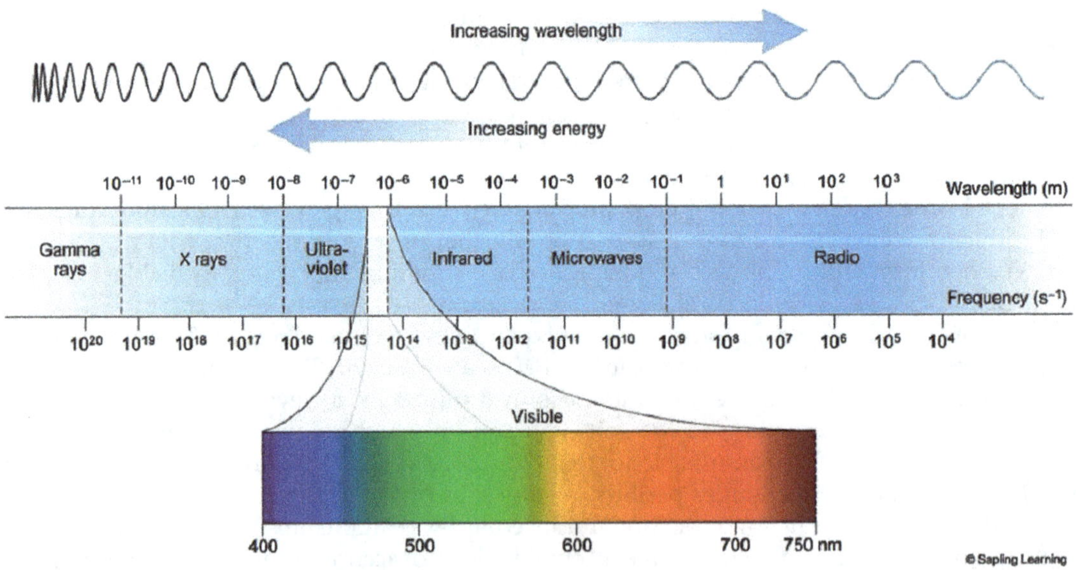

FIGURE 4.1 The electromagnetic spectrum.

4.2 Scale dependency of soil moisture

Scale dependency is a concept that highlights the relationship between the measurement or estimation of a property and the scale of measurement. It implies that the accuracy and validity of a measurement can be influenced by the spatial or temporal resolution at which it is taken. It is critical to understand the scale dependency in soil moisture estimation as the optimal scale of measurement can differ vastly depending on the application and requirement.

The selection of an appropriate scale can greatly impact the quality and usefulness of the data obtained, and therefore, it is crucial to carefully consider the scale of measurement when designing soil moisture measurement campaigns. The determination of an optimal spatio-temporal resolution for soil moisture estimation is a critical challenge in the field. The appropriate resolution is contingent on the specific application and its requirements. Lysimetric measurements, for instance, have been observed to offer acceptable accuracy at laboratory and field scales (Stoffregen et al., 2002); however, when used for decision-making at larger scales such as catchments or basins, the results are often accompanied by significant uncertainties (Verstraeten et al., 2008). Large-scale datasets commonly utilized for climate change decision-making may suffer from biases when applied at smaller scales.

Despite attempts to directly measure soil moisture at a micro scale, these efforts are also not immune to errors and limitations. A common bias or outlier shared between large and small scales is a possibility (Hallegatte et al., 2012). Therefore, it is imperative to accurately identify the appropriate scale to ensure the validity and reliability of results.

Furthermore, it is important to understand the limitations and uncertainties associated with each soil moisture estimation tool and their respective scales. This information can inform the selection of the most suitable tool for a given application, taking into account the necessary spatial and temporal resolution, as well as the accuracy and precision of the results. The domain expanded scale for each tool is shown in Fig. 4.2. The selection of the appropriate tool and scale is critical for making informed decisions in fields such as agriculture, hydrology, and climate change.

4.3 Role of survey from space

The acquisition of soil moisture data through in situ measurement and point sampling can be a challenging and costly task, with a gap in the long-term observation data being a significant concern for researchers and experts in fields such as environment, water resources, climate change, and agriculture. The advent of new soil moisture technologies offers the potential for a global perspective, with frequent revisit times, and an effective solution to overcome data scarcity. Satellites, flying at several hundred kilometers altitude and with the capability for quick revisit times, provide an opportunity for frequent monitoring of soil moisture, as reported by (Barideh et al., 2022). The objective of this approach is to achieve a long-term, consistent data record, with the highest possible temporal and spatial resolution, allowing for the monitoring of regions that are challenging to access.

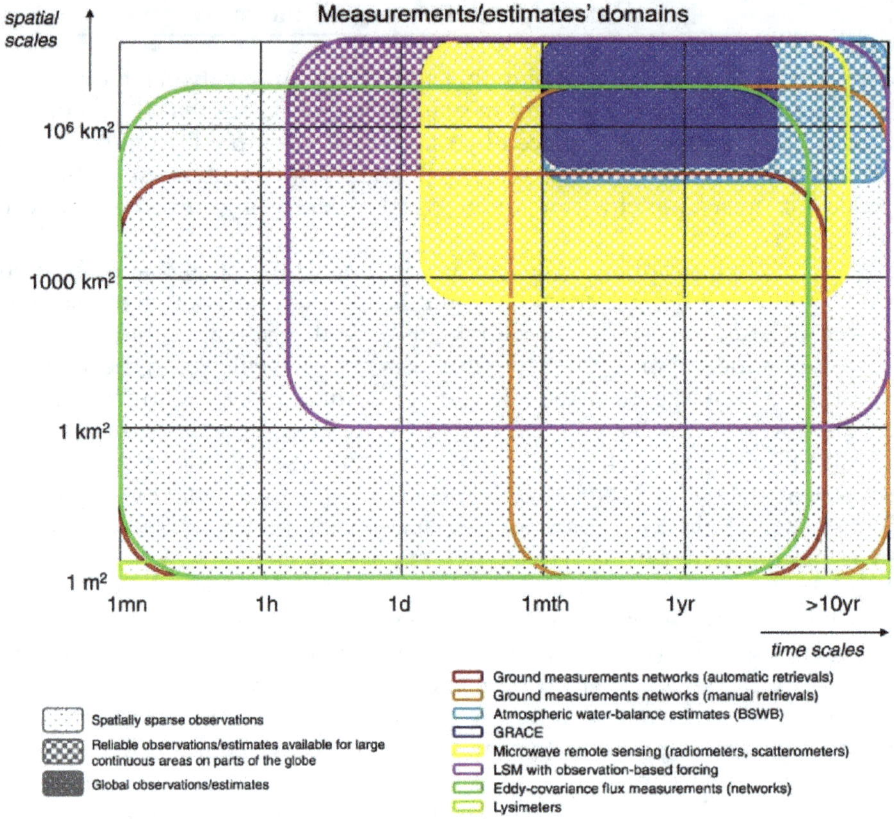

FIGURE 4.2 The schematic of scale on soil moisture measurement (Seneviratne et al., 2010).

While remote sensing offers several advantages, it is not a replacement for in situ sampling methodology, due to limitations including in situ penetration, sensing depth, and accuracy, as discussed by (West et al., 2019). In situ measurement, with its highest revisit time, provides valuable data to complement satellite-based observations and can be used for calibrating and validating current models and planning efforts, as reported by (Konings et al., 2017). By merging satellite-based observations with in situ observations, the fidelity of in situ networks can be increased, leading to the development of value-added products, as discussed by Ford and Quiring (2019).

The key to unlocking the full potential of these technologies is to strategically apply them and merge them with a land surface crop growth model, thereby improving the constraining of models from frequent satellite observations. This leads to greater information availability on topics such as dynamic agriculture and productivity indices and provides a cost-effective solution compared to in situ observation, which can be time-consuming and expensive.

4.4 Optical domain

Optical remote sensing can be used to estimate SM by measuring the reflectance of the soil surface in different spectral bands. The reflectance of soil in the optical parts is influenced by the amount of soil moisture. When soil is dry, it reflects more light in the visible spectrum and less in the NIR, whereas when soil is wet, it reflects less visible light and more NIR light. This difference in reflectance can be used to estimate the volume of water in the soil (Kaleita et al., 2005). Optical signals are affected by factors such as cloud cover and vegetation (Zhao and Li, 2013), which can make it difficult to accurately measure SM. Clouds can block the optical signals from reaching the ground, while vegetation can absorb or scatter the signals, making it challenging to distinguish the signal from the background noise. Optical sensors typically have lower spatial resolution than other remote sensing techniques such as radar, which can limit their ability to capture fine-scale variations in SM.

4.5 Combine optical and thermal

4.5.1 Triangular method

The triangular method for soil moisture estimation is a combination of both optical and thermal methods. The optical band works at the wavelength range of 400–2500 nm and is limited by the obscuring of the surface by clouds and vegetation, as well as having a shallow penetration depth over bare soil, which is only a fraction of a micrometer (Klemas and Pieterse, 2015). On the other hand, the thermal domain operates at wavelengths between 3500 and 14,000 nm, based on the principle that dry soil has a higher temperature than moist soil (Van den Hurk et al., 1997).

The triangular method was proposed by (Carlson and Bufftim, 1989; Nemani et al., 1993), who found through the Land Surface Temperature-Vegetation Index (LST-VI) curve that surface soil moisture can be determined by the relationship between LST from thermal data and vegetation index (VI) from the optical infrared band. This method allows for monitoring a top layer (0–5 cm) of soil moisture.

The correlation between LST and Normalized Difference Vegetation Index (NDVI) is higher for areas with bare soil compared to areas covered with vegetation. A positive correlation between LST and NDVI indicates energy-limited conditions, while a negative correlation indicates water-limited conditions (Fig. 4.3). These findings highlight the potential of the triangular method in providing valuable insights into soil moisture conditions and their impact on vegetation and water resources management.

One of the main challenges is the heterogeneity of surface, which means that different parts of the land surface have different physical and biological characteristics that affect their temperature and reflectance properties. This spatial heterogeneity can make it difficult to accurately determine the boundaries between dry and wet areas in the LST-VI space, which are important for characterizing land surface conditions (Wang et al., 2011). For smaller scales, such as individual pixels in satellite imagery, it is often easier to establish the LST-VI relationship because the surface properties are more homogeneous. However, at larger scales, the variability in surface properties can make it harder to accurately interpret the

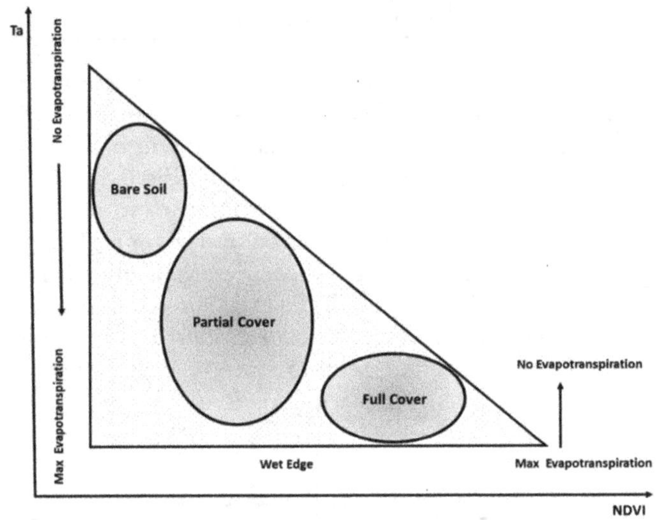

FIGURE 4.3 Schematic triangular method.

relationship between LST and VI. The triangular method of describing soil moisture content may not always be appropriate, especially in the presence of irrigation. (Boegh et al., 1999) found that the triangular curve does not accurately represent all states of soil moisture and suggested a more appropriate method, the trapezoidal method. In the trapezoidal method, soil moisture content near the dry edge indicates that crops are under stress, while moisture content near the wet edge indicates a more favorable state for plant growth. The triangular method assumes that as vegetation grows, the canopy temperature decreases. However, it is important to note that in irrigated fields, where water is applied to the soil surface, the soil moisture content may increase, leading to a reduction in soil temperature. The trapezoidal method is better suited for monitoring soil moisture content in irrigated areas, while the triangular method is more suitable for large-scale monitoring of drought and forests. Overall, it is important to choose the appropriate method of monitoring based on the each conditions (Fig. 4.4).

The comparison between triangular and trapezoidal methods of soil moisture monitoring has been extensively studied by researchers such as (Sandholt et al., 2002; Wang et al., 2007). These studies found that soil moisture oscillation is more sensitive to the variation of LST than to vegetation indices, indicating that the response and spectral reflectance in thermal bands is faster than in optical bands. The vegetation indices, showed a delay in response in humid regions, with a delay of 10 days observed after a lack of soil moisture in crops in humid regions and 5 days in arid and semi-arid regions (Wang et al., 2007). In regions with low precipitation, the trapezoidal and triangular methods performed better than in regions with high precipitation (Li et al., 2008). As a result, the combined use of optic and thermal bands in the spring and summer seasons is recommended. An example of this approach is the Temperature Vegetation Dryness Index (TVDI), which is used to estimate surface soil moisture in sugarcane fields based on the LST-VI principle. The TVDI is demonstrated in Fig. 4.5.

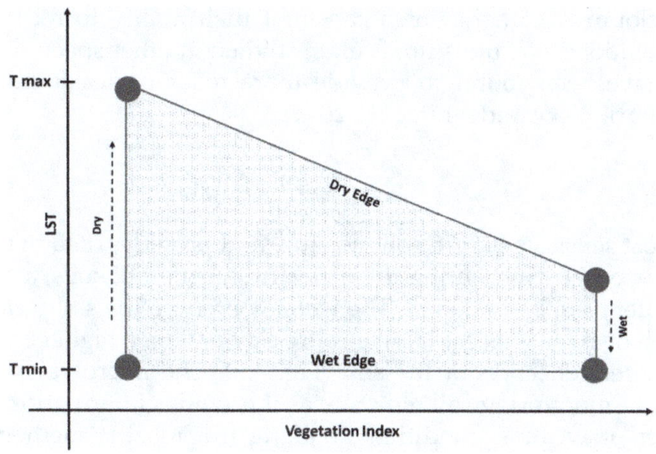

FIGURE 4.4 Schematic trapezoidal method.

The depth of soil that contributes to measurement is an essential idea in remote sensing using the electromagnetic spectrum, and it is referred to as penetration depth. This concept takes into account the extent to which soil moisture and wavelength can affect the depth of soil that is involved in the measurement. The optical bands have limitations in terms of penetration depth and thus may not provide an accurate representation of soil moisture status at deeper levels of the soil profile. Longer wavelengths tend to have better penetration depth compared to shorter wavelengths. However, both optical and thermal bands face limitations in the presence of vegetation cover, as soil moisture status can only be reported based on the condition of the crops. Another limitation of these bands is that they are subject to

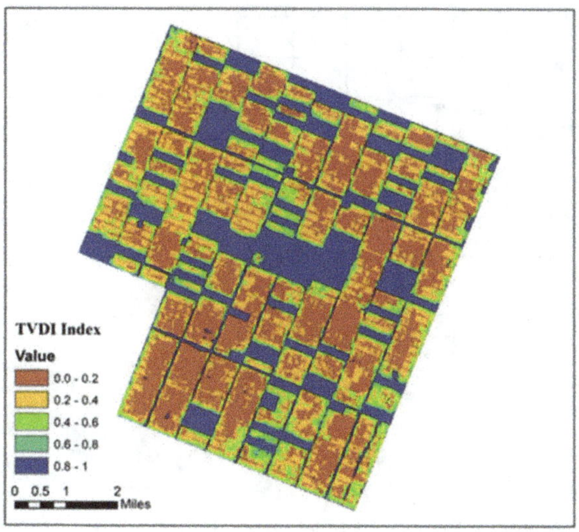

FIGURE 4.5 Temperature vegetation dryness index (TVDI) (Veysi et al., 2020).

interference from cloudy weather, which can limit their ability to receive information. To overcome these limitations, soil moisture is often studied in other spectral ranges, such as microwaves, which have been found to provide more reliable measurements (Engman and Chauhan, 1995; Behari, 2005; Sadeghi et al., 2015).

4.5.2 Microwave

Microwave is most sensitive part to soil moisture and provides data in the frequency range of 1–5 GHz. Microwaves by measuring the dielectric constant of soils, a direct measurement of soil moisture content can be obtained. The water content in the soil profile varies based on the soil texture, with heavier soils containing more clay having a higher capacity to retain water, leading to a lower reflection of the soil (Fig. 4.6). The microwave frequency range is particularly useful for monitoring soil moisture as it provides information that is not affected by vegetation cover or weather conditions, making it a reliable method for soil moisture measurement.

Also, active microwave remote sensing, the system emits its own energy in the form of radar pulses, and the backscattered radiation from the soil–vegetation interface is measured. This approach provides more precise estimates of soil moisture compared to the passive method as it is not limited by the variability of thermal emission from the surface. However, the active approach is typically more expensive and requires more sophisticated instrumentation.

Furthermore, microwave remote sensing is not only limited to soil moisture estimation but also has several other applications such as monitoring snow cover, mapping soil properties, and estimating vegetation biomass. The ability of microwaves to penetrate through vegetation and clouds, combined with their sensitivity to soil moisture, makes them an ideal tool for monitoring large-scale natural resources.

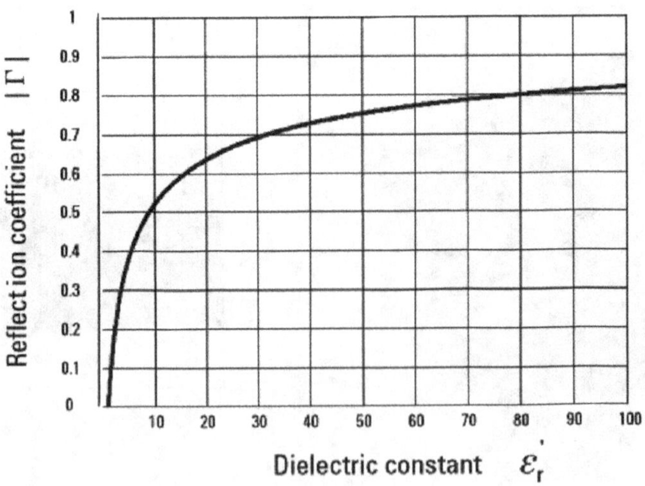

FIGURE 4.6 Relationship between spectral reflectance and dielectric constant.

In conclusion, the use of microwave remote sensing has significant potential for improving our understanding of soil moisture and its variability. Although it is still an evolving field with ongoing research, the development of advanced microwave remote sensing techniques provides an exciting opportunity to gain new insights into the water cycle and the role it plays in the Earth's climate system.

4.6 Soil moisture using passive microwave

Passive microwave is a technique used to measure the thermal emission from the Earth's surface, which is directly proportional to soil moisture content. The technique involves using a microwave radiometer to measure brightness temperature, which can provide continuous and large-scale observations with high accuracy (Njoku et al., 2002). However, this method is limited to providing information about the top layer of the soil and requires significant data processing to obtain accurate results (Jackson et al., 2011).

The emission of microwaves at different wavelengths is a function of the dielectric constant of the soil–water structure and soil moisture (Walker, 1999). The change in the spectral reflectance of the soil in the microwave region is relatively weak, which is why the emitted energy is usually introduced as brightness temperature. The brightness temperature increases as the soil goes from wet to dry (Walker, 1999).

Passive microwave remote sensing has a lower spatial resolution due to the low emitted energy at microwave wavelengths, which require large pixels on the ground to receive enough energy to record a signal (Njoku et al., 2003; Singh et al., 2005). The L-band is the most promising for soil moisture retrieval, but the spatial resolution that can be obtained is limited due to the inverse relationship between wavelength and antenna size (Jackson et al., 2011). Increasing the antenna size to improve spatial resolution presents engineering problems that cannot be solved with conventional technologies (Jackson et al., 2011).

The correlation between soil moisture and brightness temperature can vary based on factors like vegetation conditions and soil physics properties (Owe et al., 2001; Panciera et al., 2009; Rüdiger et al., 2009). The passive microwave limitations include low temporal resolution, and the need for significant data processing (Njoku et al., 2002, 2003; Singh et al., 2005) recommended the combination of passive and active microwave remote sensing to obtain a more accurate representation of soil moisture.

4.6.1 SMOS

One of the passive microwave satellites collecting data on soil moisture and ocean salinity is SMOS. SMOS launched in November 2009, and its microwave imaging radiometer with aperture synthesis (MIRAS) sensor was developed in response to the need for global observation data on these two factors. Utilizing a satellite in polar orbit, SMOS gathers two-dimensional interferometry radiometer measurements of L-band (1.4 GHz) brightness temperature. As the low frequency reduces atmospheric interference, the Earth's surface emission can be accurately measured. This measurement is strongly indicative of soil moisture over land, as well as providing information on the salinity and state of the oceans and the thickness of sea ice. The spatial resolution of the SMOS measurements range from 30 to 50 km. The expected accuracy of SMOS in estimating soil moisture is (0.04 m^3/m^3) and 0.1 g of salt in water. The main challenge of passive microwave satellites is their resolution, which can be addressed through fusion with optical satellites (Portal et al., 2018) or downscaling algorithms (Piles et al., 2012) that have produced acceptable results. In many cases, SMOS has shown itself to be capable of detecting floods (Louvet et al., 2015), and its continuous data at a large scale is useful for a better understanding of climate change (Srivastava et al., 2015). Another advantage of SMOS is that it provides soil moisture data during both day and night time (Entekhabi et al., 2010). With global coverage every 3 days, SMOS' time resolution is ideal for near real-time investigation of various phenomena.

4.7 Soil moisture using active microwave

Since passive microwaves are limited by low spatial resolution, some other remote sensing techniques have been proposed to provide observations at higher spatial resolution. Active microwave techniques can be used to retrieve soil moisture information from 10 m to 3 km resolution. Currently, two types of radars (high spatial resolution but lower soil-moisture accuracy) are being used for soil moisture retrieval: Synthetic aperture radar (SAR) and scatterometer. SAR is a coherent radar system installed on a mobile platform such as a satellite or an airborne aircraft. It creates two-dimensional images with high spatial resolution independent of light (recording data in day and night) and weather conditions from the received backscatter signals. The extent of information acquired through the antenna from the soil's depth is contingent on the soil profile's temperature. Polarizations used in SAR, depending on the direction of sending and receiving electromagnetic waves, include double polarizations (VV and HV or HH). For example, the European Space Agency (ESA) has used Earth Remote Sensing to remotely sense soil moisture using C-band SAR observations (Wagner et al., 2003, 2007). Other SAR observation sensors include the C-band polarimetric (McNairn and Brisco, 2004), the C-band SAR Sentinel-1 sensor for monitoring frozen with VH polarization (Baghdadi et al., 2018), L-band PALSAR-2 (Huang et al., 2018), X-band TerraSAR (Nunziata et al., 2019), X-band COSMO/SkyMed (Bovenga et al., 2012), and the SMAP radar sensor from L-band with a spatial resolution of 3 km (Entekhabi et al., 2010).

Microwave scatterometers are radar sensors that measure surface backscatter using an aperture of reality (Dubois et al., 1995). Scatterometer sensors are used for the measurement of near-surface ocean wind speeds, polar ice, and soil moisture measurements. Various scatterometer sensors have been worked in 2–8 GHz frequency on satellites (Wagner et al., 2003, 2007). In active microwaves, the values of the signal depend on soil moisture across the dielectric constant of water and soil. Although in radar observations, the soil signal is strongly influenced by surface roughness (Magagi and Kerr, 2001) and direct scattering from vegetation. For a smooth surface, the highest oscillation in elevation is less than 25% of the wavelength. When studying backscattering from vegetation canopies, roughness is a critical factor as any crop above a certain frequency tends to be rough.

4.8 Soil moisture active passive (SMAP)

SMAP was launched in January 2015, and its main objective is to observe soil moisture and freeze-thaw in the microwave frequency. SMAP orbits at an altitude of 685 km and utilizes a large mesh reflector, 6 m in diameter, to continuously scan the earth at 14.6 revolutions per minute (Wigneron et al., 2017). SMAP covers a swath on the Earth's surface that is a 1000 kilometers wide, enabling it to observe the entire Earth every 3 days.

SMAP has two instruments: radar observations and radiometer observations. The SMAP uses radar observations to obtain high-quality spatial resolution data; however, it can lead to lower soil moisture estimate accuracies. On the other hand, radiometer observations offer higher accuracy for soil-moisture estimates but with coarse spatial resolution. SMAP is a satellite mission that uses both active and passive sensors. The active sensor, similar to ultrasound, sends a signal to the Earth's surface and some of it bounces back to the satellite, depending on the characteristics of the surface. This signal contains information about the moisture content of the top 5 cm (2 inches) of soil everywhere. Thus, the radar is considered active, and the radiometer is considered passive. It is important to note that the radar suddenly stopped operating on July 7, 2015, after collecting data for 2.5 months and now only the radiometer product is available with a resolution of 36 km and the dimension of data is volumetric water to volumetric of soil (cm^3/cm^3). The radiometer component of SMAP continues to operate as planned (Chan et al., 2016). The SMAP mission is to generate time-series data of different levels of products, ranging from Level 1 to Level 4, for the global land area at 3-day average intervals and for land regions above 45°N at 2-day average intervals.

4.8.1 SMAP products

The public has access to 24 data products generated by the SMAP mission, with each product reflecting one of four processing levels. The data generated by the mission are useful in areas with constant cloud cover or in regions with short, dark days, as the instruments can measure soil moisture regardless of weather or nighttime conditions. The SMAP radar has a high resolution of 3 km, while the radiometer has a coarser resolution of 36 km. By combining both instruments, the mission is able to generate a soil moisture product with a resolution of 9 km. Table 4.1 lists SMAP's fundamental scientific data products, which can be obtained free of charge through two NASA-designated data centers, namely, the Alaska Satellite Facility (ASF) and the National Snow and Ice Data Center (NSIDC). Tragically, the SMAP radar stopped functioning in July 2015, limiting soil moisture product resolution to the radiometer's 40-km capability (https://smap.jpl.nasa.gov/).

4.8.1.1 Level 1: Brightness temperature

Level 1 data in the SMAP mission consist of both Level 1B (L1B) and Level 1C (L1C) data products, which are calibrated and geo-located measurements of the surface radar backscatter cross-section and brightness temperatures (Kim et al., 2016). The L1B_TB_E data product, for instance, is a calibrated brightness temperature that has been interpolated on the EASE-2 grid and acquired by the SMAP radiometer with a spatial resolution of 36 km.

TABLE 4.1 The list of products that are being generated.

Product type	Product description	Resolution	Granule extent
L1A_Radar	Parsed SMAP radar telemetry (start-July 7, 2015)		Half orbit
L1B_SC_LoRes	Low-resolution radar signal in time order (start-July 7, 2015)	5 × 30 km	Half orbit
LIC_SO_HiRes	High-resolution radar sigma on swath grid (start-July 7, 2015)	1 km	Half orbit
L1A_Radiometer	Parsed radiometer telemetry		Half orbit
LIB_TB	Geolocated, calibrated brightness temperature in time order	36 km	Half orbit
L1B_TB_E	Backus–Gilbert interpolated, calibrated brightness temperature in time order	9 km	Half orbit
LIC_TB	Geolocated, calibrated brightness temperature on EASE2 grid	36 km	Half orbit
LIC_TB_E	Backus–Gilbert interpolated, calibrated brightness temperature on EASE2 grid	9 km	Half orbit
L2_SM_A	Radar soil moisture (start-July 7, 2015)	3 km	Half orbit
L2_SM_P	Radiometer soil moisture	36 km	Half orbit
L2_SM_P_E	Radiometer soil moisture	9 km	Half orbit
L2_SM_AP	SMAP active-passive soil moisture	9 km	Half orbit
L2_SM_SP	SMAP radiometer/Copernicus Sentinel-1 soil moisture	3 km	Sentinel-1
L3_SM_P	Daily global composite radiometer soil moisture	36 km	Daily-global
L3_SM_P_E	Daily global composite radiometer soil moisture	9 km	Daily-global
L3_FT_A	Daily global composite radar freeze/thaw state (start-July 7, 2015)	3 km	Daily-North of 45 deg N
L3_SM_A	Daily global composite radar soil moisture (start-July 7, 2015)	3 km	Daily-global
L3_SM_AP	Daily global composite active-passive soil moisture (start-July 7, 2015)	9 km	Daily-global
L3_FT_P	Daily composite freeze/thaw state	36 km	Daily-global
L3_FT_P_E	Daily composite freeze/thaw state	9 km	Daily-global
L4_SM	Surface and root zone soil moisture	9 km	3 Hours-global
L4_C	Carbon net ecosystem exchange	9 km	Daily-North of 45 deg N
L1B_TB_NRT	Near real time geolocated, calibrated brightness temperature in time order	36 km	Half orbit
L2_SM_P_NRT	Near real-time radiometer soil moisture	36 km	Half orbit

This product is a gridded version of another level 1 brightness temperature product, which is based on the Single Channel (SC) Algorithm. The SMAP radiometer measures the brightness temperature of the surface, which is then used to infer soil moisture and freeze/thaw information. The level 1 data provide the foundation for the generation of higher-level products, such as level 2 and level 3 data, which provide more advanced information about soil moisture and freeze/thaw conditions (https://smap.jpl.nasa.gov/instrument-level-1-data/level-1-radiometer-data-product/).

4.8.1.2 Level 2: Soil moisture

The SMAP satellite generates five Level 2 soil moisture products that provide geophysical retrievals of soil moisture on a fixed Earth grid using Level 1 data and ancillary information. These products are updated on a half-orbit basis (Entekhabi et al., 2010).

The L2_SM_P product, identified at 36 km, is formulated from radiometer brightness temperature measurements. In comparison, the L2_SM_P_E product, available at 9 km, utilizes Backus–Gilbert's interpolated radiometer brightness temperature measurements. Furthermore, by merging Sentinel-1 active and SMAP passive data, the L2_SM_SP product produces soil moisture estimates at a highly detailed 3 km resolution (Colliander et al., 2021).

The L2_SM_A product solely utilizes high-resolution radar backscatter data to estimate soil moisture at a fine resolution. Although it may not be as accurate as other products, it provides valuable information at a higher spatial resolution. Consequently, the product is combined with ancillary data on water bodies and freeze/thaw indicators to form input for other SMAP products, enhancing their overall accuracy.

With a consistent grid size of 9 km across other SMAP products at 36 and 3 km, the L2_SM_AP product employs a 9 km Equal-Area Scalable Earth-2 grid, enabling it to produce brightness temperatures at a resolution of 9 km, as mentioned in the study by Chan et al. (2018).

While the retrieval algorithm employed for soil moisture estimation in the L2_SM_AP product is similar to that of the L2_SM_P product, its implementation may differ due to the spatial resolution variability.

4.8.1.3 Level 3: Soil moisture

Level 3 SMAP soil moisture data are a global composite of daily estimates of soil moisture derived from the SMAP passive microwave radiometer and resampled to the EASE-Grid 2.0 resolution. The data are generated by combining and processing the Level 2 soil moisture data, which provide high-resolution, near-real-time observations of soil moisture, to produce a more accurate and consistent global soil moisture product. The Level 3 data aare widely used in a variety of scientific and practical applications, as it provides valuable information about the Earth's water and energy cycles and helps to advance our understanding of the Earth's hydrological systems.

4.8.1.4 Level 4: Root zone soil moisture (RZSM)

Level 4 SMAP RZSM data are a higher-level product derived from the Level 3 SMAP soil moisture data. The data provide a detailed and accurate estimate of soil moisture at a 1-m depth in the root zone, which is the layer of soil where most plant roots are located and where most of the water uptake by plants occurs. The soil moisture information at this depth is important for understanding the water availability for plants and the overall health of vegetation (Reichle et al., 2019).

The Level 4 SMAP RZSM data have a spatial resolution of 9 km and a temporal resolution of 3 h, making it possible to track changes in soil moisture over time at a fine scale. The data are generated by assimilating the brightness temperature from SMAP Level 1C_TB soil moisture observations into a land surface model simulation and then gridding the results using the EASE-Grid 2.0 projection (Dong et al., 2019). The SMAP RZSM data are generated using an ensemble Kalman filter algorithm, which is a data assimilation method that merges the SMAP brightness temperature observations with soil moisture estimates from the NASA Catchment land surface model. The algorithm takes into account various physical processes, such as precipitation, evapotranspiration, and runoff, to produce a more accurate and consistent estimate of soil moisture at the root zone.

The accuracy of the SMAP RZSM data are further improved by including forcing data, such as precipitation and brightness temperature, which have a direct effect on the RZSM in the simulation. By providing high-resolution and accurate soil moisture information at the root zone, the SMAP RZSM data help to advance our understanding of the Earth's water and energy cycles and support a wide range of applications in fields such as agriculture, water resource management, and ecosystem monitoring (Fig. 4.7).

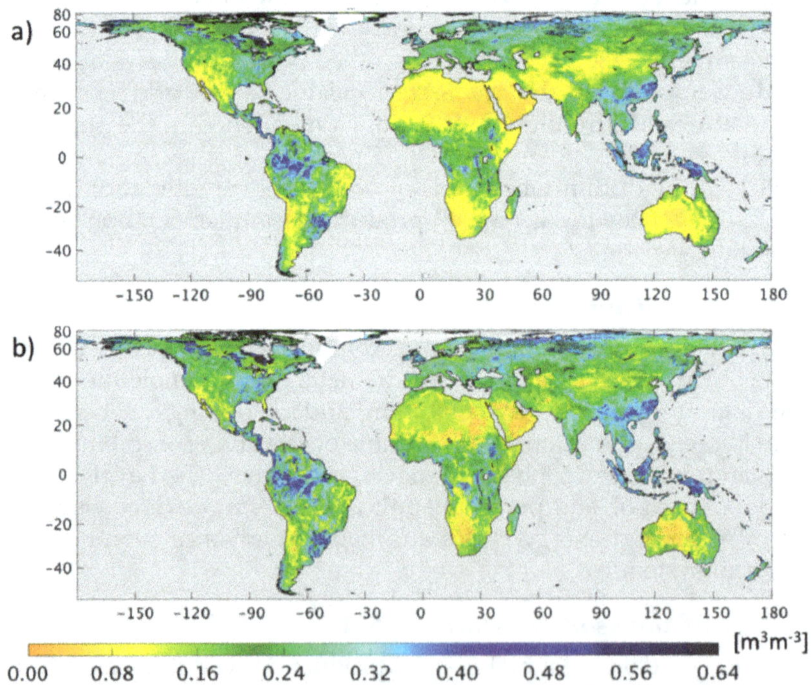

FIGURE 4.7 Time-average of Version 4 L4_SM (A) surface and (B) root zone soil moisture for 1 April 2015, to 1 April 2018, 0z. The white patches over the Greenland ice sheet and the adjacent North Atlantic are a plotting artifact that lies outside of the L4_SM domain.

4.9 Soil moisture using reanalysis dataset

Reanalysis is a sophisticated product that generates a numerical representation of climatic and hydrological variables by assimilating various types of observations such as historical meteorological stations, ships, balloons, airplanes, and satellites. This utilization of reanalysis models ensures that all observations are interpreted consistently and that spatiotemporal data gaps are filled based on physical and dynamic conditions (Hersbach et al., 2020; Snyder et al., 2005). The integration of multiple-source observations reduces ambiguities and even allows for the recognition and correction of incomplete observations. Reanalysis provides access to the recent history of the atmosphere, land surface, and oceans (Saha et al., 2010; Balsamo et al., 2009). As a result, reanalysis is a valuable tool for regenerating observed soil moisture data.

The ERA reanalysis is one of the commonly used datasets for monitoring and forecasting weather and climate conditions. Developed by the European Center for Medium-Range Weather Forecasts (ECMWF), it offers shared soil moisture products and a range of spatial resolutions for free access (Balsamo et al., 2015). ECMWF offers data and information to develop solutions that are tailored to users' needs. Fig. 4.8 provides a schematic illustration of the process used to produce the ERA5 reanalysis dataset.

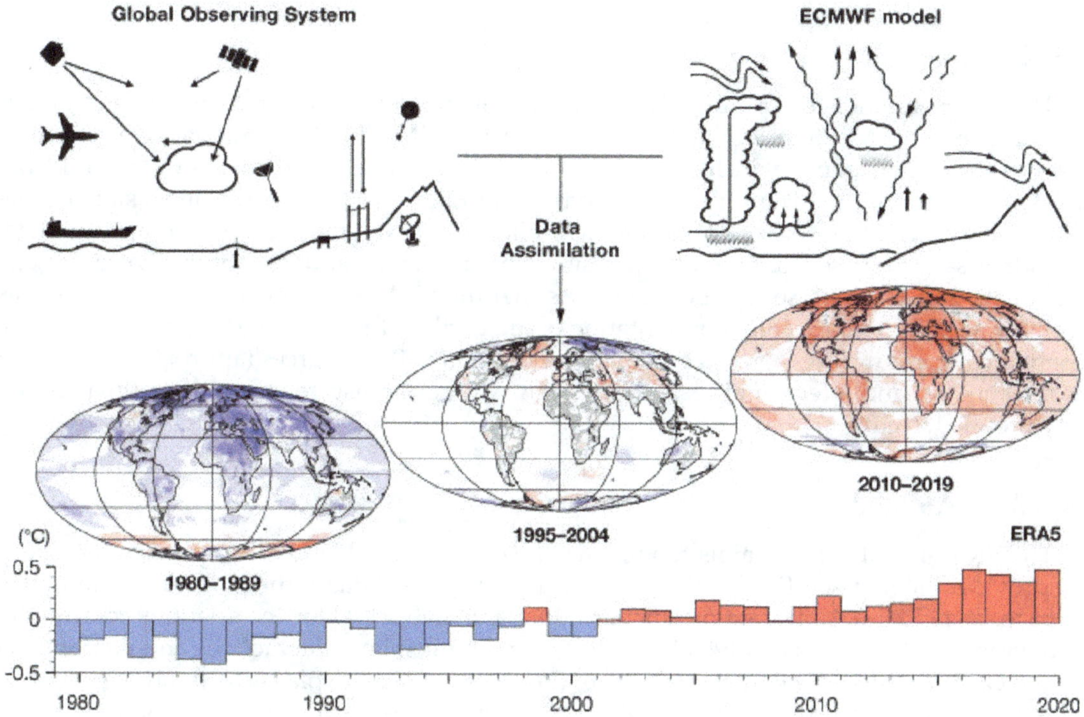

FIGURE 4.8 Schematic of producing ERA5 reanalysis dataset.

4.9.1 ERA interim

The ERA Interim soil moisture product is a highly valuable tool for many applications, including weather forecasting, hydrological modeling, climate research, and agriculture. The ERA Interim is a reanalysis data set, meaning that it combines observational data with numerical models to produce a comprehensive representation of the atmosphere, ocean, and land surface. The ERA Interim provides global coverage of soil moisture, making it useful for studying large-scale patterns and trends in soil moisture. The ERA Interim data set has a long-term record, allowing for the analysis of soil moisture variability over time and the detection of trends and patterns. Despite its advantages, there are also some limitations to the product that should be considered.

In terms of spatial resolution, the grid size of approximately 0.75° × 0.75° may not be sufficient to resolve fine-scale variability in soil moisture. As a reanalysis dataset, the ERA Interim product is subject to uncertainties associated with the numerical models used to produce it. This can limit its accuracy and reliability, especially in regions where there is a lack of observational data. The temporal resolution of the ERA Interim data are also limited to 6 h, which may not be sufficient to capture rapid changes in soil moisture. In conclusion, the ERA Interim soil moisture product is a valuable tool that provides a comprehensive view of soil moisture at a global scale. Despite its limitations, it remains a useful resource for many applications and continues to be widely used in research and practical applications. Fig. 4.9 shows the map of soil moisture results of ERA-Interim for china.

4.9.2 ERA5

The ERA5 soil moisture product is the global soil moisture product developed by the European Center for Medium-Range Weather Forecasts (ECMWF). It provides a comprehensive view of the Earth's climate, including soil moisture information, and covers the period from 1979 to the present day. The soil moisture data from ERA5 are produced using a state-of-the-art atmospheric reanalysis system that assimilates a wide range of observational data, including satellite-based observations, in-situ measurements, and meteorological data. The assimilation of this diverse set of data ensures that the ERA5 soil moisture product provides a highly accurate and reliable representation of the Earth's climate, including its soil moisture content. The product is generated at a spatial resolution of 0.25 degrees (approximately 28 km at the equator) and a temporal resolution of 6 h. The soil moisture content is estimated at a depth of 0–100 cm.

4.9.3 ERA5-land

The ERA5-LAND project aims to improve the quality of the land-atmosphere interactions by providing high-resolution, multivariation, and frequent data on the land surface and subsurface. The ERA5-LAND data product provides a valuable tool for the scientific community and various stakeholders involved in studying land-atmosphere interactions and their impact on the climate system (Malardel et al., 2016) (Fig. 4.10), which displays the trend of soil moisture from the year 2000, highlights the utility of ERA5-LAND data in monitoring and analyzing long-term changes in soil moisture conditions. The availability of the data from

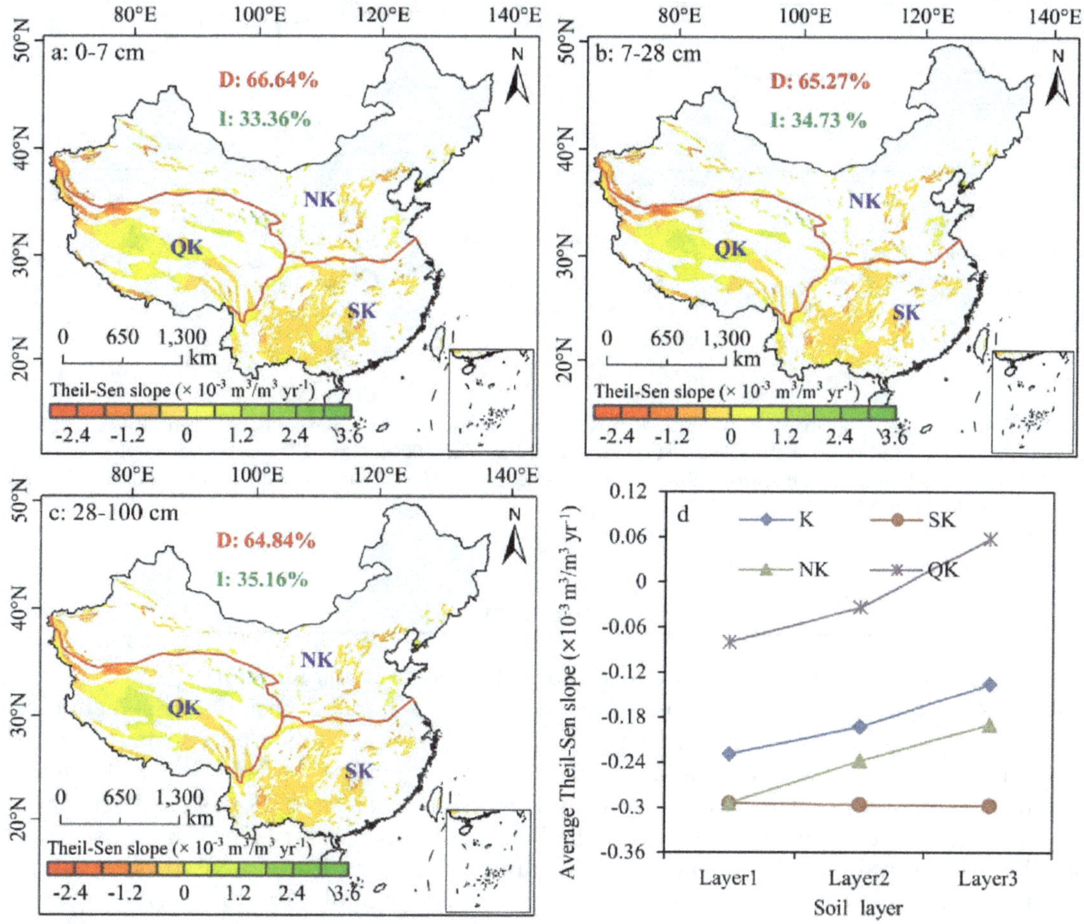

FIGURE 4.9 Illustrates the spatial distribution of soil moisture (SM) trends between 1982 and 2015 in the karst region, divided into three subfigures, (A–C), as well as the regional average of SM trends in subfigure (D). The red and green colors in subfigures A–C denote SM decreases (D) and increases (I), respectively. Deng et al. (2020).

1981, with hourly time resolution and 9 km spatial resolution, provides a comprehensive picture of the relationship among water, soil, plant, and atmosphere.

Additionally, the high-resolution and multivariant nature of the ERA5-LAND data representation of soil moisture patterns and variations across different regions and landscapes. This level of detail is critical for various applications such as drought and flood monitoring, water resources management, and agricultural planning. The frequent availability of the data with hourly time resolution provides the ability to monitor rapid changes in soil moisture conditions, making it an important resource for early warning systems. Furthermore, the integration of ERA5-LAND data into land-surface models can improve the accuracy of climate predictions and simulations and provide a valuable resource for climate researchers and practitioners.

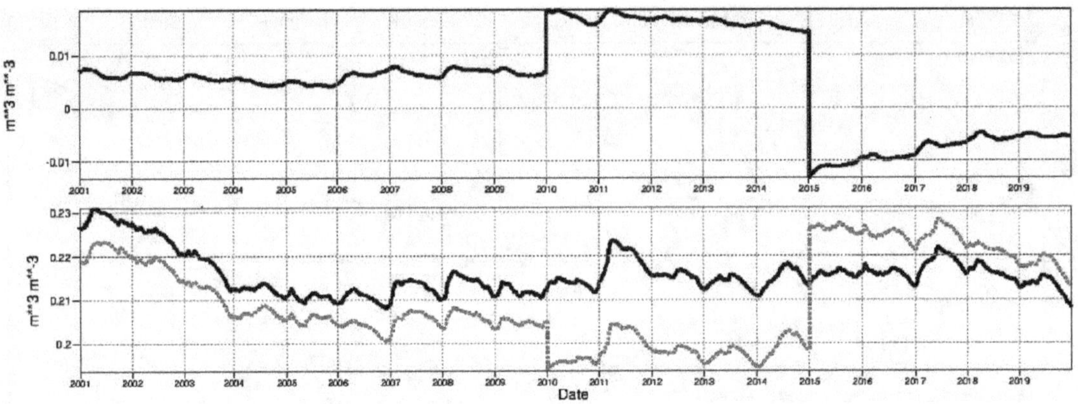

FIGURE 4.10 The image displays the mean differences in the soil moisture time series between ERA5 and ERA5-Land for the fourth soil layer of the CHTESSEL land surface model (100–289 cm). The temporal resolution for the data is 12 h. The bottom part of the image shows the raw time series of ERA5 as a dashed light gray curve and ERA5-Land as a dark curve. Muñoz et al. (2021).

4.10 Future mission

The future mission of soil moisture in space will go beyond just monitoring and managing soil moisture on a global scale. It will aim to provide actionable insights to stakeholders, including farmers, water resource managers, and policymakers, to support sustainable agricultural practices and decision-making in resource management. The mission will leverage advanced remote sensing technologies, such as satellite-based and drone-based sensors, to provide soil moisture with high-resolution and frequent observations over large areas. Additionally, it will integrate data from various sources, including ground-based observations, reanalysis datasets, weather forecasts, and other environmental data, to provide a general and realization of soil moisture status. The mission will also work toward developing user-friendly tools and platforms to help stakeholders access and use the data to support their decision-making needs. Overall, the future mission of soil moisture in space will strive to achieve a more sustainable, resilient, and productive future for our global food systems.

4.11 Conclusion

The following chapter offers a thorough review of the most recent soil moisture estimation techniques and provides a comprehensive summary of the current state-of-the-art in this field, including those based on the visible, thermal, and microwave subcomponents of the electromagnetic spectrum, as well as reanalysis datasets. A major challenge in the field of soil moisture data from space is the provision of time series observations each month or season to validate Earth observations from space and assess the performance of remote sensing and reanalysis datasets under various conditions. Despite limitations in the current data sources, it is hoped that advances in science and technology will resolve these issues in the future.

The analysis shows that, with the exception of reanalysis datasets, most remote sensing estimations using microwave data can provide soil moisture information up to a depth of 5 cm. Even when soil moisture levels are increased through precipitation or irrigation, the depth of information depicted remains limited to less than 5 cm. Microwave remote sensing methods are more effective in agricultural lands compared to optical and thermal remote sensing methods, as the depth of penetration under vegetation cover is greater with microwave data.

Designing applications based on soil moisture remote sensing and reanalysis datasets is crucial as these datasets and products are available over extended periods and can provide significant benefits for a wide range of stakeholders, including irrigation managers, flood and drought monitoring, and early warning systems to support decision-making and resource management. At the same time, it is important for remote sensing designers to consider the reliability and quality concerns of users, particularly in less-developed countries where observed data are scarce. Reliable surrogates have the potential to address major international challenges, such as water scarcity, which is a major concern in the Middle East and North Africa (MENA) region. Overcoming these challenges can not only improve the quality of life at a regional level but also alleviate global issues such as mass migration. To achieve a comprehensive and integrated solution for monitoring and managing soil moisture, it is imperative to take a proactive and comprehensive approach.

References

Baghdadi, N., Bazzi, H., El Hajj, M., Zribi, M., 2018. Detection of frozen soil using Sentinel-1 SAR data. Remote Sens. 10 (8), 1182.

Balsamo, G., Beljaars, A., Scipal, K., Viterbo, P., van den Hurk, B., Hirschi, M., Betts, A.K., 2009. A revised hydrology for the ECMWF model: verification from field site to terrestrial water storage and impact in the integrated forecast system. J. Hydrometeorol. 10 (3), 623–643.

Barideh, R., Veysi, S., Ebrahimipak, N., Davatgar, N., 2022. The challenge of reference evapotranspiration between the WaPOR data set and geostatistical methods. Irrigat. Drain. 71 (5), 1268–1279.

Behari, J., 2005. Dielectric Behavior of Soil. Springer Netherlands, pp. 22–40.

Boegh, E., Soegaard, H., Hanan, N., Kabat, P., Lesch, L., 1999. A remote sensing study of the NDVI–Ts relationship and the transpiration from sparse vegetation in the Sahel based on high-resolution satellite data. Remote Sens. Environ. 69 (3), 224–240.

Bovenga, F., Wasowski, J., Nitti, D.O., Nutricato, R., Chiaradia, M.T., 2012. Using COSMO/SkyMed X-band and ENVISAT C-band SAR interferometry for landslides analysis. Remote Sens. Environ. 119, 272–285.

Brusseau, M.L., Walker, D.B., Fitzsimmons, K., 2019. Physical-chemical characteristics of water. In: Environmental and Pollution Science. Academic Press, pp. 23–45.

Carlson, T.N., Buffum, M.J., 1989. On estimating total daily evapotranspiration from remote surface temperature measurements. Remote Sens. Environ. 29 (2), 197–207.

Chan, S.K., Bindlish, R., O'Neill, P., Jackson, T., Njoku, E., Dunbar, S., et al., 2018. Development and assessment of the SMAP enhanced passive soil moisture product. Remote Sens. Environ 204, 931–941.

Chang, D.H., Islam, S., 2003. Effects of topography, soil properties and mean soil moisture on the spatial distribution of soil moisture: a stochastic analysis. Scaling Methods in Soil Physics, pp. 193–225.

Chan, S.K., Bindlish, R., O'Neill, P.E., Njoku, E., Jackson, T., Colliander, A., Chen, F., Burgin, M., Dunbar, S., Piepmeier, J., Yueh, S., 2016. Assessment of the SMAP passive soil moisture product. IEEE Trans. Geosci. Remote Sens. 54 (8), 4994–5007.

Colliander, A., Reichle, R.H., Crow, W.T., Cosh, M.H., Chen, F., Chan, S., et al., 2021. Validation of soil moisture data products from the NASA SMAP mission. IEEE J. Sel. Top. Appl. Earth Obs. Remote Sens 15, 364–392.

Deng, Y., Wang, S., Bai, X., Luo, G., Wu, L., Chen, F., Wang, J., Li, Q., Li, C., Yang, Y., Hu, Z., 2020. Spatiotemporal dynamics of soil moisture in the karst areas of China based on reanalysis and observations data. J. Hydrol. 585, 124744.

Dong, J., Crow, W., Reichle, R., Liu, Q., Lei, F., Cosh, M.H., 2019. A global assessment of added value in the SMAP Level 4 soil moisture product relative to its baseline land surface model. Geophys. Res. Lett. 46 (12), 6604–6613.

Dubois, P.C., Van Zyl, J., Engman, T., 1995. Measuring soil moisture with imaging radars. IEEE Trans. Geosci. Rem. Sens. 33 (4), 915–926.

Engman, E.T., Chauhan, N., 1995. Status of microwave soil moisture measurements with remote sensing. Remote Sens. Environ. 51 (1), 189–198.

Entekhabi, D., Njoku, E.G., O'neill, P.E., Kellogg, K.H., Crow, W.T., Edelstein, W.N., et al., 2010. The soil moisture active passive (SMAP) mission. Proc. IEEE 98 (5), 704–716.

Ford, T.W., Quiring, S.M., 2019. Comparison of contemporary in situ, model, and satellite remote sensing soil moisture with a focus on drought monitoring. Water Resour. Res. 55 (2), 1565–1582.

Hallegatte, S., Shah, A., Brown, C., Lempert, R., Gill, S., 2012. Investment Decision Making under Deep Uncertainty-Application to Climate Change. World Bank Policy Research Working Paper, (6193).

Hersbach, H., Bell, B., Berrisford, P., Hirahara, S., Horányi, A., Muñoz-Sabater, J., Nicolas, J., Peubey, C., Radu, R., Schepers, D., Simmons, A., 2020. The ERA5 global reanalysis. Q. J. R. Meteorol. Soc. 146 (730), 1999–2049.

Huang, X., Ziniti, B., Torbick, N., Ducey, M.J., 2018. Assessment of forest above ground biomass estimation using multi-temporal C-band Sentinel-1 and polarimetric L-band PALSAR-2 data. Remote Sens. 10 (9), 1424.

Jackson, T.J., Bindlish, R., Cosh, M.H., Zhao, T., Starks, P.J., Bosch, D.D., Seyfried, M., Moran, M.S., Goodrich, D.C., Kerr, Y.H., Leroux, D., 2011. Validation of soil moisture and ocean salinity (SMOS) soil moisture over watershed networks in the US. IEEE Trans. Geosci. Rem. Sens. 50 (5), 1530–1543.

Kaleita, A.L., Tian, L.F., Hirschi, M.C., 2005. Relationship between soil moisture content and soil surface reflectance. Trans. ASAE 48 (5), 1979–1986.

Kim, S., Van Zyl, J., Dunbar, R.S., Njoku, E.G., Johnson, J.T., Moghaddam, M., Tsang, L., 2016. SMAP L2 Radar Half-Orbit 3 km EASE-Grid Soil Moisture, Version 3. Boulder, Colorado USA. NASA National Snow and Ice Data Center Distributed Active Archive Center. https://doi.org/10.5067/J8SGO1E0Y9XZ. Recuperado el, 2.

Klemas, V., Pieterse, A., 2015. Using remote sensing to map and monitor water resources in arid and semiarid regions. Advances in Watershed Science and Assessment, pp. 33–60.

Konings, A.G., Piles, M., Das, N., Entekhabi, D., 2017. L-band vegetation optical depth and effective scattering albedo estimation from SMAP. Remote Sens. Environ. 198, 460–470.

Li, Z., Wang, Y., Zhou, Q., Wu, J., Peng, J., Chang, H., 2008. Spatiotemporal variability of land surface moisture based on vegetation and temperature characteristics in Northern Shaanxi Loess Plateau, China. J. Arid Environ. 72 (6), 974–985.

Louvet, S., Pellarin, T., Al Bitar, A., Cappelaere, B., Galle, S., Grippa, M., de Rosnay, P., 2015. SMOS soil moisture product evaluation over West-Africa from local to regional scale. Remote Sens. Environ 156, 383–394.

Magagi, R.D., Kerr, Y.H., 2001. Estimating surface soil moisture and soil roughness over semiarid areas from the use of the copolarization ratio. Remote Sens. Environ. 75 (3), 432–445.

McNairn, H., Brisco, B., 2004. The application of C-band polarimetric SAR for agriculture: a review. Can. J. Rem. Sens. 30 (3), 525–542.

Malardel, S., Wedi, N.P., 2016. How does subgrid-scale parametrization influence nonlinear spectral energy fluxes in global NWP models? J. Geophys. Res. Atmos. 121 (10), 5395–5410.

Muñoz-Sabater, J., Dutra, E., Agustí-Panareda, A., Albergel, C., Arduini, G., Balsamo, G., Boussetta, S., Choulga, M., Harrigan, S., Hersbach, H., Martens, B., 2021. ERA5-Land: a state-of-the-art global reanalysis dataset for land applications. Earth Syst. Sci. Data 13 (9), 4349–4383.

Nemani, R., Pierce, L., Running, S., Goward, S., 1993. Developing satellite-derived estimates of surface moisture status. J. Appl. Meteorol. Climatol. 32 (3), 548–557.

Njoku, E.G., Wilson, W.J., Yueh, S.H., Dinardo, S.J., Li, F.K., Jackson, T.J., et al., 2002. Observations of soil moisture using a passive and active low-frequency microwave airborne sensor during SGP99. IEEE Trans. Geosci. Rem. Sens. 40 (12), 2659–2673.

Njoku, E.G., Jackson, T.J., Lakshmi, V., Chan, T.K., Nghiem, S.V., 2003. Soil moisture retrieval from AMSR-E. IEEE Trans. Geosci. Rem. Sens. 41 (2), 215–229.

Nunziata, F., de Macedo, C.R., Buono, A., Velotto, D., Migliaccio, M., 2019. On the analysis of a time series of X–band TerraSAR–X SAR imagery over oil seepages. Int. J. Remote Sens. 40 (9), 3623–3646.

Owe, M., de Jeu, R., Walker, J., 2001. A methodology for surface soil moisture and vegetation optical depth retrieval using the microwave polarization difference index. IEEE Trans. Geosci. Rem. Sens. 39 (8), 1643–1654.

Panciera, R., Walker, J.P., Kalma, J.D., Kim, E.J., Saleh, K., Wigneron, J.P., 2009. Evaluation of the SMOS L-MEB passive microwave soil moisture retrieval algorithm. Remote Sens. Environ. 113 (2), 435–444.

Petropoulos, G.P., Ireland, G., Barrett, B., 2015. Surface soil moisture retrievals from remote sensing: current status, products & future trends. Phys. Chem. Earth, Parts A/B/C 83, 36–56.

Piles, M., Vall-Llossera, M., Laguna, L., Camps, A., July 2012. A downscaling approach to combine SMOS multi-angular and full-polarimetric observations with MODIS VIS/IR data into high resolution soil moisture maps. In: 2012 IEEE International Geoscience and Remote Sensing Symposium. IEEE, pp. 1247–1250.

Portal, G., Vall-Llosscra, M., Piles, M., Camps, A., Chaparro, D., Pablos, M., Rossato, L., Aabouch, K., 2018. July. Microwave and optical data fusion for global mapping of soil moisture at high resolution. In: IGARSS 2018-2018 IEEE International Geoscience and Remote Sensing Symposium. IEEE, pp. 341–344.

Reichle, R.H., Liu, Q., Koster, R.D., Crow, W.T., De Lannoy, G.J., Kimball, J.S., et al., 2019. Version 4 of the SMAP level-4 soil moisture algorithm and data product. J. Adv. Model. Earth Sys 11 (10), 3106–3130.

Rüdiger, C., Calvet, J.C., Gruhier, C., Holmes, T.R., De Jeu, R.A., Wagner, W., 2009. An intercomparison of ERS-Scat and AMSR-E soil moisture observations with model simulations over France. J. Hydrometeorol. 10 (2), 431–447.

Sadeghi, M., Jones, S.B., Philpot, W.D., 2015. A linear physically-based model for remote sensing of soil moisture using short wave infrared bands. Remote Sens. Environ. 164, 66–76.

Saha, S., Moorthi, S., Pan, H.L., Wu, X., Wang, J., Nadiga, S., Tripp, P., Kistler, R., Woollen, J., Behringer, D., Liu, H., 2010. The NCEP climate forecast system reanalysis. Bull. Am. Meteorol. Soc. 91 (8), 1015–1058.

Sandholt, I., Rasmussen, K., Andersen, J., 2002. A simple interpretation of the surface temperature/vegetation index space for assessment of surface moisture status. Remote Sens. Environ. 79 (2–3), 213–224.

Seneviratne, S.I., Corti, T., Davin, E.L., Hirschi, M., Jaeger, E.B., Lehner, I., Orlowsky, B., Teuling, A.J., 2010. Investigating soil moisture–climate interactions in a changing climate: a review. Earth Sci. Rev. 99 (3–4), 125–161.

Singh, R.P., Oza, S.R., Chaudhari, K.N., Dadhwal, V.K., 2005. Spatial and temporal patterns of surface soil moisture over India estimated using surface wetness index from SSM/I microwave radiometer. Int. J. Rem. Sens. 26 (6), 1269–1276.

Snyder, R.L., Orang, M., Matyac, S., Grismer, M.E., 2005. Simplified estimation of reference evapotranspiration from pan evaporation data in California. J. Irrigat. Drain. Eng. 131 (3), 249–253.

Srivastava, P.K., Han, D., Rico-Ramirez, M.A., O'Neill, P., Islam, T., Gupta, M., Dai, Q., 2015. Performance evaluation of WRF-Noah Land surface model estimated soil moisture for hydrological application: synergistic evaluation using SMOS retrieved soil moisture. J. Hydrol. 529, 200–212.

Stoffregen, H., Zenker, T., Wessolek, G., 2002. Accuracy of soil water content measurements using ground penetrating radar: comparison of ground penetrating radar and lysimeter data. J. Hydrol. 267 (3–4), 201–206.

Van den Hurk, B.J., Bastiaanssen, W.G., Pelgrum, H., van Meijgaard, E., 1997. A new methodology for assimilation of initial soil moisture fields in weather prediction models using Meteosat and NOAA data. J. Appl. Meteorol. 36 (9), 1271–1283.

Verstraeten, W.W., Veroustraete, F., Feyen, J., 2008. Assessment of evapotranspiration and soil moisture content across different scales of observation. Sensors 8 (1), 70–117.

Veysi, S., Naseri, A.A., Hamzeh, S., 2020. Relationship between field measurement of soil moisture in the effective depth of sugarcane root zone and extracted indices from spectral reflectance of optical/thermal bands of multispectral satellite images. Journal of the Indian Society of Remote Sensing 48, 1035–1044.

Wagner, W., Scipal, K., Pathe, C., Gerten, D., Lucht, W., Rudolf, B., 2003. Evaluation of the agreement between the first global remotely sensed soil moisture data with model and precipitation data. J. Geophys. Res. Atmos. 108 (D19).

Wagner, W., Naeimi, V., Scipal, K., De Jeu, R., Martínez-Fernández, J., 2007. Soil moisture from operational meteorological satellites. Hydrogeol. J. 15, 121–131.

Walker, J.P., 1999. Estimating Soil Moisture Profile Dynamics from Near-Surface Soil Moisture Measurements and Standard Meteorological Data (Doctoral Dissertation, University of Newcastle).

Wang, G., Kim, Y., Wang, D., 2007. Quantifying the strength of soil moisture–precipitation coupling and its sensitivity to changes in surface water budget. J. Hydrometeorol. 8 (3), 551–570.

Wang, P., Reinelt, G., Gao, P., Tan, Y., 2011. A model, a heuristic and a decision support system to solve the scheduling problem of an earth observing satellite constellation. Comput. Ind. Eng. 61 (2), 322–335.

Wigneron, J.P., Jackson, T.J., O'neill, P., De Lannoy, G., de Rosnay, P., Walker, J.P., Ferrazzoli, P., Mironov, V., Bircher, S., Grant, J.P., Kurum, M., 2017. Modelling the passive microwave signature from land surfaces: a review of recent results and application to the L-band SMOS & SMAP soil moisture retrieval algorithms. Remote Sens. Environ. 192, 238–262.

West, H., Quinn, N., Horswell, M., 2019. Remote sensing for drought monitoring & impact assessment: progress, past challenges and future opportunities. Remote Sens. Environ. 232, 111291.

Yang, X., Yu, Y., Li, M., 2019. Estimating soil moisture content using laboratory spectral data. J. For. Res. 30, 1073–1080.

Zhao, W., Li, A., 2013. A downscaling method for improving the spatial resolution of AMSR-E derived soil moisture product based on MSG-SEVIRI data. Rem. Sens. 5 (12), 6790–6811.

Further reading

Abbaszadeh, P., Moradkhani, H., Zhan, X., 2019. Downscaling SMAP radiometer soil moisture over the CONUS using an ensemble learning method. Water Resour. Res. 55 (1), 324–344.

Klotzsche, A., Jonard, F., Looms, M.C., van der Kruk, J., Huisman, J.A., 2018. Measuring soil water content with ground penetrating radar: a decade of progress. Vadose Zone J. 17 (1), 1–9.

Njoku, E.G., Entekhabi, D., 1996. Passive microwave remote sensing of soil moisture. J. Hydrol. 184 (1–2), 101–129.

Mohanty, B.P., Skaggs, T.H., 2001. Spatio-temporal evolution and time-stable characteristics of soil moisture within remote sensing footprints with varying soil, slope, and vegetation. Adv. Water Resour. 24 (9–10), 1051–1067.

Setia, R., Pandey, D.K., Putrevu, D., Misra, A., Pateriya, B., 2021. Soil Moisture Retrieval Techniques Using Satellite Remote Sensing. Geospatial Technologies for Crops and Soils, pp. 357–385.

Srivastava, P.K., 2017. Satellite soil moisture: review of theory and applications in water resources. Water Resour. Manag. 31 (10), 3161–3176.

Xu, L., Chen, N., Zhang, X., Moradkhani, H., Zhang, C., Hu, C., 2021. In-situ and triple-collocation based evaluations of eight global root zone soil moisture products. Remote Sens. Environ. 254, 112248.

CHAPTER

5

Dust-source monitoring using remote sensing techniques

Esmaeil Heydari Alamdarloo, Hassan Khosravi and Azam Abolhasani

Department of Reclamation of Arid and Mountainous Regions, Faculty of Natural Resource, University of Tehran, Tehran, Iran

5.1 Introduction

In recent years, climate change and land degradation have intensified the processes of dust emission into the atmosphere and have caused increasing concern (IPCC, 2018). This phenomenon involves about one-third of the Earth's area (Shao et al., 2011), mostly located in semi-arid and arid environments (Jebali et al., 2021; Qor-el-aine et al., 2022). The phenomenon is caused by strong winds and as a result of the separation of loose and unstable particles of the soil surface (Rayegani et al., 2020). In various studies, the amount of dust that entered the atmosphere is estimated between 100 and 10,000 tons annually (Shao, 2008). Although a significant amount of dust is deposited in the oceans (Ashrafi et al., 2017), the most abundant particle in the atmosphere is dust worldwide (Albugami et al., 2019).

The frequency of dust occurrence in an environment depends on various variables including the extent of bare soil, wind speed, soil moisture, vegetation density, the intensity and extent of desertification, the intensity and duration of drought, and land use changes (Qu et al., 2006; Effati et al., 2019). This phenomenon creates a wide range of environmental and economic hazards for human society (Middleton and Kang, 2017) and combating it requires multilateral political, economic, and environmental methods. Therefore, monitoring dust storms and their sources is one of the first steps in managing and controlling this phenomenon. Due to the potential of dust transport over long distances and the spatio-temporal limitations of field studies, remote sensing technology is the only truthful approach to monitor and recognize dust emission events (Bakker et al., 2018). Currently, existing satellites with different spatio-temporal resolutions are a unique tool for monitoring and identifying the characteristics of dust (Fang et al., 2016). Significant progresses have been recently

Remote Sensing of Soil and Land Surface Processes
https://doi.org/10.1016/B978-0-443-15341-9.00015-0

occurred in remote sensing technology to record dust storms and their sources (Gherboudj et al., 2017). The purpose of this chapter of the book is to evaluate the use of remote sensing in monitoring dust storms and their production sources.

5.2 Aerosol and dust

The aerosol term generally refers to solid or liquid particles that are suspended in a gaseous environment such as air (Wei et al., 2020). Therefore, aerosols are characterized by two-stage processes that consist of a suspended liquid or solid stage and a neighboring gas stage, whose suspended solid or liquid phase is formed by converting gases into particles or by breaking down liquids or solids into smaller components (Kulkarni et al., 2011). Aerosol particles are produced by the Earth's environment in different forms. As these particles are affected by Earth systems, they have significant effects on the key components of these systems (Cao, 2017). Aerosols are seen in many common forms including suspended particulate matter, dust, mist, clouds, fume, smoke, fog, smog, haze, etc (Kulkarni et al., 2011).

Among different types of aerosols in the troposphere and stratosphere, dust is the most numerous in terms of both constitutes and mass (Gkikas et al., 2021). Kulkarni et al. (2011) defined dust as solid particles created by mechanical methods (e.g., crushing) that lead to the physical disintegration of parent materials. They also stated that the shape of dust particles is irregular and their size is mostly more than 0.5 μm.

5.3 Wind erosion and dust storms

The transport process of dust emissions to the atmosphere is complicated that directly affected by the interaction between surface soil particles and wind (Gherboudj et al., 2017). The movement of particles of land surface by wind happens when the wind speed trespasses the threshold velocity of friction (Fécan et al., 1998; Astitha et al., 2012). According to the soil properties and diameter of soil particles (d), this process, which is called wind erosion, occurs in three various shapes (Shao and Lu, 2000) including creeping ($d > \sim 500\ \mu m$), saltating ($\sim 70\ \mu m \leq d \leq \sim 500\ \mu m$), and suspension ($d < 20\ \mu m$).

Various variables affecting wind erosion phenomenon are categorized into some classes (Zou et al., 2015) including soil antierodibility factors (SAF), wind erosivity factors (WEF), and roughness interference factors (RIF). WEF refers to factors (e.g., wind turbulence, speed, direction, etc.) that indicate the potential of storm to employ a shear strength on land and separate particles. SAF is defined as an effective factor in creating surface soil resistance against wind shear strength (e.g., size of soil particles, percentage of organic matter, concentration of soil salts, etc). Woodruff and Siddoway (1965) have considered soil erodibility as the possible loss of soil from a smooth and uncovered land. This soil property depends on soil cohesion and has an inverse relationship with the decrease in the portion of soil particles with a size greater than 0.84 mm (Zou et al., 2015). RIF focuses on the features (e.g., percentage of stubble and vegetation, height, distance of surface roughness, etc.) related to the force of the wind on the land surface. In conditions with the same WEF, the land cover characteristics are not the same. In addition, as roughness elements and vegetation change, the wind shear strength also changes. The shearing strength of the wind that enters the Earth's surface is divided into two portions, one of them applies on the rough basics and the second acts on the soil surfaces that can hardly be eroded (Raupach, 1992).

Most of the time, the effect of wind erosion is visible and tangible only when accompanied by dust (Shao, 2008). Dust storms are one of the most tangible and important forms of wind erosion that have adverse effects on human health and the ecosystem. According to the World Meteorological Organization, strong winds that can pick up an extensive amount of dust particles and decrease vision to less than 1000 m are called dust storms (Middleton et al., 2019). Dust storms influences wide regions and dust can be transported to far distances (Bin Abdulwahed et al., 2019).

5.4 Application of remote sensing in dust monitoring

The use of ground data, as well as satellite images, are two important methods of data collection to monitor the behavior of aerosols and dust events. The heterogeneity of the data collection network as well as the insufficient number of samples are among the most important weaknesses of the method of using ground data (Hu et al., 2007; Chen et al., 2014), which makes these data not comprehensive and sufficient. For example, Huang et al. (2007) recorded more than 50 dust events from June to September 2006 using CALIPSO data, which was 5 times more than the known events by traditional terrestrial methods. Due to these limitations, the use of remote sensing technology is much more effective than ground-based data in monitoring and analyzing aerosols and dust storms (Chen et al., 2014). In the following, we will explain the role of remote sensing in dust monitoring from different perspectives.

5.4.1 Satellite and sensors

The use of satellites to monitor aerosols started almost 4 decades ago (Mhawish et al., 2018). In the previous decades, images of Landsat and GOES were used to monitor fog and air pollution (e.g., Lyons and Husar, 1976; Todd et al., 1979). Since the launch of the first Landsat satellite in 1972 by the National Aeronautics and Space Administration (NASA), several aerosol monitoring satellites have been launched. In their study, Sayer et al. (2020) also categorized the sensors that are used in the field of aerosol studies. Fig. 5.1 shows the sensors used in aerosol studies based on the mentioned research.

From the table above, it can be concluded that there are many remote sensing tools for studying aerosols. But in the case of dust storms, it is slightly different. Because dust has specific and different spectral characteristics compared to other aerosols. There have been few studies about satellites and sensors used in dust storm monitoring and assessment including Li et al. (2021) and Akhlaq et al. (2012). In this part, we try to introduce the most important ones that have been used in most studies. It is necessary to remember that satellites rotate in orbits around the Earth. Generally, these orbits are either geostationary or polar orbiting and their variants (such as Sun-synchronous). The return period in polar satellites is high, but they have a high spatial and spectral resolution, unlike geostationary orbiting satellites that have low spatial and spectral resolution and suitable temporal resolution. These satellites are

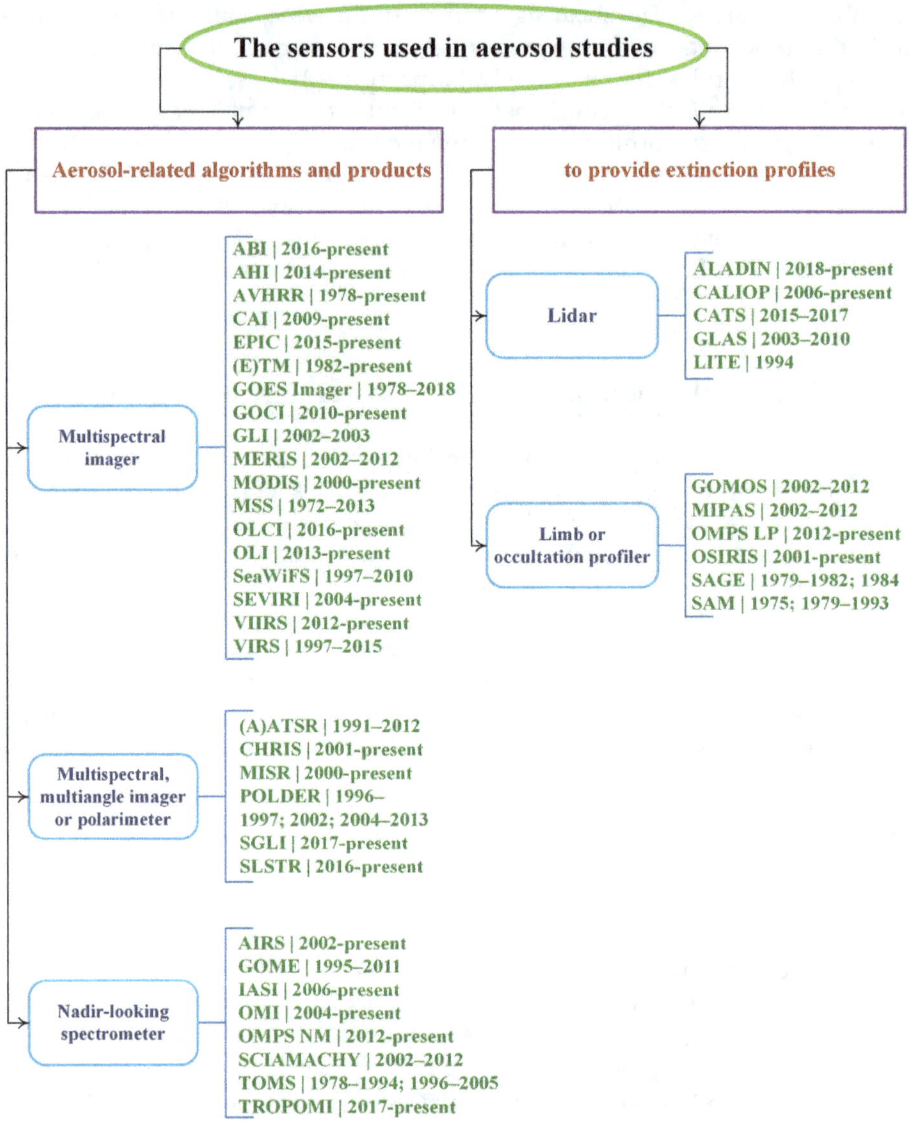

FIGURE 5.1 The sensors used in aerosol studies.

located in an orbit at a higher altitude and stay constant relative to a point on the Earth. According to Li et al. (2021), since geostationary satellites capacitate continuing observation, they can record the creation and transportation of dust storms.

5.4.1.1 MODIS, Aqua, and Terra satellites

The moderate resolution spectroradiometer (MODIS) sensor was launched on the Sun-synchronous Terra and Aqua satellites in 1999 and 2002, respectively. This sensor has a viewing swath width of 2330 km from an altitude of 700 km and observes the whole Earth's surface with a return period of 1–2 days. MODIS consists of 36 spectral bands with a wide wavelength rich at three different pixel-size (250 m, 500 m, and 1 km). This sensor is very reliable for dust monitoring and can reveal the dust in the air well (Taheri Shahraiyni et al., 2015). In addition, aerosol products of MODIS provide suitable temporal and spatial coverage for whole the Earth. In particular, MODIS detects aerosol optical depth (AOD) over land and ocean (El-Askary et al., 2004). To retrieve the aerosol optical properties, the MODIS sensor uses three algorithms including dark target over the ocean, dark target over land, and deep blue on land (Levy et al., 2007).

5.4.1.2 The AVHRR sensor of the NOAA satellite

The advanced very high-resolution radiometer (AVHRR) is used to determine the cloud cover and the temperature of the land, the upper levels of the clouds, or the surface of a water body. The latest version of AVHRR has six bands, which was first used on the NOAA-15 satellite that was launched in May 1988. The advantage of AVHRR is its daily global coverage and high spatial resolution (approximately 1.1 km) in spectral channels with nearly integrated atmospheric transmission (Evan et al., 2006).

5.4.1.3 The SEVIRI sensor of the METEOSAT

METEOSAT satellites were launched to monitor the weather (Murphy, 2013). METEOSAT satellites include two generations. The first generation of METEOSAT (MFG: Meteosat First Generation) is equipped with three visible bands and an infrared imager and a repetition cycle of 30 min (Murphy, 2013). The second-generation satellites (MSG: Meteosat Second Generation) are applied to observe the current status of the weather and numerical weather prediction (numerical weather prediction [NWP]) in addition to important information for weather monitoring. The MSG system has made major advances through its 12 spectral band radiometer. The time resolution of 15 min and the spatial power of about 1 kilometer in the high-resolution visible band (HRV) made it possible to monitor the changes in the atmosphere and ground surface rapidly. Atmospheric dust is scrutinized using MSG during the day and night in high spatial and temporal resolution using IR bands. For this purpose, the brightness temperature (16) is used, which is calculated from the radiances measured by the SEVIRI instrument on the MSG geostationary satellite. With a sampling rate of 15 min and a spatial resolution of 3 km, SEVIRI brightness temperatures are available at a high spatial and temporal resolution that can be used to detect active dust sources.

EUMETSAT organization provides a dust index product using MSG images on its website based on brightness temperature difference (BTD) (Schepanski et al., 2007). EUMETSAT MSG dust product is RGB composition based on SEVIRI infrared band which is planned to estimate the dust evolution over the desert.

5.4.1.4 The sensor OMI of Aura satellite

NASA's Aura satellite was launched in 2004 that has a transit time of about 13:45 min over the equator and a 16-day repetition cycle. This satellite is in flight with other Earth observation satellites called A-Train. The Aura satellite flies about 15 min behind Aqua in the A-Train satellite constellation, which consists of several satellites (https://aura.gsfc.nasa.gov/about.html).

Aura satellite has sensors including TES, HIRDLS, MLS, and OMI, which uses them to measure gases with small amounts in the atmosphere. Among these instruments, the ozone monitoring instrument (OMI) can monitor cloud cover and pressure and detect various types of aerosols, for example, dust, smoke, and sulfates. One of the weaknesses of OMI aerial retrieval is that OMI crosses the equator at 1:45 p.m. in a Sun-synchronous orbit. But most dust sources are active during the afternoon and night (storms associated with convective clouds). Therefore, during the passage of OMI, the dust in the air has already been mixed in the atmosphere and transported by local and regional wind systems.

5.4.1.5 *The sensor MISR of the Terra satellite*

MISR measures shortwave radiation from the ground in four spectral bands with central wavelengths of 446, 558, 672, and 866 (Carboni et al., 2012). This 9-directional observation is done to detect the angular changes of sunlight reflected by aerosols, clouds, and the surface (Kalashnikova et al., 2004). In addition, it is used to retrieve several physical aerosols parameters as well as their microphysical properties (size and shape) (Marey et al., 2011). Using the special features of MISR with several observation angles, the detection of dust events has been enhanced. MISR also is able to monitor local dust emissions. Therefore, it may be useful in the early detection of dust storms. In addition, the combination of information from different viewing angles can be used to distinguish between dust clouds and normal clouds. This can be beneficial in reducing background effects for desert areas by choosing appropriate viewing angles (El-Askary et al., 2004).

5.4.2 Dust study methods using remote sensing data

In general, the study of dust is done with two general goals: detecting dust and monitoring its concentration. Empirical physics-based and machine learning-based methods are two methods for dust detection (Shahrisvand and Akhoondzadeh, 2013). Experimental approaches are physical and radiation-based concepts that have been made based on the spectral and physical features of dust particles in thedifferent types of bands (Yan et al., 2020) that their most important weakness is threshold dependence (Li et al., 2021). The approach based on machine learning is more feasible and does not need a hypothetical threshold, but it has its complexities and uses aerosol optical properties to evaluate the dust concentration. Therefore, the methods of studying dust using remote sensing data can be classified into three categories: using optical properties of aerosols, methods based on experimental physics, and methods based on machine learning.

5.4.2.1 *Indicators of optical properties of aerosols*

Some of the key optical characteristics of aerosols are the single-scattering albedo (SSA), AOD, the asymmetry parameter (g), the absorption Angstrom exponent (AAE), the refractive index, the phase function P(θ), the phase matrix, the turbidity coefficient (β), the absorption and extinction coefficients, and the backscatter features (Mhawish et al., 2018).

5.4.2.1.1 Aerosol optical depth (AOD)

AOD, the extinction coefficient of aerosols, demonstrates the reduction of solar radiation via aerosols due to the absorbing or scattering. According to Mhawish et al. (2018), chemical composition, morphology, and the number of aerosols are affecting the overall reduction of solar radiation, and the value of this index is directly relevant to the number of particles in the column of the atmosphere. This index is obtained from ground measurements or satellite imagery. Although ground observations and field surveys are characterized by a high accuracy, this method only considers a small zone in the study area. Unlike ground measurements, satellite imagery has been extensively used in research due to its high spatio-temporal resolution as well as monitoring large areas (Namdari et al., 2016).

5.4.2.1.2 Absorption Angstrom exponent

Absorption Angstrom exponent (AAE) demonstrates the spectral properties of radiation absorbed by aerosols (Helin et al., 2021). This index is typically utilized to distinguish between various aerosol types. Mineral dust particles show a high amount of AAE because of absorbing more light in the near-UV wavelengths (Moosmüller et al., 2009; Caponi et al., 2017; Pokhrel et al., 2016). This index has become one of the most widely used optical indices for differentiating aerosol types due to its suitability for describing the composition of aerosols (Goetz et al., 2018; Olson et al., 2015).

5.4.2.1.3 Turbidity coefficient (β)

The opalescence of a column of the atmosphere is determined by the turbidity coefficient (β), which is a dimensionless index. This index illustrates the weakness of solar radiation in 1 μm (Trabelsi and Masmoudi, 2011) as well as the combined effect of scattering and absorbing caused by aerosols.

5.4.2.1.4 Asymmetry parameter (g)

The asymmetry parameter (g) is an index that shows the asymmetry degree of the light scattering angular and its direction. The value of this index is between +1 and −1. The value +1 represents forward scattering by larger particles. The values 0 and −1 denote respectively uniform scattering under the influence of smaller particles and backward scattering (Mhawish et al., 2018).

5.4.2.1.5 Single-scattering albedo (ω)

Single-scattering albedo (SSA) is a dimensionless index that represents the amount of light scattering caused by aerosols relative to the sum of scattering and absorbing in a specific wavelength. Its value ranges from zero to one. Values near zero show the dominance of light absorption by aerosols, and values near one imply the dominance of light scattering by them.

5.4.2.1.6 Mass extinction coefficient (bext)

This index shows how much aerosols absorb or scatter solar radiation at a certain wavelength per unit mass in a fixed volume.

5.4.2.2 Physical-based methods

Physical-based methods employ spectral, temporal, and spatial data to identify dust from clear-day surfaces and clouds. The dust storm reflectance is mostly higher than the reflectance of the water surface or clear-day land, and its temperature is usually cooler than the beneath area. In both thermal infrared (TIR) and visible near-infrared (VIR) images, dust storms are more homogeneous than land surfaces. According to this property, many tests have been established to differ dust and clear-day surfaces and clouds.

5.4.2.2.1 Dust spectral index

The dust spectral index considers the spectral signature of dust storms. Different spectral signatures of various attributes show their ability to absorb or scatter at several wavelengths. The main advantages and applications that can be determined include aerosols (e.g., dust storms), clouds, and clear-sky surfaces (ocean or land). According to Li et al. (2021), the dust indicators isolate various features by the significant differences in the features' spectral signature in both infrared emissivity and solar reflectance bands.

The Normalized Dust Difference Index (NDDI) and brightness temperature difference (BTD) are two major algorithms applied to monitor dust emission. The NDDI and BTD work based on the spectral signature in the visible to near-infrared and thermal, respectively (Prata, 1989; Qu et al., 2006).

To use the merits of NDDI and BTDs, enhanced methods are created including the Normalized Dust Layer Index (NDLI) (Kazi et al., 2019), Three-band Volcanic Ash Product (TVAP) (Ellrod et al., 2003), and Brightness Temperature Adjusted Dust Index (BADI) (Yue et al., 2017).

5.4.2.2.1.1 Brightness temperature difference The most widely used method is the BTD. It employs the dust spectral signature in the TIR region. This method was first created for volcanic aerosol discovery (Prata, 1989). Because of the likeness between volcanic aerosol and mineral dust aerosol, this technique was later used for dust detection. The principle that includes the basis of this technique is that, unlike the clouds, dust particles extinct more light at 11 μm than at 12 μm. This fact leads to a positive value for clouds and a negative value of BTD11-12 for dust (Potts, 1993; Prata, 1989). The value and sign of BTD are specified by dust optical thickness, dust mineral composition, and dust layer altitude, as well as the view climax angle and surface emissivity (Bin Abdulwahed et al., 2019).

5.4.2.2.1.2 Normalized Difference Dust Index NDDI is also a dust index that employs the spectral features in the near-infrared and the visible bands (Qu et al., 2006). NDDI is commonly positive, negative, and near zero for dust pixels, clouds, and the clear-sky surface, respectively. Since NDDI is not sensitive to dust altitude and density, it shows low values even in dust storms with high density. Due to their similar spectral pattern with dust, NDDI is not able to identify bright surfaces from dust (Xu et al., 2011). Generally, NDDI is utilized to discover dust along with BTD (Kazi et al., 2019; Xu et al., 2011).

5.4.2.2.1.3 Thermal infrared dust index The TDI index can detect nighttime dust and dust density cause it can gain AOT at 550 nm. Four MODIS thermal bands are used by this algorithm for detecting dust (Hao and Qu, 2007) including BT11 with the largest weight followed by BT12, BT.3, and BT9.7. Coefficients are evaluated by creating a regression connection between BTs and AOD. Since BT9.7 is also susceptible to ozone (Landgraf and Hasekamp, 2007), it achieves the least weight in TDI.

5.4.2.2.1.4 Middle East Dust Index (MEDI) The MEDI was created for dust discovery over the Middle East. The wavelengths at around 8.55, 11, and 12 μm are combined by MEDI to recognize dust from desert surfaces (Karimi et al., 2012), which has been the major challenge (Ackerman, 1997; Roskovensky and Liou, 2005).

5.4.2.2.1.5 Three-band volcanic ash product This index was created for volcanic ash identification (Ellrod et al., 2003). Since the index unifies BTD12-11 and BTD3-11, It can demonstrate both dust intensity and dust presence.

5.4.2.2.1.6 Brightness temperature adjusted difference index As we know, BTD11-12 can identify the dust extent, and BTD3-11 can show dust density. By merging BTD3-11 and BTD11-12, the BADI boosts BTD methods (Yue et al., 2017). So, this index can simultaneously determine dust storm density and extent.

5.4.2.2.1.7 D-parameter algorithms This index was first created to identify thin cirrus clouds (Roskovensky and Liou, 2003) and then was developed to isolate cirrus clouds from dust storms (Roskovensky and Liou, 2005).

5.4.2.2.2 The anomalous dust detection

Disorders including smoke, clouds, and hurricane can be well-identified by anomalous detection techniques and through ascertaining different thresholds. The time-series images are utilized by anomalous identification algorithms to consider the arena at the same period and site. The algorithm supposes a stable Earth-atmosphere system. So, anomalies in the images are likely caused by the unstable atmosphere, natural disasters, dust storms, and cloudiness. These algorithms can be used for indices susceptible to dust storms as well as all wavelengths (Li et al., 2021).

5.4.2.2.3 Spatial coherence test method

One of the most important challenges in separating dust storms from clouds using information-based algorithms is the similarity of thin clouds (e.g., cirrus) and dust storms (Martin et al., 2002).

Determining the homogeneity (dust) or heterogeneity (cloud) of the spatial structure of these phenomena is one of the solutions to this challenge (Coakley and Bretherton, 1982). This difference has led to the creation of a method called the spatial coherence, which is also acknowledged as a test based on the spatial uniformity or the spatial variability. The basis of this method is the standard deviation, a high standard deviation value show clouds, and a low standard deviation represents the presence of dust (Martins et al., 2002). The simplicity, low influence of image calibration, and low dependence on wavelength are some advantages of this method (Li et al., 2021). The spatial coherence test method is a complementary method that should be used with other methods such as the spectral index.

5.4.2.3 Machine learning-based algorithms

Defining the threshold for dust storms identification is one of the important challenges of physical threshold-based methods (Rivas-Perea et al., 2010a,b). The ability of machine learning techniques to solve this challenge has been confirmed. It should be noted that using machine learning techniques is a new approach in dust storm studies and needs scientific development. The most significant artificial intelligence procedures used in the field of dust events studies are random forest, support vector machine, probabilistic neural network, logistic regression, artificial neural network, and maximum likelihood.

5.5 Conclusion

Remote sensing has been used as one of the most important tools for spatiotemporal monitoring of dust storms and their production resources. Here, we tried to review the general application of remote sensing in the monitoring of dust storms and detecting their production sources by using the research done so far. According to various studies, there are different satellite platforms and sensors that are employed to study dust. Therefore, one of the most significant techniques in the study of dust storms is remote sensing, which can be the only way to study large areas. Since the development speed of dust study using remote sensing data is very high, the efficiency of new methods should be periodically evaluated and a report of this evaluation should be presented.

Acknowledgment

This chapter forms part of Esmaeil Heydari Alamdarloo's postdoctoral project (Number: 99017799) with the title of Feasibility Study of Proposing an Early Warning System for Desertification of IRAN using Remote Sensing Data. The Iran National Science Foundation (INS) is thanked for the financial support.

References

Ackerman, S.A., 1997. Remote sensing aerosols using satellite infrared observations. J. Geophys. Res. Atmos. 102 (D14), 17069–17079.

Akhlaq, M., Sheltami, T.R., Mouftah, H.T., 2012. A review of techniques and technologies for sand and dust storm detection. Rev. Environ. Sci. Biotechnol. 11, 305–322.

Albugami, S., Palmer, S., Cinnamon, J., Meersmans, J., 2019. Spatial and temporal variations in the incidence of dust storms in Saudi Arabia revealed from in situ observations. Geosciences 9 (4), 162.

Ashrafi, K., Motlagh, M.S., Neyestani, S.E., 2017. Dust storms modeling and their impacts on air quality and radiation budget over Iran using WRF-Chem. Air Qual., Atmos. Health 10, 1059–1076.

Astitha, M., Lelieveld, J., Abdel Kader, M., Pozzer, A., De Meij, A., 2012. Parameterization of dust emissions in the global atmospheric chemistry-climate model EMAC: impact of nudging and soil properties. Atmos. Chem. Phys. 12 (22), 11057–11083.

Bakker, N., Drake, N., Bristow, C., 2018. Identifying African mineral dust sources and tracking dust events towards the Amazon Basin. In: EGU General Assembly Conference Abstracts, p. 334.

Bin Abdulwahed, A., Dash, J., Roberts, G., 2019. An evaluation of satellite dust-detection algorithms in the Middle East region. Int. J. Rem. Sens. 40 (4), 1331–1356.

Cao, J., 2017. The importance of aerosols in the earth system: science and engineering perspectives. Aerosol. Sci. Eng. 1 (1), 1–6.

Caponi, L., Formenti, P., Massabó, D., Di Biagio, C., Cazaunau, M., Pangui, E., Chevaillier, S., Landrot, G., Andreae, M.O., Kandler, K., Piketh, S., 2017. Spectral-and size-resolved mass absorption efficiency of mineral dust aerosols in the shortwave spectrum: a simulation chamber study. Atmos. Chem. Phys. 17 (11), 7175–7191.

Carboni, E., Thomas, G.E., Sayer, A.M., Siddans, R., Poulsen, C.A., Grainger, R.G., Ahn, C., Antoine, D., Bevan, S., Braak, R., Brindley, H., 2012. Intercomparison of desert dust optical depth from satellite measurements. Atmos. Meas. Tech. 5 (8), 1973–2002.

Chen, B., Zhang, P., Zhang, B., Jia, R., Zhang, Z., Wang, T., Zhou, T., 2014. An overview of passive and active dust detection methods using satellite measurements. J. Meteorol. Res. 28 (6), 1029–1040.

Coakley Jr., J.A., Bretherton, F.P., 1982. Cloud cover from high-resolution scanner data: detecting and allowing for partially filled fields of view. J. Geophys. Res.: Oceans 87 (C7), 4917–4932.

Effati, M., Bahrami, H.A., Gohardoust, M., Babaeian, E., Tuller, M., 2019. Application of satellite remote sensing for estimation of dust emission probability in the Urmia Lake Basin in Iran. Soil Sci. Soc. Am. J. 83 (4), 993–1002.

El-Askary, H., Gautam, R., Kafatos, M., 2004. Remote sensing of dust storms over the Indo-Gangetic basin. J. Indian Soc. Remote Sens. 32, 121–124.

Ellrod, G.P., Connell, B.H., Hillger, D.W., 2003. Improved detection of airborne volcanic ash using multispectral infrared satellite data. J. Geophys. Res. Atmos. 108 (D12).

Evan, A.T., Heidinger, A.K., Pavolonis, M.J., 2006. Development of a new over-water advanced very high resolution radiometer dust detection algorithm. Int. J. Rem. Sens. 27 (18), 3903–3924.

Fang, L., Wang, S., Yu, T., Gu, X., Zhang, X., Wang, W., Ren, S., 2016. Changes in aerosol optical and micro-physical properties over Northeast Asia from a severe dust storm in April 2014. Rem. Sens. 8 (5), 394.

Fécan, F., Marticorena, B., Bergametti, G., December 1998. Parametrization of the increase of the aeolian erosion threshold wind friction velocity due to soil moisture for arid and semi-arid areas. Ann. Geophys. 17 (1), 149–157 (Berlin/Heidelberg: Springer-Verlag).

Gherboudj, I., Beegum, S.N., Ghedira, H., 2017. Identifying natural dust source regions over the Middle-East and North-Africa: estimation of dust emission potential. Earth Sci. Rev. 165, 342–355.

Gkikas, A., Proestakis, E., Amiridis, V., Kazadzis, S., Di Tomaso, E., Tsekeri, A., Marinou, E., Hatzianastassiou, N., Pérez García-Pando, C., 2021. ModIs Dust AeroSol (MIDAS): a global fine-resolution dust optical depth data set. Atmos. Meas. Tech. 14 (1), 309–334.

Goetz, J.D., Giordano, M.R., Stockwell, C.E., Christian, T.J., Maharjan, R., Adhikari, S., Bhave, P.V., Praveen, P.S., Panday, A.K., Jayarathne, T., Stone, E.A., 2018. Speciated online PM 1 from South Asian combustion sources—part 1: fuel-based emission factors and size distributions. Atmos. Chem. Phys. 18 (19), 14653–14679.

Hao, X., Qu, J.J., 2007. Saharan dust storm detection using moderate resolution imaging spectroradiometer thermal infrared bands. J. Appl. Remote Sens. 1 (1), 013510.

Helin, A., Virkkula, A., Backman, J., Pirjola, L., Sippula, O., Aakko-Saksa, P., Väätäinen, S., Mylläri, F., Järvinen, A., Bloss, M., Aurela, M., 2021. Variation of absorption Ångström exponent in aerosols from different emission sources. J. Geophys. Res. Atmos. 126 (10) e2020JD034094.

Hu, X.Q., Lu, N.M., Zhang, P., 2007. Remote sensing and detection of dust storm in China using the thermal bands of geostationary meteorological satellite. J. Appl. Meteorol. Sci. 18 (3), 266–275.

Huang, J., Minnis, P., Yi, Y., Tang, Q., Wang, X., Hu, Y., Liu, Z., Ayers, K., Trepte, C., Winker, D., 2007. Summer dust aerosols detected from CALIPSO over the Tibetan Plateau. Geophys. Res. Lett. 34 (18).

IPCC, 2018. Special Report on Global Warming of 1.5°C. Cambridge University Press, UK.

Jebali, A., Zare, M., Ekhtesasi, M.R., Jafari, R., 2021. Detection of areas prone to wind erosion and air pollution using DSI and PDSI indices. Nat. Hazard. 108, 1221–1235.

Kalashnikova, O.V., Diner, D.J., Kahn, R.A., Gaitley, B.J., December 2004. Dust aerosol retrieval results from MISR. Pass. Opt. Remote Sens. Atmos. Cloud. IV 5652, 309–317 (SPIE).

Karimi, N., Moridnejad, A., Golian, S., Vali Samani, J.M., Karimi, D., Javadi, S., 2012. Comparison of dust source identification techniques over land in the Middle East region using MODIS data. Can. J. Rem. Sens. 38 (5), 586–599.

Kazi, A.,K., Nagatani, I., Kawano, K., Kudoh, J.I., 2019. Development of a new dust index NDLI for Asian dust extraction system based on Aqua MODIS data and monitoring of trans-boundary Asian dust events in Japan. Int. J. Rem. Sens. 40 (3), 1030–1047.

Kulkarni, P., Baron, P.A., Willeke, K. (Eds.), 2011. Aerosol Measurement: Principles, Techniques, and Applications. John Wiley and Sons.

Landgraf, J., Hasekamp, O.P., 2007. Retrieval of tropospheric ozone: the synergistic use of thermal infrared emission and ultraviolet reflectivity measurements from space. J. Geophys. Res. Atmos. 112 (D8).

Levy, R.C., Remer, L.A., Dubovik, O., 2007. Global aerosol optical properties and application to Moderate resolution imaging Spectroradiometer aerosol retrieval over land. J. Geophys. Res. Atmos. 112 (D13).

Li, J., Wong, M.S., Lee, K.H., Nichol, J., Chan, P.W., 2021. Review of dust storm detection algorithms for multispectral satellite sensors. Atmos. Res. 250, 105398.

Lyons, W.A., Husar, R.B., 1976. SMS/GOES visible images detect a synoptic-scale air pollution episode. Mon. Weather Rev. 104 (12), 1623–1626.

Marey, H.S., Gille, J.C., El-Askary, H.M., Shalaby, E.A., El-Raey, M.E., 2011. Aerosol climatology over Nile delta based on MODIS, MISR and OMI satellite data. Atmos. Chem. Phys. 11 (20), 10637–10648.

Martins, J.V., Tanré, D., Remer, L., Kaufman, Y., Mattoo, S., Levy, R., 2002. MODIS cloud screening for remote sensing of aerosols over oceans using spatial variability. Geophys. Res. Lett. 29 (12), 1–MOD4.

Mhawish, A., Kumar, M., Mishra, A.K., Srivastava, P.K., Banerjee, T., 2018. Remote sensing of aerosols from space: retrieval of properties and applications. In: Remote Sensing of Aerosols, Clouds, and Precipitation. Elsevier, pp. 45–83.

Middleton, N., Kang, U., 2017. Sand and dust storms: impact mitigation. Sustainability 9 (6), 1053.

Middleton, N., Tozer, P., Tozer, B., 2019. Sand and dust storms: underrated natural hazards. Disasters 43 (2), 390–409.

Moosmüller, H., Chakrabarty, R.K., Arnott, W.P., 2009. Aerosol light absorption and its measurement: a review. J. Quant. Spectrosc. Radiat. Transf. 110 (11), 844–878.

Murphy, D., 2013. EUMETSAT Geostationary Meteorological Satellite Programs. Handbook of Satellite Applications. Springer, New York, pp. 991–1019.

Namdari, S., Valizade, K.K., Rasuly, A.A., Sari Sarraf, B., 2016. Spatio-temporal analysis of MODIS AOD over western part of Iran. Arabian J. Geosci. 9, 1–11.

Olson, M.R., Victoria Garcia, M., Robinson, M.A., Van Rooy, P., Dietenberger, M.A., Bergin, M., Schauer, J.J., 2015. Investigation of black and brown carbon multiple-wavelength-dependent light absorption from biomass and fossil fuel combustion source emissions. J. Geophys. Res. Atmos. 120 (13), 6682–6697.

Pokhrel, R.P., Wagner, N.L., Langridge, J.M., Lack, D.A., Jayarathne, T., Stone, E.A., Stockwell, C.E., Yokelson, R.J., Murphy, S.M., 2016. Parameterization of single-scattering albedo (SSA) and absorption Ångström exponent (AAE) with EC/OC for aerosol emissions from biomass burning. Atmos. Chem. Phys. 16 (15), 9549–9561.

Potts, R.J., 1993. Satellite observations of Mt pinatubo ash clouds. Aust. Meteorol. Mag. 42 (2), 59–68.

Prata, A.J., 1989. Observations of volcanic ash clouds in the 10-12 μm window using AVHRR/2 data. Int. J. Rem. Sens. 10 (4–5), 751–761.

Qor-el-aine, A., Beres, A., Geczi, G., 2022. Dust storm simulation over the Sahara Desert (Moroccan and Mauritanian regions) using HYSPLIT. Atmos. Sci. Lett. 23 (4), e1076.

Qu, J.J., Hao, X., Kafatos, M., Wang, L., 2006. Asian dust storm monitoring combining Terra and Aqua MODIS SRB measurements. Geosci. Rem. Sens. Lett. IEEE 3 (4), 484–486.

Raupach, M., 1992. Drag and drag partition on rough surfaces. Boundary-Layer Meteorol. 60 (4), 375–395.

Rayegani, B., Barati, S., Goshtasb, H., Gachpaz, S., Ramezani, J., Sarkheil, H., 2020. Sand and dust storm sources identification: a remote sensing approach. Ecol. Indicat. 112, 106099.

Rivas-Perea, P., Rosiles, J.G., Chacon, M.I.M., 2010. Traditional and neural probabilistic multispectral image processing for the dust aerosol detection problem. In: 2010 IEEE Southwest Symposium on Image Analysis and Interpretation (SSIAI). IEEE, pp. 169–172.

Rivas-Perea, P., Rosiles, J.G., Murguia, M.I.C., Tilton, J.J., 2010b. Automatic dust storm detection based on supervised classification of multispectral data. Soft Comput. Recognit. Based Biomet. 443–454.

Roskovensky, J.K., Liou, K.N., 2003. Detection of thin cirrus from 1.38 μm/0.65 μm reflectance ratio combined with 8.6–11 μm brightness temperature difference. Geophys. Res. Lett. 30 (19).

Roskovensky, J.K., Liou, K.N., 2005. Differentiating airborne dust from cirrus clouds using MODIS data. Geophys. Res. Lett. 32 (12).

Sayer, A.M., Govaerts, Y., Kolmonen, P., Lipponen, A., Luffarelli, M., Mielonen, T., Patadia, F., Popp, T., Povey, A.C., Stebel, K., Witek, M.L., 2020. A review and framework for the evaluation of pixel-level uncertainty estimates in satellite aerosol remote sensing. Atmos. Meas. Tech. 13 (2), 373–404.

Schepanski, K., Tegen, I., Laurent, B., Heinold, B., Macke, A., 2007. A new Saharan dust source activation frequency map derived from MSG-SEVIRI IR-channels. Geophys. Res. Lett. 34 (18).

Shahrisvand, M., Akhoondzadeh, M., 2013. A comparison of empirical and inteligent methods for dust detection using modis satellite data. Int. Arch. Photogram. Rem. Sens. Spatial Inf. Sci. 1 (W3), 371–375.

Shao, Y., Lu, H., 2000. A simple expression for wind erosion threshold friction velocity. J. Geophys. Res. Atmos. 105 (D17), 22437–22443.

Shao, Y. (Ed.), 2008. Physics and Modelling of Wind Erosion. Springer Netherlands, Dordrecht.

Shao, Y., Shao, Y., Wyrwoll, K.H., Chappell, A., Huang, J., Lin, Z., McTainsh, G.H., Mikami, M., Tanaka, T.Y., Wang, X., Yoon, S., 2011. Dust cycle: an emerging core theme in Earth system science. Aeolian Res. 2 (4), 181–204.

Taheri Shahraiyni, H., Karimi, K., Habibi Nokhandan, M., Hafezi Moghadas, N., 2015. Monitoring of dust storm and estimation of aerosol concentration in the Middle East using remotely sensed images. Arabian J. Geosci. 8, 2095–2110.

Todd, W., George Jr., A., Bryant, N., 1979. Satellite-aided evaluation of population exposure to air pollution. Environ. Sci. Technol. 13 (8), 970–974.

Trabelsi, A., Masmoudi, M., 2011. An investigation of atmospheric turbidity over Kerkennah Island in Tunisia. Atmos. Res. 101 (1–2), 22–30.

Wei, X., Chang, N.B., Bai, K., Gao, W., 2020. Satellite remote sensing of aerosol optical depth: advances, challenges, and perspectives. Crit. Rev. Environ. Sci. Technol. 50 (16), 1640–1725.

Woodruff, N.P., Siddoway, F.H., 1965. A wind erosion equation. Soil Sci. Soc. Am. J. 29 (5), 602–608.

Xu, D., Qu, J.J., Niu, S., Hao, X., 2011. Sand and dust storm detection over desert regions in China with MODIS measurements. Int. J. Rem. Sens. 32 (24), 9365–9373.

Yan, X., Luo, N., Liang, C., Zang, Z., Zhao, W., Shi, W., 2020. Simplified and fast atmospheric radiative transfer model for satellite-based aerosol optical depth retrieval. Atmos. Environ. 224, 117362.

Yue, H., He, C., Zhao, Y., Ma, Q., Zhang, Q., 2017. The brightness temperature adjusted dust index: an improved approach to detect dust storms using MODIS imagery. Int. J. Appl. Earth Obs. Geoinf. 57, 166–176.

Zou, X., Zhang, C., Cheng, H., Kang, L., Wu, Y., 2015. Cogitation on developing a dynamic model of soil wind erosion. Sci. China Earth Sci. 58, 462–473.

CHAPTER

6

Land surface temperature and related issues

Shadman Veysi[1] *and Eslam Galehban*[2]

[1]Soil and Water Research Institute (SWRI), Agricultural Research, Education and Extension Organization (AREEO), Karaj, Iran; [2]Department of Remote Sensing & Geographical Information System, Faculty of Geography, University of Tehran, Tehran, Iran

6.1 Introduction

Land surface temperature (LST) refers to the temperature of the Earth's land surface, usually measured in degrees Celsius or Kelvin. Every object with a temperature above absolute zero (273 Celsius) emits electromagnetic waves (Rees, 2013). Additionally, the Earth, with an average temperature of 300 K, emits electromagnetic waves with a peak in the infrared region and a range close to 9.7 μm (Kuenzer and Dech, 2013). The Earth absorbs a significant portion of the incoming solar radiation and then emits a corresponding amount in longer wavelengths (Iqbal, 2012). It is a critical variable in understanding and modeling many Earth system processes, such as the energy and water cycles (Veysi et al., 2016), as well as the carbon cycle and ecosystems (Sen, 2008).

LST can be measured using a variety of methods, including ground-based sensors, airborne sensors, and satellite remote sensing (Fabrizi et al., 2010). Remote sensing is particularly used for mapping and monitoring LST over large areas and at high temporal resolutions. Satellites can provide global coverage and frequent observations, which are essential for studying LST dynamics and variability at different spatial and temporal scales (Zhou et al., 2018).

LST is influenced by a variety of factors, including solar radiation, atmospheric conditions, surface properties (such as vegetation cover and soil moisture), and land use and land cover changes (Weng et al., 2004). Changes in LST can have important impacts on ecological and societal systems, including agriculture, water resources, and human health (Anderson et al., 2012). Therefore, accurate and timely LST measurements and modeling are essential for a wide range of applications, such as climate monitoring and forecasting, weather prediction, urban planning, and disaster management (Kerr et al., 1992).

Remote Sensing of Soil and Land Surface Processes
https://doi.org/10.1016/B978-0-443-15341-9.00017-4

The combined effects of land surface characteristics and atmospheric conditions control energy transfer from the atmosphere to the Earth's surface or vice versa, making LST a crucial factor in predicting environmental phenomena such as snowmelt, drought, forest fires, and urban heat islands (Hao et al., 2018).

Recent advancements in remote sensing platforms, such as satellites, have revolutionized environmental science studies by providing extensive coverage of the Earth's surface and enabling LST measurements over large areas and at regular intervals (Gupta, 2017). These data are essential for climate modeling, environmental monitoring, and land management, and they can help identify areas at risk of environmental stresses, such as drought (Wan and Li, 2004). LST influences air temperature through sensible heat flux, which transfers heat from the Earth's surface to the atmosphere (Weng and Schubring, 2004). Therefore, monitoring changes in LST can help us better understand and manage the Earth's environment and inform efforts to mitigate and adapt to the impacts of climate change (Wulder et al., 2019).

The advancements in remote sensing platforms have opened new avenues for environmental science studies and provided a way to measure LST over large areas and at regular intervals, creating a great revolution in environmental science studies.

It should be noted that the LST and Ta of the air are conceptually, methodologically, and quantitatively different (Jin and Dickinson, 2010). LST is obtained from satellite sensors through direct radiometric temperature measurements, which involve some assumptions about surface radiation power and atmospheric conditions (Norman and Becker, 1995). On the other hand, the Ta is the thermodynamic temperature measured in contact with the air at a height of about 2 m from the Earth's surface (Mildrexler and Running, 2011). For homogeneous and isothermal surfaces, the radiometric temperature is approximately equal to the thermodynamic temperature (Li et al., 2013a,b). Since monitoring the trend of Ta and LST is crucial for studying local and global changes, many researchers have used LST to estimate Ta, and according to statistical methods, a significant relationship has been found between them (Cresswell et al., 1999; Florio et al., 2004; Benali et al., 2012). Fig. 6.1 shows the trend of daily LST and air temperature variation.

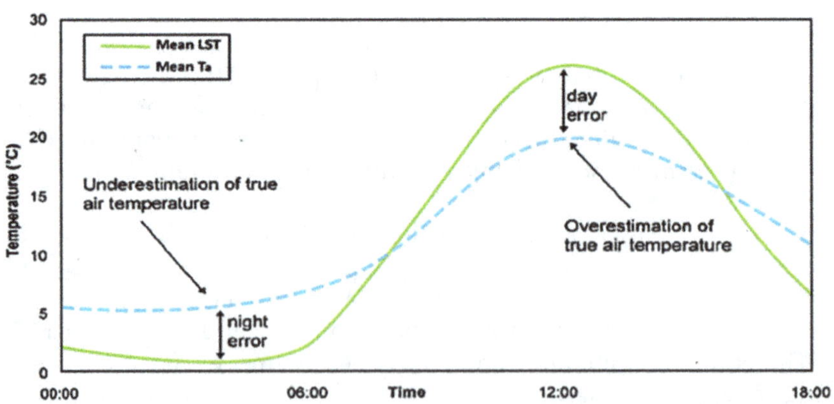

FIGURE 6.1 The trend of daily air temperature and land surface temperature.

6.2 Types of temperature

6.2.1 Kinetic temperature

Thermodynamic temperature is typically measured using a thermometer, which is a device that responds to changes in temperature through changes in some physical property, such as the length of a column of liquid or the electrical resistance of a metal wire (Becker and Li, 1995). The International System of Units (SI) defines the unit of thermodynamic temperature as the kelvin (K), which is based on the triple point of water, the temperature at which water can exist in all three phases (solid, liquid, and gas) in equilibrium (Kalinin and Kononogov, 2010). Kinetic temperature, on the other hand, is a measure of the average kinetic energy of the particles in a system, which is related to the temperature through the kinetic theory of gases (Jeans and Jeans, 1982). In a gas, for example, the temperature is related to the average speed of the gas molecules, which is determined by their mass and the temperature of the gas.

Although the fundamental nature of kinetic and thermodynamic temperature is different, they are usually defined as equivalent in most practical applications (Norman and Becker, 1995). This equivalence is based on the ideal gas law, which relates the pressure, volume, and temperature of a gas to the number of gas molecules and their average kinetic energy (Loeb, 2004). This relationship allows for the conversion of temperature between the kinetic and thermodynamic scales and is used in many areas of physics, engineering, and chemistry.

6.2.2 Radiometric temperature

Radiometric temperature, also known as brightness temperature, is a temperature derived from the radiance emitted by an object in a particular wavelength band (Norman and Becker, 1995; Dash et al., 2002). It is defined as the temperature of a blackbody that would emit the same amount of radiance in the same wavelength band (Artis et al., 1982). This makes radiometric temperature a powerful tool in remote sensing applications, allowing us to estimate the temperature of objects without direct contact (Berdahl and Fromberg, 1982). If ε_λ is the emissivity of an object and R_λ is its radiance in a given wavelength band, then the radiometric temperature (T_{sr}) can be calculated as follows (Norman and Becker, 1995):

$$T_{sr} = B^{-1}\left[\frac{R_\lambda}{\varepsilon_\lambda}\right] \tag{6.1}$$

In practice, the radiometric temperature is often used as an estimate of the skin temperature of an object, since it is a noncontact measurement that can be made from a distance using remote sensing techniques. Skin temperature is the temperature of the surface layer of an object, such as the surface of the human body, and is an important parameter in many fields, including medicine, physiology, and environmental science. Fig. 6.2 shows the pattern of the daily trend of the radiant temperature of different objects.

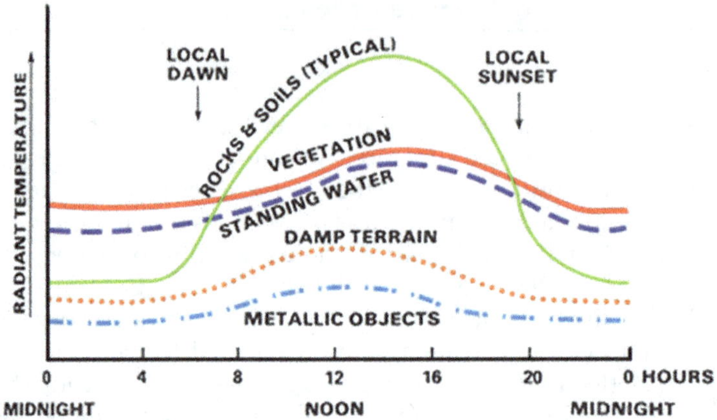

FIGURE 6.2 Daily changes in rock and soil temperature compared to vegetation, wet ground, and metal objects.

6.2.3 Brightness temperature (TB)

Brightness temperature is a measure of the temperature of an object based on the radiance it emits in a particular wavelength range. It is derived from the radiance equation, which describes the power per unit area emitted by an object at a given temperature and wavelength (Norman and Becker, 1995; DeWitt and Nutter, 1991). The *TB* of an object is obtained by integrating the radiance equation over all wavelengths in a particular wavelength range and taking into account the response of the detector used to measure the radiance. This is done using the Planck law, which describes the spectral radiance of a black body as a function of its temperature and wavelength. The resulting *TB* is the temperature of a black body that would emit the same amount of radiance as what was actually measured by the detector. In other words, it is the temperature at which a black body would have the same radiance in the same wavelength range as the object being measured (Ruff et al., 1997).

One important consideration in using *TB* to estimate the temperature of an object is that it assumes the object behaves like a black body, which emits radiation at all wavelengths and absorbs all incident radiation. In reality, most objects are not perfect black bodies and may have varying degrees of emissivity and reflectivity at different wavelengths. Therefore, the measured *TB* may differ from the true temperature of the object, and corrections may be necessary to account for these effects (Howell et al., 2020).

$$T_b = B^{-1}\left[\frac{\int_{\lambda_1}^{\lambda_2} f_i(\lambda) R_\lambda d\lambda}{\int_{\lambda_1}^{\lambda_2} f_i(\lambda) d\lambda}\right] \tag{6.2}$$

Where B^{-1} represents the inverse of Planck's law function. λ denotes the wavelength, while λ_1 and λ_2 refer to the higher and lower wavelengths associated with the sensitivity of the detector in channel *i*. Additionally, R_λ represents the measured radiance at a given wavelength and the spectral response function of the detector in channel *i*.

6.2.4 Land surface temperature

Infrared sensors are capable of recording thermal radiation images of objects on the Earth's surface. One of the products extracted from thermal infrared images is the LST, which can be used to study spatiotemporal changes on the Earth's surface (Dech et al., 1998). Due to the heterogeneity of surface characteristics such as vegetation cover, topography, and soil, LST values change rapidly with respect to location and time, which highlights the need for measuring LST with suitable spatial and temporal resolutions (Prata et al., 1995). In order to accurately estimate LST in the thermal infrared region, atmospheric effects must also be taken into account in addition to parameters such as LST and radiance. Therefore, in addition to radiometric correction and cloud removal, atmospheric correction and radiance should be corrected to determine LST from satellite thermal infrared data (Li and Becker, 1993). In Figs. 6.3 and 6.4, the *TB*, *LST* (*Ts*), aerodynamic temperature (TAero), radiometric temperature (*TR*), and air temperature (TAir) at height Z_1 are shown. Fig. 6.3 illustrates the *TB* observed by a satellite, which is influenced by the atmosphere, the Earth, and the Sun. Additionally, the air temperature according to the energy balance parameters and taking into account intermediate media environments is related to the aerodynamic temperature, radiometric temperature, and *LST*, as shown in Fig. 6.4.

(1) The radiance observed from the Earth's surface and attenuated by the atmosphere, (2) the upward thermal radiance of the atmosphere (3), the upward scattered solar radiance (4), the direct radiance from the surface of the Earth (5), the downward thermal radiance of the atmosphere (6), the upward scattered and reflected solar radiance from the surface of the Earth (6), and the reflected direct solar radiance from the surface of the Earth (7).

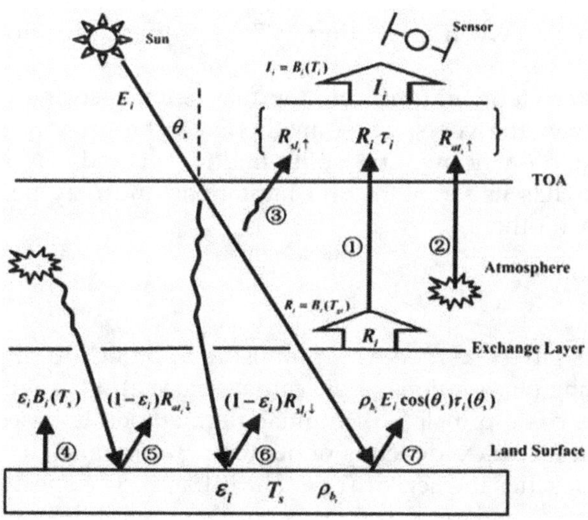

FIGURE 6.3 Energy exchange between the atmosphere and the earth.

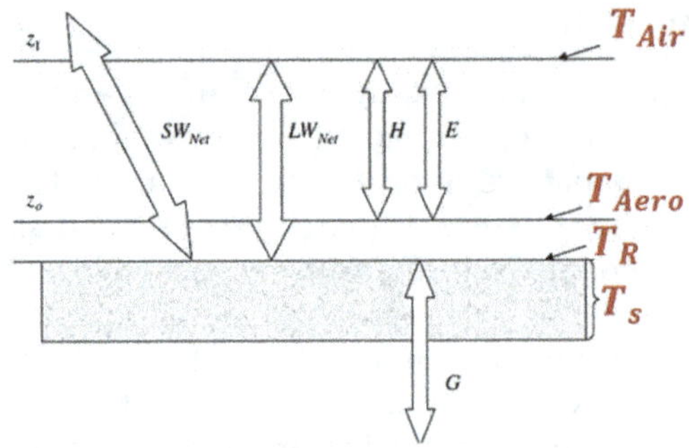

FIGURE 6.4 Display of boundary interface layers between the surface of the earth and air temperature at the height z_1 above the surface of the earth.

6.3 Methods to retrieve the land surface temperature

Thermal remote sensing is a complex field, and researchers have not yet identified a single approach that can be universally applied to estimate *LST* (Neteler, 2010). Current research is focused on addressing the limitations of existing methods and improving the accuracy and applicability of thermal remote sensing. To this end, numerous studies have been conducted to investigate the relationship between object temperature and thermal infrared waves as they pass through the atmosphere.

However, the accuracy of thermal remote sensing is influenced by various factors that can affect the relationship between thermal radiation and energy scattering. These factors include solar radiation, wind speed, temperature, atmospheric humidity, vegetation cover structure and type, soil characteristics, accessible humidity, and sensor status. Therefore, it is important to consider all these factors when estimating *LST* using thermal remote sensing (Weng and Schubring, 2004).

To determine a surface temperature from thermal remote sensing data, various methods have been proposed over the years. These methods can be broadly categorized into three groups: multiangle, mono-window, and split-window methods (Alavipanah et al., 2018). Each of these methods has its strengths and limitations and may be more appropriate for certain applications than others.

6.3.1 Multiangle

Multiangle methods have emerged as a promising approach to improve the accuracy of *LST* retrieval by utilizing observations of the same surface at different viewing angles (Houborg et al., 2007). The basic principle of multiangle methods is based on the fact that the emitted radiation from a surface depends on its temperature and the amount of direct and diffuse radiation that reaches the sensor (Li et al., 2013a,b). The spectral emissivity of a surface is typically uniform at a given wavelength but varies with the viewing angle due to differences in the amount of direct and diffuse radiation. By observing a surface at multiple angles, the differences in the observed radiation can be used to estimate the surface temperature (Norman and Becker, 1995). Multiangle methods have several advantages over other

LST retrieval techniques. Firstly, they can better correct for atmospheric effects such as water vapor and aerosols by using observations at multiple angles and spectral bands. Secondly, multiangle methods can be more sensitive to small temperature variations than other methods, making them useful for applications such as urban heat island mapping. Thirdly, they can be applied to a wide range of surfaces and environmental conditions, making them useful for a variety of applications. Finally, multiangle methods can provide information on the anisotropy of the surface, which is important for modeling energy and water balance in complex terrain (Sobrino et al., 1996).

However, multiangle methods also have limitations that must be considered. Firstly, the advanced instrumentation required for multiangle methods can be more expensive than other methods. Secondly, multiangle methods may not be suitable for continuous monitoring of temperature over long periods due to the need for multiple observations at different angles. Thirdly, multiangle methods may be limited in their spatial coverage due to the need for multiple observations at different angles, which may not be feasible in some areas. Finally, multiangle methods require careful calibration and validation to ensure accurate results (Dash et al., 2001).

6.3.2 Split window (SW)

The split window algorithm is a widely used method for estimating *LST* from satellite data. It is based on the principle that different parts of the electromagnetic spectrum are absorbed and emitted differently by the Earth's surface, and that these differences can be used to estimate the temperature of the surface (Snyder and Wan, 1998).

In the split-window algorithm, two channels in the thermal infrared region of the spectrum are used to estimate the *LST*: one at a wavelength of around 11 μm and another at a wavelength of around 12 μm (Jiménez-Muñoz et al., 2014). These two channels are chosen because they are sensitive to different levels of atmospheric water vapor, which can affect the accuracy of the temperature estimate. For example, based on thermal bands of MODIS is following Eq. (6.3):

$$LST = a + bTB_{31} + c(TB_{31} - TB_{32}) \tag{6.3}$$

Where *LST* is the land surface temperature, TB_{31} and TB_{32} are the *TB*s measured by channels 31 and 32, and a, b, and c are coefficients that are determined through calibration and validation using ground-based measurements (Garajeh and Feizizadeh, 2021). This form of the split-window algorithm now enhances the use of atmospheric transmissivity and land surface emissivity (*LSE*). For example, for Landsat 8, the split-window equation is based on the following equation (Wan and Dozier, 1996).

$$LST = \left(TB_{11} - (1-\varepsilon)TB_{12}\frac{\tau_{12}}{\tau_{11}}\right) / \left(\varepsilon\sigma\left(1 - \frac{\tau_{12}}{\tau_{11}}\right)\right)^{1/4} \tag{6.4}$$

Where TB_{11} and TB_{12} are the *TB*s at channels 11 and 12, respectively. ε is the surface emissivity, which is a measure of the surface's ability to emit thermal radiation, and σ is the Stefan–Boltzmann constant. τ_{11} *and* τ_{12} are the atmospheric transmittances at channels 11 and 12, respectively. The term $(1-\varepsilon)TB_{12}\frac{\tau_{12}}{\tau_{11}}$ represents the atmospheric contribution to the

TB at band 12, which is caused by the absorption and emission of thermal radiation by atmospheric gases and aerosols. By subtracting this term from TB_{11}, the atmospheric effect can be removed to estimate the *LST*. The Price equation is a widely used method for *LST* estimation and has been applied to a variety of satellite sensors, including Landsat and MODIS. However, like all *LST* estimation algorithms, the accuracy of the results can be affected by uncertainties in the input data and assumptions made in the algorithm.

The split-window algorithm includes its simplicity and speed, which make it well-suited for use in operational remote sensing applications. It also requires only two channels, which make it more robust to errors in atmospheric correction than more complex algorithms (Yu et al., 2007). The split-window algorithm includes its sensitivity to errors in atmospheric water vapor content, which can affect the accuracy of the temperature estimate (Qin et al., 2001). In terms of disadvantages, the split-window algorithm may not be suitable for some surfaces, such as urban areas, where the surface temperature can be affected by factors such as building materials and the urban heat island effect (Weng and Schubring, 2004). Finally, the accuracy of the algorithm may be affected by the quality of the calibration and validation data used to determine the coefficients a, b, and c. Fig. 6.5 shows an overall schematic of the SW method. It also shows a schematic of SW methods to retrieve *LST*.

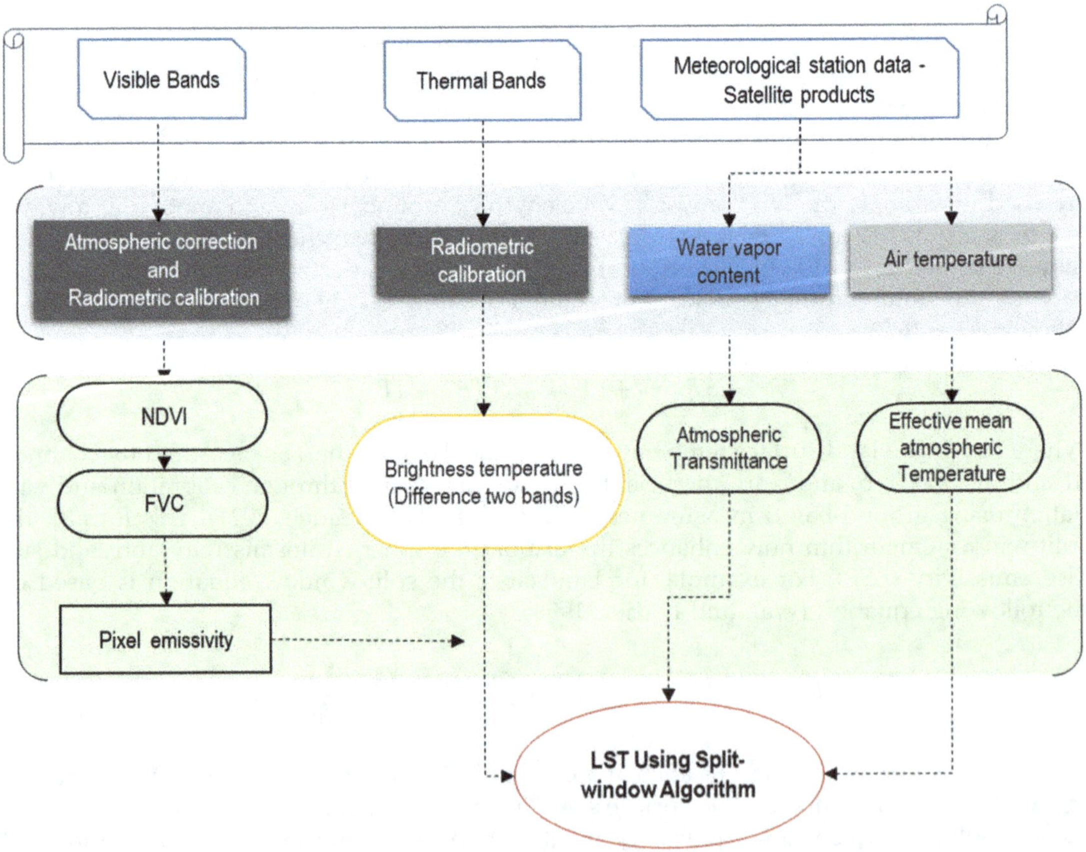

FIGURE 6.5 Schematic of split window methods to retrieve land surface temperature (*LST*).

6.3.3 Mono Window method

The Mono Window (MW) algorithm uses a single band of thermal radiance to calculate the *LST*, which makes it computationally simple and efficient (Wan and Dozier, 1996). The advantage of this algorithm is its simplicity, which allows for the fast and easy processing of large datasets. However, the MW algorithm has some limitations. It assumes a uniform atmospheric profile across the scene, which can cause errors in temperature estimation. Additionally, it is sensitive to the emissivity of the surface, which must be known or assumed to estimate the temperature accurately (Wan and Dozier, 1996). The equation for the MW algorithm using MODIS data is:

$$LST = (TB_{31} + (\lambda * C_1) / (C_2 * \lambda) - 273.15 \tag{6.5}$$

Where *LST* is the land surface temperature in degrees Celsius, *BT* is the measured *TB*, C_1, and C_2 are MODIS calibration constants, and λ is the wavelength of the MODIS band 31 for *LST* estimation.

To account for atmospheric effects, the *MW* algorithm also includes atmospheric transmissivity and surface emissivity (Jiménez-Muñoz and Sobrino, 2003). These can be estimated from ancillary data sources or assumed based on prior knowledge of the surface type. The estimated transmissivity and emissivity values are then used to adjust the *TB* in the equation above. The *MW* algorithm has several advantages, including its simplicity and computational efficiency (Wan and Dozier, 1996). However, like any algorithm, it also has limitations. The main limitation is that it assumes a uniform atmospheric profile across the scene, which can cause errors in the temperature estimation. Additionally, it is sensitive to surface emissivity, which must be known or assumed to estimate the temperature accurately. Fig. 6.6 shows a schematic of *MW* methods to retrieve *LST*.

6.4 Land surface emissivity

Land surface emissivity is a critical parameter used to retrieve LST from thermal infrared satellite data (Peres and DaCamara, 2005). It is defined as the ratio of the radiation emitted by the Earth's surface to that of a blackbody at the same temperature (Dash et al., 2002). The *LSE* value varies depending on the surface type and condition, and hence, accurate retrieval of *LSE* is essential for precise *LST* estimation. Various methods can be employed to retrieve *LSE*.

6.4.1 Spectral libraries

Spectral libraries are collections of spectra of materials with known emissivity values. Spectral libraries contain a wide range of spectra for different land surface types, such as vegetation, soil, water bodies, and urban areas (Herold et al., 2004). The spectral library method is based on the principle that every land surface type has a unique emissivity spectrum, which can be used to retrieve *LSE*. One example of a spectral library for *LSE* retrieval is the ASTER spectral library, which was developed using data from the advanced spaceborne thermal emission and reflection radiometer (ASTER) sensor onboard NASA's Terra satellite. The ASTER spectral library contains more than 2400 spectra for various land surface types,

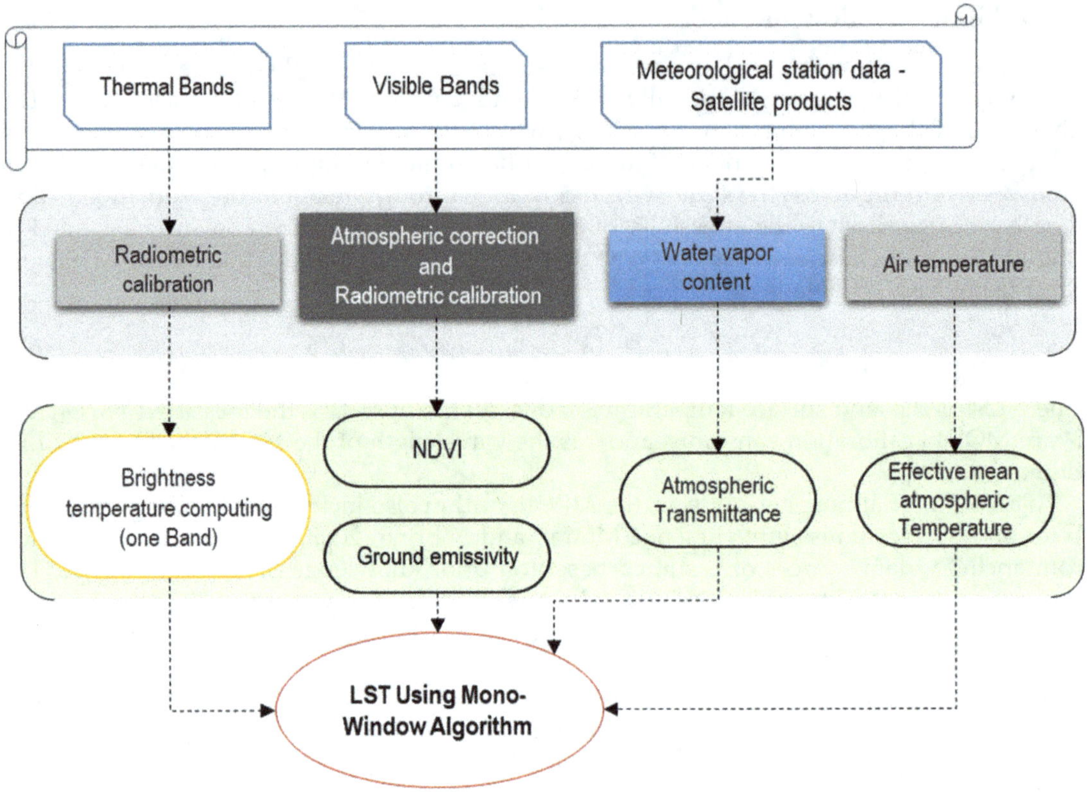

FIGURE 6.6 Schematic of mono window methods to retrieve land surface temperature (*LST*).

including minerals, soils, vegetation, and human-made materials (Mohamed and Mutanga, 2017). Another one is the MODIS MCD43C1 product, which is a global *LSE* product generated from moderate resolution imaging spectroradiometer (MODIS) satellite data (Ghent et al., 2010). The MODIS spectral library includes emissivity values for various land surface types, and it is updated periodically to reflect changes in land cover and surface conditions (Weng and Gao, 2014). The spectral library method is a fast and efficient way to retrieve *LSE* from thermal infrared satellite data (Wang et al., 2013). It is particularly useful for large-scale applications, as it can provide a quick estimate of *LSE* over a wide area. The accuracy of spectral library-based *LSE* retrieval depends on the spatial resolution of the thermal infrared satellite data. Spectral libraries have limited accuracy for small-scale features or mixed pixels. Spectral libraries may not account for spectral variability within land surface types, which can lead to errors in *LSE* retrieval. For example, different vegetation types may have different emissivity spectra depending on their moisture content and structure (Li et al., 2013a,b). Spectral libraries are based on the assumption that the emissivity spectra of land surface types are constant across different locations and conditions. However, emissivity spectra can vary with time and location, particularly for dynamic land surface types such as vegetation. Spectral libraries are accurate, efficient, and flexible, but they have limitations related to spatial resolution, spectral variability, and generalization. Spectral libraries are best used in combination with other methods, such as empirical and physical models, to improve the accuracy and reliability of *LSE* retrieval.

6.4.2 Empirical models

Empirical models are widely used for the retrieval of *LSE* from thermal infrared satellite data (Jiang et al., 2006). These models are based on a statistical analysis of field measurements or other empirical data to find the most significant variables that affect *LSE* and to derive mathematical equations that estimate *LSE* from these variables. The Vegetation Emissivity Index (VEI) is an example of an empirical model that is based on the relationship between *LSE* and vegetation cover.

One of the advantages of empirical models is their flexibility. These models can be tailored to specific land surface types and conditions, making them more adaptable than spectral libraries, which provide a limited set of emissivity spectra for specific materials (Wegener, 2004). Empirical models can also provide more accurate *LSE* estimates than spectral libraries for small-scale features or mixed pixels, as they can account for the effects of multiple variables on *LSE*. In addition, empirical models require fewer resources and data than physical models, as they do not require detailed knowledge of the physical properties of the land surface (Kalma et al., 2008).

However, empirical models also have limitations that must be taken into account. One limitation is their limited applicability, as empirical models are developed based on specific datasets and conditions and may not apply to other locations or time periods. Another limitation is their complexity, as empirical models require extensive statistical analysis to identify the most significant variables and to develop the mathematical equations that estimate *LSE*. This complexity may limit the ability of nonexperts to develop or use these models. Finally, empirical models are based on assumptions about the relationships between variables and *LSE*, which may not hold true for all land surface types or conditions, leading to errors in *LSE* retrieval (Li et al., 2013a,b). In conclusion, their flexibility, improved accuracy, and cost-effectiveness make them a useful complement to other methods, such as spectral libraries and physical models. However, the limitations of empirical models must also be taken into account, and further research is needed to develop more accurate and reliable methods for *LSE* retrieval.

6.4.3 Physical models

Physical models are based on the physical properties of the land surface and the physics of the interaction between electromagnetic radiation and the surface (Liang, 2005). These models use radiative transfer equations to simulate the interaction between radiation and the land surface and to calculate *LSE* from satellite data. Physical models require detailed knowledge of the physical properties of the land surface, such as the geometry, thermal properties, and spectral properties of the surface.

Several physical models are commonly used for *LSE* retrieval, including the Radiative Transfer Equation (RTE) and the Hapke model. The RTE is a fundamental equation that describes the propagation of electromagnetic radiation through a medium, such as the Earth's atmosphere and land surface (Howell et al., 2020). The RTE is solved numerically to simulate the radiative transfer process and to calculate the radiation intensity at different wavelengths. The Hapke model is a widely used a physical model for *LSE* retrieval that is based on the reflectance and bidirectional reflectance distribution function (BRDF) of the land surface (Cao et al., 2019). The Hapke model assumes that the surface is composed of a mixture of Lambertian and specular reflectors and that the surface roughness and the orientation of surface features affect the BRDF. The model uses a set of equations to calculate the reflectance and BRDF of the surface and to derive *LSE* from these values. Physical models require validation against ground truth measurements, which can be challenging and may introduce additional sources of uncertainty. Physical models can be sensitive to input parameters, such as atmospheric conditions and surface properties, which can affect the accuracy of *LSE* estimates. Physical models may have limitations in spatial resolution due to computational constraints or data availability, which can limit their usefulness for certain applications (Wood et al., 2011). Physical models may not be applicable to all types of remote sensing data, such as data from sensors with limited spectral bands or spatial resolution. Overall, physical models have the potential to provide accurate and interpretable estimates of *LSE*, but their use may be limited by computational and data availability constraints, as well as uncertainties and limitations in their application to remote sensing data. In conclusion, physical models are a powerful tool for estimating LSE from thermal infrared satellite data, as they provide accuracy, versatility, and interpretability. However, they have limitations related to their computational intensity, complexity, and uncertainty. Physical models are best used in combination with other methods, such as empirical models and spectral libraries, to improve the accuracy and reliability of LSE retrieval.

6.5 Atmospheric transmissivity

Atmospheric transmissivity refers to the ability of the Earth's atmosphere to transmit solar radiation from the Sun to the Earth's surface (Leckner, 1978; Monteith, 1972; Ångström, 1929; Ricchiazzi et al., 1998). Atmospheric transmissivity plays an important role in the retrieval of *LST* (Jiménez-Muñoz et al., 2014; Vlassova et al., 2014; Yu et al., 2014). It is a measure of the fraction of solar radiation that is able to pass through the atmosphere without being absorbed or scattered by gases, particles, and clouds in the atmosphere. Atmospheric transmissivity is affected by various atmospheric conditions such as the concentration of gases like water vapor, ozone, and carbon dioxide, as well as the amount and size of particles in the atmosphere such as dust, smoke, and pollutants (Baumbach, 2012). It is usually expressed as a percentage or decimal value ranging from 0 to 1, where a value of 1 means that all of the incoming solar radiation is transmitted through the atmosphere without any loss (Katan, 1981; Ghassemlooy and Popoola, 2010). In areas with low atmospheric transmissivity, such as locations with high levels of air pollution or dense cloud cover, less solar radiation can reach the Earth's surface. This can result in lower *LST*s due to reduced solar heating. On the other hand, in areas with high atmospheric transmissivity, such as locations with clear skies and low levels of atmospheric pollution, more solar radiation is able to reach the Earth's surface. This can result in higher *LST*s due to increased solar heating. If the atmospheric transmissivity is overestimated, the retrieved *LST* may be biased toward cooler values, while if it is underestimated, the retrieved *LST* may be biased toward warmer values (Windahl and de Beurs, 2016; Duan et al., 2019). Therefore, accurate estimation of atmospheric transmissivity is critical to ensure reliable and accurate retrieval of *LST*. Various types of retrieval of atmospheric transmissivity can be performed using different approaches, including the experience-based equation and satellite products (Zhou et al., 2010; Huang et al., 2019).

6.5.1 Experience-based equation

Experience-based equations are statistical models that use historical data to estimate atmospheric transmissivity based on factors such as temperature, humidity, and air pressure (Huang et al., 2019). They can be derived from ground-based measurements or satellite data. The Beer–Lambert law is a well-known empirical equation used to describe the attenuation of light as it passes through a material (Weingartner et al., 2003; Sun and Goldys, 2008; Mayerhöfer et al., 2020) The equation is expressed as:

$$A = \varepsilon l c \tag{6.6}$$

Where A is the absorbance, ε is the molar absorptivity coefficient, l is the path length of the light through the material, and c is the concentration of the absorbing species. This equation was first introduced by August Beer in 1852 and later modified by Johann Lambert in 1760 (Binkley et al., 2013). Strengths Beer–Lambert Law is a simple and widely used equation for describing the attenuation of light as it passes through a material (Mayerhöfer et al., 2020). It can be used to estimate the concentration of absorbing species in a sample, such as pollutants in the atmosphere and its limitation is it assumes that the sample is homogeneous, which may not be true for the atmosphere where the composition of air can vary spatially and temporally. Additionally, the equation only accounts for absorption and does not consider other processes such as scattering. The Kasten–Young model is an empirical equation used to estimate the atmospheric transmittance of solar radiation (Bason, 2007; Norwood et al., 2014). The equation is expressed as:

$$\tau = \exp\left(-kSec\theta\right) \tag{6.7}$$

Where τ the transmittance, k is the atmospheric extinction coefficient, and $Sec\theta$ is the secant of the solar zenith angle. This equation was proposed by the German meteorologist Erich Kasten and the American physicist Andrew Young in 1989. The Kasten–Young model is a widely used empirical equation for estimating atmospheric transmittance for solar radiation. It is based on a simple physical model and can be easily implemented in atmospheric radiation transfer models (Kittler et al., 2011; Norwood et al., 2014). Assumes a homogeneous atmosphere with a constant atmospheric extinction coefficient, which may not be true in reality. It also assumes that the atmosphere is horizontally uniform, which may not be the case in regions with high topographical variability.

Another empirical equation is the Angstrom Equation which makes an empirical relationship between the wavelength dependence of aerosol optical depth and the size distribution of atmospheric particles (Eck et al., 1999; Schuster et al., 2006). The equation is expressed as:

$$\tau(\lambda) = \tau_0(\lambda) - \alpha \tag{6.8}$$

Where $\tau(\lambda)$ is the aerosol optical depth at a given wavelength λ, $\tau_0(\lambda)$ is the aerosol optical depth at a reference wavelength, and α is the Angstrom exponent. The equation was first proposed by the Swedish physicist Anders Angstrom in 1929. The Angstrom equation provides a simple and intuitive way to relate the aerosol optical depth to the size distribution of atmospheric particles (O'Neill et al., 2001; Giles et al., 2019). It is widely used in atmospheric remote sensing applications to estimate aerosol loading and size distribution. The Angstrom equation assumes that the size distribution of atmospheric particles is constant over time and space, which may not be true in reality. It also assumes that the optical properties of aerosols are wavelength-independent, which is not true for some types of aerosols. As well as Nir Krakauer, a climate scientist at City College of New York, developed an equation that estimates the atmospheric transmissivity using only temperature and relative humidity data. The equation is expressed as:

$$\tau = 1 - 0.0094(T - 23)(1 - RH) \tag{6.9}$$

Where τ is the atmospheric transmissivity, T is the temperature in Celsius, and RH is the relative humidity as a fraction between 0 and 1. This equation was published by Krakauer in 2013. Nir Krakauer's equation provides a simple and easy-to-implement method for estimating atmospheric transmissivity using only temperature and relative humidity data, which are commonly available from weather stations. It can be useful for estimating transmissivity in data-sparse regions where other methods may not be applicable. Nir Krakauer's equation is based on statistical relationships between temperature, relative humidity, and transmissivity, which may not be valid for all atmospheric conditions. It also assumes that the atmosphere is horizontally homogeneous and does not account for the effects of aerosols or other atmospheric constituents on transmissivity. Overall, each empirical equation has its own strengths and limitations, and the choice of which equation to use depends on the specific application and available data. It is important to evaluate the validity and accuracy of each equation for a given situation before using it for atmospheric transmissivity estimation.

6.5.2 Satellite products

Products that estimate derived from satellite measurements of radiation emitted or reflected by the Earth's surface and atmosphere (Loeb et al., 2003; Wang et al., 2005; Duveiller et al., 2018). These products can provide global coverage and high temporal resolution but may have limitations in accuracy and spatial resolution. Several satellite products that can be introduced are:

MODerate resolution atmospheric TRANsmission (MODTRAN)—MODTRAN is a widely used atmospheric radiative transfer code that can calculate the transmittance, path radiance, and radiance for satellite data. The advantages of MODTRAN include its ability to handle a wide range of atmospheric conditions and its high accuracy. However, MODTRAN requires input data such as atmospheric profiles, which can be time-consuming to obtain, and the program can be computationally expensive (Berk et al., 1987; Duveiller et al., 2018).

Atmospheric Correction (ATCOR)—ATCOR is a commercially available atmospheric correction software that can estimate surface reflectance from satellite images by correcting for atmospheric effects (Gao et al., 2009; Chrysoulakis et al., 2010). The advantages of ATCOR include its ability to handle complex atmospheric conditions and its user-friendly interface. However, ATCOR can be costly and requires calibration and validation data.

Atmospheric Correction Now (ACORN)—ACORN is an open-source atmospheric correction software that can estimate surface reflectance from satellite images by correcting for atmospheric effects (Bassani et al., 2010; Lee et al., 2020). The advantages of ACORN include its open-source nature and its ease of use. However, ACORN may not be suitable for very high spatial resolution satellite data and may require input data such as atmospheric profiles.

Fast Line-of-sight Atmospheric Analysis of Spectral Hypercubes (FLAASH)—FLAASH is a commercially available atmospheric correction software that can estimate surface reflectance from hyperspectral satellite data by correcting for atmospheric effects (Cooley et al., 2002; Matthew et al., 2002; Anderson et al., 2002). The advantages of FLAASH include its ability to handle complex atmospheric conditions and its high accuracy. However, FLAASH can be computationally expensive and may require input data such as atmospheric profiles. Overall, the choice of retrieval method depends on the specific application and the trade-offs between accuracy, temporal and spatial resolution, and data availability.

6.6 Validation of LST

Land surface temperature validation is the process of assessing the accuracy and reliability of the estimated *LST* values obtained from remote sensing data against ground-based measurements or other independent sources (Li et al., 2009; Wang et al., 2007; Guillevic et al., 2014). *LST* validation is critical for ensuring the quality and usefulness of *LST* data in a wide range of applications, including climate modeling, agriculture, urban planning, and public health (Anderson et al., 2012).

6.6.1 Field measurements

This involves collecting ground-based measurements of *LST* using instruments such as infrared thermometers or thermal cameras at different locations and times that correspond to satellite overpasses (Li et al., 2009; Wang et al., 2007; Guillevic et al., 2014). The collected *LST* values are then compared with the corresponding satellite-derived *LST* values to assess the agreement and accuracy of the satellite data. They provide direct and accurate measurements of *LST* values at specific locations and times. Field measurement validation can capture small-scale variations in *LST* due to local environmental factors and be used to validate the accuracy of other validation methods. This method of validation has two important limitations spatial coverage and representativeness due to the high cost and time-consuming nature of field campaigns and vulnerable to human error and instrument calibration issues.

6.6.2 Radiative transfer models

Radiative transfer models simulate the interaction of radiation with the land surface and atmosphere to estimate *LST* values (Wan and Dozier, 1996; Zhang et al., 2004; Cao et al., 2019). These models require input data such as atmospheric conditions, land cover types, and surface properties. The *LST* values estimated from radiative transfer models can be compared with satellite-derived *LST* values to validate the accuracy of the satellite data. Radiative transfer models can provide estimates of *LST* values at a high spatial resolution and over a large area and are able to simulate different atmospheric and surface conditions to evaluate their effects on *LST* values (Chauhan et al., 2003; Rosas et al., 2017; Ermida et al., 2020). This method of validation is useful for generating long-term datasets for trend analysis and modeling purposes.

6.6.3 Intercomparison

This involves comparing *LST* data from different satellites or sensors to evaluate the consistency and accuracy of the data (Vancutsem et al., 2010). This approach can also help identify potential errors or biases in the *LST* data from different sources and establish a consistent and standardized approach to *LST* validation. Also, provide insights into the strengths and limitations of different *LST* products. Intercomparison may not be able to determine the true *LST* values due to the lack of a reliable reference dataset.

6.6.4 In situ networks

In situ networks are ground-based monitoring systems that collect continuous measurements of *LST* at multiple locations (Albergel et al., 2012; Luzi et al., 2009). The collected data can be used to validate the accuracy and consistency of satellite-derived *LST* data. In situ validation of *LST* refers to the process of comparing satellite-derived *LST* data with *LST* measurements taken at the same location on the ground (Yu et al., 2011; Hall et al., 2008). This type of validation involves using instruments that are placed in the same location as the satellite sensor to measure the *LST* directly. In contrast, field measurement validation methods involve taking *LST* measurements using portable instruments that are carried to various locations on the ground (Guillevic et al., 2018). In situ validation requires specialized instruments that are similar to the satellite sensor, while field measurement validation can be conducted using portable instruments that measure *LST* directly (Guillevic et al., 2018; Krishnan et al., 2020). In situ validation provides high-quality data that are directly comparable to satellite data, while field measurement validation provides data that may be affected by factors such as atmospheric conditions, vegetation cover, and surface properties.

In summary, validation of *LST* data is essential to ensure the accuracy and reliability of remote sensing-derived temperature information (Brocca et al., 2011). Different validation methods can be used, including ground-based measurements, radiative transfer models, intercomparison, and in situ networks (Jackson et al., 2010; Takala et al., 2011; Haeffelin et al., 2005; Al-Yaari et al., 2019). The choice of validation method depends on the specific application, data availability, and research objectives.

6.7 Future mission

Global in-situ measurements of *LST* are crucial for understanding and monitoring changes in the Earth's climate. As such, it is imperative to establish a set of standardized protocols for measuring *LST* that can be followed consistently by researchers and practitioners. To ensure the accuracy and reliability of *LST* data, it is important to implement robust quality control procedures that can identify and correct errors and outliers in the data. This can help to mitigate issues that may arise due to data inconsistencies and improve our understanding of climate patterns.

Moreover, it is necessary to increase the number of standard measurement stations, especially in areas where data are currently sparse, such as in remote regions. By doing so, we can obtain a more comprehensive and accurate understanding of the global distribution of *LST*, which can help us better understand and predict changes in the Earth's climate and ecosystems.

A unified archiving system for *LST* product storing and distribution would be an essential tool in this regard. It would involve a centralized platform that allows for the storage and distribution of *LST* data products in a standardized and accessible manner. The system would need to have a robust data management infrastructure that can store and manage large volumes of *LST* data effectively. It would also require a standardized format for *LST* data products to ensure consistency and accessibility across different systems and platforms.

The unified archiving system would need to provide easy access to *LST* data products for users, including researchers, policymakers, and the general public. To achieve this, a user-friendly interface should be created that allows users to search, browse, and download *LST* data products easily. It is also essential to ensure that the system can distribute *LST* data products to users in a timely and efficient manner, utilizing various distribution methods such as cloud-based platforms or FTP servers.

Overall, the design of a unified archiving system for *LST* product storing and distribution should prioritize scalability, accessibility, and quality. By implementing these key components, the system could provide a valuable resource for researchers, policymakers, and other users who rely on *LST* data for their work. This would help advance our understanding of the Earth's climate and ecosystem, leading to more informed decisions and policies that can benefit both the environment and society.

6.8 Conclusion

The accuracy and quality of LST products are often limited by various factors that can introduce uncertainties and biases. Atmospheric factors, such as clouds, aerosols, and water vapor, can significantly impact satellite sensors' measurements, leading to errors and inaccuracies in *LST* products. The temporal and spatial resolution of *LST* products can also affect the detection of small-scale temperature variations and the monitoring of rapid changes over time. Sensor calibration and retrieval algorithms are also significant sources of error that can introduce biases and uncertainties into *LST* products. Moreover, the limited temporal coverage of available *LST* products presents a challenge for researchers and data users, as most *LST* products are only available from the year 2000 onwards. This restricts the ability to conduct a long time series analysis that is essential for understanding changes in *LST* over time.

Current *LST* products can only provide instantaneous measurements at the satellite overpass time. Consequently, researchers are tasked with estimating these mean values from the limited instantaneous *LST*s available. *LST* is only measured when the satellite is passing overhead. However, it is possible to estimate daily mean *LST* using a combination of satellite observations and modeling techniques. One approach is to use a statistical model, such as a generalized linear mixed model (GLMM), to estimate daily mean *LST* based on a time series of satellite *LST* measurements. The GLMM can account for the random error associated with satellite measurements and incorporate additional information, such as land cover, soil moisture, and topography, to improve the accuracy of the estimates. Another approach is to use a land surface model, such as the Community Land Model (CLM), to simulate daily *LST* based on inputs such as surface radiation, soil moisture, and vegetation cover. The model can be calibrated and validated using satellite observations and can provide estimates of daily mean *LST* at high spatial resolutions.

In addition, some reanalysis land surfaces products, such as GLDAS and NLDAS, provide estimates of daily mean *LST* based on a combination of satellite observations and modeling techniques. These products incorporate satellite observations of *LST* with land surface models to estimate daily mean *LST* at high spatial resolutions and over extended periods.

While there are limitations and uncertainties associated with estimating daily mean *LST* from satellite observations and modeling techniques, these approaches provide a useful tool for studying land surface processes and understanding the impacts of climate and land use changes on surface temperature.

To address these challenges, researchers are developing new techniques and methods to improve the quality and accuracy of *LST* products. These methods include integrating multiple satellite data sources, such as using data from different sensors to account for atmospheric conditions and incorporating ground-based observations to validate and refine *LST* products. Refining atmospheric correction algorithms is another area of focus, as this can help reduce the impact of atmospheric factors on *LST* measurements. Increasing the spatial and temporal resolution of *LST* products is also a priority, as this can improve the detection of small-scale temperature variations and the monitoring of rapid changes over time.

References

Al-Yaari, A., Wigneron, J.P., Dorigo, W., Colliander, A., Pellarin, T., Hahn, S., Mialon, A., Richaume, P., Fernandez-Moran, R., Fan, L., Kerr, Y.H., 2019. Assessment and inter-comparison of recently developed/reprocessed microwave satellite soil moisture products using ISMN ground-based measurements. Rem. Sens. Environ. 224, 289–303.

Alavipanah, S., Schreyer, J., Haase, D., Lakes, T., Qureshi, S., 2018. The effect of multi-dimensional indicators on urban thermal conditions. J. Clean. Prod. 177, 115–123.

Albergel, C., De Rosnay, P., Gruhier, C., Muñoz-Sabater, J., Hasenauer, S., Isaksen, L., Kerr, Y., Wagner, W., 2012. Evaluation of remotely sensed and modelled soil moisture products using global ground-based in situ observations. Rem. Sens. Environ. 118, 215–226.

Anderson, G.P., Felde, G.W., Hoke, M.L., Ratkowski, A.J., Cooley, T.W., Chetwynd Jr., J.H., Gardner, J.A., Adler-Golden, S.M., Matthew, M.W., Berk, A., Bernstein, L.S., 2002. August. MODTRAN4-based atmospheric correction algorithm: FLAASH (fast line-of-sight atmospheric analysis of spectral hypercubes). Algorith. Technol. Multispect., Hyperspect. Ultraspect. Imag. VIII 4725, 65–71 (SPIE).

Anderson, M.C., Allen, R.G., Morse, A., Kustas, W.P., 2012. Use of Landsat thermal imagery in monitoring evapotranspiration and managing water resources. Rem. Sens. Environ. 122, 50–65.

Ångström, A., 1929. On the atmospheric transmission of sun radiation and on dust in the air. Geogr. Ann. 11 (2), 156–166.

Artis, D.A., Carnahan, W.H., 1982. Survey of emissivity variability in thermography of urban areas. Rem. Sens. Environ. 12 (4), 313–329.

Bason, F., 2007. Linke's turbidity factor applied to worldwide global horizontal irradiance measurements. Aarhus Univers. Denmar. Meteorol. Inst. 2 (7).

Bassani, C., Cavalli, R.M., Pignatti, S., 2010. Aerosol optical retrieval and surface reflectance from airborne remote sensing data over land. Sensors 10 (7), 6421–6438.

Baumbach, G., 2012. Air Quality Control: Formation and Sources, Dispersion, Characteristics and Impact of Air Pollutants—Measuring Methods, Techniques for Reduction of Emissions and Regulations for Air Quality Control. Springer Science and Business Media.

Becker, F., Li, Z.L., 1995. Surface temperature and emissivity at various scales: definition, measurement and related problems. Rem. Sens. Rev. 12 (3–4), 225–253.

Benali, A., Carvalho, A.C., Nunes, J.P., Carvalhais, N., Santos, A., 2012. Estimating air surface temperature in Portugal using MODIS LST data. Rem. Sens. Environ. 124, 108–121.

Berdahl, P., Fromberg, R., 1982. The thermal radiance of clear skies. Sol. Energy 29 (4), 299–314.

Berk, A., Bernstein, L.S., Robertson, D.C., 1987. MODTRAN: A Moderate Resolution Model for LOWTRAN. Spectral Sciences Inc Burlington MA.

Binkley, D., Campoe, O.C., Gspaltl, M., Forrester, D.I., 2013. Light absorption and use efficiency in forests: why patterns differ for trees and stands. For. Ecol. Manag. 288, 5–13.

Brocca, L., Hasenauer, S., Lacava, T., Melone, F., Moramarco, T., Wagner, W., Dorigo, W., Matgen, P., Martínez-Fernández, J., Llorens, P., Latron, J., 2011. Soil moisture estimation through ASCAT and AMSR-E sensors: an intercomparison and validation study across Europe. Rem. Sens. Environ. 115 (12), 3390–3408.

Cao, B., Liu, Q., Du, Y., Roujean, J.L., Gastellu-Etchegorry, J.P., Trigo, I.F., Zhan, W., Yu, Y., Cheng, J., Jacob, F., Lagouarde, J.P., 2019. A review of earth surface thermal radiation directionality observing and modeling: historical development, current status and perspectives. Rem. Sens. Environ. 232, 111304.

Chauhan, N.S., Miller, S., Ardanuy, P., 2003. Spaceborne soil moisture estimation at high resolution: a microwave-optical/IR synergistic approach. Int. J. Rem. Sens. 24 (22), 4599–4622.

Chrysoulakis, N., Abrams, M., Feidas, H., Arai, K., 2010. Comparison of atmospheric correction methods using ASTER data for the area of Crete, Greece. Int. J. Rem. Sens. 31 (24), 6347–6385.

Cooley, T., Anderson, G.P., Felde, G.W., Hoke, M.L., Ratkowski, A.J., Chetwynd, J.H., Gardner, J.A., Adler-Golden, S.M., Matthew, M.W., Berk, A., Bernstein, L.S., 2002. June. FLAASH, a MODTRAN4-based atmospheric correction algorithm, its application and validation. In IEEE Int. Geosci. Rem. Sens. Sym. 3, 1414–1418 (IEEE).

Cresswell, M.P., Morse, A.P., Thomson, M.C., Connor, S.J., 1999. Estimating surface air temperatures, from Meteosat land surface temperatures, using an empirical solar zenith angle model. Int. J. Rem. Sens. 20 (6), 1125–1132.

Dash, P., Göttsche, F.M., Olesen, F.S., Fischer, H., 2001. Retrieval of land surface temperature and emissivity from satellite data: physics, theoretical limitations and current methods. J. Ind. Soci. Rem. Sens. 29, 23–30.

Dash, P., Göttsche, F.M., Olesen, F.S., Fischer, H., 2002. Land surface temperature and emissivity estimation from passive sensor data: theory and practice-current trends. Int. J. Rem. Sens. 23 (13), 2563–2594.

Dech, S.W., Tungalagsaikhan, P., Preusser, C., Meisner, R.E., 1998. Operational value-adding to AVHRR data over Europe: methods, results, and prospects. Aero. Sci. Technol. 2 (5), 335–346.

DeWitt, D.P., Nutter, G.D. (Eds.), 1991. Theory and Practice of Radiation Thermometry. John Wiley and Sons.

Duan, S.B., Li, Z.L., Li, H., Göttsche, F.M., Wu, H., Zhao, W., Leng, P., Zhang, X., Coll, C., 2019. Validation of Collection 6 MODIS land surface temperature product using in situ measurements. Rem. Sens. Environ. 225, 16–29.

Duveiller, G., Hooker, J., Cescatti, A., 2018. The mark of vegetation change on Earth's surface energy balance. Nat. Commun. 9 (1), 679.

Eck, T.F., Holben, B.N., Reid, J.S., Dubovik, O., Smirnov, A., O'neill, N.T., Slutsker, I., Kinne, S., 1999. Wavelength dependence of the optical depth of biomass burning, urban, and desert dust aerosols. J. Geophys. Res. Atmos. 104 (D24), 31333–31349.

Ermida, S.L., Soares, P., Mantas, V., Göttsche, F.M., Trigo, I.F., 2020. Google earth engine open-source code for land surface temperature estimation from the landsat series. Rem. Sens. 12 (9), 1471.

Fabrizi, R., Bonafoni, S., Biondi, R., 2010. Satellite and ground-based sensors for the urban heat island analysis in the city of Rome. Rem. Sens. 2 (5), 1400–1415.

Florio, E.N., Lele, S.R., Chi Chang, Y., Sterner, R., Glass, G.E., 2004. Integrating AVHRR satellite data and NOAA ground observations to predict surface air temperature: a statistical approach. Int. J. Rem. Sens. 25 (15), 2979–2994.

Gao, B.C., Montes, M.J., Davis, C.O., Goetz, A.F., 2009. Atmospheric correction algorithms for hyperspectral remote sensing data of land and ocean. Rem. Sens. Environ. 113, S17–S24.

Garajeh, M.K., Feizizadeh, B., 2021. A comparative approach of data-driven split-window algorithms and MODIS products for land surface temperature retrieval. Appli. Geomat. 13, 715–733.

Ghassemlooy, Z., Popoola, W.O., 2010. Terrestrial free-space optical communications. Mobile Wirel. Communicat. Network layer Circ. Lev. Desi. 17, 355–391.

Ghent, D., Kaduk, J., Remedios, J., Ardö, J., Balzter, H., 2010. Assimilation of land surface temperature into the land surface model JULES with an ensemble Kalman filter. J. Geophys. Res. Atmos. 115 (D19).

Giles, D.M., Sinyuk, A., Sorokin, M.G., Schafer, J.S., Smirnov, A., Slutsker, I., Eck, T.F., Holben, B.N., Lewis, J.R., Campbell, J.R., Welton, E.J., 2019. Advancements in the Aerosol Robotic Network (AERONET) Version 3 database—automated near-real-time quality control algorithm with improved cloud screening for Sun photometer aerosol optical depth (AOD) measurements. Atmos. Meas. Tech. 12 (1), 169–209.

Guillevic, P.C., Biard, J.C., Hulley, G.C., Privette, J.L., Hook, S.J., Olioso, A., Göttsche, F.M., Radocinski, R., Román, M.O., Yu, Y., Csiszar, I., 2014. Validation of land surface temperature products derived from the visible infrared imaging radiometer suite (VIIRS) using ground-based and heritage satellite measurements. Rem. Sens. Environ. 154, 19–37.

Guillevic, P., Göttsche, F., Nickeson, J., Hulley, G., Ghent, D., Yu, Y., Trigo, I., Hook, S., Sobrino, J.A., Remedios, J., Román, M., 2018. Land surface temperature product validation best practice protocol. Version 1.1. Best Pract. Satell. Deriv. Land Prod. validat. 60.

Gupta, R.P., 2017. Remote Sensing Geology. Springer.

Haeffelin, M., Barthès, L., Bock, O., Boitel, C., Bony, S., Bouniol, D., Chepfer, H., Chiriaco, M., Cuesta, J., Delanoë, J., Drobinski, P., 2005. February. SIRTA, a ground-based atmospheric observatory for cloud and aerosol research. Ann. Geophys. 23 (2), 253–275 (Copernicus GmbH).

Hall, D.K., Box, J.E., Casey, K.A., Hook, S.J., Shuman, C.A., Steffen, K., 2008. Comparison of satellite-derived and in-situ observations of ice and snow surface temperatures over Greenland. Rem. Sens. Environ. 112 (10), 3739–3749.

Hao, Z., Singh, V.P., Xia, Y., 2018. Seasonal drought prediction: advances, challenges, and future prospects. Rev. Geophys. 56 (1), 108–141.

Herold, M., Roberts, D.A., Gardner, M.E., Dennison, P.E., 2004. Spectrometry for urban area remote sensing—development and analysis of a spectral library from 350 to 2400 nm. Rem. Sens. Environ. 91 (3–4), 304–319.

Houborg, R., Soegaard, H., Boegh, E., 2007. Combining vegetation index and model inversion methods for the extraction of key vegetation biophysical parameters using Terra and Aqua MODIS reflectance data. Rem. Sens. Environ. 106 (1), 39–58.

Howell, J.R., Mengüç, M.P., Daun, K., Siegel, R., 2020. Thermal Radiation Heat Transfer. CRC press.

Huang, G., Li, Z., Li, X., Liang, S., Yang, K., Wang, D., Zhang, Y., 2019. Estimating surface solar irradiance from satellites: past, present, and future perspectives. Rem. Sens. Environ. 233, 111371.

Iqbal, M., 2012. An Introduction to Solar Radiation. Elsevier.

Jackson, T.J., Cosh, M.H., Bindlish, R., Starks, P.J., Bosch, D.D., Seyfried, M., Goodrich, D.C., Moran, M.S., Du, J., 2010. Validation of advanced microwave scanning radiometer soil moisture products. IEEE Trans. Geosci. Rem. Sens. 48 (12), 4256–4272.

Jeans, J., Jeans, J.H., 1982. An Introduction to the Kinetic Theory of Gases. CUP Archive.

Jiang, G.M., Li, Z.L., Nerry, F., 2006. Land surface emissivity retrieval from combined mid-infrared and thermal infrared data of MSG-SEVIRI. Rem. Sens. Environ. 105 (4), 326–340.

Jiménez-Muñoz, J.C., Sobrino, J.A., 2003. A generalized single-channel method for retrieving land surface temperature from remote sensing data. J. Geophys. Res. Atmos. 108 (D22).

Jiménez-Muñoz, J.C., Sobrino, J.A., Skoković, D., Mattar, C., Cristobal, J., 2014. Land surface temperature retrieval methods from Landsat-8 thermal infrared sensor data. Geosci. Rem. Sens. Lett. IEEE 11 (10), 1840–1843.

Jin, M., Dickinson, R.E., 2010. Land surface skin temperature climatology: benefitting from the strengths of satellite observations. Environ. Res. Lett. 5 (4), 044004.

Kalinin, M.I., Kononogov, S.A., 2010. Redefinition of the unit of thermodynamic temperature in the international system of units. High Temp. 48, 23–28.

Kalma, J.D., McVicar, T.R., McCabe, M.F., 2008. Estimating land surface evaporation: a review of methods using remotely sensed surface temperature data. Surv. Geophys. 29, 421–469.

Katan, J., 1981. Solar heating (solarization) of soil for control of soilborne pests. Annu. Rev. Phytopathol. 19 (1), 211–236.

Kerr, Y.H., Lagouarde, J.P., Imbernon, J., 1992. Accurate land surface temperature retrieval from AVHRR data with use of an improved split window algorithm. Rem. Sens. Environ. 41 (2–3), 197–209.

Kittler, R., Kocifaj, M., Darula, S., 2011. Daylight Science and Daylighting Technology. Springer Science and Business Media.

Krishnan, P., Meyers, T.P., Hook, S.J., Heuer, M., Senn, D., Dumas, E.J., 2020. Intercomparison of in situ sensors for ground-based land surface temperature measurements. Sensors 20 (18), 5268.

Kuenzer, C., Dech, S., 2013. Theoretical background of thermal infrared remote sensing. Thermal Infrar. Rem. Sens.: Sens., Meth., Applicat. 1–26.

Leckner, B., 1978. The spectral distribution of solar radiation at the earth's surface—elements of a model. Sol. Energy 20 (2), 143–150.

Lee, K., Kim, K., Lee, S.G., Kim, Y., 2020. Determination of the normalized difference vegetation index (NDVI) with top-of-canopy (TOC) reflectance from a KOMPSAT-3A image using orfeo ToolBox (OTB) extension. ISPRS Int. J. Geo-Inf. 9 (4), 257.

Li, X., Li, X., Li, Z., Ma, M., Wang, J., Xiao, Q., Liu, Q., Che, T., Chen, E., Yan, G., Hu, Z., 2009. Watershed allied telemetry experimental research. J. Geophys. Res. Atmos. 114 (D22).

Li, Z.L., Becker, F., 1993. Feasibility of land surface temperature and emissivity determination from AVHRR data. Rem. Sens. Environ. 43 (1), 67–85.

Li, Z.L., Tang, B.H., Wu, H., Ren, H., Yan, G., Wan, Z., Trigo, I.F., Sobrino, J.A., 2013a. Satellite-derived land surface temperature: current status and perspectives. Rem. Sens. Environ. 131, 14–37.

Li, Z.L., Wu, H., Wang, N., Qiu, S., Sobrino, J.A., Wan, Z., Tang, B.H., Yan, G., 2013b. Land surface emissivity retrieval from satellite data. Int. J. Rem. Sens. 34 (9–10), 3084–3127.

Liang, S., 2005. Quantitative Remote Sensing of Land Surfaces. John Wiley and Sons.

Loeb, L.B., 2004. The Kinetic Theory of Gases. Courier Corporation.

Loeb, N.G., Manalo-Smith, N., Kato, S., Miller, W.F., Gupta, S.K., Minnis, P., Wielicki, B.A., 2003. Angular distribution models for top-of-atmosphere radiative flux estimation from the clouds and the earth's radiant energy system instrument on the tropical rainfall measuring mission satellite. Part I: methodology. J. Appl. Meteorol. 42 (2), 240–265.

Luzi, G., Noferini, L., Mecatti, D., Macaluso, G., Pieraccini, M., Atzeni, C., Schaffhauser, A., Fromm, R., Nagler, T., 2009. Using a ground-based SAR interferometer and a terrestrial laser scanner to monitor a snow-covered slope: results from an experimental data collection in Tyrol (Austria). IEEE Trans. Geosci. Rem. Sens. 47 (2), 382–393.

Matthew, M.W., Adler-Golden, S.M., Berk, A., Felde, G., Anderson, G.P., Gorodetzky, D., Paswaters, S., Shippert, M., October 2002. Atmospheric correction of spectral imagery: evaluation of the FLAASH algorithm with AVIRIS data. In: Applied Imagery Pattern Recognition Workshop, 2002. Proceedings. IEEE, pp. 157–163.

Mayerhöfer, T.G., Pahlow, S., Popp, J., 2020. The bouguer-beer-Lambert law: shining light on the obscure. ChemPhysChem 21 (18), 2029–2046.

Mildrexler, D.J., Zhao, M., Running, S.W., 2011. A global comparison between station air temperatures and MODIS land surface temperatures reveals the cooling role of forests. J. Geophys. Res.: Biogeosciences 116 (G3).

Mohamed, A.A., Odindi, J., Mutanga, O., 2017. Land surface temperature and emissivity estimation for Urban Heat Island assessment using medium-and low-resolution space-borne sensors: a review. Geocarto Int. 32 (4), 455–470.

Monteith, J.L., 1972. Solar radiation and productivity in tropical ecosystems. J. Appl. Ecol. 9 (3), 747–766.

Neteler, M., 2010. Estimating daily land surface temperatures in mountainous environments by reconstructed MODIS LST data. Rem. Sens. 2 (1), 333–351.

Norman, J.M., Becker, F., 1995. Terminology in thermal infrared remote sensing of natural surfaces. Rem. Sens. Rev. 12 (3–4), 159–173.

Norwood, Z., Nyholm, E., Otanicar, T., Johnsson, F., 2014. A geospatial comparison of distributed solar heat and power in Europe and the US. PLoS One 9 (12), e112442.

O'Neill, N.T., Eck, T.F., Holben, B.N., Smirnov, A., Dubovik, O., Royer, A., 2001. Bimodal size distribution influences on the variation of Angstrom derivatives in spectral and optical depth space. J. Geophys. Res. Atmos. 106 (D9), 9787–9806.

Peres, L.F., DaCamara, C.C., 2005. Emissivity maps to retrieve land-surface temperature from MSG/SEVIRI. IEEE Trans. Geosci. Rem. Sens. 43 (8), 1834–1844.

Prata, A.J., Caselles, V., Coll, C., Sobrino, J.A., Ottle, C., 1995. Thermal remote sensing of land surface temperature from satellites: current status and future prospects. Rem. Sens. Rev. 12 (3–4), 175–224.

Qin, Z., Dall'Olmo, G., Karnieli, A., Berliner, P., 2001. Derivation of split window algorithm and its sensitivity analysis for retrieving land surface temperature. J. Geophys. Res. 106 (22), 655–722.

Rees, W.G., 2013. Physical Principles of Remote Sensing. Cambridge university press.

Ricchiazzi, P., Yang, S., Gautier, C., Sowle, D., 1998. SBDART: a research and teaching software tool for plane-parallel radiative transfer in the Earth's atmosphere. Bull. Am. Meteorol. Soc. 79 (10), 2101–2114.

Rosas, J., Houborg, R., McCabe, M.F., 2017. Sensitivity of Landsat 8 surface temperature estimates to atmospheric profile data: a study using MODTRAN in dryland irrigated systems. Rem. Sens. 9 (10), 988.

Ruff, S.W., Christensen, P.R., Barbera, P.W., Anderson, D.L., 1997. Quantitative thermal emission spectroscopy of minerals: a laboratory technique for measurement and calibration. J. Geophys. Res. Solid Earth 102 (B7), 14899–14913.

Schuster, G.L., Dubovik, O., Holben, B.N., 2006. Angstrom exponent and bimodal aerosol size distributions. J. Geophys. Res. Atmos. 111 (D7).

Sen, Z., 2008. Solar Energy Fundamentals and Modeling Techniques: Atmosphere, Environment, Climate Change and Renewable Energy. Springer Science and Business Media.

Snyder, W.C., Wan, Z., 1998. BRDF models to predict spectral reflectance and emissivity in the thermal infrared. IEEE Trans. Geosci. Rem. Sens. 36 (1), 214–225.

Sobrino, J.A., Li, Z.L., Stoll, M.P., Becker, F., 1996. Multi-channel and multi-angle algorithms for estimating sea and land surface temperature with ATSR data. Int. J. Rem. Sens. 17 (11), 2089–2114.

Sun, J., Goldys, E.M., 2008. Linear absorption and molar extinction coefficients in direct semiconductor quantum dots. J. Phys. Chem. C 112 (25), 9261–9266.

Takala, M., Luojus, K., Pulliainen, J., Derksen, C., Lemmetyinen, J., Kärnä, J.P., Koskinen, J., Bojkov, B., 2011. Estimating northern hemisphere snow water equivalent for climate research through assimilation of space-borne radiometer data and ground-based measurements. Rem. Sens. Environ. 115 (12), 3517–3529.

Vancutsem, C., Ceccato, P., Dinku, T., Connor, S.J., 2010. Evaluation of MODIS land surface temperature data to estimate air temperature in different ecosystems over Africa. Rem. Sens. Environ. 114 (2), 449–465.

Veysi, S., Naseri, A.A., Hamzeh, S., Moradi, P., 2016. Estimation of sugarcane field temperature using Split Window Algorithm and OLI LandSat 8 satellite images. J. RS GIS Nat. Res. 7 (1), 27–40.

Vlassova, L., Perez-Cabello, F., Nieto, H., Martín, P., Riaño, D., De La Riva, J., 2014. Assessment of methods for land surface temperature retrieval from Landsat-5 TM images applicable to multiscale tree-grass ecosystem modeling. Rem. Sens. 6 (5), 4345–4368.

Wan, Z., Dozier, J., 1996. A generalized split-window algorithm for retrieving land-surface temperature from space. IEEE Trans. Geosci. Rem. Sens. 34 (4), 892–905.

Wan, Z., Wang, P., Li, X., 2004. Using MODIS land surface temperature and normalized difference vegetation index products for monitoring drought in the southern Great Plains, USA. Int. J. Rem. Sens. 25 (1), 61–72.

Wang, K., Wan, Z., Wang, P., Sparrow, M., Liu, J., Haginoya, S., 2007. Evaluation and improvement of the MODIS land surface temperature/emissivity products using ground-based measurements at a semi-desert site on the western Tibetan Plateau. Int. J. Rem. Sens. 28 (11), 2549–2565.

Wang, K., Wan, Z., Wang, P., Sparrow, M., Liu, J., Zhou, X., Haginoya, S., 2005. Estimation of surface long wave radiation and broadband emissivity using Moderate Resolution Imaging Spectroradiometer (MODIS) land surface temperature/emissivity products. J. Geophys. Res. Atmos. 110 (D11).

Wang, N., Li, Z.L., Tang, B.H., Zeng, F., Li, C., 2013. Retrieval of atmospheric and land surface parameters from satellite-based thermal infrared hyperspectral data using a neural network technique. Int. J. Rem. Sens. 34 (9–10), 3485–3502.

Wegener, M., 2004. Overview of land use transport models. In: Handbook of Transport Geography and Spatial Systems, vol. 5. Emerald Group Publishing Limited, pp. 127–146.

Weingartner, E., Saathoff, H., Schnaiter, M., Streit, N., Bitnar, B., Baltensperger, U., 2003. Absorption of light by soot particles: determination of the absorption coefficient by means of aethalometers. J. Aerosol Sci. 34 (10), 1445–1463.

Weng, Q., Fu, P., Gao, F., 2014. Generating daily land surface temperature at Landsat resolution by fusing Landsat and MODIS data. Rem. Sens. Environ. 145, 55–67.

Weng, Q., Lu, D., Schubring, J., 2004. Estimation of land surface temperature–vegetation abundance relationship for urban heat island studies. Rem. Sens. Environ. 89 (4), 467–483.

Windahl, E., de Beurs, K., 2016. An intercomparison of Landsat land surface temperature retrieval methods under variable atmospheric conditions using in situ skin temperature. Int. J. Appl. Earth Obs. Geoinf. 51, 11–27.

Wood, E.F., Roundy, J.K., Troy, T.J., Van Beek, L.P.H., Bierkens, M.F., Blyth, E., de Roo, A., Döll, P., Ek, M., Famiglietti, J., Gochis, D., 2011. Hyperresolution global land surface modeling: meeting a grand challenge for monitoring Earth's terrestrial water. Water Resour. Res. 47 (5).

Wulder, M.A., Loveland, T.R., Roy, D.P., Crawford, C.J., Masek, J.G., Woodcock, C.E., Allen, R.G., Anderson, M.C., Belward, A.S., Cohen, W.B., Dwyer, J., 2019. Current status of Landsat program, science, and applications. Rem. Sens. Environ. 225, 127–147.

Yu, X., Guo, X., Wu, Z., 2014. Land surface temperature retrieval from Landsat 8 TIRS—comparison between radiative transfer equation-based method, split window algorithm and single channel method. Rem. Sens. 6 (10), 9829–9852.

Yu, Y., Privette, J.L., Pinheiro, A.C., 2007. Evaluation of split-window land surface temperature algorithms for generating climate data records. IEEE Trans. Geosci. Rem. Sens. 46 (1), 179–192.

Yu, Y., Tarpley, D., Privette, J.L., Flynn, L.E., Xu, H., Chen, M., Vinnikov, K.Y., Sun, D., Tian, Y., 2011. Validation of GOES-R satellite land surface temperature algorithm using SURFRAD ground measurements and statistical estimates of error properties. IEEE Trans. Geosci. Rem. Sens. 50 (3), 704–713.

Zhang, Y., Rossow, W.B., Lacis, A.A., Oinas, V., Mishchenko, M.I., 2004. Calculation of radiative fluxes from the surface to top of atmosphere based on ISCCP and other global data sets: refinements of the radiative transfer model and the input data. J. Geophys. Res. Atmos. 109 (D19).

Zhou, D.K., Larar, A.M., Liu, X., Smith, W.L., Strow, L.L., Yang, P., Schlüssel, P., Calbet, X., 2010. Global land surface emissivity retrieved from satellite ultraspectral IR measurements. IEEE Trans. Geosci. Rem. Sens. 49 (4), 1277–1290.

Zhou, D., Xiao, J., Bonafoni, S., Berger, C., Deilami, K., Zhou, Y., Frolking, S., Yao, R., Qiao, Z., Sobrino, J.A., 2018. Satellite remote sensing of surface urban heat islands: progress, challenges, and perspectives. Rem. Sens. 11 (1), 48.

CHAPTER

7

Unraveling the changes in soil properties availed by UAV-derivative data in an arid floodplain: Lessons learned and things to fathom

Sedigheh Maleki[1,2], *Aiding Kornejady*[3] *and Minoo Haghani Shirazi*[4]

[1]Department of Plant Production, Faculty of Agriculture, University of Torbat Heydarieh, Torbat Heydarieh, Iran; [2]Saffron Institute, University of Torbat Heydarieh, Torbat Heydarieh, Iran; [3]Research Department of Natural Resources, Golestan Agricultural and Natural Resources Research and Education Center, Agricultural Research, Education and Extension Organization (AREEO), Gorgan, Iran; [4]Spatial Sciences Innovators Consulting Engineering Company, Tehran, Iran

7.1 Introduction

Realizing the potential land suitability and designating the suitable land use classes is the key to setting a logical and adaptable balance between two competing interests, most notably, societies' activities and demands, and what actually environment can offer and supply. There is no doubt that attaining sustainable development would be insurmountable without paying due attention to land suitability and environmental capacities. Hence, land planning missions can facilitate regulating the current use from environmental resources and securing the needs of next generations in future. Land suitability analysis is the prerequisite for any land planning attempt. Land suitability is the art of conglomerating land's opportunities and limitations in order to lead the current exploitation plans toward an optimal

Remote Sensing of Soil and Land Surface Processes
https://doi.org/10.1016/B978-0-443-15341-9.00012-5

use considering land capability (Bandyopadhya et al., 2009). In general, land suitability is conducted for some predefined land uses. Overall, it is an inseparable part of sustainable land management.

Soil, as one of the substantial prerequisites of attaining sustainable development, plays an important role in solving serious environmental issues. Therefore, creating thematic maps by addressing spatial variation of soil's key characteristics and different classes can pave the way toward sustainable development goals in a cost-effective fashion (Bouma et al., 2019). Also, understanding one particular soil characteristic can facilitate managing others with similar patterns (Van Huyssteen et al., 2013; Edokpayi et al., 2017). In contrast with climatic features that are almost intact and rarely change in a certain region, soil characteristics can significantly change in a small area. Hence, despite the fact that climate precedes all environmental factors and permits the lands to be home to certain vegetation covers, soil can also play the role of a decisive factor particularly in small areas where the climate is intact (Amini et al., 2016).

Satellite images can be used over large areas, while using unmanned aerial vehicle (UAV) in small regions offers a flexible routine and eases to produce multiple photogrammetry products with high spatial resolutions in a reasonable time (Sishodia et al., 2020; Maleki et al., 2020; Zhu et al., 2021). In a recent publication, Watts et al. (2012) studied in detail the chronological development of unmanned aircraft systems (UAS) (i.e., UAV and all the operating systems and crews). Satellites and UAVs can be adopted to indirectly estimate soil attributes; an important role that has been only partially addressed in the literature, especially in regions with limited ground data on soil properties. Kariminejad et al. (2022a) reported that the application of UAV and its derivatives has increased the accuracy of the erosion models in Golestan province. Also, they were able to calculate the geometrical characteristics of the collapsed piping and assess the susceptibility of their study area to erosional processes, all availed by UAV. Still, only a handful of studies have tackled the question that how actually UAVs can assist in reducing the need for soil samples (i.e., handling limited soil data) or better delimitation of soil boundary maps (Oltmanns et al., 2012; Krenz et al., 2019; Maleki et al., 2020; Kariminejad et al., 2022). The application of UAV-mounted multispectral sensors for calculating vegetation indices to extract soil properties has been previously addressed in the literature (De Almeida et al., 2019; Zhu et al., 2020). Such endeavors have also gone beyond to examine soil texture from UAV products, which can be accompanied by field surveys to attain estimations with satisfactory accuracy in a reasonable time (Yue et al., 2019).

Arid and semi-arid regions often exhibit a heterogeneous spatial pattern of bare and vegetation cover (Deblauwe et al., 2008). Most common vegetation patterns are banded (Saco et al., 2007) or spotted (Deblauwe et al., 2008). Other heterogeneous chemicals, structural, and textural soil properties such as spatial variations of infiltration capacity (Krenz et al., 2019), soil nutrient content (Schlesinger et al., 1996; Rietkerk et al., 2000), and soil erodibility (Castillo et al., 1997; Wynn and Mostaghimi, 2006) also attest to the patchwork behavior of vegetation (Schlesinger et al., 1996). There is no doubt that the prediction performance of soil mapping models relies tremendously on the spatial scale of the input data (Krenz et al., 2019). UAVs can overcome the deficiency of rigorous, time- and budget-consuming field assessments and insufficient data quality from coarse resolution imagery obtained from satellite images. Moreover, producing high-resolution 3D terrain models with multilateral characteristics of the landscape can be achieved by applying UAVs (Krenz et al., 2019).

Based on these premises, the application of UAV in producing soil series maps and land suitability classification bearing on limited soil samples was investigated in the Niatak region in the southeast of Iran. The goal of the study can be listed as enriching the base maps and data required for a more holistic detailed study, aiming at flood spreading, reclamation of vegetation cover, and stabilization of dunes. This study marks one of the first land suitability classification attempts in Iran that employs UAV products.

7.2 Sampling design and soil analysis

The conditioned Latin hypercube sampling (cLHS; Minasny and McBratney, 2006) is an intuitive approach to circumvent the tedious process of soil sampling in large areas. The UAV-derived digital elevation model (DEM) with a 1 m spatial resolution was used to extract morphometric indices (Table 7.1) in SAGA software (Conrad et al., 2015) and then used along with other soil/vegetation factors as inputs in the cLHS R package developed by Roudier et al. (2020) to identify the soil pedons location for the study. Fig. 7.1 represents the spatial distribution of 20 soil pedons throughout the study area.

After air-drying, a total of 68 soil samples of different pedons were analyzed in a laboratory. The calcium carbonate equivalent (CCE) was determined using the titration method (Page et al., 1982). Soil organic carbon (SOC) was measured by Walkley–Black method (Nelson and Sommers, 1982). The soil pH and electrical conductivity (EC) were determined in saturated soil paste and extracted from saturated soil paste using a pH meter (Thomas, 1996) and an EC meter (Rhoades, 1983), respectively. The saturation percentage (SP) was measured in soil paste (Rhoades, 1983). The soil texture was determined based on the hydrometer method (Gee and Bauder, 1986). Total nitrogen (TN) was measured by the Kjeldahl approach (Bremner and Mulvaney, 1982), and the available phosphorus (P_{av}) was determined by the colorimetric method (Olsen and Sommer, 1982). Available potassium (K_{av}) was extracted with 1 M NH4OAC (pH 7) and measured via flame photometer (Knudsen et al., 1982). Soluble calcium (Ca_{aq}) and magnesium (Mg_{aq}) were measured by titration method with EDTA and soluble sodium (Na_{aq}) by flame photometry (Tucker and Kurtz, 1961). Cation exchange capacity (CEC) was measured by the method of saturation with ammonium acetate (Chapman, 1965). Sodium adsorption ratio (SAR; Vasu et al., 2016) and exchangeable sodium percentage (ESP) were calculated following Eqs. (7.1) and (7.2). Additionally, water permeability in soil was measured using a cylinder infiltrometer (Bouwer, 1986).

$$\mathrm{SAR} = \frac{\mathrm{Na}^{+}}{\sqrt{\frac{\mathrm{Ca}^{2+} + \mathrm{Mg}^{2+}}{2}}} \tag{7.1}$$

$$\mathrm{ESP} = \frac{\mathrm{Na}^{+}}{\mathrm{CEC}} \times 100 \tag{7.2}$$

TABLE 7.1 Environmental covariates used in the conditioned Latin hypercube sampling (cLHS) method in the study area.

Covariates	Covariate name (abbreviation)	Definition	References
Topographic attributes	Elevation (DEM)	Height over sea level	Gallant and Dowling (2003)
	Slope (Slope)	Gradient of line is changing of elevation in the direction and steepness	Gallant and Dowling (2003)
	Aspect (ASP)	Slope direction	
	Curvature (Curvature)	The degree to which a curve deviates from a straight line	
	Plan curvature (Plancurv)	Perpendicular line to the direction of maximum slope	
	Profile curvature (Profcurv)	Direction of maximum slope	
	Convergence (Convergence)	Highlights the convergent areas as channels and divergent areas as ridges.	
	Analytical hillshade	Angle between the surface and the incoming light beams	
	Channel network	The interpolated channel network base level elevations	Conrad et al. (2015)
	Vertical distance to channel network (VDTCN)	Calculates the vertical distance to a channel network base level	Olaya (2004)
	Topographic roughness index (TRI)	Measures terrain ruggedness	Moore et al. (1993)
	Stream power index (SPI)	An indirect connotation of sediment transport	
	Topographic wetness index (TWI)	A measure of the topographic control on soil wetness or the ratio	
	Multiresolution valley bottom flatness index (MRVBF)	Flat areas in low altitudes and valleys	Gallant and Dowling (2003)
	Multiresolution of ridge top flatness index (MRRTF)	Flat areas in high altitudes and ridges	

	Topographic position index (TPI)	Difference between a cell elevation value and the average elevation of the neighborhood around that cell	Olaya (2004)
	Mass balance index (MBI)	Soil erosion and deposition balance	
		Position Openness	
		Represents areas visible from a line of sight	
	Wind effect	Dimensionless index indicating areas exposed to wind	Bohner and Antonic (2009)
	Valley depth	The vertical distance to a channel network base level	
	Relative slope position (RSP)	The position of one point relative	Olaya (2004)
Terrain surface texture (Texure)	The geometrical irregularities present at a surface or the nested-means terrain classification		
Organism	Land use map	Prepared using orthophoto of unmanned aerial vehicles (UAVs)	–
Soil, vegetation, topography	Curve number (CN)	Prediction of direct runoff or amount of infiltration resulting from excess precipitation	Natural resources protection organization (USDA)
Parent material, topography, soil	Land unit map	–	FAO (1976)

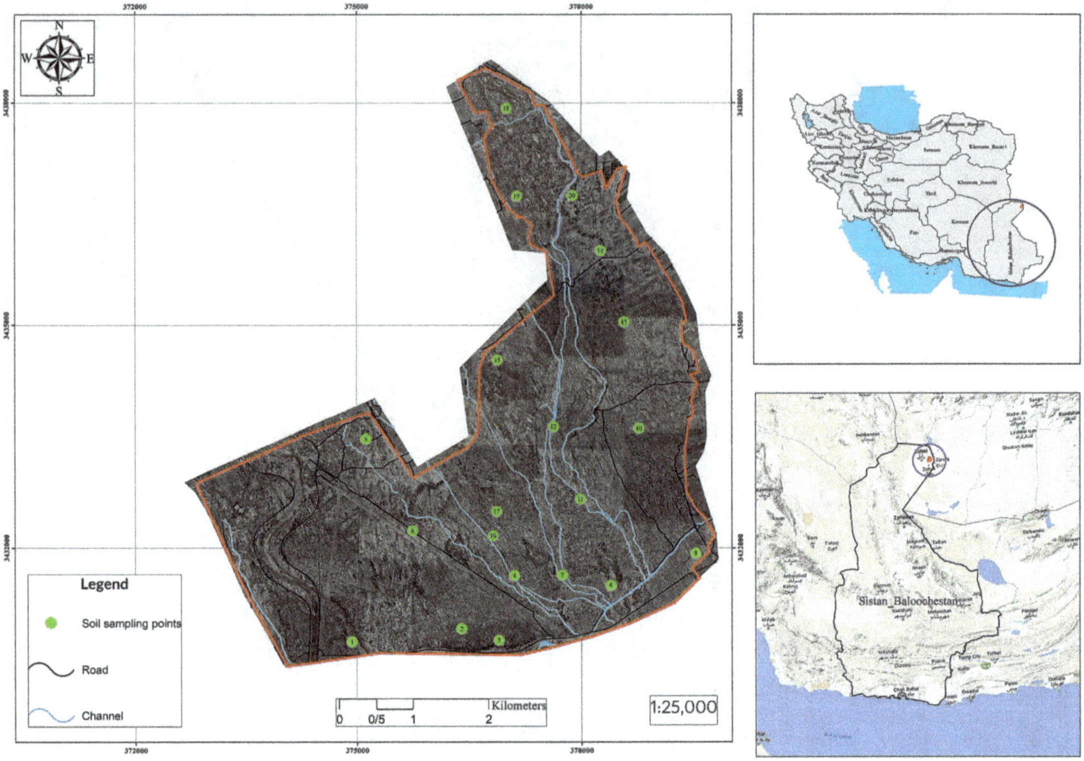

FIGURE 7.1 Geographical location of the study area, stream network, roads, and distribution of the excavated soil profiles.

7.3 Aerial images obtained from unmanned aerial vehicles and environmental covariates

A multirouter UAV, Phantom 4 pro v2, was employed with the following configuration: a 70%–80% longitudinal and transverse overlap, supplied with a DJI FC6310 camera with 5472 × 3648 resolution, 8.8 mm focal length, and the study scale of 1:500. The UAV images were processed using different photogrammetry software and different photogrammetry produced such as 3D model, orthophoto, and most importantly DEM. Previous soil studies have also proved the successful application of UAV-derived products for digital soil mapping and soil erosion studies (Maleki et al., 2020; Kariminejad et al., 2022).

7.4 Preparing of land suitability evaluation map

FAO framework for land suitability evaluation was adopted as follows:

1. Describing promising land-use type
2. Defining the main requirements for each land use class.

3. Comparing the acquired qualities and the required prerequisites for each land use class and further syntheses for assigning suitability ranks to each of the land units/components.

According to the FAO land suitability framework (FAO, 1976), suitability ranks are classified into four categories of order, class, subclass, and unit. Each order is classified into suitable (S) and not suitable (N). Suitable order is categorized into three classes suitable (S1), fairly suitable (S2), and critically/poorly suitable (S3). Not suitable order is categorized into temporarily/impermanently unsuitable and permanently unsuitable.

7.5 Case study

7.5.1 Study area

Niatak region geographically lies within 375,000–378,000 m longitudes and 3,432,000–3,438,000 m latitudes, UTM zone 41N (Fig. 7.1). It extends for an area of about 25.60 km^2 and is considered a flood plain with active flood channels. The area is topologically homogenous with gentle slopes situated in Sistan and Baluchestan Province, south-east of Iran; 19% of which is covered by small moderately active-to-stabilized sand dunes (reaching up to a few meters tall) that occasionally exhibit as Nebka. Niatak is selected as a pilot area to test different mechanical and biomechanical flood spreading measures for stabilizing active sand dunes and restoring vegetation cover. Soil cover is mostly associated with alluvial and aeolian sediments. Soil texture varied between silty loam to sandy soils with poorly developed soil pedons, and some hydromorphic features summarize the overall soil development of the region.

According to Emberger's climate classification, Niatak is classified as a hyper-arid, hyper-warm, and dusty region. The average annual precipitation and temperature steepestreach 56 mm and 22.5 ° C. Soil pedon moisture and temperature regimes indicate an aridic and hyperthermic status. Geologically, approximately 8 km^2 of the area is covered by fine-grained flood sediments (Q_f); in which aeolian sediments have lower contribution and areal extension, instead, most sediments are originated from river overflows during significant flood events. Most notably, signatures of rill and river erosion signify sequential concentrated surface runoff. Some areas with fertile soils have gone under agricultural activities, while the sporadic presence of *Tamarix* and *Haloxylon* species can be seen in other areas (Fig. 7.2A), which has led to the formation of the current Nebkas (Fig. 7.2B). Moreover, 12.22 km^2 is covered by the mixed aeolian and flood sediments (Q_{fs}). On the other hand, 19.12% of the entire region is covered by aeolian sediments that are sporadically scattered across western and eastern parts. The constituent granules of the aeolian sediments in the region vary from fine sands to coarse silts, forming sand dunes reaching up to 3 m tall (Fig. 7.2C).

7.5.2 Identifying land types/component in Niatak

Geologic processes above and below the Earth's surface have shaped the current landscape and physiographic units. In this regard, the land types of the study area can be categorized into floodplains (7), miscellaneous lands (sand dunes, X), and complex (C, 7) (Table 7.2, Fig. 7.3). Floodplain land type, as the name implies, lays on an almost flat surface with varied

FIGURE 7.2 Sediment and vegetation status in the study area (A): presence of *Tamarix* and *Haloxylon* species, (B): formation of Nebkas around local species, and (C): aeolian sediments signifying severe wind erosion status in the region). *photo taken by authors.*

TABLE 7.2 The characteristics of land units in the study area.

Land type	Land unit	Land component	Description	Area (ha)
7	7.1.1	7.1.1	Sparse vegetation cover with moderate developed soil (Typic Haplocambids), loamy to sandy soil texture, lithological discontinuity, alluvial sediments with different particle sizes	441,516.13
		7.1.2	Different ranges of vegetation density with moderate to well-developed soil (Typic Torifluvents to Sodic Aquicambids and Fluventic Haplocambids), loamy to sandy soil texture, lithological discontinuity, alluvial sediments with different particle sizes	219,089.231
		7.1.3	Dense vegetation cover with low erosion, loamy soil texture with well-developed soil pedon (Typic Natrargids, Sodic Aquicambids and Fluventic Haplocambids)	197,991.416
		7.1.4	Cultivated lands with deep to moderately developed soils and evidence of disturbance	66,038.14
X	X.1	X.1.1	Unstable sand dunes with shallow to semi-deep soils and sandy soil texture (Typic Torripsamment to Typic Torifluvents)	907,741.477
		X.1.2	Stabilized sand dunes treated by vegetation and mulch, shallow to semi-deep soils with sandy soil texture (Typic Torripsamment to Typic Torifluvents)	713,805.11
C (7, X)			Mixed sand dunes and floodplain land units with different sediments, medium to high soil depth, and loamy to sandy soil texture (typic Haplocambids)	795,018.1221

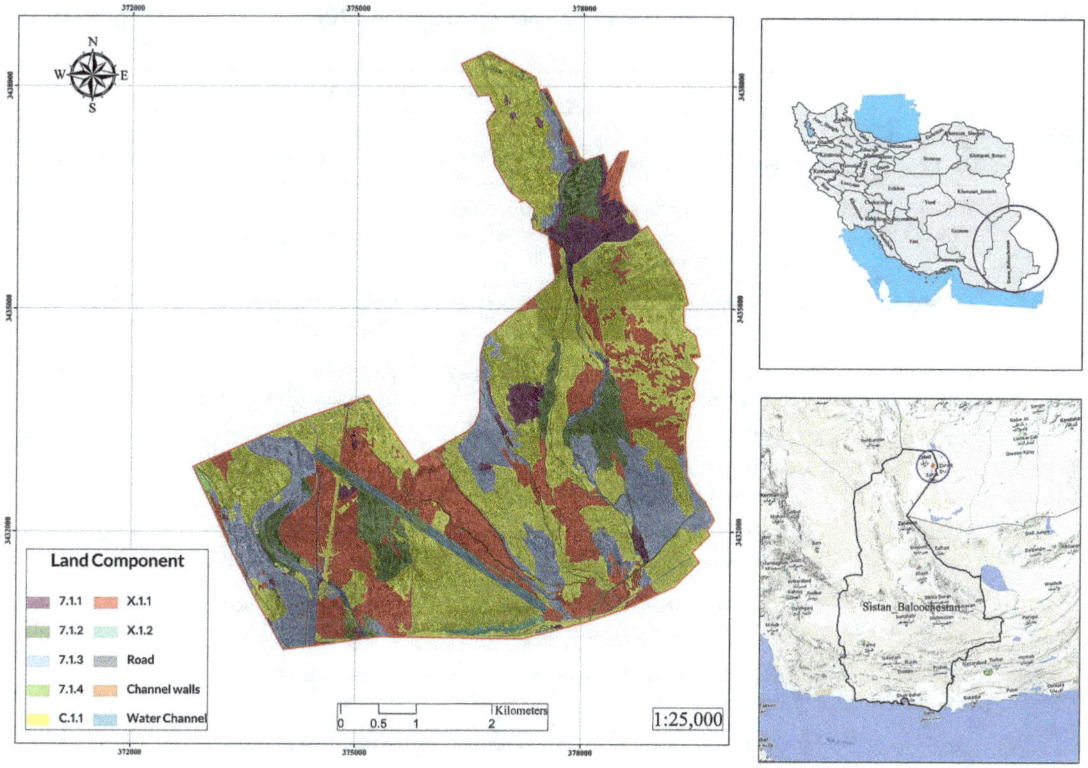

FIGURE 7.3 Land component map of the study area.

shapes and sizes, through which active streams transmit the flood water across the area offering lands for agriculture as far as the soil/water quality allows it to do so. Some areas with sparse vegetation may also be exposed to rill and channel erosion. Miscellaneous lands (X) are the units that characteristically do not fall within the provided definition categories for physiographic units and components, which includes sand dunes (Fig. 7.3, Table 7.2). Complex land type (C) includes both sand dunes and floodplains where a mixture of wind-driven sands and fine-grained flood sediments is discernible.

We used orthophoto obtained from UAV at the scale of 1:500 to extract highly differentiated units, based on land types, land units, and land components. The latter units were scrutinized through multiple field visits and soil pedons were excavated accordingly.

7.5.3 Soil classification in Niatak

The results of soil classification are presented in Table 7.3, summarizing the appearance of the land, soil properties, pedon's characteristics, that is, color, texture, structure, concentration of materials (sodium, gypsum, and lime), and soil moisture and temperature regimes (Table 7.3). Based on this, the soils of the studied area are classified into Aridisols and Entisols orders. Fig. 7.4 shows some examples of soil pedons in the region. The soil infiltration rate in

TABLE 7.3 Soil classification of the study area based on soil taxonomy (2014) and WRB (2014, Updated 2015).

Pedon no.	UTM coordinates (X, Y)	Soil classification		
		Soil taxonomy (Soil Survey Staff, 2014)	**WRB (2014, Updated, 2015)**	**Soil series**
1	374,934.0 3,430,732.2	Coarse loamy layer laid on sandy, aniso, mixed, active, calcarouse, hyperthermic, typic Haplocambids	Fluvic Cambisols (Loamic, Raptic)	Mixed flood
2	376,405.8 3,430,907	Mixed, active, calcarouse, hyperthermic, typic Torripsamments	Calcaric Arenosols (Raptic, Aridic)	Sand dune
3	376,902.0 3,430,752.8	Coarse loamy, mixed, active, calcarouse, hyperthermic, typic Haplocambids	Fluvic Cambisols (Loamic, Aridic, Turbic)	Mixed flood
4	377,112.5 3,431,624.2	Fine loamy, mixed, semiactive, calcarouse, hyperthermic, fluventic Haplocambids	Fluvic Cambisols (Loamic, Aridic, Turbic)	Mixed flood
5	375,120.0 3,433,465.0	Coarse loamy, mixed, active, calcarouse, hyperthermic, fluventic Haplocambids	Fluvic Cambisols (Loamic, Aridic, Turbic)	Mixed flood
6	375,756.8 3,432,230.2	Coarse- loamy, mixed, active, nonacid, hyperthermic, Sodic Aquicambids	Gleyic, Fluvic, Sodic Cambisols (Siltic, Aridic)	Water channel- alkali
7	377,743.0 3,431,628.7	Mixed, active, nonacid, hyperthermic, typic Torripsamments	Arenosols (Aridic)	Sand dune
8	378,388.3 3,431,492.4	Mixed, active, nonacid, hyperthermic, typic Torripsamments	Arenosols (Aridic)	Sand dune
9	379,520.0 3,431,928.0	Coarse loamy, mixed, active, calcarouse, hyperthermic, typic Haplocambids	Fluvic Cambisols (Loamic, Aridic, Turbic)	Mixed flood
10	378,763.6 3,433,608.6	Coarse loamy, mixed, semiactive, hyperthermic, typic Torifluvents	Sodic Fluvisols (Loamic, Aridic)	Young sandy soil and alluvial
11	378,763.6 3,433,608.6	Coarse loamy, mixed, active, hyperthermic, typic Torifluvents	Fluvisols (Loamic, Aridic)	Flood alluvial
12	377,627.6 3,433,630.9	Coarse loamy, mixed, active, calcarouse, hyperthermic, typic Haplocambids	Cambisols (Loamic, Aridic, Turbic)	Mixed flood
13	378,565.9 3,435,043.3	Coarse loamy, mixed, superactive, calcarouse, hyperthermic, Sodic Haplocambids	Sodic Cambisols (Loamic, Aridic)	Alkali
14	378,253.2 3,436,006.1	Coarse loamy, mixed, superactive, calcarouse, hyperthermic, typic Haplocambids	Fluvic Cambisols (Loamic, Aridic, Turbic)	Mixed flood
15	376,884.2 3,434,533.6	Mixed, active, nonacid, hyperthermic, typic Torripsamments	Arenosols (Aridic, Turbic)	Stabilized dune

TABLE 7.3 Soil classification of the study area based on soil taxonomy (2014) and WRB (2014, Updated 2015).—cont'd

Pedon no.	UTM coordinates (X, Y)	Soil classification		
		Soil taxonomy (Soil Survey Staff, 2014)	WRB (2014, Updated, 2015)	Soil series
16	376,824.2 3,432,162.5	Fine loamy, mixed, active, hyperthermic, typic Torifluvents	Fluvisols (Loamic, Aridic)	Flood alluvial
17	376,873.7 3,432,490.6	Silt loamy soil laid on sandy loam, aniso, mixed, active, hyperthermic, typic Torifluvents	Fluvisols (Loamic, Turbic, Aridic)	Flood alluvial
18	377,006.4 3,437,925.5	Fine loamy, mixed, active, calcarouse, hyperthermic, typic Haplocambids	Fluvic Cambisols (Loamic, Aridic)	Mixed flood
19	377,145.5 3,436,744.0	Fine- loamy, mixed, semiactive, calcarouse, hyperthermic, typic Natrargids	Haplic Solonetz (Loamic, Aridic)	Natric
20	377,878.0 3,436,749.0	Loamy soil laid on sandy, aniso, mixed, calcareous, hyperthermic typic Torifluvents	Fluvisols (Loamic, Aridic)	Flood alluvial

the study area varied from I0 (infiltration rate >2 cm h^{-1}) in the Young sand dune soil series (pedon NO. 8) to mixed flood soil series (pedon NO. 12) with the infiltration rate <0.2 cm h^{-1}.

7.5.4 Land suitability in Niatak

It is noteworthy that Niatak is considered a geomorphologically and physiographically homogeneous region. Therefore, pedological attributes (i.e., soil series) were substituted to differentiate the area into different units. Considering the overall field surveys, visual interpretations on the orthophoto images, physiographic units, and the changes in pedogenic and vegetation variables, eight soil series were identified and their corresponding boundaries were delimited based on the auxiliary laboratory data of the control soil pedons (Fig. 7.5). Results attest to the reliability of the interpretations conducted on UAV orthophoto as the obtained land unit and soil series boundaries are highly consistent with the reality and soil pedon data. In the denomination of soil series, it was strived to adopt the elements and features observed during field visits and UAV images as well as the inferences acquired from laboratory analysis. Hence, the assigned names to soil series borrow some distinct features such as soil development, vegetation status (e.g., density), and accumulation status of the sediments (Fig. 7.5).

Every land bears some capabilities and limitations based on its distinct characteristics. As the number of restrictions increases and the quality conditions become more and more severe, the capability of the land will be lower. Our findings revealed that the most important limitations in the Niatak region are the lack of irrigation water sources (surface and groundwater), climatic conditions (succession of droughts, high temperature, and evaporation, and

FIGURE 7.4 Examples of the studied pedons in different soil series across the study area. *photo taken by authors.*

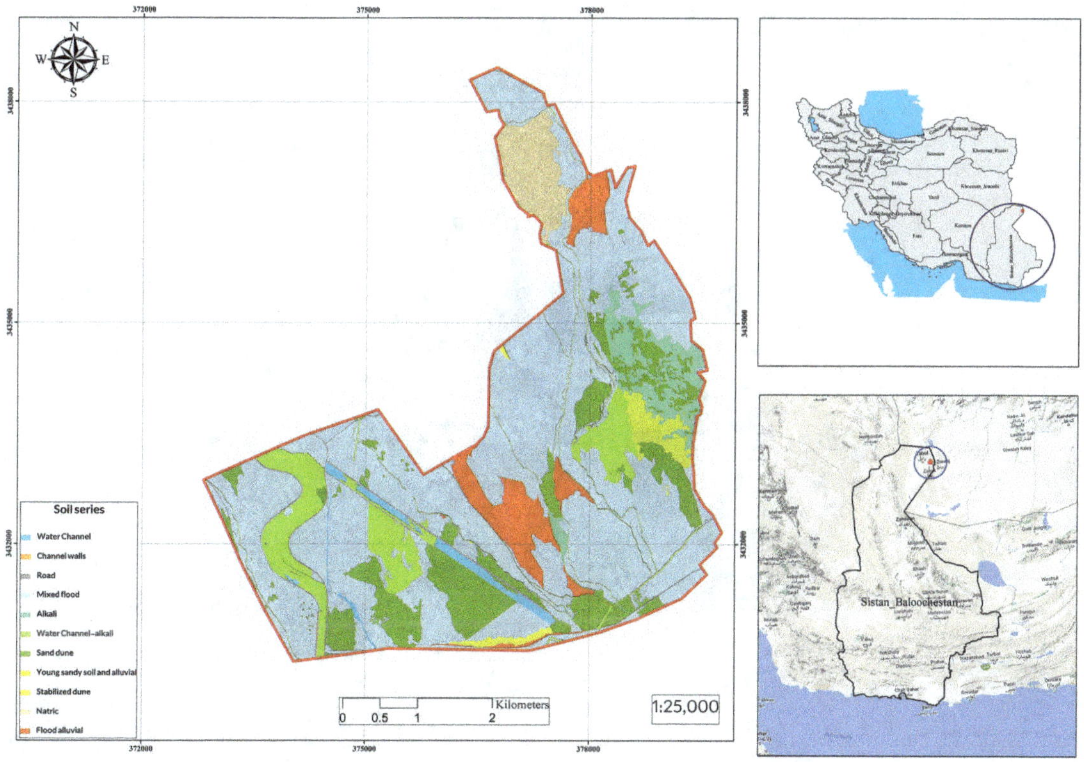

FIGURE 7.5 Soil series map of the study area.

reduction of annual rainfall), irregular stratification of sediments, and high salinity and alkalinity in some parts. The area is mostly characterized by light-textured soil and aeolian deposits. Therefore, the aforementioned classification can be reiterated as: climatic limitations, soil type limitations, salinity and alkalinity limitations, floods, soil erosion, and food poverty, among which are some recoverable instances such as water erosion and flood control, while climate and some soil restrictions are not recoverable or their improvement is not cost-effective whatsoever. Ultimately, the soil restriction map was produced, based on which different parts of the study area are susceptible to moderate-to-severe restrictions (Fig. 7.6).

The most important advantage of using the UAV derivatives (orthophoto image and DEM) is that they enable us to well differentiate the area into different distinct units within which studying only one pedon would suffice. In other words, the surveyor should once and for all visit the representative units, and that will cover the entire area in the shortest possible operating time. Hence, field surveys can be conducted in a more cost- and time-effective manner, and more time can be devoted to desk studies. This would also facilitate the assignment of management strategies and plans and the required budget. In general, since the area is exposed to severe droughts and water shortage, supplying the irrigation water would not be a wise decision and an in-depth analysis would dictate that land utilization should be synced to each unit's inherent potential. In such a way, the current excessive and unbridled

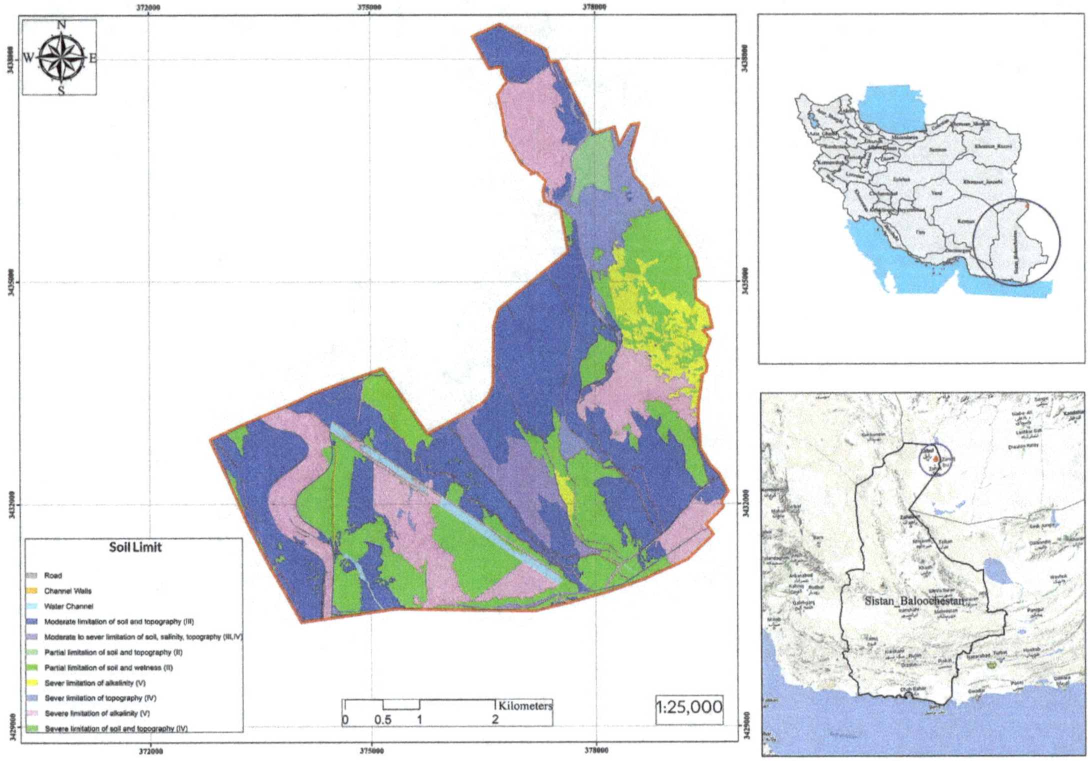

FIGURE 7.6 Soil limitation map based on the excavated soil pedons in the study area.

exploitation of the area would be decentralized from one specific area and land use to different locations and various types of activities.

The limitations of soil type mostly include soil texture and permeability. These limitations are often not recoverable, or it is not cost-effective to be improved. Still, the alternative methods for obviating or ameliorating these limitations should be always investigated. It is noteworthy that the state of layering and granularity of the soils in the region, with due respect to different alluvial and aeolian sediments in lower layers, needs to be examined in principle and taken into account in management decision-making. On the other hand, improving soil texture in large areas is not feasible nor cost-effective. It should be noted that in order to stabilize the soil and reduce soil erosion, it is advised to plant local plant species that require less water. Also, using soils of heavier texture with higher clay content would somewhat obviate such limitations. Inorganic or organic mulching methods should also be used to increase soil stabilization, especially in areas covered with loose sands.

The limitation of salinity and alkalinity in the study area is strongly discernible in the alkali and water channel-alkali soil series. Since this part of the land has many other limitations including erosion and sandy soil texture, circumventing or eliminating such a limitation is not possible nor justified. Erosion problems are one of the most effective limitations in the region, especially in areas with severe wind erosion. Therefore, impeding and decelerating erosion and sedimentation cycle are one of the most important goals of mitigation operations in the study area. By doing so, the land class would be more suitable for the main applications. Due to the fact that many of the lands in the study area have mixed aeolian and alluvial sediments, it is very important to consider the risks of soil erosion in the entire study area before undertaking any sort of executive project. For example, in areas with a high erodibility coefficient (erosion-prone areas), manipulation of natural soil or plant cover can be extremely problematic and cause irrecoverable damages. Opting for the most resistant plant species and the means of water harvesting that best suit the chosen species and the climatic conditions of hyper-arid regions such as Niatak would need more in-depth analysis and extensive studies. A combination of mechanical measures such as small dams and divergent weirs in pursuit of flood spreading as well as biological measures can also be advised, where both strategies should adhere scrupulously to a nature-friendly and -adaptive designation and construction rule. Based on these premises, the land suitability map was produced (Fig. 7.7).

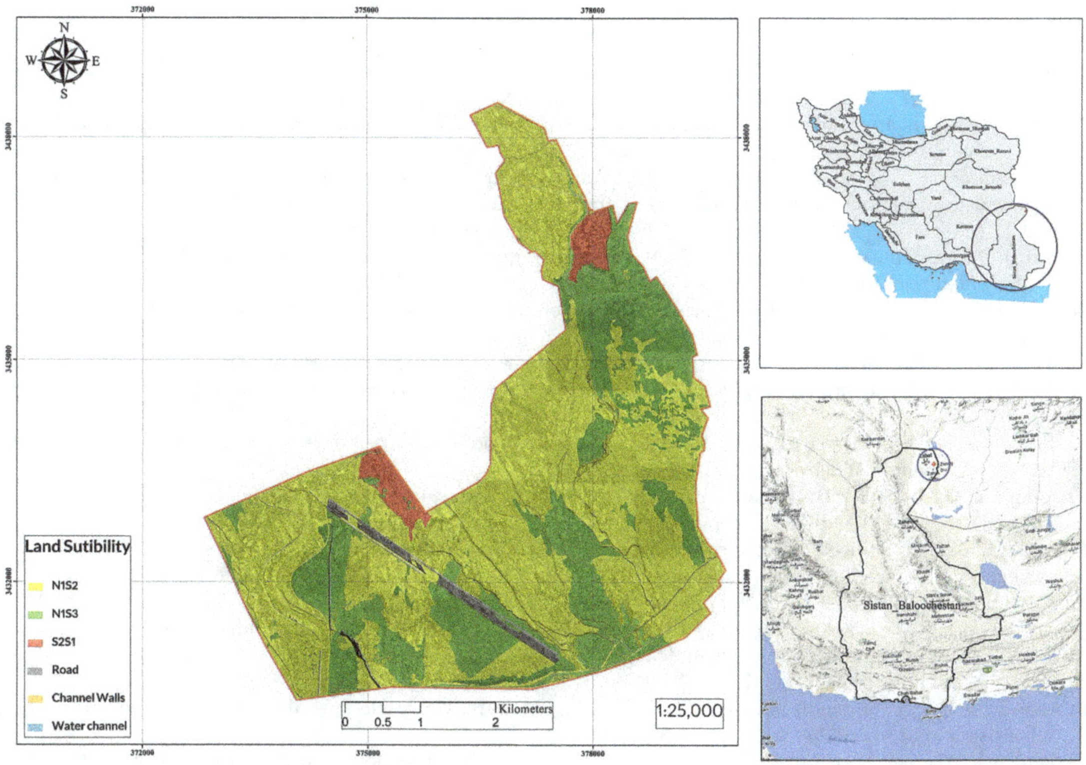

FIGURE 7.7 Soil suitability map in the study area (N1S2: temporarily unsuitable for irrigated agriculture, moderate suitability for revitalizing vegetation; N1S3: temporarily unsuitable for irrigated agriculture, low suitability for the revitalization of vegetation; S2S1: moderate suitability for agriculture, good suitability for revitalizing vegetation).

7.6 Conclusions

The results of preparing soil series and land suitability maps availed by UAV-derived orthophoto and terrain attributes at a spatial resolution of 1 m showed that superimposition of thematic layers and delimiting the boundaries, despite the limited soil pedons, were carried out with high accuracy and high agreement with reality. Hence, the preparation of a land suitability map by conglomerating ground- and UAV-derived data can better (i.e., more accurately and swiftly) reveal the limitations of the region and expedite the process of sustainable development by leading the experts in the right direction. Also, the use of cLHS methods for soil sampling based on the existence of data and accurate auxiliary variables proved to give a cost- and time-effective assessment of the region. Therefore, combining the use of accurate data and expertise in soil science can lead to better results.

Acknowledgments

We would like to thank Abbas Goli Jirandeh, Hadi Alizadeh, Amir Alizadeh, and Mehdi Karimi of *Spatial Science Innovators Co.* for putting meticulous effort for the UAV flight missions and providing us with high-resolution photogrammetry products.

References

Amini, S., Gandomkar, A., Bagheri Bodaghabadi, M., 2016. Ecological capability evaluation of Ghara-Ghach dam area to establish a tourism region. Desert 21 (2), 193–204.

Bandyopadhyay, S., Jaiswal, R.K., Hegde, V.S., Jayaraman, V., 2009. Assessment of land suitability potentials for agriculture using a remote sensing and GIS based approach. Int. J. Remote Sens. 30 (4), 879–895.

Bohner, J., Antonic, O., 2009. Land-surface parameters specific to topo-climatology. Dev. Soil Sci. 33, 95–226.

Bouma, J., Montanarella, L., Evanylo, G., 2019. The challenge for the soil science community to contribute to the implementation of the UN Sustainable Development Goals. Soil Use Manag. 35 (4), 538–546.

Bouwer, H., 1986. Intake rate: cylinder infiltrometer. Methods of Soil Analysis: Part 1 Physical and Mineralogical Methods, vol. 5, pp. 825–844.

Bremner, J.M., Mulvaney, C.S., 1982. Nitrogen-total. In: Page, A.L., Miller, R.H., Keeney, D.R. (Eds.), Methods of Soil Analysis. Part 2. Chemical and Microbiological Properties. Soil Science Society of America, Madison, WI, USA, pp. 595–624.

Castillo, V.M., Martinez-Mena, M., Albaladejo, J., 1997. Runoff and soil loss response to vegetation removal in a semi-arid environment. Soil Sci. Soc. Am. J. 61 (4), 1116–1121.

Chapman, H.D., 1965. Cation-exchange capacity. Methods of Soil Analysis: Part 2 Chemical and Microbiological Properties, 9, pp. 891–901.

Conrad, O., Bechtel, B., Bock, M., Dietrich, H., Fischer, E., Gerlitz, L., Wehberg, J., Wichmann, V., Böhner, J., 2015. System for Automated Geoscientific Analyses (SAGA) v. 2.1.4. Geosci. Model Dev. 8, 1991–2007.

de Almeida, C.T., Galvao, L.S., Ometto, J.P.H.B., Jacon, A.D., de Souza Pereira, F.R., Sato, L.Y., Lopes, A.P., de Alencastro Graça, P.M.L., de Jesus Silva, C.V., Ferreira-Ferreira, J., Longo, M., 2019. Combining LiDAR and hyperspectral data for aboveground biomass modeling in the Brazilian Amazon using different regression algorithms. Remote Sens. Environ. 232, 111323.

dOleire-Oltmanns, S., Marzolff, I., Peter, K.D., dan Ries, J.B., 2012. Unmanned aerial vehicle (UAV) for monitoring soil erosion in Morocco. Remote Sens. 4, 3390–3416.

Deblauwe, V., Barbier, N., Couteron, P., Lejeune, O., Bogaert, J., 2008. The global biogeography of semi-arid periodic vegetation patterns. Global Ecol. Biogeogr. 17 (6), 715–723.

Edokpayi, A.A., Agho, C.A., Adeh, S.A., Okpamen, S.U., 2017. Comparison of the different hierarchical clustering techniques for the classification of soils under oil palm in Nigeria. Int. J. Basic Sci. Technol. 3 (1), 37–46.

[FAO] UN Food and Agriculture Organization, 1976. A framework for land evaluation. Soils Bull. 32, 71.

Gallant, J.C., Dowling, T.I., 2003. A multi resolution index of valley bottom flatness for mapping depositional areas. Water Resour. Res. 39, 1347–1360.

Gee, G.W., Bauder, J.W., 1986. Particle-size analysis. Methods of Soil Analysis. Part 1, pp. 383–411.

Kariminejad, N., Shariati, M., Hosseinalizadeh, M., Pourghasemi, H.R., Maleki, S., 2022a. Investigating geometrical characteristics of collapsed pipes and the changing role of driving factors. J. Environ. Manag. 312, 114910.

Kariminejad, N., Pourghasemi, H.R., Maleki, S., Hosseinalizadeh, M., 2022b. Digital soil mapping and modeling in Loess-derived soils of Iranian Loess Plateau. Geocarto Int. 37 (26), 11633–11651.

Knudsen, D., Peterson, G.A., Pratt, P.F., 1982. Lithium, sodium and potassium. Chapter 13. Methods of Soil Analysis, Part 2. Chemicals and Microbiological Properties, pp. 225–246.

Krenz, J., Greenwood, P., Kuhn, N.J., 2019. Soil degradation mapping in drylands using unmanned aerial vehicle (UAV) data. Soil Syst. 3 (33).

Maleki, S., Khormali, F., Bodaghabadi, M.B., Mohammadi, J., Hoffmeister, D., Kehl, M., 2020. Role of geomorphic surface on the above-ground biomass and soil organic carbon storage in a semi-arid region of Iranian loess plateau. Quat. Int. 552, 111–121.

Minasny, B., McBratney, A.B., 2006. A conditioned Latin hypercube method for sampling in the presence of ancillary information. Comput. Geosci. 32 (9), 1378–1388.

Moore, I.D., Gessler, P., Nielsen, G., Peterson, G., 1993. Soil attribute prediction using terrain analysis. Soil Sci. Soc. Am. J. 57, 443–452.

Nelson, D.W., Sommers, L.W., 1982. In: Page, A.L., Miller, R.H., Keeney, P.R. (Eds.), Organic Carbon in: Method of Soil Analysis Part 2. Agronomy Monograph No. 9. American Society of Agronomy, pp. 561–573.

Olaya, V.F., 2004. A Gentle Introduction to Saga GIS. The SAGA User Group e.V, Göttingen, Germany, p. 208.

Olsen, S., 1982. Phosphorus. Chemical and Microbiological Properties, pp. 416–418.

Page, A.L., Miller, R.H., Keeney, D.R., 1982. Methods of Soil Analysis. Part 2. American Society of Agronomy, vol. 4. Soil Science Society of America, Madison, WI, USA, pp. 167–179.

Rhoades, J.D., 1983. Soluble salts. Methods of Soil Analysis: Part 2 Chemical and Microbiological Properties, vol 9, pp. 167–179.

Rietkerk, M., Ketner, P., Burger, J., Hoorens, B., Olff, H., 2000. Multiscale soil and vegetation patchiness along a gradient of herbivore impact in a semi-arid grazing system in West Africa. Plant Ecol. 148, 207–224.

Roudier, P., Brugnard, C., Beaudette, D., Louis, B., 2020. Conditioned Latin Hypercube Sampling. R Package Version 0.7-3. https://github.com/pierreroudier/clhs/.

Saco, P.M., Willgoose, G.R., Hancock, G.R., 2007. Eco-geomorphology of banded vegetation patterns in arid and semi-arid regions. Hydrol. Earth Syst. Sci. 11 (6), 1717–1730.

Schlesinger, W.H., Raikes, J.A., Hartley, A.E., Cross, A.F., 1996. On the spatial pattern of soil nutrients in desert ecosystems: ecological archives E077-002. Ecology 77 (2), 364–374.

Sishodia, R.P., Ray, R.L., Singh, S.K., 2020. Applications of remote sensing in precision agriculture: a review. Remote Sens. 12 (3136), 1–31.

Soil Survey Staff, 2014. Keys to Soil Taxonomy, 12th. USDA-Natural Resources Conservation Service, Washington, DC, USA.

Thomas, G.W., 1996. Soil pH and soil acidity. Methods of Soil Analysis: Part 3 Chemical Methods, vol 5, pp. 475–490.

Tucker, B.B., Kurtz, L.T., 1961. Calcium and magnesium determinations by EDTA titrations. Soil Sci. Soc. Am. J. 25 (1), 27–29.

Van Huyssteen, C.W., Turner, D.P., Le Roux, P.A.L., 2013. Principles of soil classification and the future of the South African system. S. Afr. J. Plant Soil 30 (1), 23–32.

Vasu, D., Singh, S.K., Ray, S.K., Duraisami, V.P., Tiwary, P., Chandran, P., Nimkar, A.M., Anantwar, S.G., 2016. Soil quality index (SQI) as a tool to evaluate crop productivity in semi-arid Deccan plateau, India. Geoderma 282, 70–79.

Watts, A.C., Ambrosia, V.G., Hinkley, E.A., 2012. Unmanned aircraft systems in remote sensing and scientific research: classification and considerations of use. Remote Sens. 4 (6), 1671–1692.

World Reference Base for Soil Resources (WRB), 2014. Update 2015. Food and Agriculture Organization of the United Nations, p. 204.

Wynn, T., Mostaghimi, S., 2006. The effects of vegetation and soil type on streambank erosion, southwestern Virginia, USA 1. J. Am. Water Resour. Assoc. 42 (1), 69–82.

Yue, J., Yang, G., Tian, Q., Feng, H., Xu, K., Zhou, C., 2019. Estimate of winter-wheat above-ground biomass based on UAV ultrahigh-ground-resolution image textures and vegetation indices. ISPRS J. Photogrammetry Remote Sens. 150, 226–244.

Zhu, W., Sun, Z., Yang, T., Li, J., Peng, J., Zhu, K., Li, S., Gong, H., Lyu, Y., Li, B., Liao, X., 2020. Estimating leaf chlorophyll content of crops via optimal unmanned aerial vehicle hyperspectral data at multi-scales. Comput. Electron. Agric. 178, 105786. https://doi.org/10.1016/j.compag.2020.105786.

Zhu, W., Rezaei, E.E., Nouri, H., Yang, T., Li, B., Gong, H., Lyu, Y., Peng, J., Sun, Z., 2021. Quick detection of field-scale soil comprehensive attributes via the integration of UAV and sentinel-2B remote sensing data. Rem. Sens. 13 (22), 4716. https://doi.org/10.3390/rs13224716.

CHAPTER

8

Investigating the land use changes effects on the surface temperature using Landsat satellite data

Hadi Eskandari Damaneh, Hassan Khosravi and Hamed Eskandari Damaneh

Department of Reclamation of Arid and Mountainous Regions, Faculty of Natural Resource, University of Tehran, Tehran, Iran

8.1 Introduction

Urban heat islands, the result of urban climate, are one of the most critical environmental challenges in the current century (Wang, 2022). So that this phenomenon is considered one of the emerging environmental hazards caused by the expansion of urbanization. Due to the formation of these heat islands, cities usually experience higher temperatures than their surrounding areas. Cities usually experience higher air temperatures than their surrounding areas due to the construction of these heat islands (Zhu et al., 2022; Guo et al., 2020). In recent decades, heat island effects have been one of the reasons for significant changes in the weather (Fenner et al., 2019). This phenomenon occurs when a large percentage of natural surface covers are destroyed and replaced by roads and other urban facilities (Rahaman et al., 2022). This extensive change in the Earth's surface causes extensive changes in surface reflection, heat retention, and heat transfer or, in other words, significant effects on the local climate (Duveiller et al., 2018). So, an urban area temperature becomes higher than in the surrounding areas. One of the effects of this phenomenon is the temperature difference of several degrees Celsius compared to the surrounding regions (Lazzarini et al., 2013). The change in use and increased construction by humans prevent wind flow, which helps maintain hot air masses and causes events like urban heat islands (Molina et al., 2022). According to this, the urban heat island is a closed isothermal line that represents a hot area, especially areas such as cities and towns where humans have disturbed the natural state of the Earth with their actions. Urbanization increases the intensity and frequency of urban heat islands effects on

Remote Sensing of Soil and Land Surface Processes
https://doi.org/10.1016/B978-0-443-15341-9.00007-1

highly developed urbans and cities (Jato-Espino et al., 2022). In other words, the urban heat island is a phenomenon in which the land surface temperature (LST) in an urban area is significantly higher than the local climate of big cities (Naikoo et al., 2022).

Assessing changes in the surface temperature of each region can reflect changes in urbanization, industrial activities, population changes, and natural factors (Eludoyin et al., 2019). Rapid urbanization causes severe problems in the urban area. The urban heat island effects have attracted a plenty attention due to their enlarged impact on their negative role in urban viability, climate changes, and sustainable development (Chen et al., 2022).

The main reason for the special attention of urban climatologists to this phenomenon is its unfavorable consequences. In other words, this phenomenon not only affects the peace and health of city residents but also affects environmental conditions, such as the quality of air, biodiversity, and energy consumption. Investigating the dispersion of thermal radiation and its relationship with the land use type is very important in understanding the microclimate of urban areas (Kang et al., 2022). Therefore, special considerations have been paid for controlling land use and land cover changes.

Land use change means a change in the land use type, which is not necessarily a change in the surface of the land but also a change in the density and management of the land. In other words, the land use change is considered to include the change in land use type and the distribution and spatial patterns of activities and work (Eskandari Damaneh et al., 2022). Therefore, the study of these islands is very important. In other words, the change in land use is considered to include the change in land use type and the distribution and spatial patterns of activities and work (Eskandari Dameneh et al., 2022). Therefore, the study of these islands is very important.

Investigation of urban heat islands by remote sensing data to estimate LST is a relatively new method that significantly reduces the cost of temperature estimation compared to traditional methods (Lyu et al., 2022). Thermal remote sensing data have enabled it to dynamically monitor and evaluate urban heat islands (Halder et al., 2022). These data can continuously have a general view of the whole city, which is especially important for the detailed investigation of the city's climate (Cocci Grifoni et al., 2022).

Investigating the effects of increasing the urban heat island considered as one of the main studies concerning environmental changes caused by urbanization using remote sensing. In this regard, researchers have conducted various studies, some of which are mentioned below. Chen et al. (2006), by investigating the relationship between land use change and urban heat island in Guangdong province, found that heat islands have become more evident in urban areas. Jiang and Tian (2010) investigated the changes in land use/land cover on surface temperature. Its results showed that the evolution of land use is an essential factor in increasing the Earth's surface temperature. There is a high temperature in areas with scattered vegetation and low in areas with dense vegetation. Ning (2011) investigated the coastal land use changes effect on LST using Landsat's TM sensor. The results showed that the temperature of the Earth's surface increases with distance from the sea. Polydoros et al. (2018) quantified the trend of ground surface temperature and surface urban heat islands severity in Mediterranean area. The results of their research showed that the trend of surface temperature in urban areas is increasing over time. Eludoyin et al. (2019) assessed the temperature of land surface by remote sensing in southwestern Nigeria. Their research results showed that there was an overall increase in mean LST, with an average rise of 1.43°C between 2004 and 2015.

This increase in LST has occurred due to changes in land use/land cover of cities. Also, their research results showed that different land uses experienced various degrees of LST in both wet and dry seasons. Their temperature difference is approximately 2°C in dry seasons compared to wet seasons. Moisa et al. (2022) investigated heat islands caused by land cover/land use changes in the city of Jima in southwest Ethiopia; their research results showed that with the reduction of vegetation cover, the surface temperature of the Earth has increased from 12.3% in 2000 to 10.8% in 2020. Also, the results showed the negative relation between vegetation cover and temperature of ground surface. The average surface temperature in this region has increased from 24.5°C in 2000 to 26.3°C in 2020. Chen et al. (2022) investigated land use change and its effect on Wuhan, China's spatial and temporal distribution of heat islands. Based on their research the built-up areas increased from 32.81% to 46.01% between 2007 and 2020, and agricultural land decreased by 12.14% during this period. This land use has changed to the land use of residential and artificial areas. The average of the earth's surface temperature has increased from 25.92°C to 31.71°C having an upward trend. The relationship between vegetation and the Earth's surface temperature is negative. Carrillo et al. (2022) investigated deforestation's temporal and spatial changes and their effects on the formation of urban heat islands using Landsat satellite images. Their research showed that after deforestation, the surface temperature of the Earth has increased by 3.94°C. Most of the heat islands occurred in areas converted from forest to residential areas. In Khan et al. (2022) investigated the effects of land use change on the surface temperature of Pakistan. They stated that the changes in land use from 1990 to 2019 have significantly affected the LST of this region. So that LST has increased from 20.1, 21.5, and 17.1°C to 32.1, 35.5, and 28.2°C in residential areas, barren land, and vegetation, respectively.

Due to the availability of amenities in the city of Tehran, the capital of Iran, in recent years, we have seen an increase in immigration from many regions and neighboring provinces to this province. A significant change of use has been made in this city to provide comfortable facilities for people. This change in land use has caused a decrease in range lands, barren lands, and agricultural lands. The use of urban and manufactured areas has increased, while urban and manufactured areas have increased. The consequences of these extensive land use changes in this area are becoming more evident daily. Therefore, the present study tries to evaluate the process of land use changes in Tehran and its effect on the formation of heat islands.

8.2 Remote sensing

Different environmental phenomena provide different responses to electromagnetic radiation. But what is clear is that the optical interaction of phenomena is summarized in the absorption, passage, and reflection of electromagnetic energy. The basis of remote sensing in the detection of phenomena is based on these characteristics and mainly based on the amount of reflected energy. Remote sensing is the art of obtaining information about phenomena from a distance and in other words, collecting data from areas that cannot be directly accessed due to being dangerous, high cost, and distance (Abdollahi et al., 2021). This technology is one of the effective tools in the field of environmental studies and Earth sciences. Knowing many sources of soil, water, and vegetation and monitoring hazardous phenomena such as floods, desertification, water and wind erosion, dust storms, movement of dunes, water and soil salinity, destruction of forests and range lands, etc., is possible using remote sensing. Application of new technology such as remote sensing most of the times increases accuracy and speed and reduces costs. Day by day, the importance of this technology is growing in line with sustainable development. Satellites and sensors can provide information in different dimensions, scales, and electromagnetic spectrums (bands) (AlaviPanah, 2010; Arekhi et al., 2019).

The primary purpose of remote sensing is to extract useful information by remote sensing data. Extraction of information can be done by interpretation, digital analysis of data, and combining both methods. Useful digital information can be obtained through classification, identification of complications, examination of changes, extraction of physical quantities (such as temperature measurement, amount of atmospheric compounds, altitude, slope, etc.), and also an examination of statistical relationships between natural phenomena and their spectral values. One of the conventional methods in the field of information extraction from satellite images is the calculation of spectral indices. In simple words, the spectral index consists of combining bands using mathematical relationships and extracting new information from satellite images. This information cannot be obtained from any of the bars alone.

One of the most essential and prominent sensors is ETM+, TM, and OLI in the world today. These sensors were put into orbit on the Landsat satellite in 1982, 1999, and 2013, respectively. Landsat plays a vital role in developing Earth systems models. It enables us to accurately predict global changes, that can be useful for policymakers and managers in environmental conservation and management of optimal resource (Khosravi et al., 2017).

8.3 Application of remote sensing in land use evaluation

The increase in population and the provision of housing and food for this growing population have caused changes in land use in different regions (Shirazi and Kazemi, 2017). Land use changes affect diverse ecosystems, including arid and semi-arid regions (Eskandari et al., 2020). Among these effects in these areas, we can mention the change in local and regional temperature regimes, the increase in the amount of solar radiation and the emission of greenhouse gases (Bazgir et al., 2020), the loss of biodiversity (Amini parsa et al., 2020; Hosseini et al., 2020), reducing the surface quantity and quality and underground water, disrupting the physical and chemical properties of the soil (Eskandari Damaneh et al., 2020). Planning for the protection and development of these areas requires monitoring their changes over time. Monitoring land use changes in ecosystems leads to a correct understanding of how humans and the environment interact (Etemadi et al., 2018). This issue is very important in the case of cities due to their continuous changes for multiple uses.

The use of remote sensing techniques is one of the new and efficient techniques in the field of monitoring changes in the use of ecosystems (Abbaszadeh Tehrani et al., 2022). Remote sensing data provide researchers with up-to-date information on land cover and land use. This information is very useful in spatial analysis and simulation of land changes in different time frames (Pan et al., 2020). Compared with traditional methods, these data provide more significant amounts of land use information at a geographic location that is cost-effective in terms of time and cost at the regional scale (Islam et al., 2018). Compared with traditional methods, these data provide more significant amounts of land use information at a geographic location that is cost-effective in terms of time and cost at the regional scale (Islam et al., 2018; Hoque et al., 2022). This technology is a useful science that can obtain various information about how land is used. It has advantages such as creating a wide and holistic view of an area, accessing information easily, repeating the process, and producing high-quality data. These advantages make this technology more desirable than other methods for studying and managing land use changes (Wicke et al., 2012; Panuju et al., 2020). The new common point of view in land use investigation and monitoring is using various spectral indices. Land-use change indices are mathematical transformations designed to evaluate the spectral distribution of multiple land uses in multispectral satellite observations.

Multiple sensors with different temporal and spatial resolutions allow systematic monitoring of ecosystems' other temporal characteristics, providing multilingual measurements at the ground level. In this context, Landsat 7, 8, etc. are sensors that can be used to monitor land use dynamics (Holden et al., 2016; Deng et al., 2019). The main advantage of these sensors is their time resolution, which can be very useful in time series research (Zhu, 2017). The time series products of land use in those sensors differ in terms of temporal, spatial, and spectral resolution. Time series data analysis of vegetation, water, and soil indicators significantly contributes to our understanding of land use changes over time (Girma et al., 2023). A long-term study of land use changes can provide a correct understanding of the response of these land uses to stimuli such as human and climate changes (Mishra et al., 2019). Changes related to the increase and decrease of land use types (Moody and Johnson, 2001) and assessment of habitat destruction (Pettorelli et al., 2005) are among the applications related to the evaluation of time series of land use and vegetation change indicators.

8.4 The application of remote sensing in the assessment of land surface temperature

In the last few decades, due to population growth and technological development, the exploitation of primary resources has increased significantly, which causes excessive pressure on land and can ultimately lead to land destruction. Land degradation, which includes a series of processes, can destroy water resources (Eskandari et al., 2016), soil (Khosravi et al., 2017), and vegetation (Khosravi et al., 2017; Eskandari Damaneh et al., 2018, 2022). One of the most important effects of land degradation is the temperature increase in the Earth's surface, which leads to very harmful effects that may be irreparable. Therefore, this issue is a wake-up call for more protection of ecosystems against destruction and productivity reduction. Although the increase in the Earth's surface temperature may not be as dangerous as earthquakes or landslides, it is considered a great environmental risk.

The increase of the Earth's surface temperature in arid and semi-arid areas is a problem in the direction of unreasonable and unsustainable use of soil and water resources. To continue sustainable development, this phenomenon must be identified, monitored, and managed. An increase in the Earth's surface temperature not only destroys the fertility and dynamism of the environment but also endangers the safety of the residents of an area. In this regard, there are various ways to investigate the rising temperature of the Earth's surface, among which one of these important and influential tools is the technology of remote sensing technology and satellite data. The use of remote sensing technology, in addition to reducing costs compared to traditional methods, increases accuracy and speed; for this reason, the use of this technology is increasing day by day. In general, remote sensing as a powerful tool can play an influential role in identifying and preparing a map of the Earth's surface temperature in a region. In general, the science of remote sensing provides worthy information about the features of the Earth's surface temperature and other related properties by evaluating and processing satellite image data. Satellite images are helpful due to being digital, providing information while providing an all-round view, using different parts of the electromagnetic spectrum to record the characteristics of phenomena, periodicity, transmission speed, and variety of data. The change in land use and, as a result, the increase in the temperature of the Earth's surface can be recognized and interpreted in satellite images due to the high reflections.

8.5 The land use change effects on the land surface temperature of Tehran city: As a case study

Tehran province is located between 35 degrees 14′ to 36 degrees 17′ north latitude and 50 degrees 14′ to 53 degrees 6′east longitude. The Alborz mountain range separates this province from its northern neighbor, Mazandaran, like a long wall. With more than 13 million people, Tehran province has 17.5% of the total population of Iran. Regarding topography, about 57% of the area of Tehran is in mountainous regions. Due to the unique geographical conditions and the slope, direction, and microclimates, it has diverse climatic conditions. The air currents that affect Tehran province include the Mediterranean and Siberian winds and are rarely covered by the influence of the Arabian and Indian Ocean currents. The average temperature of Tehran province is 17°C, reaching 22.1°C in summer and 11.5°C in winter. In terms of precipitation, atmospheric precipitation is the most critical input to catchment basins; on the other hand, precipitation in Tehran province has many changes due to the presence of northern elevations. The average annual rainfall reaches 213.8 mm. The general source of rain in the Iranian plateau is moist currents. The low-pressure centers associated with the fronts pass over the country in the west–east direction for 7 months of the year until the middle of May. Fig. 8.1 shows the studied area, a part of Tehran province with an area of 9943.40 square kilometers.

8.5.1 Satellite data and preprocessing

Determine the changes of land use in Tehran province over 30 years (1990–2020), Landsat sensor images were used. This section uses ENVI5.3, GIS10.3, Excel, and remote sensing data, including satellite images of Landsat 5 and 8, presented in Table 8.1. The images were downloaded from the American Geology website, earthexplorer.usgs.gov, then prepared for preprocessing. After downloading the images, radiometric and atmospheric corrections were performed (Eskandari et al., 2016).

8.5.2 Land use land change

Table 8.1 earthexplorer.usgs.govEskandari et al., 2016 After applying the preprocessing, we prepared the maximum likelihood method of land use classification for the urban area

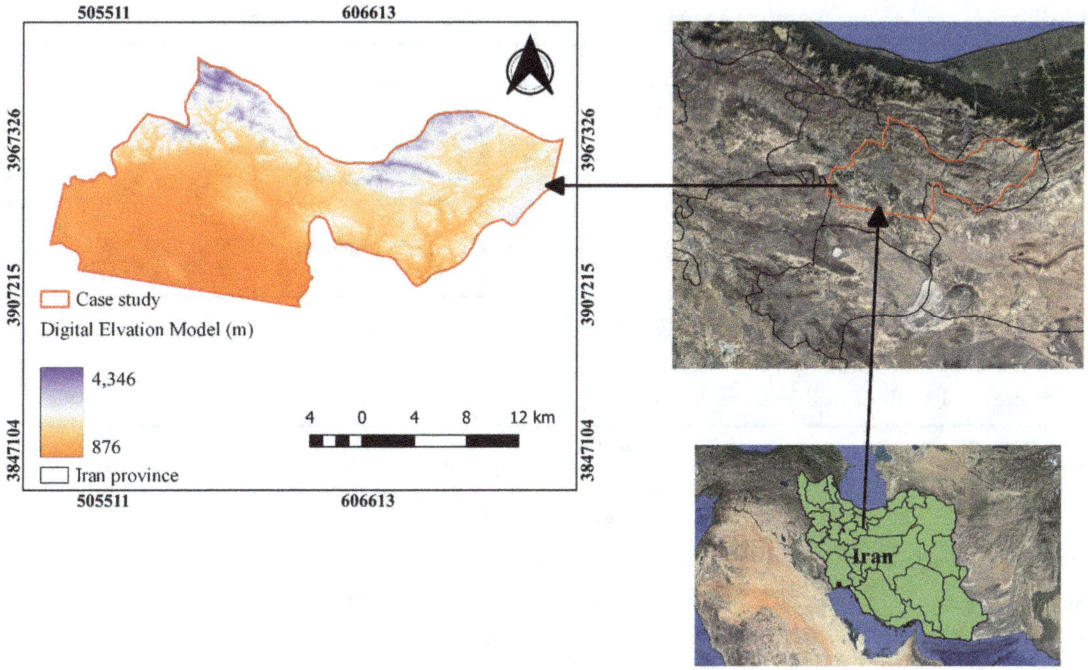

FIGURE 8.1 Location map of the case study.

TABLE 8.1 Specifications related to the images used in this research.

Data	Satellite	Sensor	Path/row	Spatial resolution (m)
1990/05/28	Landsat5	TM	164/35	30
2000/06/08	Landsat5	TM	164/35	30
2010/06/04	Landsat5	TM	164/35	30
2020/06/15	Landsat5	OLI	164/35	30

of Tehran between 1990, 2000, 2010, and 2020 (Eskandari Damaneh et al., 2021). In addition, the NDVI, NDBI, and NDWI auxiliary indices * were utilized (Rugel et al., 2017; Eskandari Damaneh et al., 2020; Eskandari Damaneh et al., 2019) (Table 8.2).

Finally, a map depicting land use was prepared using the maximum likelihood method and the mentioned auxiliary index, dividing the range of land use into five major areas: urban, agricultural land, barren, water bodies, and other lands. The details of each of these users are presented in Table 8.3.

The classified maps were compared with ground reality and accuracy criteria to quantify the classification results. An error matrix was formed, and based on that, the accuracy of the producer and the user, the overall accuracy, and the Kappa coefficient of the classified maps

TABLE 8.2 A description of the Landsat indices that were used in the present research.

Description	Range	Index
Normalized Difference Vegetation Index (NDVI) (Tucker et al., 1986)	Ranges from −1 (negative) to +1 (positive)	$NDVI = \frac{NIR - R}{NIR + R}$
Normalized Difference Water Index (NDWI) (Gao et al., 1986)	Ranges from −1 (negative) to +1 (positive)	$NDWI = \frac{NIR - SWR}{NIR + SWR}$
Normalized Difference Built-up Index (NDBI) (Zha et al., 2003)	Ranges from −1 (negative) to +1 (positive)	$NDBI = \frac{SWR - NIR}{SWR + NIR}$

NIR, near infrared band; *R*, red band; *SWR*, shortwave infrared band.

TABLE 8.3 Details of used land use/land cover.

Land use/land cover	Description
Urban	Residential, commercial areas, transportation networks such as roads and rail network
Agricultural land	A densely vegetated area, an agricultural area, or a garden
Barren	It includes salty lands, sand fields, and bare lands without vegetation
Water bodies	Pools, lakes and dams
Other lands	Meadows, lands with scattered vegetation, mountainous areas and other uses not mentioned above

were calculated based on the ground truth points obtained from field studies and Google Earth images (Eskandari Damaneh et al., 2020; Kafy et al., 2020). A kappa index calculates the classification accuracy based on the pixels' accuracy. A kappa index was calculated using Eq. (8.1):

$$\text{Kappa} = \frac{P_0 - P_c}{1 - P_c} \tag{8.1}$$

where P_0 is the observed correctness and P_c is the expected agreement (Eskandari Damaneh et al., 2022).

8.5.3 Land surface temperature

The temperature of ground surface is essential for estimating physiological parameters and their processes. We can consider several parameters, including surface energy flow, such as evaporation and transpiration. We can also consider plant production, air temperature near the surface, etc. Increased land use changes, especially urbanization, and industrialization have drastically altered the region's temperature balance; these areas are hotter than their surroundings and are known as urban heat islands. In urban areas, for instance, vegetation is lost and replaced with coverings such as asphalt, metal, and concrete, causing significant changes to the Earth's surface. These areas do not experience evaporation or transpiration. Urban and rural areas are characterized by contrasts of solar radiation, which upset nature's balance. Since the introduction of remote sensing technology, satellite data can now be used to identify these islands at different scales and investigate their environmental impact. The thermal bands, 10 and 11 of Landsat 8 (OLI/TIRS) and six of Landsat (TM), used to estimate the LST. We first calculated the Landsat five thermal band's spectral radiance, then calculated the Earth's surface's temperature using Eqs. (8.3) and (8.4). These are relationships 8.2 to 8.4 (Arabi Aliabad et al., 2020; Jiménez-Muñoz and Sobrino, 2008).

$$L_{\lambda}(TM) = L_{min} + \frac{L_{max} \times L_{min}}{Qcal_{max} \times Qcal_{min}} \times DN \tag{8.2}$$

$$L_{\lambda}(OLI) = Ml \times DN + Al \tag{8.3}$$

$$LST = \frac{T_B}{1 + \left(\lambda \times \frac{T_B}{\rho}\right) \times \ln(\varepsilon)} - 273.15 \tag{8.4}$$

where L_{λ} = upper atmospheric spectral radiance, M_L = conversion factor obtained from metadata, A_L = conversion coefficient obtained from metadata.

Q_{cal} = raw band digital number (DN), L_{max} and L_{min} are obtained from Landsat five metadata, where $\rho = 1.438 \times 10^{(-2)}$ mk also ε = emissivity calculated using the NDVI method (Sobrino et al., 2004; Khosravi et al., 2017; Eskandari Damaneh et al., 2020; Ahmadaali et al., 2021), TB = satellite brightness temperature, which is calculated using Eq. (8.5).

$$T_B = \frac{K2}{\ln\left(\frac{K_1}{K_2} + 1\right)} \tag{8.5}$$

where K_1 and K_2 = thermal constant of band six in Landsat five and bands 10 and 11 in Landsat 8, which are obtained from satellite metadata (Ullah et al., 2019; Al Rakib et al., 2020)

8.5.4 Analyses of land cover and land use

Urban, agricultural land, barrens, water bodies, and other land uses were mapped for the period 1990 to 2020 (Fig. 8.2). According to Table 8.4, the overall accuracy of the use map for 1990, 2000, 2010, and 2020 is 85%, 79%, 80%, and 81%, respectively, and the Kappa coefficient for these uses is also 0.81, 0.73, 0.75, and 0.76. According to Fig. 8.2 and Table 8.5, in the period from 1990 to 2000, the trend of land use changes in the study area is significant, so

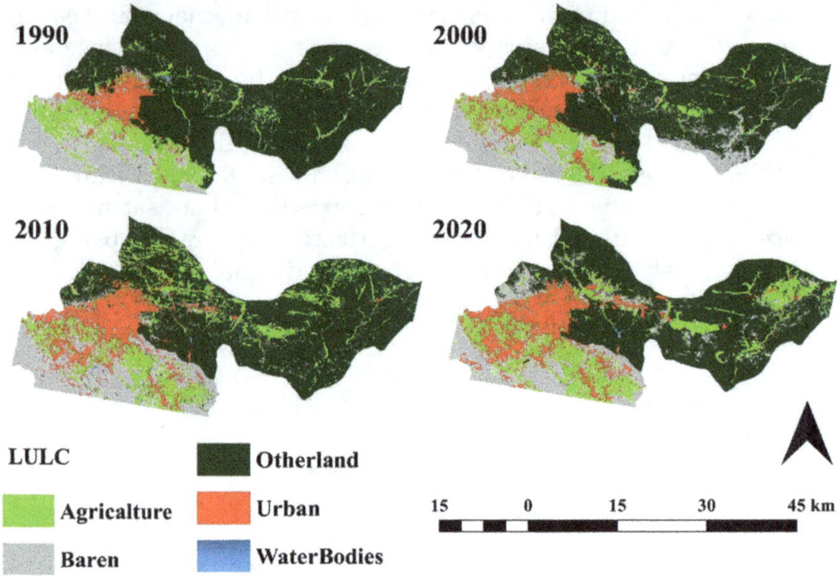

FIGURE 8.2 Map of land use/land cover changes (LULC) in 1990, 2000, 2010, and 2020.

TABLE 8.4 Accuracy assessments of the land cover types.

	Product accuracy					User accuracy						
Year	Agriculture	Baren	Other lands	Urban	Water bodies	Agriculture	Baren	Other lands	Urban	Waterbodies	Overall accuracy	Kappa
1990.00	0.81	0.86	0.87	0.84	0.90	0.90	0.87	0.77	0.88	0.81	85	0.81
2000.00	0.72	0.81	0.83	0.83	0.78	0.78	0.70	0.74	0.85	0.91	79	0.73
2010.00	0.79	0.76	0.69	0.93	0.81	0.80	0.79	0.75	0.77	0.89	80	0.75
2020.00	0.81	0.88	0.78	0.81	0.77	0.77	0.74	0.90	0.75	0.90	81	0.76

in the period from 1990 to 2000, the class of other uses decreased by 5.81%, so In 1990, these areas covered 36.99% and in 2000, about 58.17% of the region. This is while urban areas, agricultural lands, barren lands, and waterbodies showed an increase of 0.79%, 1.96%, 3.03%, and 0.04%, respectively. According to an analysis of land use trends between 2000 and 2010, barren lands and other lands decreased by 0.90% and 3.43%, respectively, whereas agricultural land uses, urban areas, and waterbodies increased by 0.93%, 3.39%, and 0.01%, respectively. Between 2010 and 2020, the trend of changes showed that barren lands and other lands decreased by 0.23% and 4.26%, respectively, whereas agricultural land, urban areas, and waterbodies increased by 2.56%, 1.88%, and 0.05%, respectively. A general trend revealed that other lands decreased by 13.51% between 1990 and 2020, while agricultural land, barren land, urban, and water bodies grew by 4.28%, 0.83%, 8.29%, and 0.11%.

TABLE 8.5 The change in land use in Tehran province between 1990 and 2020.

	1990		2000		2010		2020		1990–2000	2000–10	2010–20	1990–2020
	Area (km^2)	area (%)	Area (km^2)	area (%)	Area (km^2)	Area (%)	Area (km^2)	Area (%)	Area (%)	Area (%)	Area (%)	Area (%)
Agriculture	1421.07	14.3	1499	15.08	1592	16.01	1846.2	18.57	0.79	0.93	2.56	4.28
Baren	1625.68	16.3	1820	18.30	1731	17.41	1708.4	17.18	1.96	−0.90	−0.23	0.83
Other land	6362.68	63.9	5784	58.17	5443	54.74	5019.7	50.48	−5.81	−3.43	−4.26	−13.51
Urban	527.89	5.31	829.2	8.34	1166	11.73	1352.4	13.60	3.03	3.39	1.88	8.29
Water bodies	6.36	0.06	10.30	0.10	11.55	0.12	16.90	0.17	0.04	0.01	0.05	0.11
	9943.68	100	9943	100	9943	100	9943.7	100	100	100	100	100

8.5.5 Spatial and temporal changes of LST, NDVI, NDWI, and NDBI in the period from 1990 to 2020

The pattern of spatial and temporal changes in LST, NDVI, NDWI, and NDBI in the period from 1990 to 2020 is shown in Figs. 8.3–8.6, respectively. According to Fig. 8.2, the highest surface temperature values in 1990 and 2000 corresponded to the western to southwestern parts of the studied area. In contrast, the highest temperature values in 2010 and 2020

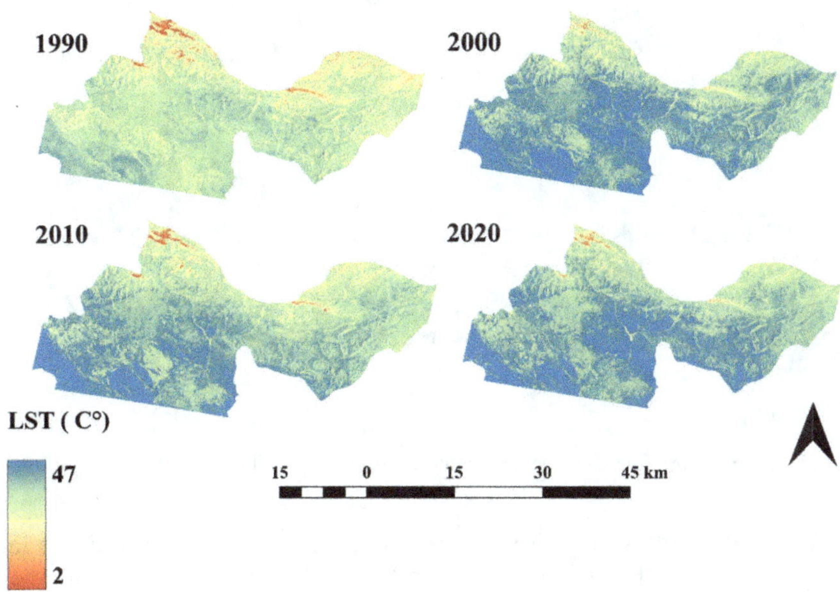

FIGURE 8.3 Map of land surface temperature (LST) changes in 1990, 2000, 2010, and 2020.

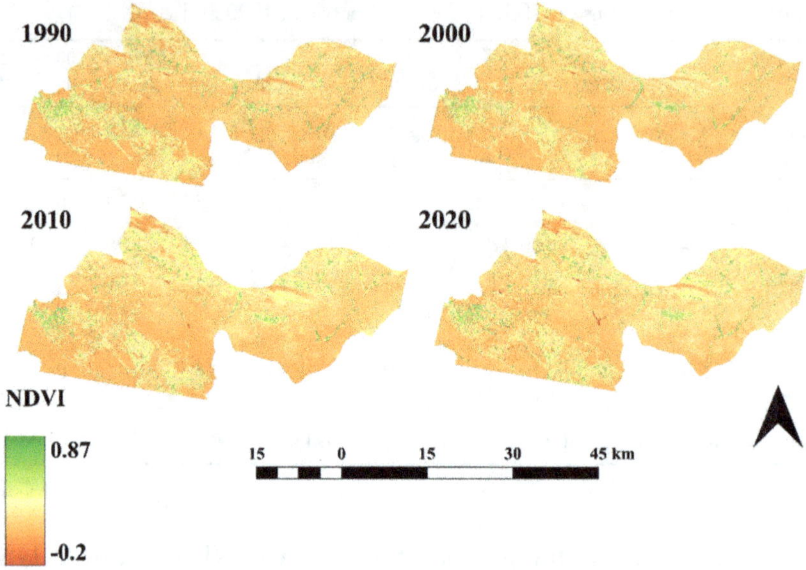

FIGURE 8.4 Map of normalized difference vegetation index changes (NDVI) in 1990, 2000, 2010, and 2020.

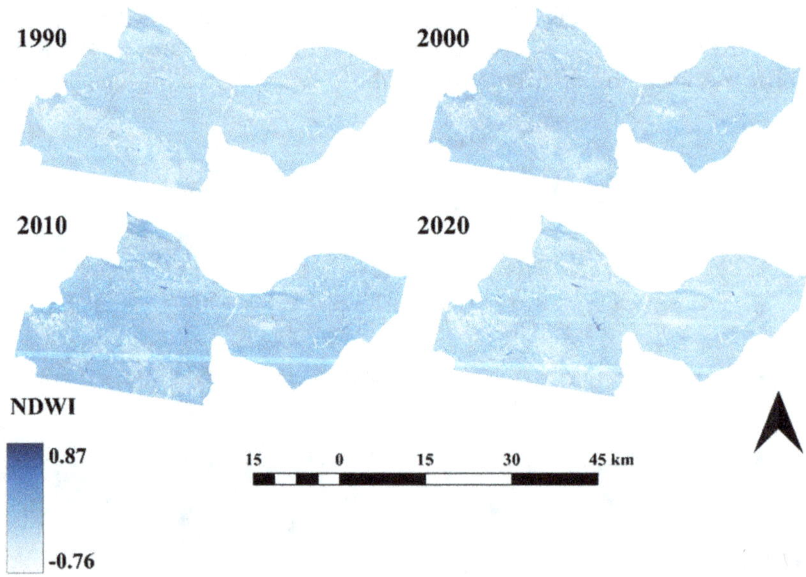

FIGURE 8.5 Map of normalized difference water index changes (NDWI) in 1990, 2000, 2010, and 2020.

compared to the southern to southwestern parts of this region, and this index is increasing in the region's east and north. However, Fig. 8.3 indicates that the maximum value of the NDVI index in all years is associated with the central to southern parts of the study area. Also, the

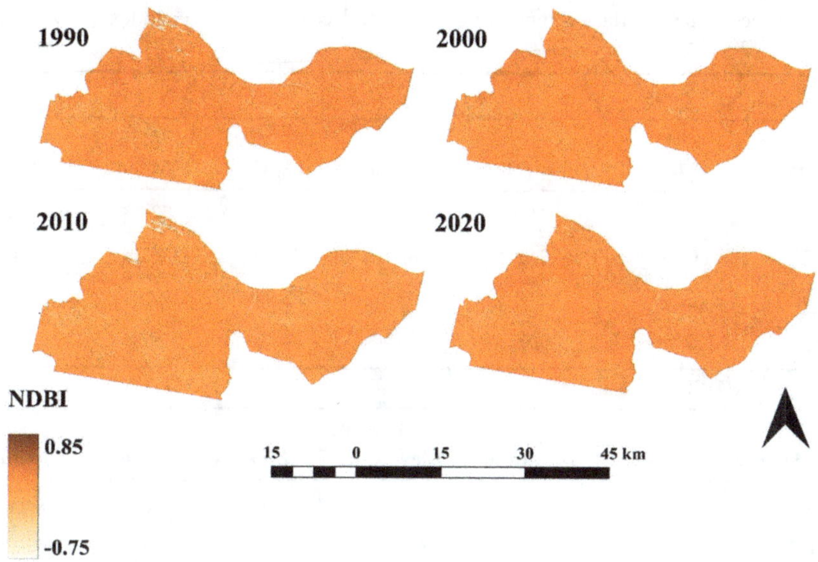

FIGURE 8.6 Map of normalized difference built-up index (NDBI) in 1990, 2000, 2010, and 2020.

trends of changes in LST and NDVI are opposite to each other. Also, the examination of the NDWI index in Fig. 8.5 showed that the maximum values of this index are related to the parts of the studied area where dams and catchment beds have been built. As shown in Fig. 8.6, the maximum values of the NDBI index are also scattered throughout the study area. The minimum, maximum, and average LST and NDVI index values increased from 1990 to 2020, so the average LST temperature increased from 32.58°C in 1990 to 39.96°C in 2020. Table 8.6 shows that the NDVI index values increased until 2000, so in 1990, the average value of this index was 0.21; in 2000, it was 0.23, and in 2010 and 2020, it was 0.20% and 19.0%. An examination of the trend of changes in the NDBI index shows that its average values in 1990 and 2000 are 0.16% and 0.14%, respectively, but in 2010 and 2020, they are 0.16% and 0.20%. NDWI index values also follow this rule.

8.5.6 Land surface temperature variations in different classes and land use in Tehran city

The investigation of the spatial and temporal changes in the Earth's surface temperature using Landsat satellite thermal bands in different layers is shown in Fig. 8.7. Based on this figure, in the period from 1990 to 2020, the most significant percentage of the area of Tehran province was in the <33°C classes, so in the years 1990, 2000, 2010, and 2020, respectively, these classes accounted for about 53.23%, 88.57%, 79.54%, and 89.38% of it covered the area. This is despite the fact that the 30–33°C range has increased during this period from 73.12% in 1990 to about 95.57% in 2020. Observing the trend of surface temperature change over different land uses in Fig. 8.8 also showed that the highest average surface temperature in 1990 was observed in the barren land. The trend of temperature changes in all land uses

TABLE 8.6 Changes in land surface temperature (LST) and vegetation cover index (NDVI) in the period of 1990–2020.

NDVI				LST				
Standard deviation	Maximum	Mean	Minimum	Standard deviation	Maximum	Mean	Minimum	Year
0.12	0.98	0.21	−0.99	5.60	47.06	32.58	0.76	1990
0.08	0.52	0.23	−0.92	5.67	54.90	40.35	2.41	2000
0.10	1.00	0.20	−0.89	6.33	53.85	37.94	8.60	2010
0.08	0.57	0.19	−0.76	5.58	54.91	39.96	5.54	2020
NDWI				**NDBI**				
Standard deviation	Maximum	Mean	Minimum	Standard deviation	Maximum	Mean	Minimum	Year
0.09	1.00	−0.22	−0.76	0.11	0.83	0.16	−0.20	1990
0.08	0.44	−0.21	−0.72	0.09	0.76	0.14	−0.20	2000
0.09	0.58	−0.22	−0.71	0.11	0.79	0.16	−0.20	2010
0.10	0.87	−0.27	−0.77	0.12	0.88	0.20	−0.20	2020

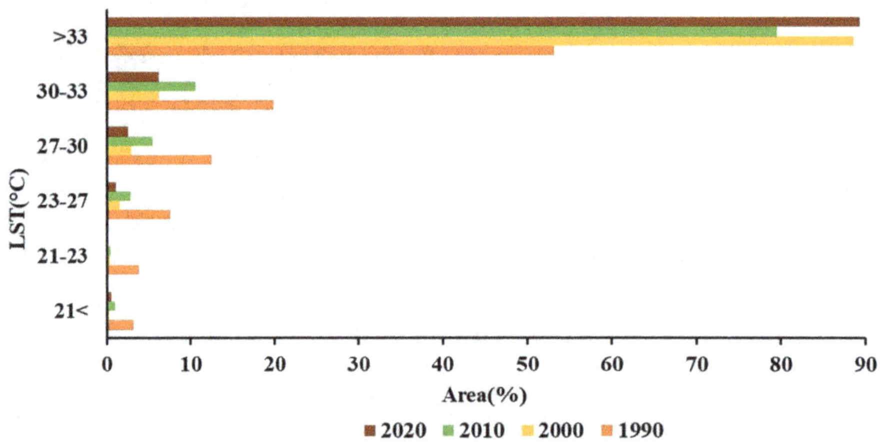

FIGURE 8.7 Changes in the land surface temperature (LST) classes of Tehran province in the period of 1990–2020.

indicates that temperatures have been rising during this period, so the average temperature in agricultural land uses, barren lands, other land uses, urban areas, and water beds in 1990 was 30.68°C and 37.18°C respectively, 31.77°C, 33.36°C, and 22.37°C, respectively. In 2020, these uses will have an average temperature of 39.10°C, 44.53°C, 38.94°C, 39.39°C, and 27.80°C.

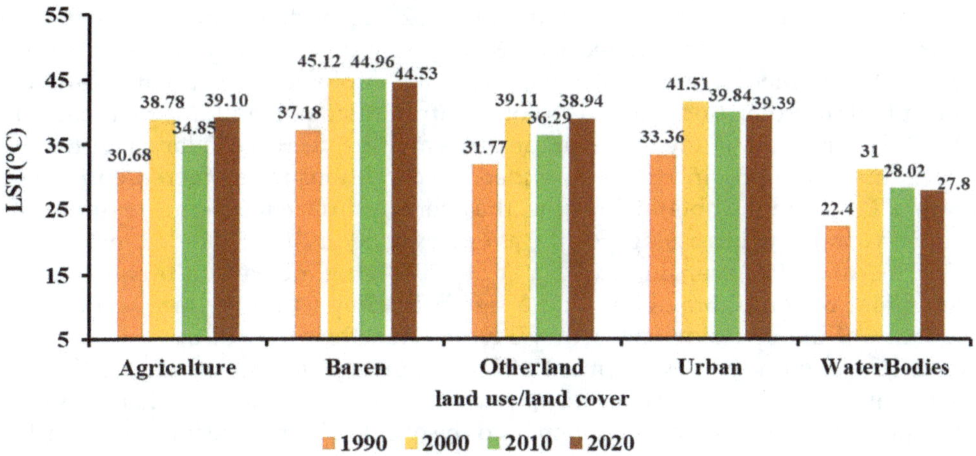

FIGURE 8.8 Changes in land surface temperature (LST) in land use of Tehran province in the period of 1990–2020.

8.6 Conclusions

The present research uses Landsat OLI and TM satellite images to investigate land use changes and LST. It also studies the mutual effects of these two factors in Tehran province from 1990 to 2020. Examining the trend of land use changes in Tehran province showed that in these 30 years, urban and manufactured areas have been increasing. In the past 20 years, the urban and manufactured portions of land use have increased by 8.29%. The land use has grown from 527.89 square kilometers in 1990 to 1352.41 square kilometers in 2020. Rousta et al. (2018), the Landsat OLI and TM satellite images were used to examine Tehran's expansion, pointed out that the process of changes in the city of Tehran is increasing irregularly. So that the land use change map resulting from their research in 1988–2018 showed that urban areas (impermeable) had expanded the most among all land use classes. Because in 1988, the area of this land use (urban and manufactured areas) was 144.72 square kilometers, which increased to 244.50 square kilometers in 1998. Subsequently, in 2008, it increased again to 343.83 square kilometers and finally; it increased again to 430.77 square kilometers in 2018. From 1988 to 2018, this region has been converted into urban and manufactured areas covering 286.04 square kilometers (27.48% of its total area). In the current situation, rapid development and excess urbanization and types of land use and land cover have undergone unprecedented global and national changes, which have significantly affected the quality of the environmental environment. Or in other words, although the increase in urban areas can have positive economic and social effects on people, on the other hand, it can also have negative effects on the natural environment (Estoque and Murayama, 2012). In Tianshan, China, Airiken et al. (2022) assessed the quality of the environment due to changes in land use and urban density. Land use and urbanization have changed the environmental quality indicators for 2001–18. The increase in urban areas has resulted in decreased natural lands, such as saline lands without vegetation, and these lands decreased by 13.51% in the period 1990–2020 respectively. This reduction in the amount of natural land causes the reduction and disruption of ecosystem services such as weather regulation, biodiversity protection (especially biological conservation and control, including plant reproduction), and erosion control (Estoque and Murayama, 2012; Li et al., 2017). This increase in the use of urban areas and the decrease in lands with low vegetation cover in Tehran province have provided the conditions for an increase in the surface temperature. Based on the obtained results, the average surface temperature (in the studied period of 30 years) was increasing. So, the average value of the LST index in 1990 reached 32.58–39.96°C in 2020. According to Silva et al. (2018), the increase in urbanization in the Paco do Lumiar region of Brazil had had a significant impact on the local climate, as well as the quality of life of residents, as they investigated the effect of spatial and temporal changes in land use on the development of heat islands. The results indicate that the city of Paco do Lumiar has experienced large growth during the 16 years of urbanization. which has led to the loss of green spaces and increased the effects of urban heat islands. Gashu et al. (2018) investigated the trend of spatial-temporal changes in urban land use/land cover in Hawassa and Bahir Dar, two cities of Ethiopian, from 1973 to 2015. The results of their research showed that vegetation cover has reduced by 14% and 30% respectively in Hawasa and Bahir Dar from 1973 to 2015. The built-up areas have increased by 24 and 10% in these two cities, respectively. which has caused numerous environmental effects in this area. According to the IPCC report (2014) and Alamgir et al.'s study (2020), the increase in urban areas, the decrease in vegetation cover, and the climatic changes in different regions have increased the LST index. According to the current research results, in this region, the upper classes of the earth's surface temperature are also rising, so the maximum temperature reached 54.91°C in 2020, from 47.06°C in 1990. Also, this increase in temperature has increased due to the use of urban and manufactured areas. In general, this study shows that using Landsat satellite remote sensing data archive, the process of dynamic changes in different areas, especially in urban areas, as well as climatic changes such as LST resulting from the expansion of evil urban areas, can be very useful. In general, our results show that the land uses of Tehran province have changed significantly in the 30 years from 1990 to 2020. The analysis of our results also shows that there has been a significant increase in agricultural land and urban land, which are about 4.28% and 8.29%, respectively; these increases have been seen along with a decrease of 13.51% in other land uses. Also, the increase in agricultural land and the decrease of 0.83% in the barren land, considering the climatic conditions prevailing in the world and especially in Iran, also indicate that this growing population has increased the need for food and the demand for food, which in itself causes an increase in the land use changes. Therefore, we conclude that the indiscriminate and unauthorized urban growth and increased nonprincipled agricultural land are critical factors in the increase of urban areas and LST in Tehran province. In summary, the new technology of remote sensing has an essential role in improving the periodic monitoring efficiency and management of urban growth in large cities such as Tehran. The information obtained from this chapter can provide valuable information to policymakers and managers of urban areas and other relevant bodies to use this information to make appropriate plans in these areas' present and future time frames.

Khan, R., Li, H., Basir, M., Chen, Y.L., Sajjad, M.M., Haq, I.U., Ullah, B., Arif, M., Hassan, W., 2022. Monitoring land use land cover changes and its impacts on land surface temperature over Mardan and Charsadda Districts, Khyber Pakhtunkhwa (KP), Pakistan. Environ. Monit. Assess. 194 (6), 409.

Khosravi, H., Azareh, A., Dameneh, H.E., Sardoii, E.R., Dameneh, H.E., 2017. Assessing the effects of the climate change on land cover changes in different time periods. Arabian J. Geosci. 10, 1–10.

Lazzarini, M., Marpu, P.R., Ghedira, H., 2013. Temperature-land cover interactions: the inversion of urban heat island phenomenon in desert city areas. Remote Sens. Environ. 130, 136–152.

Li, H., Peng, J., Yanxu, L., Yina, H., 2017. Urbanization impact on landscape patterns in Beijing City, China: a spatial heterogeneity perspective. Ecol. Indicat. 82, 50–60.

Lyu, F., Wang, S., Han, S.Y., Catlett, C., Wang, S., 2022. An integrated cyberGIS and machine learning framework for fine-scale prediction of urban Heat Island using satellite remote sensing and urban sensor network data. Urban Informatics 1 (1), 6.

Mishra, A., Appadurai, A.N., Choudhury, D., Regmi, B.R., Kelkar, U., Alam, M., Chaudhary, P., Mu, S.S., Ahmed, A.U., Lotia, H., Fu, C., 2019. Adaptation to climate change in the Hindu Kush Himalaya: stronger action urgently needed. Hindu Kush Himalaya Assess.: Mountains, Climate Change, Sust. People 457–490.

Moisa, M.B., Merga, B.B., Gemeda, D.O., 2022. Urban heat island dynamics in response to land use land cover change: a case of Jimma city, southwestern Ethiopia. Theor. Appl. Climatol. 149 (1–2), 413–423.

Molina Gomez, N.I., Varon-Bravo, L.M., Sierra-Parada, R., Lopez Jimenez, P.A., 2022. Urban growth and heat islands: a case study in micro-territories for urban sustainability. Urban Ecosyst. 25 (5), 1379–1397.

Moody, A., Johnson, D.M., 2001. Land-surface phenologies from AVHRR using the discrete fourier transform. Remote Sens. Environ. 75 (3), 305–323.

Naikoo, M.W., Islam, A.R.M.T., Mallick, J., Rahman, A., 2022. Land use/land cover change and its impact on surface urban heat island and urban thermal comfort in a metropolitan city. Urban Clim. 41, 101052.

Ning, J., 2011. Influence of coastal land use change to land surface temperature. Energy Proc. (11), 3999–4004.

Pan, Z., He, J., Liu, D., Wang, J., 2020. Predicting the joint effects of future climate and land use change on ecosystem health in the middle reaches of the Yangtze River Economic Belt, China. Appl. Geogr. 124, 102293.

Panuju, D.R., Paull, D.J., Griffin, A.L., 2020. Change detection techniques based on multispectral images for investigating land cover dynamics. Rem. Sens. 12 (11), 1781.

Pettorelli, N., Vik, J.O., Mysterud, A., Gaillard, J.M., Tucker, C.J., Stenseth, N.C., 2005. Using the satellite-derived NDVI to assess ecological responses to environmental change. Trends Ecol. Evol. 20 (9), 503–510.

Polydoros, A., Mavrakou, T., Cartalis, C., 2018. Quantifying the trends in land surface temperature and surface urban heat island intensity in mediterranean cities in view of smart urbanization. Urban Sci. 2 (1), 16.

Rahaman, Z.A., Kafy, A.A., Saha, M., Rahim, A.A., Almulhim, A.I., Rahaman, S.N., Fattah, M.A., Rahman, M.T., Kalaivani, S., Al Rakib, A., 2022. Assessing the impacts of vegetation cover loss on surface temperature, Urban Heat Island and carbon emission in Penang city, Malaysia. Build. Environ. 222, 109335.

Rousta, I., Sarif, M.O., Gupta, R.D., Olafsson, H., Ranagalage, M., Murayama, Y., Zhang, H., Mushore, T.D., 2018. Spatiotemporal analysis of land use/land cover and its effects on surface urban heat island using Landsat data: a case study of Metropolitan City Tehran (1988–2018). Sustainability 10 (12), 4433.

Rugel, E.J., Henderson, S.B., Carpiano, R.M., Brauer, M., 2017. Beyond the Normalized Difference Vegetation Index (NDVI): developing a natural space index for population-level health research. Environ. Res. 159, 474–483.

Shirazi, M.G., Kazemi, A., Kelishadi, R., Mostafavi, F., 2017. A review on determinants of nutritional behavior in teenagers. Iranian J. Pediatr. 27 (3).

Silva, J.S., da Silva, R.M., Santos, C.A.G., 2018. Spatiotemporal impact of land use/land cover changes on urban heat islands: a case study of Paço do Lumiar, Brazil. Build. Environ. 136, 279–292.

Sobrino, J.A., Jimenez Munoz, J.C., Paolini, L., 2004. Land surface temperature retrieval from LANDSAT TM 5. Remote Sens. Environ. 90 (4), 434–440.

Tucker, C.J., Fung, I.Y., Keeling, C.D., Gammon, R.H., 1986. Relationship between atmospheric CO2 variations and a satellite-derived vegetation index. Nature 319 (6050), 195–199.

Ullah, S., Ahmad, K., Sajjad, R.U., Abbasi, A.M., Nazeer, A., Tahir, A.A., 2019. Analysis and simulation of land cover changes and their impacts on land surface temperature in a lower Himalayan region. J. Environ. Manag. 245, 348–357.

Wang, Z.H., 2022. Reconceptualizing urban heat island: beyond the urban-rural dichotomy. Sustain. Cities Soc. 77, 103581.

Wicke, B., Verweij, P., Van Meijl, H., Van Vuuren, D.P., Faaij, A.P., 2012. Indirect land use change: review of existing models and strategies for mitigation. Biofuels 3 (1), 87–100.
Zha, Y., Gao, J., Ni, S., 2003. Use of normalized difference built-up index in automatically mapping urban areas from TM imagery. Int. J. Rem. Sens. 24 (3), 583–594.
Zhu, R., Dong, X., Wong, M.S., 2022. Estimation of the urban heat island effect in a reformed urban district: a scenario-based study in Hong Kong. Sustainability 14 (8), 4409.
Zhu, Z., 2017. Change detection using landsat time series: a review of frequencies, preprocessing, algorithms, and applications. ISPRS J. Photogr. Remote Sens. 130, 370–384.

CHAPTER

9

The application of remote sensing on wetlands spatio-temporal change detection

Saleh Yousefi[1], *Sayed Naeim Emami*[1], *Aleksandar Valjerević*[2], *Yaghoub Iranmanesh*[3], *Mohammad Nekoeimehr*[1] *and Nastaran Naderi*[4]

[1]Soil Conservation and Watershed Management Research Institute, Chaharmahal and Bakhtiari Agricultural and Natural Resources Research and Education Center, AREEO, Shahrekord, Iran; [2]Department of Geospatial and Environmental Science, Faculty of Geography, University of Belgrade, Belgrade, Serbia; [3]Research Division of Natural Resources, Chaharmahal and Bakhtiari Agricultural and Natural Resources Research and Education Center, AREEO, Shahrekord, Iran; [4]Islamic Azad University of Tonekabon, Mazandaran, Iran

9.1 Wetlands in our ecosystem

Wetlands are one of the most unique, beautiful, pristine, and amazing natural phenomena that have very original, interesting, unknown, and mysterious effects (Gren et al., 1994; Board, 2005). Wetlands are the most important, valuable, and complex ecosystems on Earth, and the reason for this is the diverse functions that include many economic, social, environmental, and scientific benefits (Chen et al., 2019). Wetlands are one of the most important and sensitive ecosystems on the planet and have been the shelter of an important part of the biodiversity on this planet (Rundquist et al., 2001; Batzer and Sharitz, 2007). An interconnected set of aquatic or water-dependent plants has been formed in them, and the intervention of natural and unnatural factors can disrupt this coherence and quality(Mitsch and Gosselink, 2015; Chen et al., 2019). The study of vegetation is valuable in solving ecological problems such as protecting biodiversity, managing natural resources, and evaluating environmental conditions, and based on the results, future changes can be predicted (Fraser and Keddy, 2005; Dong et al., 2016; Kandus et al., 2018). As unique, rich, and fertile ecosystems, wetlands

Remote Sensing of Soil and Land Surface Processes
https://doi.org/10.1016/B978-0-443-15341-9.00010-1

are considered to be one of the most productive environments in the world and the most vital factors for the future of mankind (Epaphras et al., 2008; Pal and Talukdar, 2018). Despite the countless functions and benefits of wetlands, unfortunately, there is still very little awareness and full knowledge of these unique habitats. It can be said that none of the world's ecosystems have experienced damage caused by human short-sightedness and selfish human-centered tendencies as much as wetlands (Rundquist et al., 2001; Kandus et al., 2018). These habitats have environmental importance from the point of view of biological diversity and aesthetics, and they also play a vital role in people's economic life. Wetlands support large communities of birds, mammals, reptiles, amphibians, fish, and invertebrates, so among the 20,000 different species of fish in the world, more than 40% of them live in freshwater (wetlands) (Majumdar et al., 1989; Chen et al., 2019). In addition, these water bodies are considered to be a very important reserve of plant genetic material (Johnston and Barson, 1993; Messina and Conner, 1997).

The change in hydrological systems, especially the construction of multiple dams during the last four decades, has caused the drying up of rivers, wetlands, and groundwater subsidence, which has a great role in reducing the diversity and extinction of plant species (Fraser and Keddy, 2005; Betbeder et al., 2014). On the other hand, with the increasing human intervention in nature and the phenomenon of global warming, plant species need stronger protection. Although wetlands are sometimes small water environments, they have great functions in the environment (Batzer and Sharitz, 2007). Wetlands provide various services to living organisms, and at the same time, they face many challenges (Chen et al., 2019; Hayri Kesikoglu et al., 2019).

9.2 Remote sensing and wetlands

The basis of remote sensing (RS) is the measurement of the reflective energy of remote phenomena (Ozesmi and Bauer, 2002). Therefore, it seems necessary to know the basic concepts of physics due to the emphasis of RS on obtaining quantitative information over qualitative information. Meanwhile, many factors affect spectral symptoms that should not be overlooked (Butera, 1983; Meseșan et al., 2019). The issue of using different parts of the electromagnetic spectrum in RS will depend on the characteristics of solar radiation, the characteristics of reflection of terrestrial phenomena, and the absorption of different wavelengths by the atmosphere (atmospheric effects) (Kavzoglu and Mather, 2003). For this reason, the basic understanding of the concepts of electromagnetic, absorption, spectral reflection, spectral characteristics, spectral sources, etc. is actually a part of RS and with the basic principles of light physics, such as particle and wave models, as well as it is related to concepts such as wavelength and frequency (Hall et al., 2016).

Global warming and climate changes caused by human activities, as some of the main factors disrupting the vital network, have attracted the attention of many responsible people in the world in the last two decades (Feng and Fu, 2013). To control and manage the devastating effects of this phenomenon, solutions have been proposed at the national, regional, and global levels (Clark et al., 2001). Meanwhile, the use of RS images as an emerging technology can play an important role in the temporal and spatial monitoring of the Earth's temperature (Thenkabail and Gamage, 2004). One of the most important methods of studying the Earth's changes is the use of information resulting from the processing of satellite images and thermal images (Hames, 2009). It seems that the potentials in RS technology and geographic information system and their integration, could be useful to create spatial data (with high temporal, spatial, and spectral frequency) and check and improve the indicators used in monitoring and evaluating ecosystem health (Johnston and Barson, 1993). Planning for the protection of wetlands requires monitoring the changes in these aquatic ecosystems. RS is an efficient and valuable tool to identify and monitor changes in wetlands over time (Sharitz, 1986; Ozesmi and Bauer, 2002).

Recent advances in sensor design techniques and data analysis have developed some RS systems to monitor changes in natural ecosystems such as wetlands (Kandus et al., 2018; Liu et al., 2020). Infrared thermal scanners can map coastal water temperatures, and microwave radiometers can measure water salinity (Dregne, 2002). Light detection and ranging (LIDAR) can map the topography of wetlands, the profile of beaches, and the depths of the sea floor (Chase and Weishampel, 2016). RS and Geographical Information System (GIS) are efficient and valuable tools for land identification and monitoring (Yousefi et al., 2015). Changes in wetlands occur over different periods. These tools are used for mapping wetland habitats, determining their biodiversity, changes in wetland land use, and monitoring protected areas (Kaplan and Avdan, 2018). The sources of RS that are used for the wetland include aerial photographs and satellite images from different years. In the past, aerial photos and sometimes satellite images were used to interpret the maps of wetlands, and recently satellite image processing is used. There are no standard methods for classifying wetlands that are based on computer software. Satellites including, Landsat, SPOT, AVHRR, IRS, and other radar systems often use satellite measurements to detect wetlands and changes in their characteristics (water, soil, and vegetation) (Geerken and Ilaiwi, 2004; Vogelmann et al., 2009; Jacqueminet et al., 2013; Yousefi et al., 2022).

9.3 Vegetation indexes

Normalized difference vegetation index: NDVI is a simple graphical index that is used in RS analysis and measurements and evaluating the presence or absence of vegetation in an area (Beck et al., 2006). The NDVI index is a vegetation health index based on how plants reflect a certain range of the electromagnetic spectrum. In simple words, the NDVI index indicates plant health based on the way a plant reflects different light waves (Nanzad et al., 2019). The value of the NDVI index will always be between -1 and $+1$; values between 0 and 1 indicate dead plants or inorganic objects such as rocks, roads, and houses (Demissie et al., 2015). NDVI index values for living plants are between 0 and 1, where 1 is the healthiest and 0 is the least healthy value.

Soil adjusted vegetation index (SAVI): In order to improve NDVI, the SAVI vegetation index was proposed, which minimizes the effects of soil illumination on the spectrum of red and infrared bands.

$$SAVI = \{(NIR - RED) \ / \ (NIR + RED + L)\}(1 + L) \tag{9.1}$$

where L is the correction factor changes from 0 for high vegetation area to 1 for very low vegetation area and 0.5 for medium vegetation area. L + 1 in this formula makes the vegetation index changes from −1 to +1, and if the L factor reaches zero, the SAVI index will be equal to the NDVI index (Haghighian et al., 2020).

9.4 Case study on wetland monitoring in Iran

Gadaman Wetland, with an area of about 1000 ha, is one of the top 10 wetlands in Iran, which is registered with the International Bureau of Waterfowl Research in London. Gandaman Wetland is located at an altitude of 2215 m above sea level and 4 km away from Gandaman city. The climate of the Gdanman wetland area is cold and semi-arid, and the average annual rainfall of the area is 566.1 mm. The scope of the study area of Gadaman Wetland is related to the prohibited hunting area of the Wetland, the area of which is 1093 ha (Fig. 9.1).

In this study, the satellite images of different sensors (TM, ETM+, and OLI) of the Landsat satellite were used from the years 1995–2022 (Table 9.1).

In the present study, the NDVI and SAVI indices for 1955, 1999, 2002, 2008, 2013, 2016, 2019, and 2022 were prepared in the ENVI 5.6 software for the study area.

Using the ENVI 5.6 software and using the infrared and red bands of the prepared images, the NDVI index was extracted (Fig. 9.2), and its statistical parameters were determined for different years of the study period in the Gandaman wetland (Table 9.2). Also, the trend of changes in the average NDVI index in the Gadaman wetland is shown in Fig. 9.3.

Using the ENVI 5.6 software and using the infrared and red bands of the prepared images, the SAVI index was extracted (Fig. 9.4) and its statistical parameters were determined for different years of the study period in the Gandaman wetland (Table 9.3). Also, the trend of changes in the average NDVI index in the Gadaman wetland is shown in Fig. 9.5.

In this study, two important indicators, NDVI and SAVI, were used to evaluate the changes in the vegetation cover of the Gandman wetland. The data from various Landsat satellite sensors during 1995–2022 were also considered as the main data of this research. The results of the NDVI index in Gadaman Wetland show that the average of this index in the prohibited hunting area of the wetland during the research period has two main periods of changes. So that from 1995 to 2008, the average values of this index had a downward trend, and from 2008 to 2022, it can be said that the changing trend of this index was positive. Also, the results of this part of the research show a downward trend in the minimum values of the NDVI index in the prohibited hunting area of Gadaman wetland. The existence of droughts in the last few decades has weakened the natural vegetation in the country's natural lands.

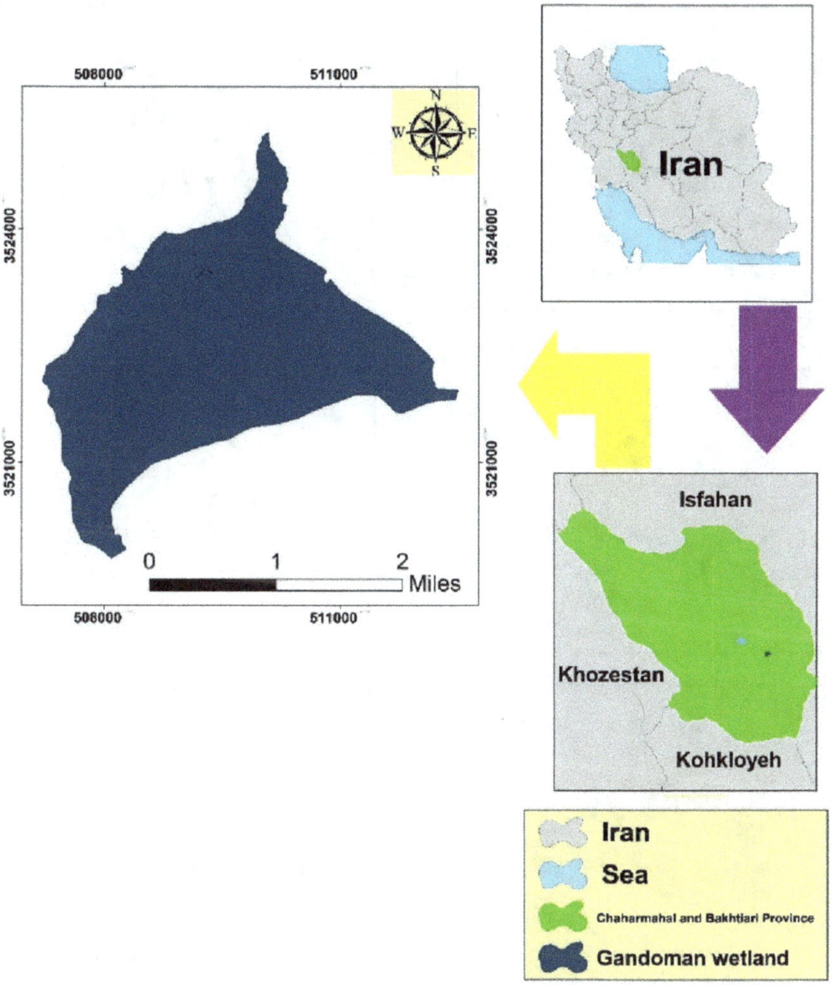

FIGURE 9.1 Geographical location of Gandoman Wetland in Iran.

TABLE 9.1 Characteristics of remote sensing data used in the present study.

Pixel size(M)	Date	Sensor
30 × 30	July 13, 1955	TM
30 × 30	August 01, 1999	ETM
30 × 30	July 24, 2002	ETM
30 × 30	July 16, 2008	TM
15 × 15	July 14, 2013	OLI
15 × 15	July 27, 2016	OLI
15 × 15	July 31, 2019	OLI
15 × 15	July 23, 2022	OLI

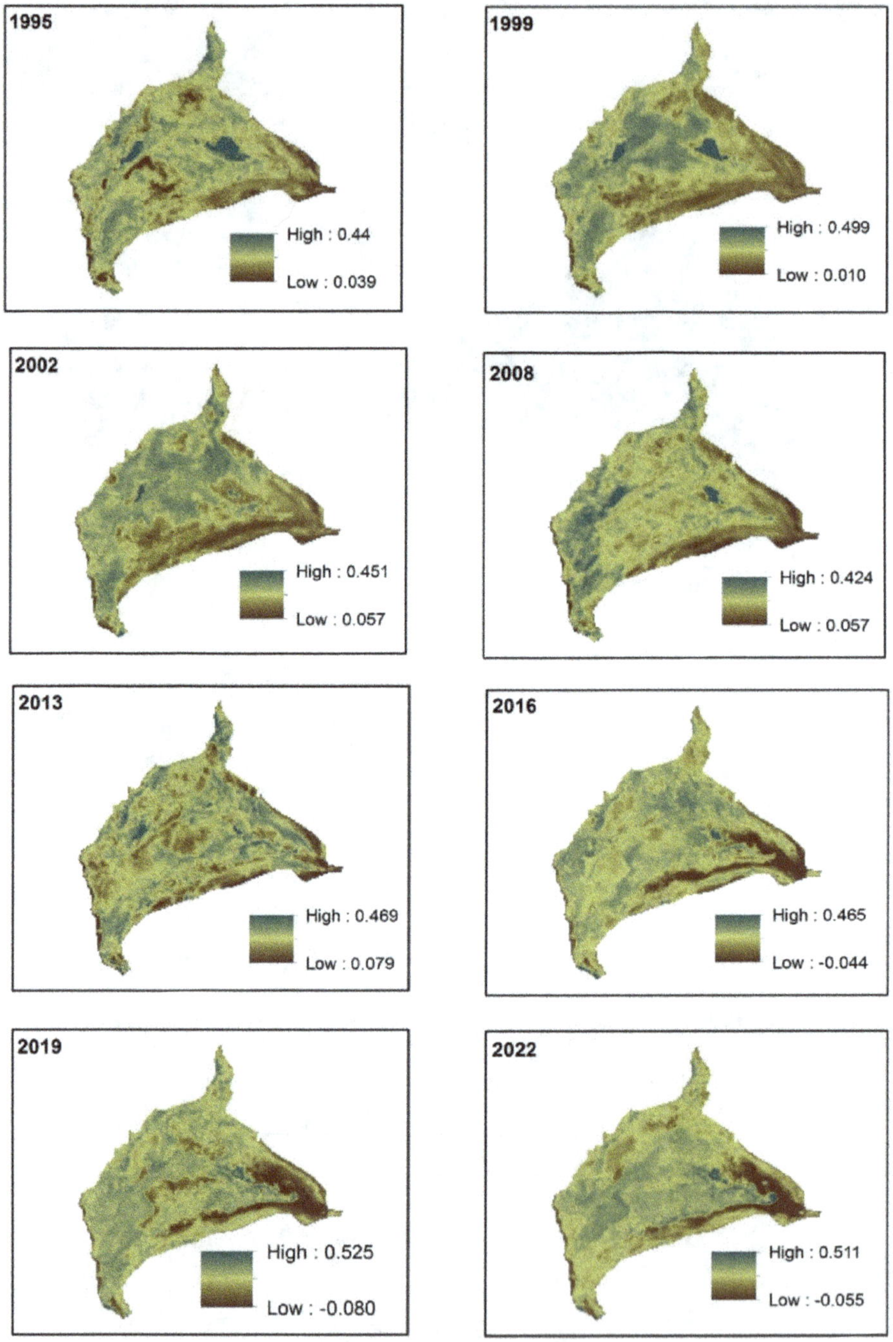

FIGURE 9.2 Normalized difference vegetation index (NDVI) index for different years, prohibited hunting area of Gadaman Wetland.

TABLE 9.2 NDVI index values in different years for Gandaman wetland.

	1995	1999	2002	2008	2013	2016	2019	2022
Min	0.039184	0.010453	0.05796	0.05723	0.079437	−0.04457	−0.08037	−0.05546
Max	0.442148	0.499373	0.451848	0.424427	0.469217	0.465678	0.525742	0.511908
Mean	0.255188	0.229701	0.227469	0.19996	0.263461	0.226968	0.252296	0.255986
SD	0.052092	0.076011	0.067303	0.047189	0.050707	0.071224	0.082468	0.079527

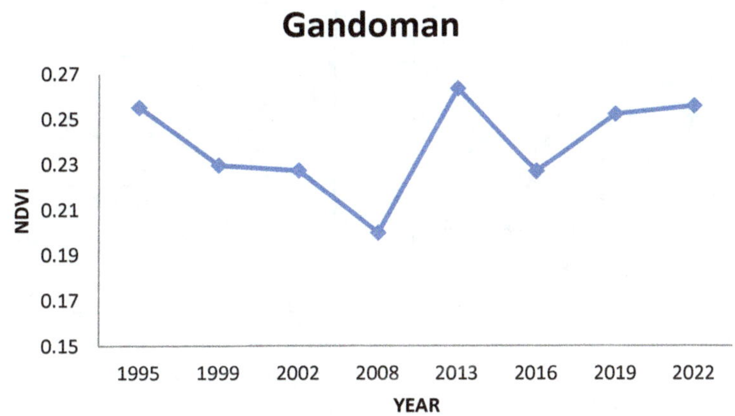

FIGURE 9.3 The trend of changes in the average normalized difference vegetation index (NDVI) values in the prohibited hunting area of Gandaman Wetland.

NDVI index is unstable and different against soil color, soil moisture, and saturation effects of high-density vegetation. Especially in areas where the vegetation cover is low (e.g., <40%) and the soil surface is exposed to light, light reflection in the red and infrared electromagnetic spectrum can affect the values of vegetation indices. In order to improve NDVI, the SAVI vegetation index was proposed, which minimizes the effects of soil illumination on the spectrum of red and infrared bands.

The results of the SAVI index in Gadaman Wetland show that the average of this index in the prohibited hunting area of the wetland during the research period has two main periods of changes. So that from 1995 to 2008, the average values of this index had a downward trend, and from 2008 to 2022, it can be said that the changing trend of this index was positive. Also, the results of this part of the study show a downward trend in the minimum values of the SAVI index in the prohibited hunting area of Gadaman Wetland. The existence of droughts in the last few decades has weakened the natural vegetation in the country's natural lands. Also, the changes in the water level and especially the aquatic environment of the wetland have caused changes and decrease in the values of this index in the years 2002–2008.

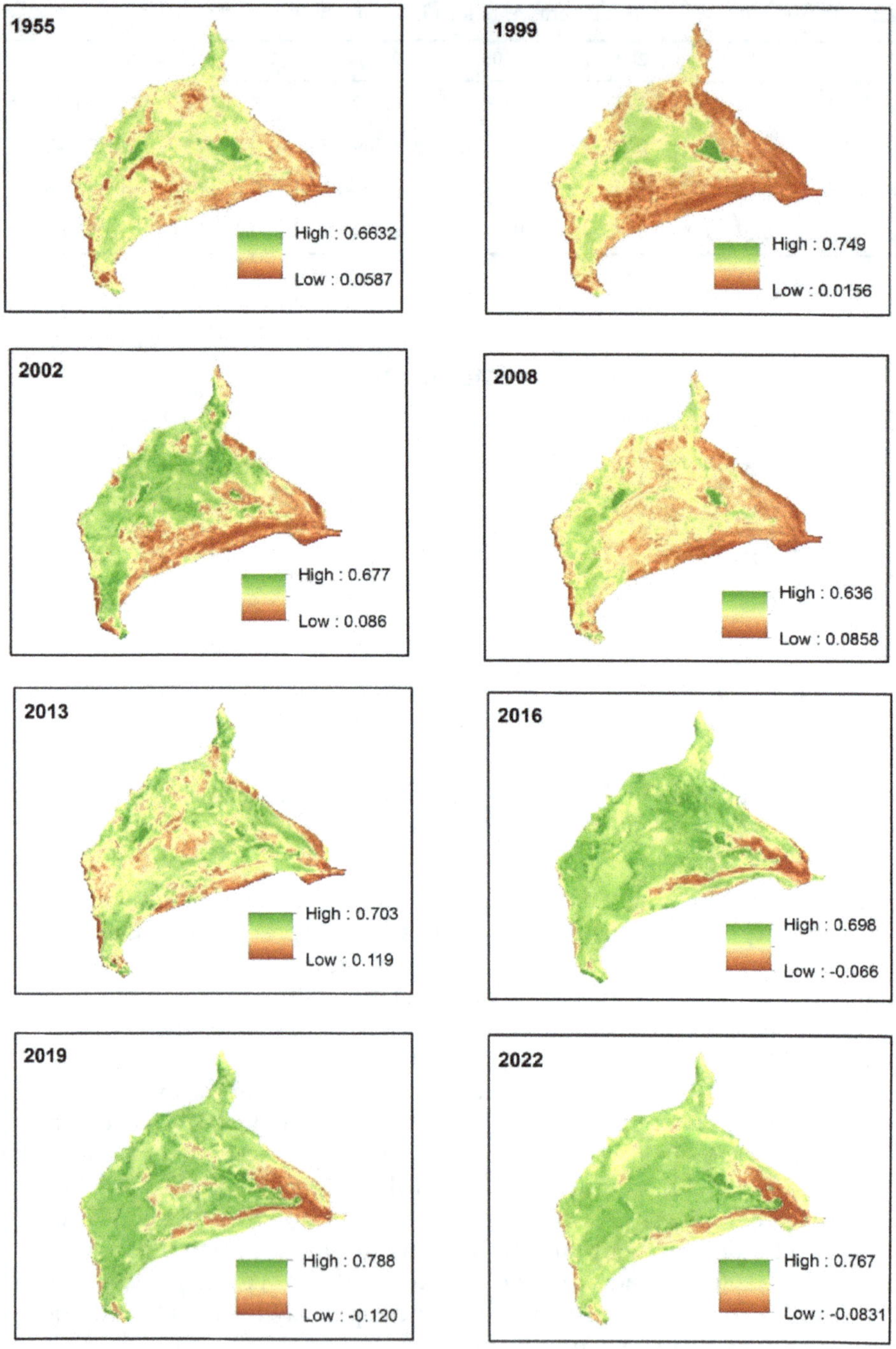

FIGURE 9.4 Soil-adjusted vegetation index (SAVI) index for different years, prohibited hunting area of Gadaman Wetland.

TABLE 9.3 SAVI index values in different years for Gandaman wetland.

	1995	1999	2002	2008	2013	2016	2019	2022
Min	0.058774	0.015678	0.086939	0.085843	0.119154	−0.06685	−0.12056	−0.08319
Max	0.663213	0.749049	0.677763	0.63663	0.703816	0.698508	0.788602	0.767852
Mean	0.382776	0.344546	0.341198	0.299934	0.395185	0.340446	0.378438	0.383972
SD	0.078136	0.114014	0.100953	0.070783	0.07606	0.106834	0.1237	0.119288

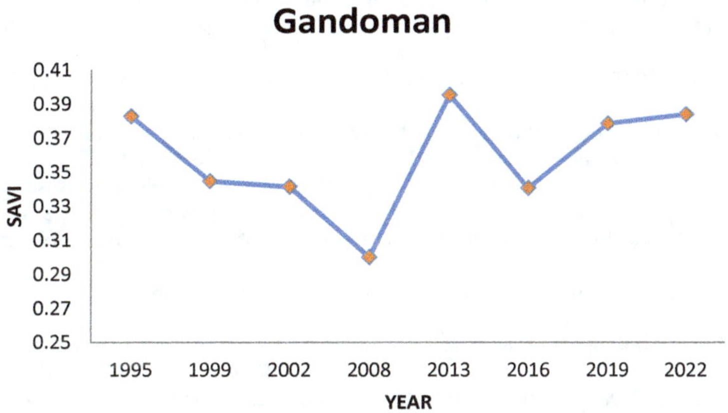

FIGURE 9.5 The trend of changes in the average soil-adjusted vegetation index (SAVI) values in the prohibited hunting area of Gandaman Wetland.

Acknowledgments

This study was funded by Agricultural Research, Education and Extension Organization, AREEO as part of the Project No. 4-42-29-005-010275.

References

Batzer, D.P., Sharitz, R.R., 2007. Ecology of Freshwater and Estuarine Wetlands, Ecology of Freshwater and Estuarine Wetlands. Univ of California Press. https://doi.org/10.1525/california/9780520247772.003.0001.

Beck, P.S.A., et al., 2006. Improved monitoring of vegetation dynamics at very high latitudes: a new method using MODIS NDVI. Remote Sens. Environ. 100 (3), 321–334.

Betbeder, J., et al., 2014. Mapping of central africa forested wetlands using remote sensing. IEEE J. Sel. Top. Appl. Earth Obs. Remote Sens. 7 (2), 531–542. https://doi.org/10.1109/JSTARS.2013.2269733.

Board, M.E.A., 2005. Ecosystems and Human Well-Being: Wetlands and Water Synthesis. Millennium Ecosystem Assessment Board.

Butera, M.K., 1983. Remote sensing of wetlands. IEEE Trans. Geosci. Remote Sens. (3), 383–392.

Chase, A.S.Z., Weishampel, J., 2016. Using Lidar and GIS to investigate water and soil management in the agricultural terracing at caracol, Belize. Adv. Archaeol. Pract. 4 (3), 357–370. https://doi.org/10.7183/2326-3768.4.3.357.

Chen, W., et al., 2019. An evaluating system for wetland ecological health: case study on nineteen major wetlands in Beijing-Tianjin-Hebei region, China. Sci. Total Environ. 666, 1080–1088. https://doi.org/10.1016/j.scitotenv.2019.02.325.

Clark, D.B., et al., 2001. Modeling the impact of land surface degradation on the climate of tropical North Africa. J. Clim. 14 (8), 1809–1822. https://doi.org/10.1175/1520-0442(2001)014<1809:MTIOLS>2.0.CO;2.

Demissie, B., et al., 2015. Biophysical controlling factors in upper catchments and Braided rivers in drylands: the case of a marginal graben of the Ethiopian Rift valley. Land Degrad. Dev. 26 (7), 748–758. https://doi.org/10.1002/ldr.2357.

Dong, X., et al., 2016. Temporal variability in hydrology modifies the influence of geomorphology on wetland distribution along a desert stream. J. Ecol. 104 (1), 18–30. https://doi.org/10.1111/1365-2745.12450.

Dregne, H.E., 2002. Land degradation in the drylands. Arid Land Res. Manag. 16 (2), 99–132. https://doi.org/10.1080/153249802317304422.

Epaphras, A.M., et al., 2008. Wildlife water utilization and importance of artificial waterholes during dry season at Ruaha National Park, Tanzania. Wetl. Ecol. Manag. 16 (3), 183–188. https://doi.org/10.1007/s11273-007-9065-3.

Feng, S., Fu, Q., 2013. Expansion of global drylands under a warming climate. Atmos. Chem. Phys. 13 (19), 10081–10094.

Fraser, L.H., Keddy, P.A., 2005. The World's Largest Wetlands: Ecology and Conservation. Cambridge University Press.

Geerken, R., Ilaiwi, M., 2004. Assessment of rangeland degradation and development of a strategy for rehabilitation. Remote Sens. Environ. 90 (4), 490–504. https://doi.org/10.1016/j.rse.2004.01.015.

Gren, I.-M., et al., 1994. Primary and secondary values of wetland ecosystems. Environ. Resour. Econ. 4 (1), 55–74.

Haghighian, F., Yousefi, S., Keesstra, S., 2020. Identifying tree health using sentinel-2 images: a case study on Tortrix viridana L. infected oak trees in Western Iran. Geocarto Int. 1–17. https://doi.org/10.1080/10106049.2020.1716397.

Hall, R.J., et al., 2016. Remote sensing of forest pest damage: a review and lessons learned from a Canadian perspective. Can. Entomol. 148 (S1), S296–S356.

Hames, A.S., 2009. Comparison of four classification methods to extract land use and land cover from raw satellite images for some remote arid areas, Kingdom of Saudi Arabia. J. King. Abdulaziz Univ. Earth Sci 20 (1), 167–191.

Hayri Kesikoglu, M., et al., 2019. Performance of ANN, SVM and MLH techniques for land use/cover change detection at Sultan Marshes wetland, Turkey. Water Sci. Technol. 80 (3), 466–477. https://doi.org/10.2166/wst.2019.290.

Jacqueminet, C., et al., 2013. Land cover mapping using aerial and VHR satellite images for distributed hydrological modelling of periurban catchments: application to the Yzeron catchment (Lyon, France). J. Hydrol. 485, 68–83. https://doi.org/10.1016/j.jhydrol.2013.01.028.

Johnston, R.M., Barson, M.M., 1993. Remote sensing of Australian wetlands: an evaluation of Landsat TM data for inventory and classification. Mar. Freshw. Res. 44 (2), 235–252.

Kandus, P., et al., 2018. Remote sensing of wetlands in South America: status and challenges. Int. J. Rem. Sens. 39 (4), 993–1016.

Kaplan, G., Avdan, U., 2018. Monthly analysis of wetlands dynamics using remote sensing data. ISPRS Int. J. Geo-Inf 7 (10), 411.

Kavzoglu, T., Mather, P.M., 2003. The use of backpropagating artificial neural networks in land cover classification. Int. J. Rem. Sens. 24 (23), 4907–4938.

Liu, L., Wang, H., Yue, Q., 2020. 'China's coastal wetlands: ecological challenges, restoration, and management suggestions. Reg. Stud. Mar. Sci. 37, 101337.

Majumdar, S.K., et al., 1989. Wetlands Ecology and Conservation Emphasis in Pennsylvania. The Pennsylvania Academy of Science Publications, Easton, PA (United States).

Meseșan, F., et al., 2019. Reconstructing snow-avalanche extent using remote sensing and dendrogeomorphology in Parâng Mountains. Cold Reg. Sci. Technol. 157, 97–109. https://doi.org/10.1016/j.coldregions.2018.10.002.

Messina, M.G., Conner, W.H., 1997. Southern Forested Wetlands: Ecology and Management. CRC Press.

Mitsch, W.J., Gosselink, J.G., 2015. Wetlands. John Wiley & Sons.

Nanzad, L., et al., 2019. NDVI anomaly for drought monitoring and its correlation with climate factors over Mongolia from 2000 to 2016. J. Arid Environ. 164, 69–77. https://doi.org/10.1016/j.jaridenv.2019.01.019.

Ozesmi, S.L., Bauer, M.E., 2002. Satellite remote sensing of wetlands. Wetl. Ecol. Manag. 10, 381–402.

Pal, S., Talukdar, S., 2018. Application of frequency ratio and logistic regression models for assessing physical wetland vulnerability in Punarbhaba river basin of Indo-Bangladesh. Hum. Ecol. Risk Assess. 24 (5), 1291–1311. https://doi.org/10.1080/10807039.2017.1411781.

Rundquist, D.C., Narumalani, S., Narayanan, R.M., 2001. A Review of Wetlands Remote Sensing and Defining New Considerations. Taylor & Francis.

Sharitz, R., 1986. Remote sensing inland wetlands: a multispectral approach. Photogramm. Eng. Rem. Sens. 52 (1), 87–100.

Thenkabail, P.S., Gamage, M.S.D.N., 2004. He Use of Remote Sensing Data for Drought Assessment and Monitoring in Southwest Asia. Iwmi.

Vogelmann, J.E., Tolk, B., Zhu, Z., 2009. Monitoring forest changes in the southwestern United States using multitemporal Landsat data. Remote Sens. Environ. https://doi.org/10.1016/j.rse.2009.04.014.

Yousefi, S., et al., 2015. Accuracy assessment of land cover/land use classifiers in dry and humid areas of Iran. Environ. Monit. Assess. 187 (10), 641. https://doi.org/10.1007/s10661-015-4847-1.

Yousefi, S., et al., 2022. Image classification and land cover mapping using Sentinel-2 imagery: optimization of SVM parameters. Land 11 (7), 993.

CHAPTER

10

Machine learning modeling of the wind-erodible fraction of soils

Bahram Choubin[1], *Nikoo Hamzehpour*[2], *Fereshteh Alizadeh*[2] *and Amir Mosavi*[3,4]

[1]Soil Conservation and Watershed Management Research Institute, West Azarbaijan Agricultural and Natural Resources Research and Education Center, AREEO, Urmia, Iran; [2]Department of Soil Sciences, Faculty of Agriculture, University of Maragheh, Maragheh, Iran; [3]Institute of Automation, Kalman Kando Faculty of Electrical Engineering, Obuda University, Budapest, Hungary; [4]Institute of Advanced Studies Koszeg (IASK), Koszeg, Hungary

10.1 Introduction

Earth processes like wind erosion and dust production play an important role in arid and semiarid regions, which can lead to increased soil salinity and decreased plant productivity (Lal, 1990). Dust consists mostly of small, fine particles that are typically from desert and arid areas due to low humidity, low vegetation coverage, and lack of soil cements. This results in weak and small soil aggregates that can easily be detached and transported by wind (Prospero et al., 2002; Colazo and Buschiazzo, 2010).

Studies have revealed that dust from dry lakes and playas can be a major source of airborne particles in the world. These particles can contain large amounts of harmful chemicals and other pollutants, which can cause health problems for people who breathe them in (e.g., Prospero et al., 2002). On the other hand, the salinization of groundwater in specific regions near Saline Lake and the use of saline water in irrigation have led to the secondary salinization of agricultural lands. Therefore, agricultural farms are becoming more economically unbeneficial for farmers, resulting in abandoning their lands. In such barren areas, soil aggregates are small due to their years of breakdown by miss-tillage practices along with a lack of soil organic matter, which is the major soil aggregating agent. Therefore, these areas (such as the Urmia Lake basin in Iran) are of great potential in becoming dust sources.

Remote Sensing of Soil and Land Surface Processes
https://doi.org/10.1016/B978-0-443-15341-9.00003-4

10.2 Wind-erodible fraction

The erodible fraction of soil is the percentage of soil that can be eroded by wind. There are several factors that affect wind erosion in different regions. These include the type of terrain, how much sand or gravel is present, and whether there is a river running through the area. In arid regions, small particles (<0.84 mm) in the soil are an important indicator of soil susceptibility to wind erosion. It has been found that aggregates smaller than 0.84 mm in diameter are easily eroded by wind, and their proportion in the upper 2.5 cm of the soil surface determines the soil's susceptibility to wind erosion (Chepil, 1953; Colazo and Buschiazzo, 2010), and thus it is considered as wind erodible fraction (WEF).

For calculating the WEF, after taking the soil samples from the soil depths of 0–5 cm, the samples are processed by passing a 2 mm sieve. The final product is then considered for physicochemical analysis. The rotary sieve stability test according to Nimmo and Perkins (2002) was used to determine the erodible fraction of soil:

$$\mathrm{WEF} = \frac{W < 0.84}{\mathrm{TW}} \times 100 \tag{10.1}$$

where WEF is wind erodible fraction of soil in (%), $W < 0.84$ is the weight (g) of aggregates <0.84 mm, and TW is the initial weight (g) of the total sample.

10.3 Machine learning

Machine learning (ML) is a type of technology that allows software applications to learn from data in order to make better predictions. This technology can be used by managers to improve their accuracy when making decisions, without being explicitly programmed to do so. Statistical methods such as multiple linear regression (MLR) and partial least squares (PLS) identify the linear relationship between the independent and dependent variables. MLR has been widely used in hydrological modeling. PLS is known as a bilinear factor model, which shares similarities with principal components regression; yet instead of using hyperplanes of maximum variance for independent variables, it uses linear regression. Unlike statistical methods, ML is a type of inquiry that seeks to improve performance on a task by learning from data. For example, Random Forest (RF) is an ensemble learning method that constructs a multitude of decision trees using the bagging technique and the number of variables available for splitting at each tree node. In the RF literature, this is referred to as the *mtry* parameter that had a strong influence on predictor variable importance estimates. However, the application of the ML models for WEF modeling is a new approach and has not been well documented.

10.4 Urmia Lake as an example

Although Playa (ULP) seems to be developed for a few decades, there have not been many studies focusing on the consequences of its drying on surrounding areas and their susceptibility to dust production. Thus, the present research designed to study the soil (WEF) in

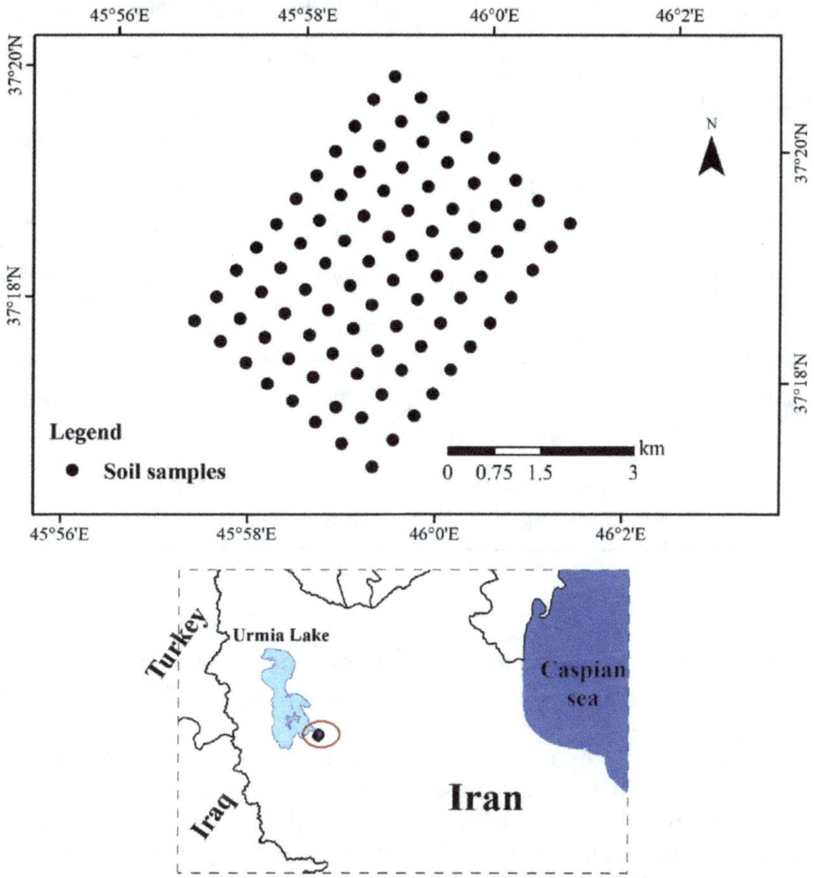

FIGURE 10.1 Study area and soil samples' locations.

southeastern Urmia Lake Playa and adjustment agricultural lands. For the first time, this research models the WEF in southeastern Urmia Lake Playa and the adjustment of agricultural lands using data-driven methods such as MLR, PLS regression, and RF.

The area of study is in north-western Iran and the southeastern segment of the Urmia Lake basin (45° 57′ 11″ to 46° 1′ 50″ E and 37° 16′ 45″ to 37° 20′ 17″ N), having an approximate area of 32.2 km^2 (Fig. 10.1). The survey area is located in the eastern edge of the Bonab Plain. The area features an annual precipitation of 265 mm (Iran Ministry of Energy, 2014). Geologically, the playa consists of alluvial deposits from the Quaternary.

10.5 Physicochemical analysis and soil sampling

In total, 88 soil samples are taken on a grid of 500 m from the depth of 0–5 cm. The samples are processed by passing a 2 mm sieve. The final product is then considered for physicochemical analysis. Table 10.1 indicates summary statistics of the predictive variables used

TABLE 10.1 Summary statistics of the predictive variables in the 88 soil samples.

Variable	Maximum	Minimum	Mean	Standard deviation
Electrical conductivity (EC)	107.50	0.33	9.34	19.01
pH	8.97	7.10	8.16	0.41
Na (meq/L)	436.92	0.92	25.65	60.16
Ca + Mg (meq/L)	108.00	4.00	25.97	15.93
Sodium adsorption ratio (SAR)	48.25	0.23	4.40	8.07
Organic carbon (OC)	1.13	0.00	0.37	0.18
Calcium carbonate equivalent (CCE)	30.58	12.45	18.04	3.83
Wet aggregate stability (WAS) (%)	87.11	0.00	16.04	18.57
Clay (%)	36.00	2.00	11.93	6.96
Silt (%)	56.00	6.00	24.89	12.61
Sand (%)	90.00	14.00	63.07	17.26

to WEF modeling. The physicochemical analysis has been arranged and implemented according to Chapman (1965), Page et al. (1982), Nelson and Sommers (1983), and Gee and Or (2002) as follows:

The soil electrical conductivity (EC), acidity (pH), Mg^{2+}, Na^{+}, and Ca^{2+} are determined in the samples' paste saturation extracts. These measurements help understand the soil's ability to hold water and its chemical composition. The back titration method is used to determine the calcium carbonate equivalent (CCE). Cation exchange capacity (CEC) was determined with ammonium acetate (1 N NH_4OAc) in pH value of 7. Soil texture was measured with a hydrometer. Soil organic carbon (OC) content was measured by wet oxidation technique. The wet sieving method was used to determine wet aggregate stability. Sodium adsorption ratio (SAR) was determined using following equation according to Soil Survey Staff (2014):

$$SAR = \frac{\text{soulable Na}(\text{mmol}(+).\text{l}^{-})}{\sqrt{\frac{\text{soulable}(\text{Ca}+\text{Mg})(\text{mmol}(+).\text{l}^{-})}{2}}} \tag{10.2}$$

10.6 Wind-erodible fraction modeling

Wind-erodible fraction modeling (EF), in this research, was conducted by the MLR,PLS, and RF using the predictive variables presented in Table 10.1. The Caret R package (Kuhn, 2015) with k-fold cross-validation methodology was used to calibrate the models. In this package, the optimum values of parameters for each model are recognized using the tuneLength with an objective function of minimizing the root mean square error (RMSE). A ratio 70%–30% of the data was considered for training and testing phases.

Modeling evaluation was conducted using the metrics of RMSE, Nash and Sutcliffe (NSE), RMSE to standard deviation ratio (RSR), index of agreement (d), and correlation (R) (Nash and Sutcliffe, 1970; Willmott, 1981; Moriasi et al., 2007) (Eqs. (10.3)–(10.6)):

$$\text{RMSE} = \sqrt{\frac{\sum_{i=1}^{N} (O_i - S_i)^2}{N}} \tag{10.3}$$

$$\text{NSE} = 1 - \frac{\sum_{i=1}^{N} (O_i - S_i)^2}{\sum_{i=1}^{n} (O_i - \overline{O})^2} \tag{10.4}$$

$$\text{RSR} = \frac{\text{RMSE}}{\text{STDEV}_{\text{obs}}} = \frac{\left[\sqrt{\sum_{i=1}^{N} (O_i - S_i)^2}\right]}{\left[\sqrt{\sum_{i=1}^{N} (O_i - \overline{O})^2}\right]} \tag{10.5}$$

$$d = 1 - \frac{\sum_{i=1}^{N} (O_i - S_i)^2}{\sum_{i=1}^{n} \left(|S_i - \overline{O}| + |O_i - \overline{O}|\right)^2} \tag{10.6}$$

where N is the number of data; O_i and S_i respectively are the observed and simulated values, and $\overline{O}$ and $\overline{S}$ are their means. A value 0 for *RMSE* and *RSR* and a value 1 for *NSE*, and *d* indicate a perfect match between observed and simulated data.

10.7 Modeling results

The WEF was modeled using 10 cross-validation methodology. Fig. 10.2 indicates the modeling error (mean RMSE across 10 cross-validations) to find the best number of principal components (*ncomp*) incorporated in the PLS model. We tested 10 different values of the tuning parameter *ncomp*. The results indicated that choosing two principal components (*ncomp* = 2) gives the smallest modeling error (RMSE = 83.7) (Fig. 10.2). Also, in the RF model, the best number of variables randomly selected for splitting at each tree node (*mtry*) was identified. The results indicated that the RF model with value equal to nine for *mtry* parameter has the smallest modeling error (RMSE = 76.7) (Fig. 10.3).

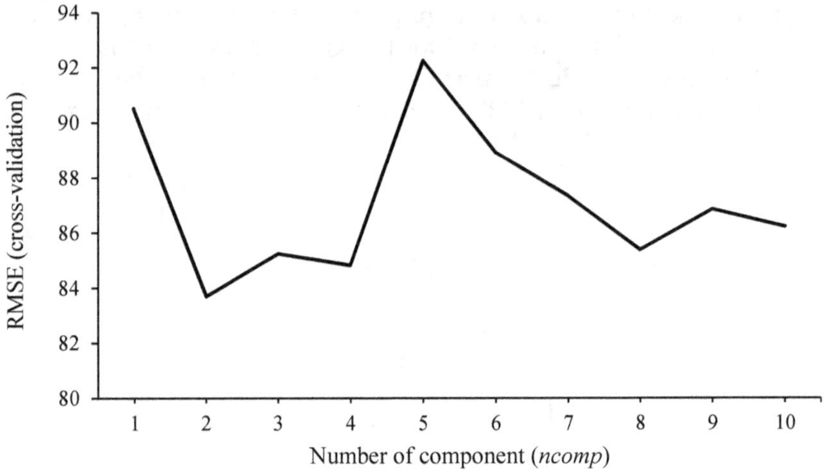

FIGURE 10.2 The optimal number of principal components (*ncomp*) in the partial least squares (PLS) model.

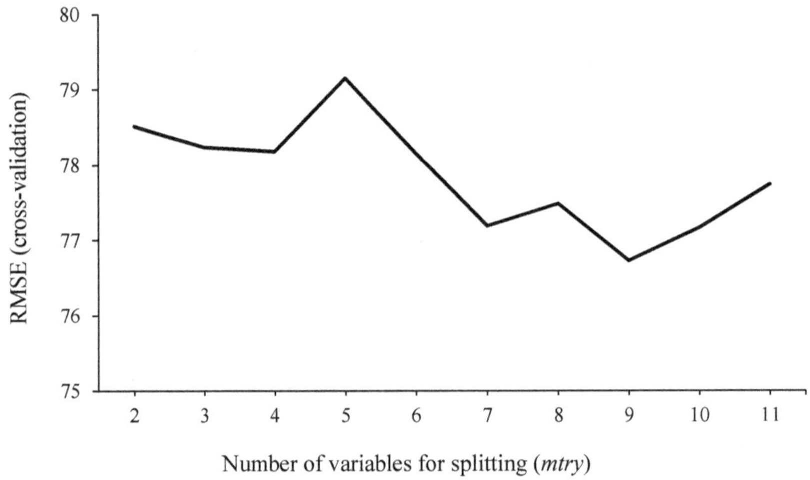

FIGURE 10.3 Number of variables randomly selected for splitting at each tree node by Random Forest (RF) model.

After tuning the parameters, the modeling process was conducted and the results for training and testing datasets were evaluated. Results of the modeling are represented in Table 10.2. A low value of RMSE (and RSR) and a high value of NSE (and d) indicate a better performance of the modeling results. As can be seen, the RF model has lower values of error (RMSE = 34.32, RSR = 0.35) and higher values of efficiency and agreement (NSE = 0.87, d = 0.96) for training dataset rather than other models (Table 10.2). Also, for testing dataset, the RF model indicated a better result (RMSE = 63.35, NSE = 0.63, RSR = 0.60, d. = 0.85), followed by the PLS (RMSE = 65.41, NSE = 0.60, RSR = 0.62, d = 0.81) and MLR (RMSE = 71.70, NSE = 0.52, RSR = 0.68, d = 0.76) models (Table 10.2). Also, scatter plots

TABLE 10.2 Results of the modeling performance.

Metric	MLR	PLS	RF
Training dataset			
RMSE	73.50	78.85	34.32
NSE	0.41	0.33	0.87
RSR	0.76	0.81	0.35
d	0.76	0.69	0.96
Testing dataset			
RMSE	71.70	65.41	63.35
NSE	0.52	0.60	0.63
RSR	0.68	0.62	0.60
d	0.76	0.81	0.85

between observed and simulated WEF data are presented in Fig. 10.4. As can be seen, the RF model has a good correlation between observed and simulated values for both training and testing datasets (respectively equal to 0.96 and 0.81).

Generally, our results indicated that the RF model, as an ensemble learning method, had a better result for WEF modeling rather than the statistical methods (i.e., PLS and MLR). Application of these models for modeling of the WEF was novel, and there is not any previous study to compare the results.

10.8 Main contributors

Contribution of the variables during the modeling process is presented in Fig. 10.5. As can be seen, contribution of the clay (100%), sand (78.6%), CCE (73.7%), and silt (64.1) is greater than other factors during modeling process by the MLR model. According to the PLS model, variables of sand (100%), silt (68.3%), NA (40.3%), and clay (33.4%) were the most important variables. Also, the RF model result indicated that the greatest contribution of inputs in the modeling process was respectively related to clay (100%), sand (49.9%), silt (33.3%), and CCE (28.2%) (Fig. 10.5). Therefore, according to the modeling results, variables of clay, sand, silt, and CCE were respectively the most important variables. Also, the Ca + Mg and OC were moderate important variables. Soil erodibility is most affected by soil texture and organic matter. These factors collect individual soil particles, which in turn helps improve soil aggregation (Fryrear et al., 1994; Zobeck et al., 2013). Sirjani et al. (2019) found that wind erosion rates were negatively related to soil properties, including clay and organic matter contents. According to Charman and Murphy (2007), there is a negative relationship between soil clay content and soil susceptibility to erosion. This means that when the levels of soil clay are high, it may be more difficult for soils to resist erosion. Soil aggregation is one of the

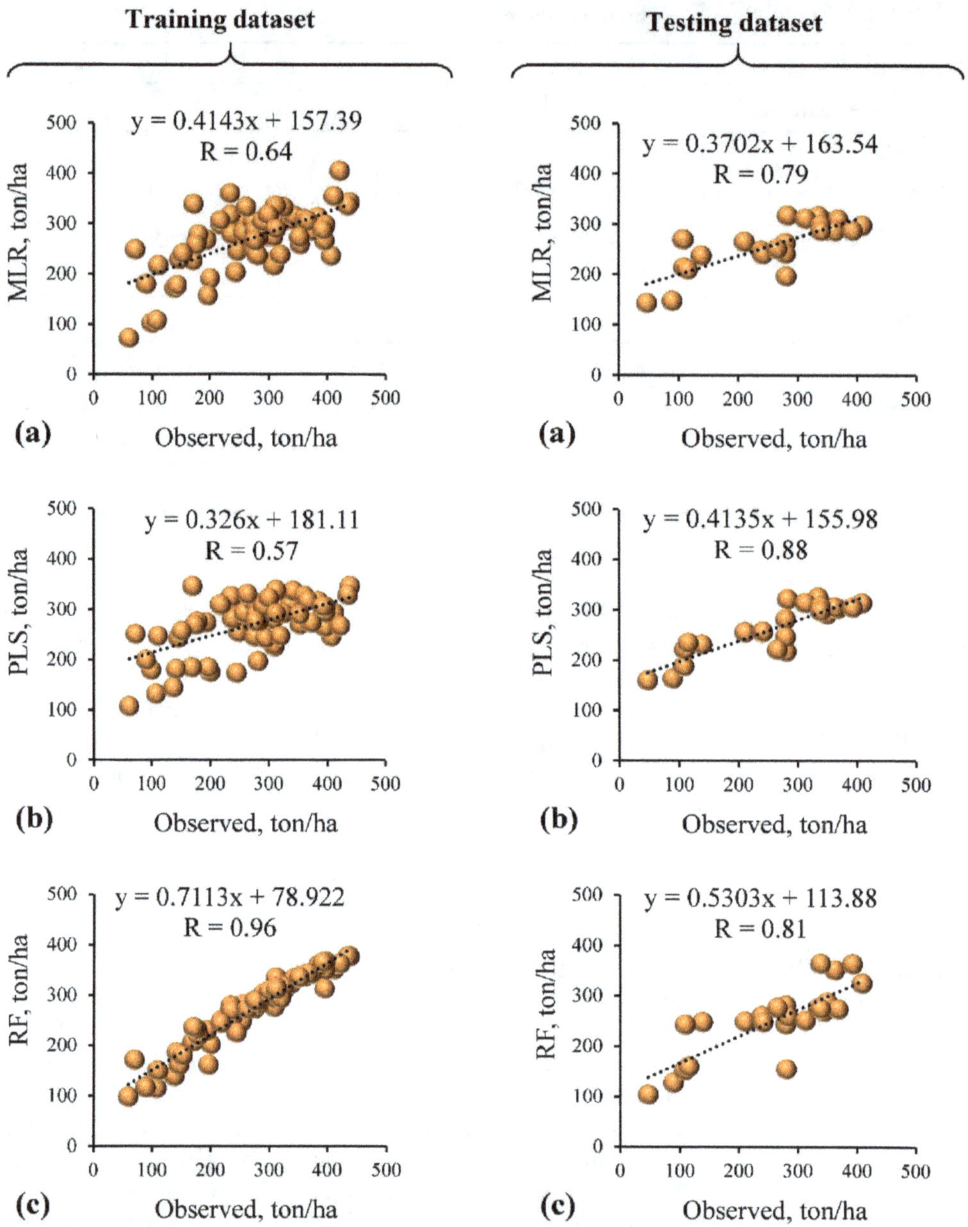

FIGURE 10.4 Scatter plot between observed and simulated wind-erodible fraction (WEF) data for training (*left*) and testing (*right*) dataset: (A) Multiple linear regression (MLR), (B) partial least squares (PLS), and (C) Random Forest (RF).

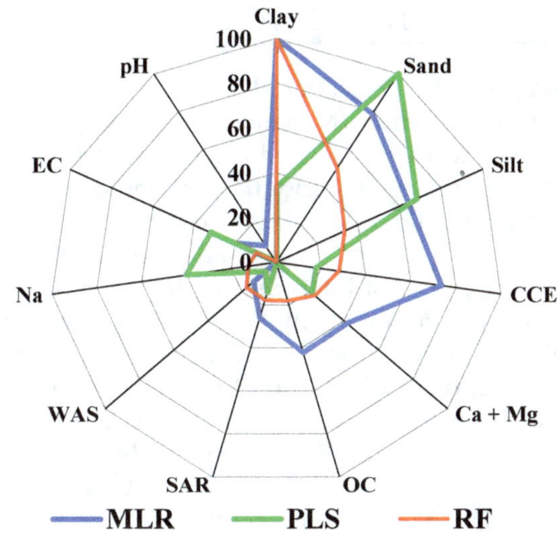

FIGURE 10.5 Importance of the predictor variables in the wind-erodible fraction modeling.

primary factors that can affect soil susceptibility to wind erosion. Clay, silt, and organic content in the soil play important roles in binding particles together into larger aggregates. A large soil aggregate allows more soil to be eroded by wind and water than smaller aggregates. This means the land is less erosion-resistant, which can lead to problems with valuable resources being lost over time. According to Shahabinejad et al. (2019), the amount of clay and silt in an area affects how quickly wind erosion takes place, while the amount of sand creates a positive relationship. They have also shown that the formation of macro-aggregates in the existence of silt and especially clay particles can reduce wind erosion while meso-aggregates with predominance of sand particles were proportional to wind erosion.

The positive relationship between sand content and WEF seems to be due to lower coherence of the sand particles because there is less salt, clay, and silt present. This can lead to the formation of larger soil aggregates (Colazo and Buschiazzo, 2010; Shahabinejad et al., 2019). Osman (2014) reported that high sand content and low organic matter in arid and semiarid regions makes them more susceptible to wind erosion.

Calcium carbonate equivalent also showed significantly negative correlation with EF. The role of calcium carbonate in soil susceptibility to wind erosion is still being investigated, but it is thought that it may play a role. In some cases, there was no correlation found between calcium carbonate levels and soil erodibility (Chepil, 1954; Colazo and Buschiazzo, 2010). There is limited evidence that calcium carbonate can improve aggregation in loamy and sandy soils. However, it could reduce wind erodibility of these soils acting as cementing agent and increasing aggregate stability (Chepil, 1954; Tatarko, 2001; Sirjani et al., 2019). The soils considered in the current study had high sand and CCE contents with mean values of 63% and 18%, respectively. CCE has played a cementing role and has improved soil aggregation and decreased soil erodibility factor.

10.9 Conclusions

This chapter introduced the WEF and its modeling using some data-driven methods. Using different soil characteristics and three (ML) models including MLR, PLS, and RF, the WEF modeling were examined in southeastern Urmia Lake Playa and the adjustment agricultural lands. Regarding to the ML modeling, contributions of the clay (100%), sand (78.6%), CCE (73.7%), and silt (64.1) were greater than other factors during modeling process by the MLR model. According to the PLS model, variables of sand (100%), silt (68.3%), NA (40.3%), and clay (33.4%) were the most important variables. Also, the RF model result indicated that the greatest contribution of inputs in the modeling process was respectively related to clay (100%), sand (49.9%), silt (33.3%), and CCE (28.2%). Therefore, according to the modeling results, variables of clay, sand, silt, and CCE were respectively the most important variables that had a higher contribution across the modeling process. The modeling results showed that the RF model outperforms (RMSE = 63.35, NSE = 0.63, RSR = 0.60, d = 0.85) other models. It had a good correlation with observed values for both training and testing datasets (respectively equal to 0.96 and 0.81). Therefore, ML modeling can help for the estimation of WEF and reduce sampling costs and time.

References

Chapman, H.D., 1965. Cation-exchange capacity. Methods of Soil Analysis: Part 2 Chemical and Microbiological Properties, vol 9, pp. 891–901.

Charman, P.E.V., Murphy, B.W. (Eds.), 2007. Soils, Their Properties and Management. Oxford University Press, USA.

Chepil, W.S., 1953. Field structure of cultivated soils with special reference to erodibility by wind. Soil Sci. Soc. Am. J. 17 (3), 185–190.

Chepil, W.S., 1954. Factors that influence clod structure and erodibility of soil by wind: III. Calcium carbonate and decomposed organic matter. Soil Sci. 77 (6), 473–480.

Colazo, J.C., Buschiazzo, D.E., 2010. Soil dry aggregate stability and wind erodible fraction in a semiarid environment of Argentina. Geoderma 159 (1–2), 228–236.

Fryrear, D.W., Krammes, C.A., Williamson, D.L., Zobeck, T.M., 1994. Computing the wind erodible fraction of soils. J. Soil Water Conserv. 49 (2), 183–188.

Gee, G.W., Or, D., 2002. 2.4 particle-size analysis. Methods of Soil Analysis: Part 4 Physical Methods, vol 5, pp. 255–293.

Iran Ministry of Energy, 2014. Deputy of water and wastewater, macro planning Bureau. The National Water Master Plan Study in the Aras, Sefidrood, between Sefidrood and Haraz, Atrac and Urmia: Agricultural Water Use Study in Urmia Lake Basin. Report Number: 2385070-4420-19464.

Kuhn, M., 2015. Caret: Classification and Regression Training. Astrophysics Source Code Library pp.ascl-1505.

Lal, R., 1990. Soil erosion and land degradation: the global risks. Adv. Soil Sci. Soil Degrad. 11, 129–172.

Moriasi, D.N., Arnold, J.G., Van Liew, M.W., Bingner, R.L., Harmel, R.D., Veith, T.L., 2007. Model evaluation guidelines for systematic quantification of accuracy in watershed simulations. Trans. ASABE 50 (3), 885–900.

Nash, J.E., Sutcliffe, J.V., 1970. River flow forecasting through conceptual models part I—a discussion of principles. J. Hydrol. 10 (3), 282–290.

Nelson, D.A., Sommers, L., 1983. Total carbon, organic carbon, and organic matter. Methods of Soil Analysis: Part 2 Chemical and Microbiological Properties, vol 9, pp. 539–579.

Nimmo, J.R., Perkins, K.S., 2002. 2.6 Aggregate stability and size distribution. Methods of Soil Analysis: Part 4 Physical Methods, vol 5, pp. 317–328.

Osman, K.T., 2014. Soil Degradation, Conservation and Remediation, vol 820. Springer Netherlands, Dordrecht.

Page, A.L., Page, A.L., 1982. Methods of Soil Analysis: Chemical and Microbiological Proerpteis. Amen Society of Agronomy.

Prospero, J.M., Ginoux, P., Torres, O., Nicholson, S.E., Gill, T.E., 2002. Environmental characterization of global sources of atmospheric soil dust identified with the Nimbus 7 Total Ozone Mapping Spectrometer (TOMS) absorbing aerosol product. Rev. Geophys. 40 (1), 1–2.

Shahabinejad, N., Mahmoodabadi, M., Jalalian, A., Chavoshi, E., 2019. The fractionation of soil aggregates associated with primary particles influencing wind erosion rates in arid to semiarid environments. Geoderma 356, 113936.

Sirjani, E., Sameni, A., Moosavi, A.A., Mahmoodabadi, M., Laurent, B., 2019. Portable wind tunnel experiments to study soil erosion by wind and its link to soil properties in the Fars province, Iran. Geoderma 333, 69–80.

Skidmore, E.L., Hagen, L.J., Armbrust, D.V., Durar, A.A., Fryrear, D.W., Potter, K.N., Wagner, L.E., Zobeck, T.M., 2017. Methods for investigating basic processes and conditions affecting wind erosion. Soil Erosion Research Methods, pp. 295–330.

Soil Survey Staff, 2014. Keys to Soil Taxonomy, twelfth ed. USDA. NRCS, p. 372.

Tatarko, J., 2001. Soil aggregation and wind erosion: processes and measurements. Ann. Arid Zone 40 (3), 251–264.

Willmott, C.J., 1981. On the validation of models. Phys. Geogr. 2 (2), 184–194.

Zobeck, T.M., Baddock, M., Van Pelt, R.S., Tatarko, J., Acosta-Martinez, V., 2013. Soil property effects on wind erosion of organic soils. Aeolian Res. 10, 43–51.

CHAPTER

11

Application of remote sensing techniques for evaluating land surface vegetation

Esmaeil Heydari Alamdarloo, Azam Abolhasani, Maliheh Behrang Manesh and Hassan Khosravi

Department of Reclamation of Arid and Mountainous Regions, Faculty of Natural Resource, University of Tehran, Tehran, Iran

11.1 Introduction

As the most important living part of the ecosystem, plants interact with soil, climate, hydrology, and other elements in the entire ecosystem (Moradi et al., 2023). Vegetation cover is not only the main part of terrestrial ecosystems but also an important intermediary for energy exchanges, water cycle, and biogeochemical cycles on the terrestrial surface (Sun et al., 2021). Especially in fragile ecosystems like semi-arid regions, vegetation provides vital products and services for the local ecosystem and has an important role in controlling the process of land degradation and desertification (Ryan et al., 2016). It is estimated that vegetation change threatens the health of one-sixth of the world's population and one-quarter of terrestrial ecosystems (Jabbar et al., 2006). Having a series of characteristics such as immobility, relatively high growth rate, large expansion, and tolerance vegetation is considered a widely used biological criterion in ecological evaluations (Raab and Bayley, 2012). Also, due to the sensitivity of vegetation to climate variability and fluctuations, its dynamic is used as a criterion for examining global changes in terrestrial ecosystems (Peng et al., 2012).

Climate fluctuations and changes and human activities are the most important factors affecting the dynamics of vegetation. Several researchers believe that climate factors may be more important in vegetation dynamics compared to human activities (Liu et al., 2016). In other words, vegetation is more sensitive to climatic factors (Mu et al., 2013). In today's era, when climate evolution has a great influence on social development (Mishra et al., 2010; Schmidhuber and Tubiello, 2007; Funatsu et al., 2019), environment, and human health (Caminade et al., 2019), and many global problems like food incertitude, water depletion,

https://doi.org/10.1016/B978-0-443-15341-9.00006-X

biodiversity loss, and health-related issues are dependent on fluctuations and changes of climate, knowing the patterns of climate fluctuations and changes and their effects is very important (Levy et al., 2019). It is predicted that with climate changes and fluctuations, the destruction of forests, pastures, and other ecosystems will happen and ultimately lead to the expansion of land degradation and desertification.

The appropriate growth of vegetation not only protects water and soil but also causes the creation and development of a microclimate, unlike weak vegetation that causes erosion and land degradation (Li et al., 2021). Therefore, vegetation monitoring, both quantitatively and qualitatively, is very necessary to plan for a sustainable ecosystem (Jiang et al., 2023).

Recently, in some studies, vegetation dynamics has been used as an indicator to reveal the result of anthropogenic factors and climate variability on land degradation and desertification (Eskandari et al., 2016). It should be noted that the relationship between land degradation and desertification and vegetation cover is a two-way relationship. In other words, the destruction of vegetation increases the speed of land destruction and desertification, and increasing the speed of these phenomena increases the intensity of vegetation destruction.

In vegetation studies, the most significant challenge is collecting the primary data related to them. One of the most important tools for solving this problem is remote sensing-based data. Data based on remote sensing provides the possibility of damaged vegetation detection and assessing the severity of the damage in different types of land (Rimkus et al., 2017). Compared to studies based on traditional methods, using remote sensing-based data for monitoring vegetation and its changes in a large area has many advantages such as saving time and money (Pei et al., 2018). Another problem in investigating vegetation dynamics is the lack of accurate spatial information from the past. Considering the accessibility of satellite data at different times and different spectral bands, methods based on time series of remote sensing data are increasingly used to monitor a wide range of remotely detectable features of vegetation (Lawley et al., 2016). Generally, the purpose of this chapter is to evaluate the general usage of remote sensing methods for the study and monitoring of vegetation cover.

11.2 Vegetation indices

Employing vegetation spectral indicators is a general view in the field of vegetation monitoring and investigation. Vegetation indices which were first developed in the 1970s are now essential tools in climate change studies, drought monitoring, land cover classification, land use detection, and ecosystem dynamics. vegetation indices are mathematical expressions that combine reflectance values in many different spectral bands to obtain a new numerical value that can be used to evaluate crop growth and establishment, as well as other plant characteristics such as chlorophyll and biomass (Ferchichi et al., 2022). The reflection of electromagnetic waves from vegetation depends on the type of them, the water extent in tissues, and other inherent features (Liu et al., 2016). Three spectral ranges including the ultraviolet spectrum (10–380 nm), the visible range including the blue, green, and red bands (450–495, 495–570, and 620–750 nm respectively), and the near and mid-infrared range (850–1700 nm) are the most used spectral ranges in vegetation assessment by remote sensing (Bin Abdul Rahim et al., 2016; Cruden et al., 2012). Of course, most researchers often combine the red and near-infrared bands in different ways related to their goals (Foley et al., 1998).

11.3 Various types of vegetation indices

Remote sensing–based vegetation indicators were first divided into distance-based and slope-based indices (Jackson and Huete, 1991). Then, another group was added to this category as orthogonal transformation vegetation indices. Here, we added another group to this classification called temporal vegetation indices. In the following, we tried to explain these groups and introduce the most important indicators of them.

11.3.1 Slope-based indices

Slope-based indices are obtained from the ratio between two bands (usually the ratio of infrared to red), and the pixel value indicates the slope of a line that passes between these two bands. The indices of this group are a combination of simple calculations that centralize the contradiction between the vegetation spectral reflex patterns, generally in the infrared and red parts of the electromagnetic spectrum. Any specific value of these indices can be created with reflectance values of red and infrared bands that construct a two-dimensional graph line of the reflectance of the red band versus infrared band (Fig. 11.1).

The most important indices of this group are as follows.

11.3.1.1 Ratio

This index was developed by Rouse et al. (1974) for characterizing vegetation using Landsat MSS images (Eq. 11.1):

$$\text{Ratio} = \frac{\text{NIR}}{R} \tag{11.1}$$

where R is the reflection of the red band and NIR is the reflection of the near-infrared band. This index shows the difference between pixels with vegetation cover. Since it is a ratio, it

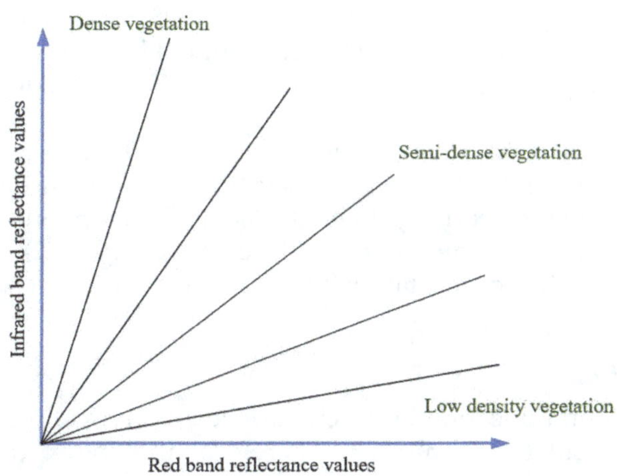

FIGURE 11.1 The basis of slope-based vegetation index.

minimizes the problems related to the topography effect. Due to being sensitive to the denominator zero, the measurement scale is not linear. Therefore, this index does not have a normal distribution.

11.3.1.2 Normalized difference vegetation index

This normalized difference vegetation index, presented by Rouse et al. (1974), is among the most famous vegetation indices that have been used in most research and satellite surveys to determine the intensity and weakness of vegetation cover (Eq. 11.2):

$$\mathrm{NDVI} = \frac{\mathrm{NIR} - R}{\mathrm{NIR} + R} \tag{11.2}$$

where NIR is the reflection of the near-infrared band and R is the reflection of the red band. Its range varies from −1 to +1. When the vegetation cover is dense, the index approaches +1, as the cover decreases, the index also decreases, so that zero and close to it represent barren soil, and −1 represents water.

Researchers stated different values for barren soil. Among them, Ding et al. (2016) announced 0.07 to 0.2, and Piao et al. (2006) determined less than 0.1 for barren lands. Therefore, it can be concluded that this value is different according to the conditions of the studied area.

11.3.1.3 Enhanced vegetation index (EVI)

In areas with sparse vegetation, the complex combination of soil type and atmospheric and vegetation effect reduces the possibility of extracting reliable information from satellite data. Therefore, efforts have always been done to provide vegetation indicators that can reduce the unwanted effects of factors such as soil. Huete et al. (1999) have proposed the EVI for use in MODIS data (Eq. 11.3):

$$\mathrm{EVI} = \frac{\mathrm{NIR} - R}{\mathrm{NIR} + \mathrm{C1} \times R - \mathrm{C2} \times B + L}(1 + L) \tag{11.3}$$

where NIR, R, and B are the reflections of the near-infrared band, the red band, and the blue band, respectively. L is equal to one, C1 is equal to 6, and C2 is equal to 7.5. This index varies from +1 to −1.

One of the advantages of this index is the correction of the background soil effects and the reduction of the atmosphere and aerosol distribution effects through blue band reflectance in the range leaf area index (LAI) (Li et al., 2009). EVI is an improved NDVI in which the soil modification agent (L) and two coefficients (C1 and C2) are considered. This method increases sensitivity to vegetation cover and reduces atmospheric effects (Huete et al., 2002).

11.3.1.4 Nonlinear index

In this index, the communication between surface biophysical parameters and vegetation indicators is considered nonlinear. Therefore, the index tries to linearize the connection between them. The following formula represents how to compute this index (Eq. 11.4):

$$\mathrm{NLI} = \frac{(\mathrm{NIR}^2 - R)}{(\mathrm{NIR}^2 + R)} \tag{11.4}$$

where NIR is the reflection of the near-infrared band and R is the reflection of the red band.

11.3.1.5 Global environmental monitoring index

This is a nonlinear index for monitoring the global environment using satellite imagery, which tries to refine atmospheric effects. Global environmental monitoring index (GEMI) is like NDVI but less susceptible to atmospheric agents and influenced by barren soil. The following equations indicate how to calculate this index (Eqs. 11.5 and 11.6):

$$\mathrm{GEMI} = \eta \times (1 - 0.25 \times \eta) - \left(\frac{R + 0.125}{1 - R}\right) \tag{11.5}$$

$$\eta = \frac{2 \times (\mathrm{NIR}^2 - R^2) + (1.5 \times \mathrm{NIR}) + (0.5 \times R)}{\mathrm{NIR} + R + 0.5} \tag{11.6}$$

11.3.2 Distance-based indices

Compared to the previous group of vegetation indices, in this group, the amount of existing vegetation is evaluated by measuring the reflection difference of each pixel from the reflection of the bare soil. In general, in the two-dimensional diagram of the reflection of the red band and the infrared band, barren soil pixels with different moisture levels form a line, which is called the soil line. In a pixel, as the vegetation cover augments, the background of the soil continuously becomes blurred and disappears, and the vertical distance of the pixel position in the two-dimensional diagram of the red band and infrared band reflectance from the soil line increases. The distance-based index is defined based on the same principle. Therefore, to calculate them, the soil line slope and its intercept are needed. The main purpose of these vegetation indices is to remove the influence of soil illumination in cases where the vegetation cover is low and the pixels are a combination of the vegetation cover and the soil background. This is very momentous in arid lands. The basis of these indices is the soil line that was described. The soil line is gained through the linear regression of the near-infrared band against the red band for a sample of barren soil pixels. Pixels close to the soil line are known as soil, while those that appear far away are water or vegetation (Fig. 11.2). Unfortunately, there is a lot of inconsistency in the logic and method for developing and calculating the soil line. Some researchers use the red band and others use the infrared band as independent variables for regression.

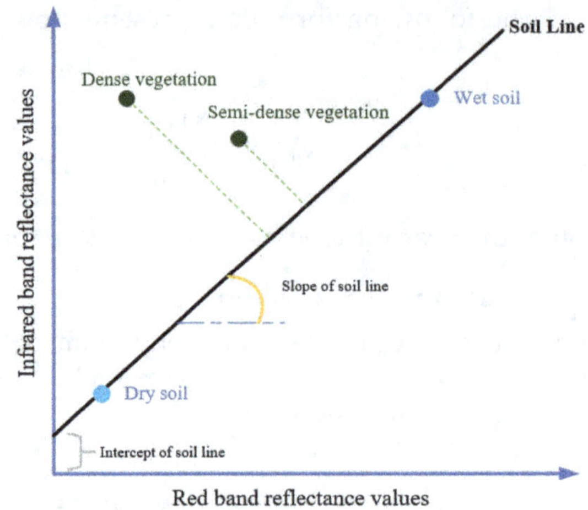

FIGURE 11.2 The basis of distance-based vegetation index.

In the first method, the values of the red band of bare lands are considered independent variables and the near-infrared band values of bare lands are considered dependent variables, unlike the second method in which the values of the near-infrared band of bare lands are considered independent variables, and the red band values of bare lands are considered dependent variables (Eastman, 2020). The most important indicators of this group are as follows.

11.3.2.1 Perpendicular vegetation index

This index which was introduced by Richardson and Wiegand (1977) is the basis of distance-based indices. According to the definition of this index, the vertical distance of each point in the diagram of the red band and infrared band from the soil line is equivalent to PVI. The efforts to improve the efficiency of this index led to the creation of three other indices.

11.3.2.1.1 PVI1

The PVI1 index was proposed by Perry and Lautenschlagert (1984) to reduce the limitation of PVI in separating vegetation from water (Eq. 11.7):

$$\mathrm{PVI1} = \frac{(a \times \mathrm{NIR} - R + b)}{\sqrt{a^2 + 1}} \tag{11.7}$$

where NIR is the reflectance of the near-infrared band, R is the reflectance of the red band, a is the slope of the soil line, and b is the intercept of the soil line.

11.3.2.1.2 PVI2

This index was also introduced by Walther and Shabaani (1991) to improve the results of the PVI index (Eq. 11.8):

$$\mathrm{PVI2} = \frac{\mathrm{NIR} - a \times R + b}{\sqrt{1 + a^2}} \tag{11.8}$$

where NIR is the reflectance of the near-infrared band, R is the reflectance of the red band, a is the slope of the soil line, and b is the intercept of the soil line.

11.3.2.1.3 PVI3

PVI3 was introduced by Qi et al. (1994) to better highlight the vegetation cover in the soil. The calculation method of this index is as follows (Eq. 11.9):

$$\mathrm{PVI3} = b \times \mathrm{NIR} - a \times R \tag{11.9}$$

where NIR and R are the reflectance of the near-infrared and the red band, respectively. a and b are the slope and the intercept of the soil line respectively.

11.3.2.2 Weighted difference vegetation index

This index which was introduced by Richardson and Wiegand (1977) is simple but effective like many slope-based indices. Weighting the weight effect of the red band with the soil line slope magnifies the vegetation signal in the near-infrared band and underestimates the effect of soil illumination. The following equation shows how to calculate this method (Eq. 11.10):

$$\mathrm{WDVI} = \mathrm{NIR} - a \times R \tag{11.10}$$

where NIR represents the reflectance of the near-infrared band, R shows the reflectance of the red band, and a is the slope of the soil line.

11.3.2.3 Soil-adjusted vegetation index

This index (Eq. 11.11) was proposed by Huet (1988). Its purpose is to lower the influence of the soil background on the vegetation signal using a constant soil modification factor (L) in the NDVI equation. This factor changes based on the reflective properties of the soil (such as color and brightness) and varies from zero to 100. When L is zero, soil-adjusted vegetation index (SAVI) is equal to NDVI, and when it is 100, SAVI is equal to PVI. The L factor depends on the vegetation congestion that will be analyzed:

$$\mathrm{SAVI} = \frac{\mathrm{NIR} - R}{\mathrm{NIR} + R + L} \times (1 + L) \tag{11.11}$$

where NIR is the reflection of the near-infrared band, R is the reflection of the red band, and L is the soil adjustment factor. To improve efficiency, this index was revised and TSAVI1, TSAVI2, MSAVI1, and MSAVI indices were extracted from it.

11.3.2.3.1 TSAVI1

Baret et al. (1989) modified SAVI based on the concept of PVI. They proposed the first modification of TSAVI as TSAVI1 (Eq. 11.12):

$$\text{TSAVI1} = \frac{a \times (\text{NIR} - a \times R - b)}{(R + (a \times \text{NIR}) - (a \times b)} \tag{11.12}$$

where NIR is the reflectance of the near-infrared band, R is the reflectance of the red band, a is the slope of the soil line, and b is the intercept. TSAVI1 with high humidity resistance can be a very suitable index in semi-arid areas. This index was created specifically for dry lands and does not have good efficiency in areas with dense vegetation.

11.3.2.3.2 TSAVI2

The SAVI index was again adjusted by Baret et al. (1991), adding an amendment factor of 0.08, to reduce the influence of the soil illumination, which is called TSAVI2 and is calculated as follows (Eq. 11.3):

$$\text{TSAVI2} = \frac{a \times (\text{NIR} - a \times R - b)}{R + (a \times \text{NIR}) - (a \times b) + 0.08 \times (1 + a^2)} \tag{11.13}$$

where NIR is the reflectance of the near-infrared band, R is the reflectance of the red band, and a and b are the soil line slope and intercept respectively.

11.3.2.3.3 MSAVI1

Based on the correction of the constant soil adjustment factor (L) in the SAVI index, this index was proposed by Qi et al. (1994). In this index, the constant factor of soil adjustment is calculated based on an empirical function. This empirical function includes NDVI and WDVI indices. This index is calculated as follows (Eqs. 11.4 and 11.5):

$$\text{MSAVI1} = \frac{\text{NIR} - R}{\text{NIR} + R + L} \times (1 + L) \tag{11.14}$$

$$L = 1 - (2 \times A \times \text{NDVI} \times \text{WDVI}) \tag{11.15}$$

where NIR is the reflectance of the near-infrared band, R is the reflectance of the red band, a is the slope of the soil line, and L is the constant factor of soil adjustment.

11.3.2.3.4 MSAVI2

This index, presented by Qi et al. (1994), uses the L factor to eliminate soil noises that are not resolved by NDVI and WDVI indices and correct values greater than 1 of MSAVI1. Therefore, it is not recommended for areas with high vegetation density. This index is calculated as follows (Eq. 11.16):

$$\text{MSAVI2} = \frac{(2 \times \text{NIR}) + 1 - \sqrt{(2 \times \text{NIR} + 1)^2 - 8 \times (\text{NIR} - R)}}{2} \tag{11.16}$$

where NIR is the reflectance of the near-infrared band, and R is the reflectance of the red band.

11.3.3 Orthogonal transformation vegetation indices

These indices are based on the conversion of existing spectral bands to a set of nondependent bands that can be called vegetation bands. The basis of these indices is based on the noncorrelation of the main bands through alignment to exploit new bands. With this procedure, a green band is produced, in which the background effects of the soil are somehow removed, and almost all the characteristics of the soil are given to another band called brightness.

Vegetation indices extracted from the principal components analysis (PCA) and Tasseled Cap transformation are among the most well-known indicators of this group.

11.3.3.1 Principal components analysis

PCA is an orthogonal transformation that uses the bands of satellite imagery to build a new set of bands or components that are not correlated and are sorted according to the changes compared to the original image. In the context of satellite images, the first component usually represents albedo and the second component indicates vegetation cover.

11.3.3.2 Tasseled Cap transformation

Tasseled Cap transformation is the transformation of the original bands of an image into a novel set of bands with specific interpretations. This method is done by taking the linear composition of the original image bands, which is conceptually similar to principal component analysis (PCA).

11.3.4 Temporal vegetation indices

These indices are generally calculated using the time series of previous indices. Vegetation indicators' time series can prepare continual information about vegetation variation and be used to assign the effects of environmental factors, such as climate, on vegetation (Zhang et al., 2010; Behrang Manesh et al., 2019). The indicators of this group are used for vegetation and agriculture drought studies. Among the most important of these indices are VCI, VHI, and PVVI.

11.3.4.1 Vegetation condition index

This index, which was first introduced by Kogan (1990), is calculated using the following equation (Eq. 11.17):

$$VCI = \frac{NDVI_i - NDVI_{min}}{NDVI_{max} - NDVI_{min}} \times 100 \quad (11.17)$$

where $NDVI_i$, $NDVI_{max}$, and $NDVI_{min}$ are respectively the NDVI value of a pixel in the ith year, the highest and the lowest NDVI values of the same pixel in the studied period. Other indices such as EVI can also be used instead of NDVI. The minimum and maximum values of NDVI represent the worst and the best conditions for vegetation, respectively, and their difference indicates the conditions of vegetation life (Sha et al., 2013; Qian et al., 2016). The VCI values close to 0 reveal less growth and the values near to 100 represent complete growth of vegetation. The most important ability of the VCI is the reduction of noise effects created by seasonal alterations and the increase of the vegetation studies' accuracy.

11.3.4.2 Vegetation health index

This index indicates vegetation health (Kundu et al., 2016) and is used to evaluate the intensity and duration of drought (Karnieli et al., 2010). This index combines TCI and VCI (Eq. 11.18):

$$\mathrm{VHI} = \alpha \times \mathrm{VCI} + (1 - \alpha) \times \mathrm{TCI} \tag{11.18}$$

α indicates the weight of VCI and TCI and its value depends on the various condition of humidity and temperature in the study area (Tran et al., 2017). This coefficient is considered equal to 0.5 if the contribution of humidity and temperature is not known (Cong et al., 2017). The TCI, which was developed based on LST and can evaluate the effects of heat stress on vegetation (Kogan, 1995), is computing using the following formula (Eq. 11.19):

$$\mathrm{TCI} = \frac{\mathrm{LST}_i - \mathrm{LST}_{\min}}{\mathrm{LST}_{\max} - \mathrm{LST}_{\min}} \times 100 \tag{11.19}$$

where LST_i, $\mathrm{LST}_{\max}$, and $\mathrm{LST}_{\min}$ are respectively the LST value of a special pixel in the ith year, the most, and the least LST value of the same pixel in the study period.

11.3.4.3 Probability of vegetation vulnerability index

This index, introduced by Heydari Alamdarloo et al. (2018), is derived from the VHI index. To calculate the probability of vegetation vulnerability index (PVVI) index, the probability percentage of occurrence of various VHI intensity classes must be calculated. Then, the percentage of occurrence probability of each class is multiplied by the weight of that class, and finally, the results of this mathematical operation for different classes are added together. The following equation shows how each class is calculated:

$$\mathrm{PVVI} = \sum_{i=1}^{n} W_i \times P_{Ci} \tag{11.20}$$

where W_i is the weight of each class and P_{Ci} is the probability of its occurrence.

11.4 Evaluating the vegetation variations trend

Vegetation cover, one of the important factors of soil protection and environmental improvement (Gu et al., 2018), has been changed for various reasons and over time due to natural or human factors. These changes affect the ecosystem condition and function (Pettorelli et al., 2005) and are considered a common indicator for measuring environmental conditions (Wang et al., 2015). Vegetation monitoring is a key factor in the sustainable management of natural resources and environmental protection (Del Valle and Jiang, 2022). Continuous degradation in dry areas makes it necessary to use monitoring and evaluation systems and also early warning systems with appropriate accuracy for determining the degradation process (Eskandari et al., 2016). Therefore, evaluating the vegetation dynamics and its reply to environmental changes is necessary to understand ecosystem alterations and the acceptable utilization of natural resources (Yang, 2022).

It is very important to determine a set of appropriate and accurate vegetation information to investigate the dynamics of vegetation and compare it in different regions. Changes in vegetation can provide essential information in vulnerable ecosystems (Liu et al., 2019). Vegetation indices are utilized to reduce the effect of unwanted factors on vegetation information. The vegetation index is a suitable factor for investigating the behavior and evolution of vegetation cover (Fatiha et al., 2013) and quantifying the greenness of vegetation. In the following, with a practical example, the trend of vegetation changes is evaluated utilizing the Theil-Sen slope, Mann-Kendall, and Hurst test.

Gavkhoni wetland is one of the most important aquatic ecosystems (Abou Zaki et al., 2020) with an area of 41,524 km^2 in central Iran (Farsi et al., 2019; Zehtabian et al., 2023). Gavkhoni wetland, with many environmental and economic benefits for the central region of Iran, has significant effects on local ecosystems and climate (Azadi et al., 2022) and has completely dried up several times in recent decades (Fig. 11.3).

To evaluate the trend of vegetation dynamics, the EVI of the MOD13Q1, the product of the MODIS sensor, has been used. First, the 4-month average of the EVI (March, April, May, and June, as the growing season of the study region) was calculated for the years 2001–2022 on the Google Earth Engine (GEE). Due to the climate of the region, the natural vegetation is at

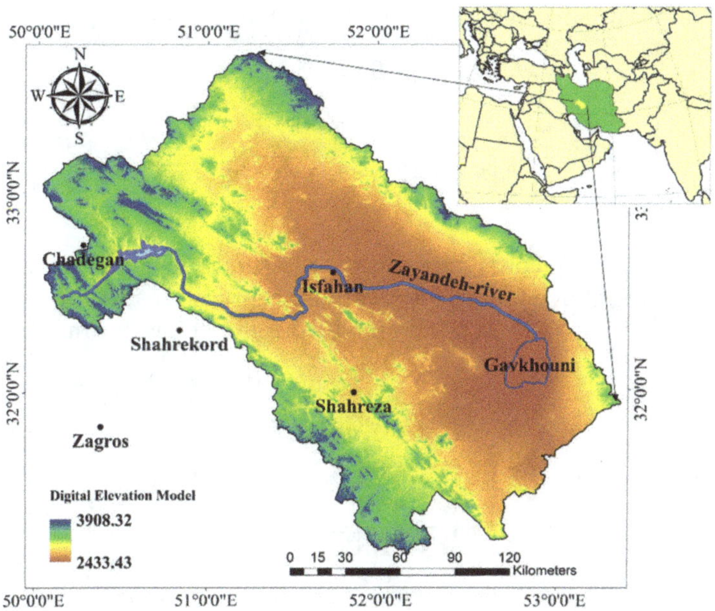

FIGURE 11.3 The location of Gavkhoni Wetland, Iran.

its peak during these 6 months (Heydari et al., 2018). Then, the Theil-Sen slope has been calculated using the time series of EVI in the TerrSet (Eq. 11.21):

$$\beta = \text{Median}\left(\frac{VI_j - VI_i}{t_j - t_i}\right), j > i \tag{11.21}$$

where β is the Theil-Sen median slope, t_i and t_j are the years, and VI_i is the value of the vegetation index in the year i. $\beta > 0$ represents the increasing trend of the index and thus the vegetation improvement unlike $\beta < 0$ which shows the decreasing trend of the index (Cao et al., 2014).

In the next step, the Mann–Kendall trend test, offered by Mann (1945) and Kendall (1975), has been done based on the following relationship (Eq. 11.22):

$$S = \sum_{i=2}^{n}\sum_{j=1}^{i-1}(\text{sign}(xi - xj)) \tag{11.22}$$

where x_i and x_j are sequential data and n is the length of the time series. The sign function is calculated as follows (Eq. 11.23):

$$\text{sign}(xi - xj) = \begin{cases} +1 & \text{if} \quad (xi - xj) > 0 \\ 0 & \text{if} \quad (xi - xj) = 0 \\ -1 & \text{if} \quad (xi - xj) < 0 \end{cases} \tag{11.23}$$

The average € and variance (Var) of S are calculated as follows (Eqs. 11.24 and 11.25):

$$E(S) = 0 \tag{11.24}$$

$$\text{VAR}(S) = \frac{1}{18}\left[n(n-1)(2n+5) - \sum_{p=1}^{q} tp(tp-1)(2tp+5)\right] \tag{11.25}$$

where tp is the number of sequences for the p-th value and q is the number of sequences. The second component in the above formula is an adjustment for sensitive sequences or data. The standardized statistic of the ZM test is obtained from the below formula (Eq. 11.26):

$$Z_M = \begin{cases} \dfrac{S-1}{\sqrt{\text{VAR}(S)}} & \text{if} \quad S > 0 \\ 0 \quad \text{if} \quad S = 0 & \\ \dfrac{S+1}{\sqrt{\text{VAR}(S)}} & \text{if} \quad S > 0 \end{cases} \tag{11.26}$$

TABLE 11.1 Trend of the EVI in Gavkhoni wetland.

S_{NDVI}	Z	The trend of the EVI	Area percentage (%)
≤0.0005	1.96≤	Significantly improved	11.37
≤0.0005	1.96–1.96	Slightly improved	6.14
0.0005–0.0005	1.96–1.96	Stable	69.54
−0.0005>	1.96–1.96	Slightly degraded	8.7
−0.0005>	−1.96>	Severely degraded	4.25

Note: the pixels with S_{NDVI} in −0.0005 and 0.0005 and $Z > 1.96$ or $Z < -1.96$ are classified as stable.

The positive amount of it represents the increasing trend and the negative amount of it exhibits the decreasing trend of time series. For performing this test, the TerrSet software is used.

The variation in the EVI trend has been classified using the combination of the Mann–Kendall trend test and the Theil-Sen median trend analysis (Jiang et al., 2015). The results are indicated in Table 11.1.

Table 11.1 shows significantly or slightly improved regions, areas with stable vegetation (or without vegetation), and areas with slight or severe vegetation deterioration covering 17.51%, 69.54%, and 12.95% of the wetland extent, respectively.

Fig. 11.4 shows the spatial dispensation of the most changes in the EVI trend in the study period. Vegetation degradation is mainly concentrated in the central and eastern sections and

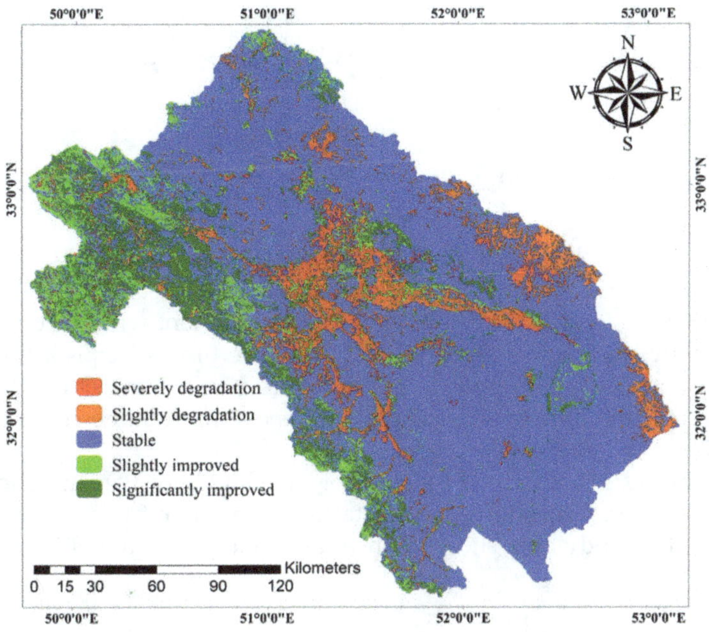

FIGURE 11.4 Changes in vegetation trend of Gavkhoni wetland.

parts of the southeast of the study area, while the vegetation improved in the northwest, west, and southwest regions.

Finally, the Hurst index has been calculated, and its results have been integrated with the previous tests. This index is widely used for the evaluation of the change persistence in time series data over long terms (Ndayisaba et al., 2016) as well as for climate and vegetation studies. This index is calculated based on the following equations in Arc GIS software.

First, the long time series $\{VI\ (\tau)\}$ ($\tau = 1, 2, \ldots, n$) are divided into τ subseries $X(t)$, and the average of EVI is defined for each time series (Eq. 11.27):

$$\overline{VI}_{(T)} = \frac{1}{T}\sum_{t}^{T} \mathrm{EVI}(T) \quad T = 1.2.\cdots.n \tag{11.27}$$

In the next step, the accumulated deviation of each EVI average is calculated (Eq. 11.28). Then, the sequence of R will be defined (Eq. 11.29):

$$X_{(T.t)} = \sum_{t=1}^{t}\left(\mathrm{EVI}_{(t)} - \overline{\mathrm{EVI}}_{(T)}\right) \quad 1 \le t \le T \tag{11.28}$$

$$R_{(T)} = \max_{1 \le t \le T} X_{(T.t)} - \min_{1 \le t \le T} X_{(T.t)} \quad T = 1.2.\cdots.n \tag{11.29}$$

After calculating the sequence of the standard deviation of (S) (Eq. 11.30), the Hurst value is computed (Eq. 11.31):

$$S_{(T)} = \left[\frac{1}{T}\sum_{t=1}^{T}\left(\mathrm{EVI}_{(t)} - \mathrm{EVI}^2_{(T)}\right)\right]^{½} \quad T = 1.2.\cdots.n \tag{11.30}$$

$$\log\left(R/S\right)_n = a + H \times \log(n) \tag{11.31}$$

A value close to 0.5 indicates that the EVI time series is an accidental series, and the future trend is independent of the research period. Values more than 0.5 represent that the time series is a consistent series and the future trend will be consistent with the current trend in the study period. Finally, values less than 0.5 show that the time series is a nonrandom series, and the past trend is likely the opposite of the future trend (Jiang et al., 2017). The results of trend analysis and the Hurst index were combined to determine consistency in the EVI trend (Table 11.2).

According to the results, 1.26% of the area of the Gavkhoni wetland showed a trend of vegetation improvement, 4.4% showed a vegetation destruction trend, 19.61% showed stability, and 73.74% of the study area showed uncertain changes in trends (Fig. 11.5).

TABLE 11.2 Statistical result of trends and the Hurst index.

S_{NDVI}	Z	H	The trend of the EVI	Area percentage (%)
<−0.0005	1.96≤	0.5<	Sustainability and severe degradation	0.47
<−0.0005	1.96 −1.96	0.5<	Sustainability and slight degradation	3.93
−0.0005−0.0005	1.96−1.96	0.5<	Sustainability and stability	19.61
≥0.0005	1.96−1.96	0.5<	Sustainability and a slight improvement	0.83
≥0.0005	−1.96>	0.5<	Sustainability and significant improvement	0.43
≥0.0005	−1.96>	<0.5	Undetermined future variation in trends	74.73

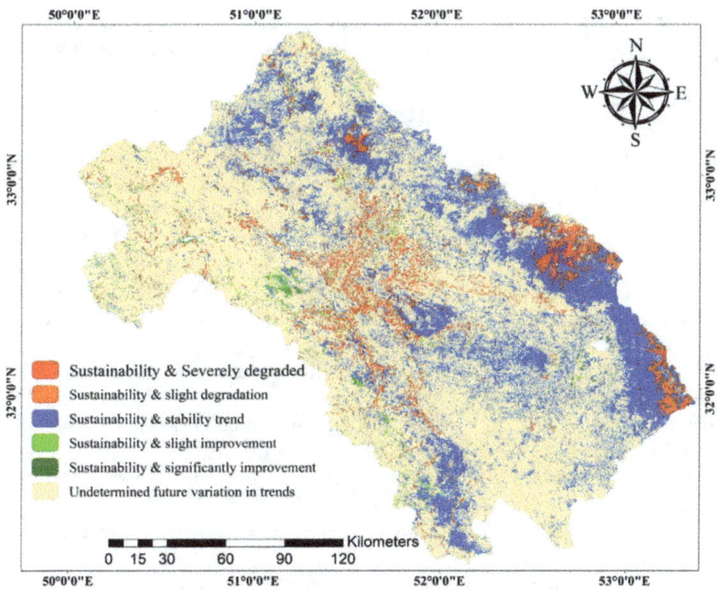

FIGURE 11.5 Stability of vegetation changes.

11.5 Conclusions

Considering the essential contribution of vegetation to the stability of ecosystems, monitoring and evaluating vegetation changes are essential. One of the most important tools for studying vegetation, especially in ecosystems with a large area, is remote sensing-based data. There are many remote sensing indices in the field of vegetation assessment. Each index has its characteristics, efficiency, and limitation. Selection of these indices should be done based on their features and the condition of the study area. In this chapter, the characteristics of vegetation indices were introduced. Due to the increasing development of remote sensing in all fields, including vegetation studies, it is necessary to evaluate the ability of vegetation indicators.

Acknowledgment

This chapter forms part of Esmaeil Heydari Alamdarloo's postdoctoral project (Number: 99017799) with the title of Feasibility Study of Proposing an Early Warning System for Desertification of IRAN using Remote Sensing Data. The Iran National Science Foundation (INS) is thanked for the financial support.

References

Abou Zaki, N., Torabi Haghighi, A., Rossi, P.M., Tourian, M.J., Bakhshaee, A., Kløve, B., 2020. Evaluating impacts of irrigation and drought on river, groundwater and a terminal wetland in the Zayanderud Basin, Iran. Water 12 (5), 1302.

Azadi, S., Yazdanpanah, H., Nasr-Esfahani, M.A., Pourmanafi, S., Dorigo, W., 2022. The Gavkhouni wetland dryness and its impact on air temperature variability in the eastern part of the Zayandeh-Rud River Basin, Iran. Water 14 (2), 172.

Baret, F., Guyot, G., 1991. Potentials and limits of vegetation indices for LAI and APAR assessment. Remote Sens. Environ. 35 (2–3), 161–173.

Baret, F., 1989. TSAVI: a vegetation index which minimizes soil brightness effects on LAI and APAR estimation. In: 12th Canadian Symp. On Remote Sensing and IGARSS'90, Vancouver, Canada, 10-14 July 1989.

Behrang Manesh, M., Khosravi, H., Heydari Alamdarloo, E., Saadi Alekasir, M., Gholami, A., Singh, V.P., 2019. Linkage of agricultural drought with meteorological drought in different climates of Iran. Theor. Appl. Climatol. 138, 1025–1033.

Bin Abdul Rahim, H.R., Bin Lokman, M.Q., Harun, S.W., Hornyak, G.L., Sterckx, K., Mohammed, W.S., Dutta, J., 2016. Applied light-side coupling with optimized spiral-patterned zinc oxide nanorod coatings for multiple optical channel alcohol vapor sensing. J. Nanophotonics 10 (3), 036009.

Caminade, C., McIntyre, K.M., Jones, A.E., 2019. Impact of recent and future climate change on vector-borne diseases. Ann. N. Y. Acad. Sci. 1436 (1), 157–173.

Cao, R., Jiang, W., Yuan, L., Wang, W., Lv, Z., Chen, Z., 2014. Inter-annual variations in vegetation and their response to climatic factors in the upper catchments of the Yellow River from 2000 to 2010. J. Geogr. Sci. 24, 963–979.

Cong, D., Zhao, S., Chen, C., Duan, Z., 2017. Characterization of droughts during 2001–2014 based on remote sensing: a case study of Northeast China. Ecol. Inf. 39, 56–67.

Cruden, B.A., Prabhu, D., Martinez, R., 2012. Absolute radiation measurement in venus and mars entry conditions. J. Spacecraft Rockets 49 (6), 1069–1079.

Del Valle, T.M., Jiang, P., 2022. Comparison of common classification strategies for large-scale vegetation mapping over the Google Earth Engine platform. Int. J. Appl. Earth Obs. Geoinf. 115, 103092.

Ding, M., Chen, Q., Xiao, X., Xin, L., Zhang, G., Li, L., 2016. Variation in cropping intensity in northern China from 1982 to 2012 based on GIMMS-NDVI data. Sustainability 8 (11), 1123.

Eastman, J.R., 2020. TerrSet 2020 Geospatial Monitoring and Modeling System. Clark Labs. Clark University, Worcester, MA, USA.

Eskandari, H., Borji, M., Khosravi, H., Mesbahzadeh, T., 2016. Desertification of forest, range and desert in Tehran province, affected by climate change. Solid Earth 7 (3), 905–915.

Farsi, N., Mahjouri, N., 2019. Evaluating the contribution of the climate change and human activities to runoff change under uncertainty. J. Hydrol. 574, 872–891.

Fatiha, B., Abdelkader, A., Latifa, H., Mohamed, E., 2013. Spatio temporal analysis of vegetation by vegetation indices from multi-dates satellite images: application to a semi arid area in Algeria. Energy Proc. 36, 667–675.

Ferchichi, A., Abbes, A.B., Barra, V., Farah, I.R., 2022. Forecasting vegetation indices from spatio-temporal remotely sensed data using deep learning-based approaches: a systematic literature review. Ecol. Inf. 101552.

Foley, W.J., McIlwee, A., Lawler, I., Aragones, L., Woolnough, A.P., Berding, N., 1998. Ecological applications of near infrared reflectance spectroscopy—a tool for rapid, cost-effective prediction of the composition of plant and animal tissues and aspects of animal performance. Oecologia 116 (3), 293–305.

Funatsu, B.M., Dubreuil, V., Racapé, A., Debortoli, N.S., Nasuti, S., Le Tourneau, F.M., 2019. Perceptions of climate and climate change by Amazonian communities. Global Environ. Change 57, 101923.

Gu, Z., Duan, X., Shi, Y., Li, Y., Pan, X., 2018. Spatiotemporal variation in vegetation coverage and its response to climatic factors in the Red River Basin, China. Ecol. Indicat. 93, 54–64.

Heydari Alamdarloo, E., Manesh, M.B., Khosravi, H., 2018. Probability assessment of vegetation vulnerability to drought based on remote sensing data. Environ. Monit. Assess. 190, 1–11.

Huete, A., Justice, C., Van Leeuwen, W., 1999. MODIS vegetation index (MOD13). Algor. Theor. Basis Doc. 3 (213), 295–309.

Huete, A., Didan, K., Miura, T., Rodriguez, E.P., Gao, X., Ferreira, L.G., 2002. Overview of the radiometric and biophysical performance of the MODIS vegetation indices. Remote Sens. Environ. 83 (1–2), 195–213.

Huete, A.R., 1988. A soil-adjusted vegetation index (SAVI). Remote Sens. Environ. 25 (3), 295–309.

Jabbar, M.T., Zhi-Hua, S.H.I., Tian-Wei, W.A.N.G., Chong-Fa, C.A.T., 2006. Vegetation change prediction with geo-information techniques in the three gorges area of China. Pedosphere 16 (4), 457–467.

Jackson, R.D., Huete, A.R., 1991. Interpreting vegetation indices. Prev. Vet. Med. 11 (3–4), 185–200.

Jiang, W., Yuan, L., Wang, W., Cao, R., Zhang, Y., Shen, W., 2015. Spatio-temporal analysis of vegetation variation in the Yellow river basin. Ecol. Indicat. 51, 117–126.

Jiang, L., Bao, A., Guo, H., Ndayisaba, F., 2017. Vegetation dynamics and responses to climate change and human activities in Central Asia. Sci. Total Environ. 599, 967–980.

Jiang, L., Liu, Y., Xu, H., 2023. Variation in vegetation quality of terrestrial ecosystems in China: coupling analysis based on remote sensing and typical stations monitoring data. Rem. Sens. 15 (9), 2276.

Karnieli, A., Agam, N., Pinker, R.T., Anderson, M., Imhoff, M.L., Gutman, G.G., Panov, N., Goldberg, A., 2010. Use of NDVI and land surface temperature for drought assessment: merits and limitations. J. Clim. 23 (3), 618–633.

Kendall, M.G., 1975. Rank Correlation Measures, vol. 202. Charles Griffin, London, p. 15.

Kogan, F.N., 1990. Remote sensing of weather impacts on vegetation in non-homogeneous areas. Int. J. Rem. Sens. 11 (8), 1405–1419.

Kogan, F.N., 1995. Application of vegetation index and brightness temperature for drought detection. Adv. Space Res. 15 (11), 91–100.

Kundu, A., Dwivedi, S., Dutta, D., 2016. Monitoring the vegetation health over India during contrasting monsoon years using satellite remote sensing indices. Arabian J. Geosci. 9, 1–15.

Lawley, V., Lewis, M., Clarke, K., Ostendorf, B., 2016. Site-based and remote sensing methods for monitoring indicators of vegetation condition: an Australian review. Ecol. Indicat. 60, 1273–1283.

Levy, O., Dayan, T., Porter, W.P., Kronfeld-Schor, N., 2019. Time and ecological resilience: can diurnal animals compensate for climate change by shifting to nocturnal activity? Ecol. Monogr. 89 (1), e01334.

Li, B., Tang, H., Chen, D., 2009. October. Drought monitoring using the modified temperature/vegetation dryness index. In: 2nd International Congress on Image and Signal Processing. IEEE, pp. 1–5.

Li, S., Wang, J., Zhang, M., Tang, Q., 2021. Characterizing and attributing the vegetation coverage changes in North Shanxi coal base of China from 1987 to 2020. Resour. Pol. 74, 102331.

Liu, C., Sun, P.S., Liu, S.R., 2016. A review of plant spectral reflectance response to water physiological changes. Chin. J. Plant Ecol. 40 (1), 80.

Liu, S., Li, W., Qiao, W., Wang, Q., Hu, Y., Wang, Z., 2019. Effect of natural conditions and mining activities on vegetation variations in arid and semiarid mining regions. Ecol. Indicat. 103, 331–345.

Mann, H.B., 1945. Nonparametric tests against trend. Econometrica. J. Econom. Soc. 245–259.

Mishra, A.K., Singh, V.P., Jain, S.K., 2010. Impact of global warming and climate change on social development. J. Comp. Soc. Welfare 26 (2–3), 239–260.

Moradi, E., Darabi, H., Alamdarloo, E.H., Karimi, M., Kløve, B., 2023. Vegetation vulnerability to hydrometeorological stresses in water-scarce areas using machine learning and remote sensing techniques. Ecol. Inf. 73, 101838.

Mu, S., Yang, H., Li, J., Chen, Y., Gang, C., Zhou, W., Ju, W., 2013. Spatio-temporal dynamics of vegetation coverage and its relationship with climate factors in Inner Mongolia, China. J. Geogr. Sci. 23, 231–246.

Ndayisaba, F., Guo, H., Bao, A., Guo, H., Karamage, F., Kayiranga, A., 2016. Understanding the spatial temporal vegetation dynamics in Rwanda. Rem. Sens. 8 (2), 129.

Pei, F., Wu, C., Liu, X., Li, X., Yang, K., Zhou, Y., Wang, K., Xu, L., Xia, G., 2018. Monitoring the vegetation activity in China using vegetation health indices. Agric. For. Meteorol. 248, 215–227.

Peng, J., Liu, Z., Liu, Y., Wu, J., Han, Y., 2012. Trend analysis of vegetation dynamics in Qinghai–Tibet plateau using Hurst Exponent. Ecol. Indicat. 14 (1), 28–39.

Perry Jr., C.R., Lautenschlager, L.F., 1984. Functional equivalence of spectral vegetation indices. Remote Sens. Environ. 14 (1–3), 169–182.

Pettorelli, N., Vik, J.O., Mysterud, A., Gaillard, J.M., Tucker, C.J., Stenseth, N.C., 2005. Using the satellite-derived NDVI to assess ecological responses to environmental change. Trends Ecol. Evol. 20 (9), 503–510.

Piao, S., Mohammat, A., Fang, J., Cai, Q., Feng, J., 2006. NDVI-based increase in growth of temperate grasslands and its responses to climate changes in China. Global Environ. Change 16 (4), 340–348.

Qi, J., Chehbouni, A., Huete, A.R., Kerr, Y.H., Sorooshian, S., 1994. A modified soil adjusted vegetation index. Remote Sens. Environ. 48 (2), 119–126.

Qian, X., Liang, L., Shen, Q., Sun, Q., Zhang, L., Liu, Z., Zhao, S., Qin, Z., 2016. Drought trends based on the VCI and its correlation with climate factors in the agricultural areas of China from 1982 to 2010. Environ. Monit. Assess. 188, 1–13.

Raab, D., Bayley, S.E., 2012. A vegetation-based Index of Biotic Integrity to assess marsh reclamation success in the Alberta oil sands, Canada. Ecol. Indicat. 15 (1), 43–51.

Richardson, A.J., Wiegand, C.L., 1977. Distinguishing vegetation from soil background information. Photogramm. Eng. Rem. Sens. 43 (12), 1541–1552.

Rimkus, E., Stonevicius, E., Kilpys, J., Maciulyte, V., Valiukas, D., 2017. Drought identification in the eastern Baltic region using NDVI. Earth Syst. Dyn. 8 (3), 627–637.

Rouse, J.W., Haas, R.H., Schell, J.A., Deering, D.W., 1974. Monitoring vegetation systems in the great plains with ERTS. NASA Spec. Publ. 351 (1), 309.

Ryan, C.M., Pritchard, R., McNicol, I., Owen, M., Fisher, J.A., Lehmann, C., 2016. Ecosystem services from southern African woodlands and their future under global change. Phil. Trans. Biol. Sci. 371 (1703), 20150312.

Schmidhuber, J., Tubiello, F.N., 2007. Global food security under climate change. Proc. Natl. Acad. Sci. USA 104 (50), 19703–19708.

Sha, S., Guo, N., Li, Y.H., Ren, Y.L., Li, Y.P., 2013. Comparison of the vegetation condition index with meteorological drought indices: a case study in henan province. J. Glaciol. Geocryol. 35 (4), 990–998.

Sun, R., Chen, S., Su, H., 2021. Climate dynamics of the spatiotemporal changes of vegetation NDVI in Northern China from 1982 to 2015. Rem. Sens. 13 (2), 187.

Tran, H.T., Campbell, J.B., Tran, T.D., Tran, H.T., 2017. Monitoring drought vulnerability using multispectral indices observed from sequential remote sensing (case study: Tuy Phong, Binh Thuan, Vietnam). GIScience Remote Sens. 54 (2), 167–184.

Walther, D., Shabaani, S., 1991. Large scale monitoring of rangelands vegetation using NOAA/AVHRR LAC data: application to the rainy seasons 1989/90 in northern Kenya. In: Range Management Handbook of Kenya, vol. 3.

Wang, J., Wang, K., Zhang, M., Zhang, C., 2015. Impacts of climate change and human activities on vegetation cover in hilly southern China. Ecol. Eng. 81, 451–461.

Yang, S., Liu, J., Wang, C., Zhang, T., Dong, X., Liu, Y., 2022. Vegetation dynamics influenced by climate change and human activities in the Hanjiang River Basin, central China. Ecol. Indicat. 145, 109586.

Zehtabian, E., Masoudi, R., Yazdandoost, F., Sedghi-Asl, M., Loáiciga, H.A., 2023. Investigation of water allocation using integrated water resource management approaches in the Zayandehroud River basin, Iran. J. Clean. Prod. 395, 136339.

Zhang, X., Goldberg, M., Tarpley, D., Friedl, M.A., Morisette, J., Kogan, F., Yu, Y., 2010. Drought-induced vegetation stress in southwestern North America. Environ. Res. Lett. 5 (2), 024008.

CHAPTER

12

A brief review of digital soil mapping in Iran

Ruhollah Taghizadeh-Mehrjardi[1], *Mojtaba Zeraatpisheh*[2], *Alireza Amirian-Chakan*[3] *and Thomas Scholten*[1]

[1]Department of Geosciences, University of Tübingen, Tübingen, Germany; [2]Rubenstein School of Environment and Natural Resources, University of Vermont, Burlington, VT, United States; [3]Faculty of Agriculture and Natural Resources, Lorestan University, Khorramabad, Iran

12.1 History of soil mapping in Iran

The first national-scale and legacy soil map of Iran was published in 1964 at a scale of 1:2,500,000 through collaboration between the Soil Institute of Iran and the FAO (Dewan and Famouri, 1964). Since then, using traditional soil survey methods, soil maps have been produced for the agricultural plains of the country at scales ranging from 1:25,000 to 1:250,000 (Zeraatpisheh et al., 2017). However, the usefulness of these maps is limited due to their coarse scale and/or qualitative nature. They rely on qualitative mental models of soil–environmental relationships that are often not described (Jafari et al., 2013). They are subjective rather than objective, difficult to update, and have limited flexibility for quantitative studies (McBratney et al., 2003). Additionally, the variation in soil across space is not illustrated within the boundaries of individual map units (i.e., polygons or delineations) of choropleth maps, and the values of soil properties may have changed since the production of these maps.

With the recent increase in the availability of digital data and robust data mining algorithms, digital soil mapping (DSM) techniques provide a cost- and time-efficient means of overcoming some of the drawbacks of legacy soil maps (Fig. 12.1). The SCORPAN framework serves as the foundation for DSM techniques (Behrens and Scholten, 2006; McBratney et al., 2003). This framework involves quantifying the correlations between soil properties and environmental factors, with the aim of predicting soil properties and classifications. Within the DSM framework, environmental covariates refer to data that aim to depict the

Remote Sensing of Soil and Land Surface Processes
https://doi.org/10.1016/B978-0-443-15341-9.00027-7

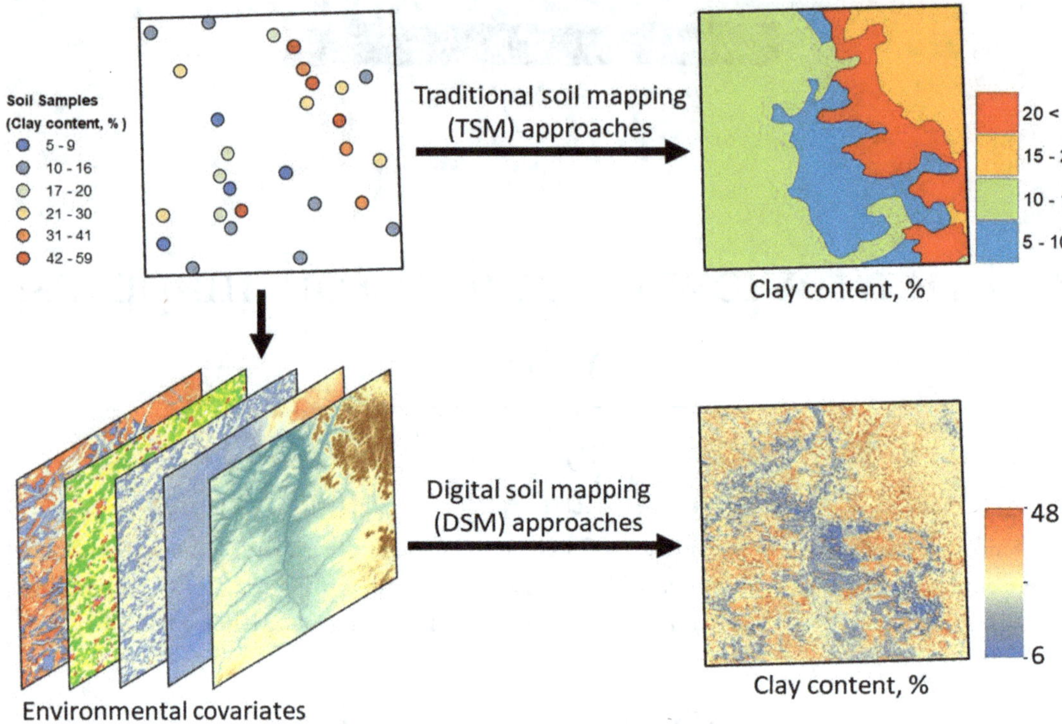

FIGURE 12.1 The steps involved in creating soil maps using traditional and digital mapping approaches.

factors responsible for soil formation. They are mostly derived from remotely sensed images, digital elevation models, climate model maps, existing digital soil maps, and digitized legacy soil maps. In Iran, the use of predictive soil mapping based on DSM techniques is still emerging, especially at the national scale. The first country-scale soil map partly based on new techniques (i.e., remote sensing and GIS) was produced by the Soil and Water Research Institute of Iran (SWRI) at a scale of 1:1,000,000 in 2000 (Balali et al., 2018). Although the map provided general information about the soil resources of Iran and their spatial distribution, its resolution was too coarse to meet the requirements for detailed management and environmental modeling purposes (Zeraatpisheh et al., 2019a).

Using 4250 soil profile observations and environmental variables at a 1000 m spatial resolution, the first soil map of Iran that employed the SCORPAN approach was developed (Hengl et al., 2007). This map provided more accurate and reliable information than the map produced by SWRI (Balali et al., 2018); however, some significant environmental variables, such as geology and climate, which vary greatly across the country, were not included in the spatial modeling of soil classes. In the SoilGrids250m project, researchers applied the DSM approach using a large set of environmental variables to model and map the global patterns of soil classes, including Iran, in the SoilGrids250m project using the World Reference Base (WRB) and Soil Taxonomy (USDA) classification systems (Hengl et al., 2007). Although these maps were the most detailed soil maps of Iran at the national scale since 2017, the final resolution of 250 m was not fine enough to meet all the requirements for detailed land management and land-use planning (Taghizadeh-Mehrjardi et al., 2012).

In contrast to national-scale studies, DSM techniques have predominantly been applied at the local and regional scales to predict soil taxonomic classes in Iran (Taghizadeh-Mehrjardi et al., 2014b, 2015, 2020c; Zeraatpisheh et al., 2019a). These localized studies are valuable for decision-makers, farmers, and local administrations to enhance sustainable soil management strategies. In countries such as Iran, where there is limited availability of quantitative data and data scarcity, these studies can be particularly beneficial.

12.2 Digital soil mapping using advanced machine learning

Since the development of the SCORPAN framework (McBratney et al., 2003), there has been a growing utilization of models worldwide that measure the connections between soil characteristics and spatially linked environmental covariates. Machine learning (ML) algorithms, which leverage spatial datasets include remotely sensed information and digital elevation models, have become particularly popular in the last two decades (Khaledian and Miller, 2020; Wadoux et al., 2020). Iran has also seen a rise in the use of both DSM techniques and ML algorithms. Table 12.1 provides a chronological overview of studies that have utilized ML algorithms in a DSM framework. The first study to use ML in a DSM approach was done in 2007 (Hengl et al., 2007), and since then, ML has increasingly been used in DSM studies across the country. Random forest (RF) was the most frequently used ML algorithm (41 times), followed by support vector machines (SVMs) (15 times) and artificial neural networks (ANNs) (12 times). Out of the 56 studies reviewed, 40 modeled continuous variables, while 16 modeled categorical variables (soil classes) using ML algorithms. The models range from simple linear regression to advanced ML algorithms such as hybridized random forest, semisupervised support vector machines, deep learners, super learners, and ensemble learners. Soil taxonomic classes, soil organic carbon, and soil texture (or texture fractions) were the most frequently modeled properties using ML algorithms.

12.3 Environmental covariates

The soil-forming factors, represented by the acronym CLORPT (climate, organism/vegetation, relief, parent material, and time), can be demonstrated using environmental covariates. By establishing the link between these environmental factors and soil properties, it becomes possible to spatially estimate the distribution of soil target variables (McBratney et al., 2003). In inferring soil properties, relief was found to have the highest level of significance, appearing in 80% of the research (McBratney et al., 2003). This was followed by soil class (S), which was present in 35% of the studies, while vegetation (O) and parent material (P) were both identified in 25% of the research. Spatial position (N) was a factor in 20% of the studies, whereas climate (C) was the least commonly considered factor, appearing in only 5% of the research. In this study, we summarize the common environmental variables used in previous studies and demonstrate their influence on soil-forming factors.

TABLE 12.1 A chronological review of the studies in DSM in Iran.

References	Location	Target variable	Model
Hengl et al. (2007)	Whole country	Soil taxonomic classes	SCML, MnLR, RK
Motaghian and Mohammadi (2011)	ChaharMahal-Va-Bakhtiari	Hydraulic conductivity	Stepwise regression, kriging, ANN
Jafari et al. (2012)	Kerman	Soil taxonomic classes	MnLR, BLR
Taghizadeh-Mehrjardi et al. (2012)	Yazd	Soil taxonomic classes	DT
Jafari et al. (2013)	Kerman	Soil taxonomic classes	ANN
Jafari et al. (2014)	Kerman	Diagnostic horizons	BRT
Pahlavan Rad et al. (2014)	Golestan	Soil taxonomic classes	RF
Taghizadeh-Mehrjardi et al. (2014a)	Yazd province	Soil salinity	RT
Bagheri Bodaghabadi et al. (2015)	ChaharMahal-Va-Bakhtiari	Soil units	ANN
Taghizadeh-Mehrjardi et al. (2015)	Kurdistan	Soil taxonomic classes	DT, ANN, RF, SVM, kNN, LR
Pahlavan-Rad et al. (2016)	Golestan	Soil taxonomic classes	RF, MnLR
Taghizadeh-Mehrjardi et al. (2016)	Kurdistan	SOC	ANN, SVM, kNN. RF, RT, GP
Rahmati et al. (2017)	Lorestan and Ilam	Gully erosion susceptibility	SVM, ANN, RF, BRT
Shiri et al. (2017)	Fars and Khorasan-e-Razavi	CEC	GEP, RF, ANN, SVM
Zeraatpisheh et al. (2017)	Chaharmahal-Va-Bakhtiari	Soil taxonomic classes	MnLR, RF
Mirakzehi et al. (2018)	Sistan and Baluchestan	Soil taxonomic classes	RF
Pahlavan-Rad et al. (2018)	Sistan and Baluchestan	SOC	RF
Pahlavan-Rad and Akbarimoghaddam (2018)	Sistan and Baluchestan	Soil texture fractions, pH	RF
Zeraatpisheh et al. (2019b)	Chaharmahal-Va-Bakhtiari	SOC, CCE, clay	MLR, RF, RT, cubist
Amirian-Chakan et al. (2019)	Khuzestan	Soil texture fractions, AWC	RF
Hamzehpour et al. (2019)	West Azerbaijan	SOC	GBM
Shahbazi et al. (2019)	East Azerbaijan	N, P, B	RF, cubist
Shahriari et al. (2019)	Sistan and Baluchestan	Soil texture fractions	RF, RKNNRK, NNRK, RK, COK

TABLE 12.1 A chronological review of the studies in DSM in Iran.—cont'd

References	Location	Target variable	Model
Sharififar et al. (2019)	Northwest of Iran	Soil taxonomic classes	DT, RF, MnLR
Tajik et al. (2020)	Golestan	Invertebrates	RF
Taghizadeh-Mehrjardi et al. (2020a)	Central Iran	Clay, silt, sand	Multitask CNN, RF
Asgari et al. (2020)	Chaharmahal-Va-Bakhtiari	Soil drainage classes	RF, SVM
Esfandiarpour-Boroujeni et al. (2020)	Chaharmahal-Va-Bakhtiari	Soil taxonomic classes	DT, ANN, RF, SVM
Fathizad et al. (2020a)	Yazd	Soil salinity	RF
Fathizad et al. (2020b)	Yazd	Soil quality index	RF
Fathololoumi et al. (2020)	Northwest of Iran	SOC, sand, CCE	RF, cubist
Mahmoudzadeh et al. (2020)	Kurdistan	SOC	Cubist, kNN, RF, XGB, SVM
Maleki et al. (2020)	Golestan	Soil taxonomic classes	RF
Mousavi et al. (2020)	West Azerbaijan	Iron forms and features	RF
Pahlavan-Rad et al. (2020)	Sistan and Baluchestan	Soil infiltration	MLR, RF
Taghizadeh-Mehrjardi et al. (2020b)	Yazd and Gilan	SOC	ANN, RF, XGB, AvNNet, DNN, cubist
Tajik et al. (2020)	Golestan	SOC	PLSR, GLM, rpart, SVM, RF, kNN, ensemble model
Tashayo et al. (2020)	East Azerbaijan	Sorptivity and hydraulic conductivity	ANN
Abedi et al. (2021)	Fars	EC, SAR	DT, kNN, SVM, RF, cubist, XGB, ensemble model
Gholami et al. (2021)	Kerman	Soil susceptibility to wind erosion	RNN, RBM
Nabiollahi et al. (2021)	Kurdistan	pH, EC, SAR	RF, hybridized RF
Neyestani et al. (2021)	Qazvin	Soil taxonomic classes	RF, C5.0, DT, ANN, SVM, kNN, LSSVMR
Omrani et al. (2021)	West Azerbaijan	Soil salt composition, major electrolytes and ionic strength	RF, cubist, DT

(*Continued*)

TABLE 12.1 A chronological review of the studies in DSM in Iran.—cont'd

References	Location	Target variable	Model
Taghizadeh-Mehrjardi et al. (2021)	West Azerbaijan	Clay, sand, $CaCO_3$, SOC, pH, EC, K, Ca + Mg, Na, SAR, EF	LR, LASSO, MARS, kNN, SVM, GP, ANN, ANFIS), cubist, RF, ERT, XGB, super learner
Alvyar et al. (2022)	West Azerbaijan	As, Cr, Cu, Ni, Pb and Zn	RF, cubist
Azizi et al. (2022)	Kurdistan	Ni, Fe, Cu, Mn	RF, cubist
Jahandideh Mahjenabadi et al. (2022)	Fars	Biological properties	RF
Mallah et al. (2022)	Mazandaran and Gilan	Soil textural classes	RF
Mousavi et al. (2022)	Alborz and Qazvin	SOC	MLR, cubist, RF, DT, KNN, SVM
Taghipour et al. (2022)	Ilam province	Soil quality index	RT, RF, ANN
Taghizadeh-Mehrjardi et al. (2022)	Isfahan and Yazd	Soil taxonomic classes	Semisupervised SVM
Azizi et al. (2023)	Kurdistan	Magnetic susceptibility	RF, SVM, cubist
Khosravi et al. (2023)	Mazandaran	Water erosion susceptibility	CNN, RNN, LSTM
Moosavi et al. (2023)	East Azerbaijan	Soil moisture	CMAC
Rezaei et al. (2023)	Fars	Wind-erodible fraction	BiGRU, BiRNN
Shahrokh et al. (2023)	Fars	Clay type	MLR, RF

ANFIS, adaptive-network-based fuzzy inference systems; *ANN*, artificial neural network; *AvNNet*, neural network ensemble based on model averaging; *AWC*, available water content; *BiGRU*, bidirectional gated recurrent unit; *BiRNN*, bidirectional recurrent neural network; *BLR*, binary logistic regression; *BRT*, boosted regression tree; *CCE*, calcium carbonate equivalent; *CEC*, cation exchange capacity; *CMAC*, Cerebellar Model Articulation Controller; *CNN*, convolutional neural network; *COK*, Cokriging; *DNN*, deep learning neural networks; *DT*, decision tree; *EC*, electrical conductivity; *EF*, erodible fraction; *ERT*, extremely randomized trees; *GBM*, gradient boosting machine; *GEP*, gene expression programming; *GLM*, generalized linear model; *GP*, genetic programming; *kNN*, k-nearest neighbors; *LASSO*, least absolute shrinkage and selection operator; *LR*, linear regression; *LSTM*, long-short term memory; *MARS*, multivariate adaptive regression splines; *MLR*, multiple linear regression; *MnLR*, multinomial logistic regression; *NF*, neuro-fuzzy; *NNRK*, neural network residual kriging; *PLSR*, partial least squares regression; *RBM*, restricted boltzmann machine; *RF*, random forest; *RK*, regression-kriging; *RKNNRK*, regression kriging-neural network residual kriging; *RNN*, recurrent neural network; *rpart*, recursive partitioning and regression trees; *SAR*, sodium adsorption ratio; *SCML*, supervised classification using maximum likelihood; *SOC*, soil organic carbon; *SVM*, support vector machines; *XGB*, extreme gradient boosting.

12.3.1 DEM-related covariates

The impact of topography on soil formation is significant, as it affects water runoff and microclimate, which in turn affects vegetation (Keshavarzi et al., 2021). Given this, several studies (Abedi et al., 2021; Taghizadeh-Mehrjardi et al., 2020a, 2021; Tajik et al., 2020) have focused on examining the relationships between soil physical properties and slope characteristics such as altitude, slope, and shape. It should be noted, however, that the effect of topography on soil properties may not always be uniform and is complex, making accurate prediction difficult. Nevertheless, topography is a critical factor in the transport of soil materials and sediments (Rahmati et al., 2017), which ultimately impacts crop production. In DSM studies, various derivatives of digital elevation models (DEMs) have been utilized to establish the relationship between soil properties and environmental covariates. These DEM derivatives include elevation, aspect, curvature (plan and profile), topographic wetness index (TWI), multiresolution valley bottom flatness index (MrVBF), topographic position index (TPI), slope, and multiresolution ridge top flatness (MrRTF).

12.3.2 Remote sensing and relevant covariates

In recent years, remote sensing has been proven to be an effective method for reconstructing the history of vegetation change, studying the properties of change, and evaluating the ecological effects of the change (Khosravi et al., 2023; Moosavi et al., 2023; Taghizadeh-Mehrjardi et al., 2022). Thus, there is a huge potential to link soil and vegetation reflectance characteristics with soil properties. It was reported that the status of vegetation cover is closely related to soil properties, and therefore, variations in soil properties can be reflected through vegetation indices (Fathizad et al., 2020b; Omrani et al., 2021). As a result, numerous vegetation indices, including the Normalized Difference Vegetation Index (NDVI), Perpendicular Vegetation Index (PVI), Leaf Area Index (LAI), Enhanced Vegetation Index (EVI), Soil Adjusted Vegetation Index (SAVI), and as well as band reflectance, a representative of soil condition, could be used to predict soil properties once a relationship was established between soil targets and remote sensing indices. Several studies have found that there is a strong relationship between vegetation indices and soil properties such as SOC (Hamzehpour et al., 2019; Pahlavan-Rad et al., 2018; Taghizadeh-Mehrjardi et al., 2016), soil pH (Moosavi et al., 2023; Taghizadeh-Mehrjardi et al., 2021), and soil EC (Abedi et al., 2021; Fathizad et al., 2020a; Taghizadeh-Mehrjardi et al., 2014a). It was also reported that remote sensing variables were not appropriate predictors for the prediction of nitrogen, phosphorous, and boron in Northwest Iran (Shahbazi et al., 2019).

12.3.3 Climatic data

The spatial variation in topsoil attributes is most influenced by precipitation and temperature, which affect chemical, physical, and biological processes at a large scale (Khalili and Rahimi, 2018). Soil moisture conditions and weathering are also influenced by precipitation and temperature (Poppiel et al., 2021). Hence, climatic data could be utilized as a predictive proxy for soil prediction in DSM studies based on an understanding of soil processes and soil-forming rates.

12.3.4 Legacy maps

The inclusion of legacy maps in this category, such as legacy soil maps (e.g., soil types and properties), geomorphology maps, geology maps, and land use/land cover maps, has significantly improved the prediction accuracy of machine learning in DSM frameworks in previous studies for legacy soil maps (Pahlavan-Rad et al., 2016) and for geomorphology and geology maps (Jafari et al., 2013; Taghizadeh-Mehrjardi et al., 2014a; Zeraatpisheh et al., 2017). The incorporation of existing legacy soil maps in DSM can be achieved through two methods. In the first approach, soil class observations are directly collected from legacy soil maps. In the second approach, field soil samples are collected, and the legacy soil survey is utilized as a covariate, or categorical covariates are derived from the original soil survey map (Pahlavan-Rad et al., 2016).

12.4 Practical use of digital soil maps

Accurate soil information is crucial for successful soil management. The advancement of technologies and equipment necessitates the use of precise soil information to enhance crop yield and meet the ever-increasing global demand for food (Roozitalab et al., 2018). However, in developing countries, the availability of digital soil fertility information is limited, and this technology is largely unknown to land managers (Farshad et al., 2018). In addition, high-resolution soil information is lacking worldwide, especially in developing countries, which could hinder agricultural intensification (Roozitalab et al., 2018). Digital soil products can improve agricultural land efficiency and production. Traditional farming practices apply fertilizer uniformly across the entire field, despite the uneven distribution of nutrients in soils (Zeraatpisheh et al., 2022). Digital soil products can show the spatial variability of soil properties at a finer resolution, providing stakeholders with information on the distribution of soil properties within landscapes and farms.

12.5 Summary and future direction

In summary, DSM is an emerging field that aims to provide accurate and high-resolution soil information for various applications, such as agriculture, environmental management, and land-use planning. DSM combines various techniques and data sources, such as remote sensing, geographic information systems, legacy soil maps, and statistical models, to generate detailed soil maps and predictions of soil properties. The accuracy of DSM depends on the quality and quantity of data sources used, as well as the expertise of the modeler in selecting appropriate predictors and modeling techniques. In recent years, there has been increasing interest in incorporating machine learning algorithms into DSM frameworks, which have shown promising results in improving prediction accuracy. The future of DSM is promising, as technological advancements and increasing availability of data sources continue to enhance the accuracy and resolution of soil maps. However, there are still many challenges to overcome, such as the need for more detailed and accurate soil data, especially in developing countries. Moreover, there is a need to develop standardized protocols and guidelines for DSM to ensure the quality and comparability of soil maps generated by different modelers and stakeholders. The integration of DSM with other emerging technologies, such as artificial intelligence, big data analytics, and precision agriculture, can further enhance the applicability and impact of DSM in various fields. Overall, DSM has the potential to revolutionize the way we understand and manage soils, which can have significant implications for food security, environmental sustainability, and socioeconomic development.

References

Abedi, F., Amirian-Chakan, A., Faraji, M., Taghizadeh-Mehrjardi, R., Kerry, R., Razmjoue, D., Scholten, T., 2021. Salt dome related soil salinity in southern Iran: prediction and mapping with averaging machine learning models. Land Degrad. Dev. 32, 1540–1554. https://doi.org/10.1002/ldr.3811.

Alvyar, Z., Shahbazi, F., Oustan, S., Dengiz, O., Minasny, B., 2022. Digital mapping of potentially toxic elements enrichment in soils of Urmia Lake due to water level decline. Sci. Total Environ. 808, 152086. https://doi.org/10.1016/j.scitotenv.2021.152086.

Amirian-Chakan, A., Minasny, B., Taghizadeh-Mehrjardi, R., Akbarifazli, R., Darvishpasand, Z., Khordehbin, S., 2019. Some practical aspects of predicting texture data in digital soil mapping. Soil Tillage Res. 194, 104289. https://doi.org/10.1016/j.still.2019.06.006.

Asgari, N., Ayoubi, S., Jafari, A., Demattê, J.A.M., 2020. Incorporating environmental variables, remote and proximal sensing data for digital soil mapping of USDA soil great groups. Int. J. Rem. Sens. 41, 7624–7648. https://doi.org/10.1080/01431161.2020.1763506.

Azizi, K., Ayoubi, S., Nabiollahi, K., Garosi, Y., Gislum, R., 2022. Predicting heavy metal contents by applying machine learning approaches and environmental covariates in west of Iran. J. Geochem. Explor. 233, 106921. https://doi.org/10.1016/j.gexplo.2021.106921.

Azizi, K., Ayoubi, S., Demattê, J.A.M., 2023. Controlling factors in the variability of soil magnetic measures by machine learning and variable importance analysis. J. Appl. Geophys. 210, 104944. https://doi.org/10.1016/j.jappgeo.2023.104944.

Bagheri Bodaghabadi, M., Martínez-casasnovas, J., Salehi, M.H., Mohammadi, J., Esfandiarpoor borujeni, I., Toomanian, N., Gandomkar, A., 2015. Digital soil mapping using artificial neural networks and Terrain-related attributes. Pedosphere 25, 580–591. https://doi.org/10.1016/S1002-0160(15)30038-2.

Balali, M.R., Siadat, H., Roozitalab, M.H., 2018. History of soil research. In: Roozitalab, M.H., Siadat, H., Farshad, A. (Eds.), The Soils of Iran, World Soils Book Series. Springer International Publishing, Cham, pp. 5–18. https://doi.org/10.1007/978-3-319-69048-3_2.

Behrens, T., Scholten, T., 2006. Digital soil mapping in Germany—a review. J. Plant Nutr. Soil Sci. 169, 434–443. https://doi.org/10.1002/jpln.200521962.

Dewan, M.L., Famouri, J., 1964. Soil fertility and response of soils to fertilizers. Soils Iran. Soil Inst. Min. Agr. Iran FAO 240–258.

Esfandiarpour-Boroujeni, I., Shahini-Shamsabadi, M., Shirani, H., Mosleh, Z., Bagheri-Bodaghabadi, M., Salehi, M.H., 2020. Assessment of different digital soil mapping methods for prediction of soil classes in the Shahrekord plain, Central Iran. Catena 193, 104648. https://doi.org/10.1016/j.catena.2020.104648.

Farshad, A., Roozitalab, M.H., Kamali, M.R.J., 2018. Land management in space and time. In: Roozitalab, M.H., Siadat, H., Farshad, A. (Eds.), The Soils of Iran, World Soils Book Series. Springer International Publishing, Cham, pp. 229–244. https://doi.org/10.1007/978-3-319-69048-3_13.

Fathizad, H., Ali Hakimzadeh Ardakani, M., Sodaiezadeh, H., Kerry, R., Taghizadeh-Mehrjardi, R., 2020a. Investigation of the spatial and temporal variation of soil salinity using random forests in the central desert of Iran. Geoderma 365, 114233. https://doi.org/10.1016/j.geoderma.2020.114233.

Fathizad, H., Ardakani, M.A.H., Heung, B., Sodaiezadeh, H., Rahmani, A., Fathabadi, A., Scholten, T., Taghizadeh-Mehrjardi, R., 2020b. Spatio-temporal dynamic of soil quality in the central Iranian desert modeled with machine learning and digital soil assessment techniques. Ecol. Indicat. 118, 106736. https://doi.org/10.1016/j.ecolind.2020.106736.

Fathololoumi, S., Vaezi, A.R., Alavipanah, S.K., Ghorbani, A., Saurette, D., Biswas, A., 2020. Improved digital soil mapping with multitemporal remotely sensed satellite data fusion: a case study in Iran. Sci. Total Environ. 721, 137703. https://doi.org/10.1016/j.scitotenv.2020.137703.

Gholami, V., Sahour, H., Hadian Amri, M.A., 2021. Soil erosion modeling using erosion pins and artificial neural networks. Catena 196, 104902. https://doi.org/10.1016/j.catena.2020.104902.

Hamzehpour, N., Shafizadeh-Moghadam, H., Valavi, R., 2019. Exploring the driving forces and digital mapping of soil organic carbon using remote sensing and soil texture. Catena 182, 104141. https://doi.org/10.1016/j.catena.2019.104141.

Hengl, T., Toomanian, N., Reuter, H.I., Malakouti, M.J., 2007. Methods to interpolate soil categorical variables from profile observations: lessons from Iran. Geoderma, Pedometrics 140, 417–427. https://doi.org/10.1016/j.geoderma.2007.04.022.

Jafari, A., Finke, P.A., Vande Wauw, J., Ayoubi, S., Khademi, H., 2012. Spatial prediction of USDA-great soil groups in the arid Zarand region, Iran: comparing logistic regression approaches to predict diagnostic horizons and soil types. Eur. J. Soil Sci. 63, 284–298.

Jafari, A., Ayoubi, S., Khademi, H., Finke, P.A., Toomanian, N., 2013. Selection of a taxonomic level for soil mapping using diversity and map purity indices: a case study from an Iranian arid region. Geomorphology 201, 86–97.

Jafari, A., Khademi, H., Finke, P.A., Van de Wauw, J., Ayoubi, S., 2014. Spatial prediction of soil great groups by boosted regression trees using a limited point dataset in an arid region, southeastern Iran. Geoderma 232–234, 148–163. https://doi.org/10.1016/j.geoderma.2014.04.029.

Jahandideh Mahjenabadi, V.A., Roohollah Mousavi, S., Rahmani, A., Karami, A., Asadi Rahmani, H., Khavazi, K., Rezaei, M., 2022. Digital mapping of soil biological properties and wheat yield using remotely sensed, soil chemical data and machine learning approaches. Comput. Electron. Agric. 197, 106978. https://doi.org/10.1016/j.compag.2022.106978.

Keshavarzi, A., Tuffour, H.O., Brevik, E.C., Ertunç, G., 2021. Spatial variability of soil mineral fractions and bulk density in Northern Ireland: assessing the influence of topography using different interpolation methods and fractal analysis. Catena 207, 105646. https://doi.org/10.1016/j.catena.2021.105646.

Khaledian, Y., Miller, B.A., 2020. Selecting appropriate machine learning methods for digital soil mapping. Appl. Math. Model. 81, 401–418. https://doi.org/10.1016/j.apm.2019.12.016.

Khalili, A., Rahimi, J., 2018. Climate. In: Roozitalab, M.H., Siadat, H., Farshad, A. (Eds.), The Soils of Iran, World Soils Book Series. Springer International Publishing, Cham, pp. 19–33. https://doi.org/10.1007/978-3-319-69048-3_3.

Khosravi, K., Rezaie, F., Cooper, J.R., Kalantari, Z., Abolfathi, S., Hatamiafkoueieh, J., 2023. Soil water erosion susceptibility assessment using deep learning algorithms. J. Hydrol. 618, 129229. https://doi.org/10.1016/j.jhydrol.2023.129229.

Mahmoudzadeh, H., Matinfar, H.R., Taghizadeh-Mehrjardi, R., Kerry, R., 2020. Spatial prediction of soil organic carbon using machine learning techniques in western Iran. Geoderma Reg. 21, e00260. https://doi.org/10.1016/j.geodrs.2020.e00260.

Maleki, S., Khormali, F., Mohammadi, J., Bogaert, P., Bagheri Bodaghabadi, M., 2020. Effect of the accuracy of topographic data on improving digital soil mapping predictions with limited soil data: an application to the Iranian loess plateau. Catena 195, 104810. https://doi.org/10.1016/j.catena.2020.104810.

Mallah, S., Delsouz Khaki, B., Davatgar, N., Scholten, T., Amirian-Chakan, A., Emadi, M., Kerry, R., Mosavi, A.H., Taghizadeh-Mehrjardi, R., 2022. Predicting soil textural classes using random forest models: learning from imbalanced dataset. Agronomy 12, 2613. https://doi.org/10.3390/agronomy12112613.

McBratney, A.B., Mendonça Santos, M.L., Minasny, B., 2003. On digital soil mapping. Geoderma 117, 3–52. https://doi.org/10.1016/S0016-7061(03)00223-4.

Mirakzehi, K., Pahlavan-Rad, M.R., Shahriari, A., Bameri, A., 2018. Digital soil mapping of deltaic soils: a case of study from Hirmand (Helmand) river delta. Geoderma 313, 233–240. https://doi.org/10.1016/j.geoderma.2017.10.048.

Moosavi, V., Karami, A., Aliramaee, R., 2023. High-resolution soil moisture mapping using PSO-based optimized cerebellar model articulation controller (CMAC). Sci. Total Environ. 857, 159493. https://doi.org/10.1016/j.scitotenv.2022.159493.

Motaghian, H.R., Mohammadi, J., 2011. Spatial estimation of saturated hydraulic conductivity from Terrain attributes using regression, kriging, and artificial neural networks. Pedosphere 21, 170–177. https://doi.org/10.1016/S1002-0160(11)60115-X.

Mousavi, A., Shahbazi, F., Oustan, S., Jafarzadeh, A.A., Minasny, B., 2020. Spatial distribution of iron forms and features in the dried lake bed of Urmia Lake of Iran. Geoderma Reg. 21, e00275. https://doi.org/10.1016/j.geodrs.2020.e00275.

Mousavi, S.R., Sarmadian, F., Omid, M., Bogaert, P., 2022. Three-dimensional mapping of soil organic carbon using soil and environmental covariates in an arid and semi-arid region of Iran. Measurement 201, 111706. https://doi.org/10.1016/j.measurement.2022.111706.

Nabiollahi, K., Taghizadeh-Mehrjardi, R., Shahabi, A., Heung, B., Amirian-Chakan, A., Davari, M., Scholten, T., 2021. Assessing agricultural salt-affected land using digital soil mapping and hybridized random forests. Geoderma 385, 114858. https://doi.org/10.1016/j.geoderma.2020.114858.

Neyestani, M., Sarmadian, F., Jafari, A., Keshavarzi, A., Sharififar, A., 2021. Digital mapping of soil classes using spatial extrapolation with imbalanced data. Geoderma Reg. 26, e00422. https://doi.org/10.1016/j.geodrs.2021.e00422.

Omrani, M., Shahbazi, F., Feizizadeh, B., Oustan, S., Najafi, N., 2021. Application of remote sensing indices to digital soil salt composition and ionic strength mapping in the east shore of Urmia Lake, Iran. Remote Sens. Appl.: Society and Environment 22, 100498. https://doi.org/10.1016/j.rsase.2021.100498.

Pahlavan Rad, M.R., Toomanian, N., Khormali, F., Brungard, C.W., Komaki, C.B., Bogaert, P., 2014. Updating soil survey maps using random forest and conditioned Latin hypercube sampling in the loess derived soils of northern Iran. Geoderma 232–234, 97–106. https://doi.org/10.1016/j.geoderma.2014.04.036.

Pahlavan-Rad, M.R., Akbarimoghaddam, A., 2018. Spatial variability of soil texture fractions and pH in a flood plain (case study from eastern Iran). Catena 160, 275–281. https://doi.org/10.1016/j.catena.2017.10.002.

Pahlavan-Rad, M.R., Khormali, F., Toomanian, N., Brungard, C.W., Kiani, F., Komaki, C.B., Bogaert, P., 2016. Legacy soil maps as a covariate in digital soil mapping: a case study from Northern Iran. Geoderma 279, 141–148. https://doi.org/10.1016/j.geoderma.2016.05.014.

Pahlavan-Rad, M.R., Dahmardeh, K., Brungard, C., 2018. Predicting soil organic carbon concentrations in a low relief landscape, eastern Iran. Geoderma Reg. 15, e00195. https://doi.org/10.1016/j.geodrs.2018.e00195.

Pahlavan-Rad, M.R., Dahmardeh, K., Hadizadeh, M., Keykha, G., Mohammadnia, N., Gangali, M., Keikha, M., Davatgar, N., Brungard, C., 2020. Prediction of soil water infiltration using multiple linear regression and random forest in a dry flood plain, eastern Iran. Catena 194, 104715. https://doi.org/10.1016/j.catena.2020.104715.

Poppiel, R.R., Demattê, J.A.M., Rosin, N.A., Campos, L.R., Tayebi, M., Bonfatti, B.R., Ayoubi, S., Tajik, S., Afshar, F.A., Jafari, A., Hamzehpour, N., Taghizadeh-Mehrjardi, R., Ostovari, Y., Asgari, N., Naimi, S., Nabiollahi, K., Fathizad, H., Zeraatpisheh, M., Javaheri, F., Doustaky, M., Naderi, M., Dehghani, S., Atash, S., Farshadirad, A., Mirzaee, S., Shahriari, A., Ghorbani, M., Rahmati, M., 2021. High resolution middle eastern soil attributes mapping via open data and cloud computing. Geoderma 385, 114890. https://doi.org/10.1016/j.geoderma.2020.114890.

Rahmati, O., Tahmasebipour, N., Haghizadeh, A., Pourghasemi, H.R., Feizizadeh, B., 2017. Evaluation of different machine learning models for predicting and mapping the susceptibility of gully erosion. Geomorphology 298, 118–137. https://doi.org/10.1016/j.geomorph.2017.09.006.

Rezaei, M., Mohammadifar, A., Gholami, H., Mina, M., Riksen, M.J.P.M., Ritsema, C., 2023. Mapping of the wind erodible fraction of soil by bidirectional gated recurrent unit (BiGRU) and bidirectional recurrent neural network (BiRNN) deep learning models. Catena 223, 106953. https://doi.org/10.1016/j.catena.2023.106953.

Roozitalab, M.H., Siadat, H., Farshad, A. (Eds.), 2018. The Soils of Iran, World Soils Book Series. Springer International Publishing, Cham. https://doi.org/10.1007/978-3-319-69048-3.

Shahbazi, F., Hughes, P., McBratney, A.B., Minasny, B., Malone, B.P., 2019. Evaluating the spatial and vertical distribution of agriculturally important nutrients—nitrogen, phosphorous and boron—in North West Iran. Catena 173, 71–82. https://doi.org/10.1016/j.catena.2018.10.005.

Shahriari, M., Delbari, M., Afrasiab, P., Pahlavan-Rad, M.R., 2019. Predicting regional spatial distribution of soil texture in floodplains using remote sensing data: a case of southeastern Iran. Catena 182, 104149. https://doi.org/10.1016/j.catena.2019.104149.

Shahrokh, V., Khademi, H., Zeraatpisheh, M., 2023. Mapping clay mineral types using easily accessible data and machine learning techniques in a scarce data region: a case study in a semi-arid area in Iran. Catena 223, 106932. https://doi.org/10.1016/j.catena.2023.106932.

Sharififar, A., Sarmadian, F., Malone, B.P., Minasny, B., 2019. Addressing the issue of digital mapping of soil classes with imbalanced class observations. Geoderma 350, 84–92. https://doi.org/10.1016/j.geoderma.2019.05.016.

Shiri, J., Keshavarzi, A., Kisi, O., Iturraran-Viveros, U., Bagherzadeh, A., Mousavi, R., Karimi, S., 2017. Modeling soil cation exchange capacity using soil parameters: assessing the heuristic models. Comput. Electron. Agric. 135, 242–251. https://doi.org/10.1016/j.compag.2017.02.016.

Taghipour, K., Heydari, M., Kooch, Y., Fathizad, H., Heung, B., Taghizadeh-Mehrjardi, R., 2022. Assessing changes in soil quality between protected and degraded forests using digital soil mapping for semiarid oak forests, Iran. Catena 213, 106204. https://doi.org/10.1016/j.catena.2022.106204.

Taghizadeh-Mehrjardi, R., Minasny, B., McBratney, A.B., Triantafilis, J., Sarmadian, F., Toomanian, N., 2012. Digital soil mapping of soil classes using decision trees in central Iran. In: Digital Soil Assessments and beyond. CRC Press.

Taghizadeh-Mehrjardi, R., Minasny, B., Sarmadian, F., Malone, B.P., 2014a. Digital mapping of soil salinity in Ardakan region, central Iran. Geoderma 213, 15–28. https://doi.org/10.1016/j.geoderma.2013.07.020.

Taghizadeh-Mehrjardi, R., Sarmadian, F., Minasny, B., Triantafilis, J., Omid, M., 2014b. Digital mapping of soil classes using decision tree and auxiliary data in the Ardakan region, Iran. Arid Land Res. Manag. 28, 147–168. https://doi.org/10.1080/15324982.2013.828801.

Taghizadeh-Mehrjardi, R., Nabiollahi, K., Minasny, B., Triantafilis, J., 2015. Comparing data mining classifiers to predict spatial distribution of USDA-family soil groups in Baneh region, Iran. Geoderma 253–254, 67–77. https://doi.org/10.1016/j.geoderma.2015.04.008.

Taghizadeh-Mehrjardi, R., Nabiollahi, K., Kerry, R., 2016. Digital mapping of soil organic carbon at multiple depths using different data mining techniques in Baneh region, Iran. Geoderma 266, 98–110. https://doi.org/10.1016/j.geoderma.2015.12.003.

Taghizadeh-Mehrjardi, R., Mahdianpari, M., Mohammadimanesh, F., Behrens, T., Toomanian, N., Scholten, T., Schmidt, K., 2020a. Multi-task convolutional neural networks outperformed random forest for mapping soil particle size fractions in central Iran. Geoderma 376, 114552. https://doi.org/10.1016/j.geoderma.2020.114552.

Taghizadeh-Mehrjardi, R., Schmidt, K., Amirian-Chakan, A., Rentschler, T., Zeraatpisheh, M., Sarmadian, F., Valavi, R., Davatgar, N., Behrens, T., Scholten, T., 2020b. Improving the spatial prediction of soil organic carbon content in two contrasting climatic regions by stacking machine learning models and rescanning covariate space. Rem. Sens. 12, 1095. https://doi.org/10.3390/rs12071095.

Taghizadeh-Mehrjardi, R., Schmidt, K., Eftekhari, K., Behrens, T., Jamshidi, M., Davatgar, N., Toomanian, N., Scholten, T., 2020c. Synthetic resampling strategies and machine learning for digital soil mapping in Iran. Eur. J. Soil Sci. 71, 352–368. https://doi.org/10.1111/ejss.12893.

Taghizadeh-Mehrjardi, R., Hamzehpour, N., Hassanzadeh, M., Heung, B., Ghebleh Goydaragh, M., Schmidt, K., Scholten, T., 2021. Enhancing the accuracy of machine learning models using the super learner technique in digital soil mapping. Geoderma 399, 115108. https://doi.org/10.1016/j.geoderma.2021.115108.

Taghizadeh-Mehrjardi, R., Sheikhpour, R., Zeraatpisheh, M., Amirian-Chakan, A., Toomanian, N., Kerry, R., Scholten, T., 2022. Semi-supervised learning for the spatial extrapolation of soil information. Geoderma 426, 116094. https://doi.org/10.1016/j.geoderma.2022.116094.

Tajik, S., Ayoubi, S., Zeraatpisheh, M., 2020. Digital mapping of soil organic carbon using ensemble learning model in Mollisols of Hyrcanian forests, northern Iran. Geoderma Reg. 20, e00256. https://doi.org/10.1016/j.geodrs.2020.e00256.

Tashayo, B., Honarbakhsh, A., Akbari, M., Ostovari, Y., 2020. Digital mapping of Philip model parameters for prediction of water infiltration at the watershed scale in a semi-arid region of Iran. Geoderma Reg. 22, e00301. https://doi.org/10.1016/j.geodrs.2020.e00301.

Wadoux, A.M.J.-C., Minasny, B., McBratney, A.B., 2020. Machine learning for digital soil mapping: applications, challenges and suggested solutions. Earth Sci. Rev. 210, 103359. https://doi.org/10.1016/j.earscirev.2020.103359.

Zeraatpisheh, M., Ayoubi, S., Jafari, A., Finke, P., 2017. Comparing the efficiency of digital and conventional soil mapping to predict soil types in a semi-arid region in Iran. Geomorphology 285, 186–204.

Zeraatpisheh, M., Ayoubi, S., Brungard, C.W., Finke, P., 2019a. Disaggregating and updating a legacy soil map using DSMART, fuzzy c-means and k-means clustering algorithms in Central Iran. Geoderma 340, 249–258.

Zeraatpisheh, M., Ayoubi, S., Jafari, A., Tajik, S., Finke, P., 2019b. Digital mapping of soil properties using multiple machine learning in a semi-arid region, central Iran. Geoderma 338, 445–452.

Zeraatpisheh, M., Bottega, E.L., Bakhshandeh, E., Owliaie, H.R., Taghizadeh-Mehrjardi, R., Kerry, R., Scholten, T., Xu, M., 2022. Spatial variability of soil quality within management zones: homogeneity and purity of delineated zones. Catena 209, 105835.

CHAPTER

13

Impacts of land use and land cover changes on soil erosion

Carla S.S. Ferreira[1,2,3], *Milica Kašanin-Grubin*[4], *Marijana Kapović Solomun*[5] *and Zahra Kalantari*[1,6]

[1]Department of Physical Geography and Bolin Centre for Climate Research, Stockholm University, Stockholm, Sweden; [2]Polytechnic Institute of Coimbra, Applied Research Institute, Coimbra, Portugal; [3]Research Centre for Natural Resources Environment and Society (CERNAS), Polytechnic Institute of Coimbra, Coimbra, Portugal; [4]Institute of Chemistry, Technology, and Metallurgy, University of Belgrade, Belgrade, Serbia; [5]Department of Forest Ecology, Faculty of Forestry, University of Banja Luka, Banja Luka, Bosnia and Herzegovina; [6]Department of Sustainable Development, Environmental Science and Engineering (SEED), KTH Royal Institute of Technology, Stockholm, Sweden

13.1 Introduction

Soil covers the Earth's surface and is vital for different forms of life since it provides valuable goods and services, such as food and biomass provision, regulation of water, nutrient cycling, filtration of contaminants, and supporting biodiversity (e.g., Panagos et al., 2020; Evans et al., 2021). Given the important ecosystem services provided, soil plays a relevant role in supporting the achievement of multiple United Nations Sustainable Development Goals (SDGs) (Keesstra et al., 2016). For example, it provides 90% of all food, feed, and fiber and production of raw materials (EEA, 2017), thus playing a vital role in SDG1 (Zero hunger). It also supports life on Earth and thus plays a fundamental role in SDG15 (Life on land). Soil is the second largest carbon pool on Earth and contributes to climate change mitigation by sequestering carbon and reducing nitrous oxide and methane emissions (Smith, 2016), making it crucial for SDG13 (climate action). The increasing recognition of soil as a nonrenewable resource at the human time scale (Rakshit et al., 2017) and of its importance to society has resulted in soil degradation being acknowledged as a key global environmental problem

https://doi.org/10.1016/B978-0-443-15341-9.00023-X

(e.g., Ma et al., 2021). Several degradation processes limit the ability of soil to provide ecosystem services (Ferreira et al., 2022). Globally, soil erosion is the dominant land degradation process and one of the most critical environmental challenges (e.g., Vanmaercke et al., 2022). Soil erosion is a physical process characterized by the transport of soil particles from the topsoil that occurs when external forces are higher than the cohesive forces that bind the soil together (FAO, 2016). In general, soil erosion is a three-phase process consisting of (i) detachment of individual soil particles from the soil mass; (ii) transport of detached particles by an erosive agent; and (iii) deposition of particles when the erosive agent lacks sufficient energy for further transport (Ferreira et al., 2019).

Soil erosion can be triggered by various agents, such as water, wind, and gravity, but water is the agent responsible for 56% of all global soil losses (Bouma and Batjes, 2000). Soil erosion by water can take four different forms: (i) splash, resulting from the impact of raindrops; (ii) sheet, caused by nonchannel uniform overland flow, often leading to low-magnitude soil losses over small distances; (iii) rill, in which water running as concentrated flow removes soil by carving out channels of increasing depth and/or width in hillslopes; and (iv) gully, comprising formation and subsequent expansion of erosional channels (that cannot be erased through tillage operations) caused by concentrated water flow. Gully erosion affects several regions worldwide and can contribute 20%–80% of the average catchment sediment yield (Vanmaercke et al., 2022).

Water-related soil erosion is often associated with a small number of extreme rainfall events, which control the annual soil loss budget (Matthews et al., 2022). In Europe, for example, 50% of rainfall erosivity is triggered by 11% of rainfall events (Bezak et al., 2021). Since the magnitude and frequency of extreme events are predicted to increase due to climate change, soil erosion should aggravate even in areas where the mean annual precipitation amount may decrease (Borrelli et al., 2017). By 2050, mean soil erosion rates in agricultural areas in the European Union (EU) are expected to increase by 13%–22.5% compared with 2016, primarily due to water erosion (Panagos et al., 2021). This increase is expected to have a skewed spatial pattern, with soil erosion being unequally distributed between EU countries (Borrelli et al., 2020).

Wind-related soil erosion is more relevant in dry areas when the soil is exposed to erosive agents. Strong winds beat upon existing landforms, cutting them, and transporting the soil particles over large distances in the direction of the wind by creeping, saltation, and suspension (depending on particle size) (Kuhwald et al., 2022).

The overall rate of soil erosion at a site is a consequence of the climate setting, particularly precipitation, wind speed and temperature, the surface environment (e.g., topography, including slope steepness and length) and soil properties (e.g., texture, structure, organic matter content, and soil depth) (e.g., Li and Fang, 2016). However, increasing global population and socioeconomic welfare have led to significant land use and land cover (LULC) changes, often associated with unsustainable human activities such as deforestation, tillage practices, overgrazing, and infrastructure construction, which have accelerated soil erosion and land degradation (Gil et al., 2021; Adams et al., 2022). Through its strong effects on soil characteristics and vegetation quality, LULC is a highly important predictor of soil erosion processes (Cunha et al., 2022), together with land management (Alem, 2022).

Globally, soil erosion rates are 10–14 times higher than soil formation rates through weathering and pedogenesis (Pimentel and Burgess, 2013). Soil erosion has been responsible for 84% of all degraded land (Lal, 2017), which currently affects 30% of the Earth's surface, and the proportion is expected to increase to 90% by 2050 (FAO, ITPS, 2015). It is estimated that global land areas experience annual soil losses of approximately 1.094 billion hm^2 and 0.549 billion hm^2 by water and wind erosion, respectively (Lal, 2003). Approximately 80% of agricultural land worldwide is subjected to moderate to severe soil erosion (International Atomic Energy Agency, 2014), with soil erosion affecting nearly 10 million hectares of agricultural land every year (Amiri et al., 2019). Soil erosion is particularly concerning in Africa but is also relatively severe in Asia and Latin America (Borrelli et al., 2017), where rates are 2–6 times higher than in North America and Europe (UNEP, 2012). Soil degradation affects more than 1.5 billion people in 110 countries, 90% of whom live in low-income countries (Nellemann, 2009). In the EU, more than 201,000 km^2 of the land area is at risk of severe soil erosion ($>10\ t\ ha^{-1}\ yr^{-1}$), largely in agricultural areas (Ma et al., 2021).

Soil erosion can cause several forms of environmental damage in a short period. Soil loss reduces the capacity for rooting space and the volume to store water. By removing the topsoil, the most fertile and organic matter-rich layer, soil erosion reduces the in situ productive capacity of the soil, resulting in lower crop yields (Adams et al., 2022). According to the FAO (2015), each 0.1 m of soil loss is associated with a 4% yield decline. In vulnerable areas with shallow soils, topsoil loss is a major cause of desertification (Kirkby et al., 2008). In addition, the loss of nutrients through erosion may exceed external inputs through fertilization, increasing the financial costs to land owners (Ferreira et al., 2018a). Erosion is often associated with soil biodiversity loss and disturbances in biogeochemical cycles (Zhao et al., 2022). For example, in the soil carbon cycle, erosion is associated with increased mineralization, increasing carbon dioxide emissions and decreasing carbon sinks (Smith, 2016). Off-site environmental consequences may include decreasing surface water quality and pervasive impacts on aquatic ecosystems (Ferreira et al., 2016a). This is because suspended sediment, the main type of sediment generated by erosion within catchments (Vercruysse et al., 2017), increases water turbidity, which reduces light penetration and affects photosynthetic activity, in turn affecting dissolved oxygen concentrations. In addition, sediments rich in nutrients, particularly phosphorus and nitrogen, can cause algae blooms and eutrophication of surface waters (Ferreira et al., 2016a). Algae blooms reduce the dissolved oxygen content in surface waters, threatening aquatic species and imposing constraints on recreational activities such as fishing and swimming (Vercruysse et al., 2017). Furthermore, suspended sediment has the capacity to adsorb potential organic and inorganic contaminants, such as heavy metals, polycyclic aromatic hydrocarbons (PAHs) and polychlorinated biphenyls, and pathogens (Ferreira et al., 2019). The magnitude of the impact of sediment on aquatic ecosystems is dependent on the physical and chemical properties of the sediment, which are determined by the characteristics of the source(s) (Ferreira et al., 2016a). These negative impacts can impair the achievement of SDG14 (life below water) and SDG6 (clean water and sanitation) (Evans et al., 2021).

Coarse sediments can also lead to siltation of downstream lakes, reservoirs, and river channels and may clog artificial drainage systems (e.g., drains and culverts) (Ferreira et al., 2019). Decreasing the drainage capacity of hydraulic infrastructure and surface water bodies can increase flood hazards, jeopardizing community and regional security (Ma et al., 2021). It can also cause constraints on travel, damage to infrastructure such as dams and transport networks (e.g., roads and rail routes) (Kalantari et al., 2019) and abrasion of hydropower turbines (Mushi et al., 2019). Globally, soil erosion is reported to cause socioeconomic losses of 40 billion dollars annually (Oldeman et al., 2017).

Accurate information on the timing and spatial distribution of erosion is vital for the development of targeted and efficient management practices. Many different models and field measurements at different scales have been used to estimate soil erosion, but large uncertainties have led to disparate results (Zhao et al., 2022). For example, the average soil erosion rate reported for China ranges from 1.4 ton ha^{-1} yr^{-1} (Teng et al., 2018) to 5 ton ha^{-1} yr^{-1} (Liu et al., 2020). Uncertainties in estimates are often associated with assumptions concerning important input variables based on case studies in other regions, and thus, it is essential to have access to local reliable information. In recent years, remote sensing has emerged as a promising tool supporting soil erosion assessments (e.g., Panahi et al., 2022).

In the remainder of this book chapter, we review the literature regarding the impact of different land uses, including agriculture, forest, and urban, on soil erosion. We then discuss the potential of remote sensing in supporting soil erosion assessments.

13.2 Soil erosion in agricultural areas

Most arable land worldwide is affected by soil erosion, and erosion rates are expected to increase in the coming century due to climate change (Borrelli et al., 2022). It is estimated that of the 1.5 billion hectares used globally for crops and the 3.5 billion hectares used for livestock, one-third of the total area is already degraded (FAO, 2016). Rapid population growth and increasing food demand, associated with changes in consumption patterns, have increased the pressure on agricultural systems, triggering the expansion of monoculture cropping using intensive practices (Barão et al., 2019). This often leads to temporary or persistent loss of productive capacity and severe soil degradation (Arif et al., 2018). For example, soil erosion is reported to cause mean annual losses of 8.2% in yield production in Africa, a reduction in cereal equivalent of 36 million tons in south Asia (Eswaran et al., 2001) and annual losses of 3 and 0.6 million tonnes of wheat and maize, respectively, in the EU (Panagos et al., 2018).

The rapid mechanisation of agriculture and increasing use of larger and heavier tractors have increased soil compaction and decreased infiltration, promoting runoff and soil erosion (Chevigny et al., 2014). Erosion rates are strongly influenced by soil management practices (Thomaz et al., 2022) (Fig. 13.1). Tillage is one of the most widely used soil management practices for preparing the seedbed for the next crop, controlling weeds and thus preventing competition with the main crop and temporarily reducing soil compaction (Evans et al., 2021). However, tillage leads to faster mineralization of soil organic matter, disrupts soil structure and aggregate stability, and modifies the soil pore system by creating large macropores that are unstable and susceptible to compaction, thus reducing the capacity of the soil for infiltration of water, resulting in runoff and water erosion (Stolte et al., 2016). Some studies have also reported surface crusting after the first rainfall events following tillage and a tendency for thicker crusts over time, contributing to increased overland flow and sediment losses (e.g., Telak et al., 2020). The overall magnitude of the impacts depends on the implement used for soil tillage (e.g., chisel plow, moldboard plow) (Kuhwald et al., 2022).

FIGURE 13.1 Examples of rill erosion in well-established (*left*) and recently established (*right*) arable crops in central Portugal.

Soil losses due to crop harvesting, although not widely investigated, may be relevant in crops that are in direct contact with the soil, such as potatoes and sugar beet (Ruysschaert et al., 2004). Loss of soil during harvesting of these crops is reported to represent <30% of all water- and wind-related soil erosion in Europe (Panagos et al., 2019). For example, it accounts for up to 22 Mg ha^{-1} per harvest in Turkey, indicating that it can be as important as water erosion (Parlak et al., 2016). This type of soil loss depends on factors such as (i) soil properties, including texture (higher losses of clayey than sandy soils), carbon content and moisture content (increasing losses with increasing soil moisture); (ii) crop characteristics, such as root density and architecture and skin roughness (a smooth skin minimizes the ability of soil to adhere to the crop); (iii) agronomic practices; and (iv) harvesting technique and the technology used, namely, the velocity and effectiveness of the harvester (Kuhwald et al., 2022).

Increasing food demand and urban sprawl have forced the expansion of agricultural activities to upland areas, which have accelerated soil erosion (Panahi et al., 2022). Crop cultivation in highland locations and on steep slopes is often associated with water erosion and can lead to the formation of gullies (Alem, 2022). Ephemeral gullies are typically found in cropland but are quickly filled in by land-leveling operations soon after their formation (Vanmaercke et al., 2022). Modern crop cultivation in sloping areas often requires engineering operations involving mechanical translocation of soil (e.g., terracing or leveling with bulldozers) to adapt the surface to mechanized agriculture. Although the bench of terraces is leveled, reducing the soil erosion risk, the back-wall component of the terrace is constructed with a steeper slope than the original land and is thus highly susceptible to soil erosion (Verheijen et al., 2009). In addition, the removal of vegetation and soil movement during terrace construction destroyed the soil structure and increased soil erodibility and losses. Over time, the lack of maintenance of terraces, particularly those on abandoned agricultural land, leads to further soil erosion problems (Djuma et al., 2017). For example, loss of the drainage function of terraces, often retained by stone walls, may lead to soil saturation in the back slope, causing the collapse of terraces and triggering shallow landslides controlled by slope and hillslope shape (Stolte et al., 2016). This soil degradation process is frequently reported, for example, in the Mediterranean region (Ferreira et al., 2022).

With increasing recognition of the consequences of soil erosion in lowering soil fertility and crop yields, multiple soil and water conservation measures have been applied in agricultural areas all over the world to mitigate soil erosion. These conservation measures include no-tillage or minimum tillage (Ferreira et al., 2018); grass cover, with a few different types of grass being used (Duan et al., 2020); cover crops, usually a legume or edible crop, which provides ground cover to reduce raindrop impact and surface runoff (Boulet et al., 2021) (Fig. 13.2); mulch, that is, organic or inorganic material that can be spread on the soil surface to protect the soil from raindrop impact, reduce surface runoff, and conserve soil moisture (Thomaz et al., 2022); vegetative filter strips (Li et al., 2020); intercropping, that is, growing more than one main crop in the same field (Xia et al., 2015) and terracing, that is, dividing a long slope into shorter slopes (Zhang et al., 2015). The effectiveness of mulching depends on the type of material (e.g., straw residues) and application rate, but worldwide, mulching at an application rate of 6–8 Mg ha^{-1} has been shown to reduce water erosion by up to 52% (Li et al., 2021). Several conservation measures have been compiled in WOCAT, a global database documenting and sharing information on sustainable land management, aiming to promote adaptation and support decision-making (https://www.wocat.net/). Experiments performed on relatively small scales have shown positive impacts of conservation measures in mitigating and reversing soil erosion (Bai et al., 2018). However, the effectiveness of these measures in protecting the soil varies according to the farming system and the agroecological zone (Kanter et al., 2018), causing uncertainties that require further investigation (Niu et al., 2021).

Overgrazing, which leads to decreasing vegetation cover and soil compaction, also exacerbates soil erosion (Ferreira et al., 2022). Lack of synchronicity between periods of greater feed demand from animals and fodder supply by vegetation is particularly influential in soil degradation (Miranda et al., 2018). Overgrazing of rangelands and various agrosilvopastoral systems has been reported to cause gully erosion in many regions worldwide, but it mainly affects arid and semiarid regions (Panahi et al., 2022).

FIGURE 13.2 Use of grass cover (*left*) and a cover crop (*right*) for soil erosion mitigation in vineyards in Sicily, Italy.

13.3 Soil erosion in forest areas

Forests occupy one-third of global land, or over 4.06 billion hectares (FAO, 2020). Their distribution varies according, for example, to climate conditions and geology, and they can be subjected to different degrees of human intervention. Some forests can regenerate spontaneously with native tree species (without noticeable human activities), whereas others are planted and only established by planting tree seedlings. Many forest plantations are intensively managed and composed of a single tree species (monoculture, usually with nonnative species), spaced regularly, for commercial/economic purposes. The use of fast-growing species allows rotation cycles of 8–10 years, a very attractive option for forestland owners (Castro et al., 2016). Globally, forest plantations have increased from eight million hectares in 1990 to 131 million hectares in 2020, with a larger incidence in South America (FAO, 2020).

Plants are a critical factor affecting soil erosion, but their ability to regulate runoff and erosion depends on vegetation type and land management practices (Gong et al., 2022). For example, a study comparing a managed dense eucalypt plantation, a sparse eucalypt stand, and an oak woodland found that the dense eucalypt stand resulted in twice as high runoff and sediment yield as the other two stands (Ferreira et al., 2016b). Nevertheless, afforestation is often considered an effective method for controlling soil erosion (FAO, 2020). However, particularly in planted forests, intensive management involves high levels of mechanization in site preparation for planting, harvesting, transport, storage, and processing (Forrester et al., 2012). For example, in eucalypt plantations (the most widespread form of plantation worldwide), establishment of a new plantation typically involves soil preparation by machinery, for example, removal of existing root systems and elimination of vegetation (Pancel, 2016). This soil preparation process can involve ripping (scraping the soil with metal blades mounted on bulldozers), disking, subsoiling, and light harrowing, involving the use of several disc types to reduce soil density and facilitate the establishment of new trees. It may also involve deep plowing (>80 cm), characterized by inversion and mixture of soil horizons and profound rupture of the soil structure. During the forest production cycle, managers may decide to clear-cut ground vegetation, leaving the soil more susceptible to erosive agents (Fig. 13.3, left). Consequently, studies have shown that forests involving trees and shrubs

FIGURE 13.3 Soil erosion in forest areas: Intensive monoculture plantations with vegetation removed during the plantation cycle (*left*) and terracing for new forest plantations (*right*).

are often more efficient in reducing runoff and erosion than forests containing only tree species (e.g., Gong et al., 2022). Harvesting also involves heavy machinery, such as caterpillar trucks for clear cutting, cranes for loading logs into trailers, and tractors that transport logs to temporary storage areas for transportation to a final destination (Castro et al., 2016). This intensive use of machinery disrupts the soil structure, making it more susceptible to erosion.

The construction of bench terraces for the plantation of fast-growing tree species (e.g., eucalypt) in sloped areas is also a widespread practice in many countries, such as Portugal. As mentioned in Section 13.2, terracing operations to convert a slope into roughly flat sections (treads) between steeper sections (risers) involve massive redistribution of topsoil (Fig. 13.3, right). Construction of terraces also involves the destruction of soil structure and exposure of soil to erosive agents, which increases splash erosion, localized interrill erosion and landslide risks (especially when the technical guidelines regarding slope angle of risers are not fulfilled). Soil erosion in these areas can be particularly high during the first years until the root system and the tree canopy provide sufficient stability and cover protection (Martins et al., 2013).

Plantation of flammable tree monocultures (e.g., pine, eucalypt) also leads to more frequent and severe wildfires. In addition, abandoned land areas are prone to wildfires due to extensive accumulation of biomass (Solomun et al., 2021). By destroying vegetation and litter cover, wildfires reduce rainfall interception and storage, evaporation, and infiltration, increasing overland flow and erosion (Esteves et al., 2012). Additionally, high-severity wildfires alter soil physical properties (e.g., increased bulk density, reduced aggregate stability, reduced soil porosity), chemical properties (e.g., reduced soil organic matter content), and biological properties (e.g., reduced soil biodiversity). These are properties that govern water flow and soil structural stability, thus increasing particle susceptibility to detachment and transportation (Gonzalez-Pelayo et al., 2015; Stoof et al., 2015). Wildfires may also consume part of the existing root system, further contributing to the loss of soil cohesion (Esteves et al., 2012). In areas where the wildfire incidence is high, declining soil fertility due to erosion impairs the ability of the vegetation to recover, thus prolonging the exposure of the soil surface to erosive agents (e.g., Shakesby et al., 2015; Solomun et al., 2021). The actual impacts of wildfires on soil degradation are complex, as they are determined by fire regime (which influences soil temperature and duration), fire history, the nature of the terrain (e.g., topography), soil properties (e.g., texture and moisture content), postfire weather conditions, and postfire management (Shakesby, 2011). Postfire erosion rates tend to decline over a period of months to years, depending on factors such as exhaustion of easily removed material and vegetation recovery. Postfire salvage logging by forest owners to recover wood and ensure some income has greater negative impacts on soil than the fire itself in increasing soil erosion (Waggenbrenner et al., 2015).

Deforestation, that is, conversion of forest areas into other land uses, is one of the most unsustainable land-use changes leading to soil erosion (Ferreira et al., 2018b). Globally, 178 million hectares of forests have been lost since 1990 (FAO, 2020). Clear-cutting and burning are the most commonly applied deforestation methods (Gharibreza et al., 2021). Gully erosion is prone to occur in deforested areas (Panahi et al., 2022).

Roads are important means of transportation within forest areas but are reported to be a major source of sediment loss within catchments (e.g., Thomaz et al., 2022) (Fig. 13.4). For

FIGURE 13.4 Unpaved roads providing access to forest areas (*left*) or within forest areas (*right*) act as sources of sediment.

example, Bezerra et al. (2020) found that even though unpaved roads comprise a small area, they can account for >90% of all sediment yield at the hillslope scale. Even in areas less affected by wildfires, unpaved roads can be the main source of sediments (Shakesby et al., 2015).

13.4 Soil erosion in urban areas

Urbanization has been a pervasive global trend in recent decades, with the rate of urban expansion reported to be even faster than the increase in urban population (EEA, 2006). The urban population is currently 3.9 billion and is expected to reach 6.3 billion, representing 70% of the world's population, by 2050 (WHO/UNICEF, 2017). The expansion of urban areas has mainly affected the agricultural land, typically adjacent to existing urban areas (Ferreira et al., 2019).

Urban areas include a mixture of sealed and unsealed surfaces and a diversity of soils under variable pressures, ranging from relatively intact soil profiles (e.g., in urban forests) to profiles different from natural soil (e.g., by covering preexisting soil with layers of soil from elsewhere or by mixing the topsoil with subsoil). Different urban land uses, such as industrial, commercial, residential, recreational, and agricultural (urban agriculture), are characterized by distinct development patterns associated with variable hydrological processes and susceptibility to erosive agents.

Within urban catchments, erosion is a serious problem at construction sites and in areas with bare soil (Fig. 13.5). At construction sites, the removal of vegetation combined with soil disturbance and compaction changes the storage and infiltration capacity of the soil, leading to more and frequency of surface runoff, which increases water erosion (Ferreira, 2021). Soil erosion at construction sites is typically 35–45 ton per ha and year, whereas in undisturbed forest areas, it is lower than 1 ton per ha and year (EPA, 1993). Erosion from construction sites can be approximately four times higher than that from clear-felled forest areas (Ferreira et al., 2012). In fact, during the clearing and grading phases of construction, erosion is reported to be 10 times higher than for other urban land uses, such residential areas (Line et al., 2002). Although soil erosion is reduced after construction activities cease, the rate is still

FIGURE 13.5 Soil erosion in different urban areas: construction sites for a large industrial park or isolated house (*top*) and bare surfaces with high soil degradation status (*bottom left*) and urban agricultural areas (*bottom right*).

often 3–6 times higher than that in agricultural and forest areas due to the greater runoff coefficient (Russel et al., 2017). In the case of bare soils (e.g., acting as parking areas), inappropriate management practices in combination with advanced stages of soil degradation increase the susceptibility of the soil to erosion. Construction sites and soils near large areas of sealed surfaces that convey storm runoff (e.g., roads), without proper artificial drainage systems, are often associated with gully formation (Panahi et al., 2022).

Erosion in established urban areas is typically associated with roads, particularly road verges (Fig. 13.6), due to inadequate planning of the urban drainage system. Several studies have demonstrated that distance from roads (or proximity to roads) and the location of drainage ditches and culverts are critical factors influencing gully erosion on bare land, especially in arid and semiarid regions (e.g., Panahi et al., 2022).

Sediment can also be lost from sealed surfaces, depending on precipitation amount and intensity (Ferreira et al., 2016c). Although these losses may represent a relatively low percentage of the total sediment load at the catchment scale (Ferreira et al., 2017), they may be large during summer storms or after long dry spells and tend to decrease in the wet season since they are carried by overland flow (Ferreira et al., 2016c). The pavement material, which influences the permeability of the sealed surface, the conservation or degradation status of the surface (e.g., road, courtyard), and the cleaning actions applied (e.g., suction, brushing) are

FIGURE 13.6 Soil erosion driven by roads, in hillslope cuts (*left*) and at road edges due to inappropriate storm drainage provision.

important factors determining the sediment load from sealed surfaces (Vaze and Chiew, 2002). The chemical properties of these sediments make them a main concern for aquatic ecosystems. Sediments from urban areas can act as a diffuse source of pollution due to their capacity to adsorb and transport pollutants, such as heavy metals and polycyclic aromatic hydrocarbons (e.g., Yu et al., 2014).

Overall, the type of urbanization conducted, characterized by the percentage of sealing, spatial urban pattern and drainage system, is highly important in determining potential sediment loss sources, transport pathways, and deposition sites. Depending on landscape connectivity, landscape disturbance, even in small parts of a catchment, can make major contributions to sediment dynamics in surface water bodies (Ferreira et al., 2020). Construction of single houses can even have an impact on suspended sediment loads in nearby streams if the site is well connected to the stream by close proximity or by high effectiveness of the drainage system in transporting runoff and sediments to downslope areas (Fig. 13.5).

Some landscape features can provide sediment disconnectivity between the source and downslope areas or even the stream network. These features include (i) buffer zones (e.g., well-grassed areas and walls), which prevent sediment from entering the channel network on hillslopes; (ii) barriers (e.g., sediment traps and sediment ponds), which impede sediment transport downstream; and (iii) blankets (e.g., mulch), which provide temporary cover and support for establishing vegetation on bare soils (Ferreira et al., 2021). Clogging of urban streams typically leads to increased sedimentation, often associated with a reduction in drainage capacity or even blockage (Ferguson and Deak, 1994). In urban catchments, various nature-based solutions can be implemented to reduce surface runoff and erosion (e.g., drainage basins, wetlands) while also providing multiple environmental, social and economic benefits (Kalantari et al., 2019).

13.5 Remote sensing in supporting soil erosion assessments

Remote sensing collects spectral, spatial, and/or temporal information about objects or areas. It is based on the principle that at temperatures above absolute zero, all objects radiate electromagnetic energy as a result of their atomic and molecular oscillations. All material objects have unique distribution of radiation, including absorbed, emitted, and reflected, meaning that spectral properties can be used to differentiate them or to acquire information regarding the size, shape, and other physical and chemical properties of the investigated object (King et al., 2005). In addition to its physical and chemical properties, the radiation emitted from an object is also influenced by the properties of the radiation reaching the object (i.e., wavelength or intensity), as well as its surface roughness. Remote sensing techniques consider a wide range of the electromagnetic spectrum, providing data as digital images characterized by arrays of pixels in several spectral bands, including visible, near-infrared, and infrared, that are detected by optical sensor systems or as microwaves (used by radar systems). Currently available satellite multiplatforms and multiscale sensors cover an extensive spectrum of electromagnetic radiation, providing different spatial, temporal, and spectral resolution that can be used within different applied domains (Dubovyk, 2017). Although remote sensing data have been available since the early 1970s, the quantity and quality of information (i.e., spectral and spatial accuracy) have increased considerably over recent years as a result of rapid technological developments. For example, a relatively recent type of radar is synthetic aperture radar (SAR), which involves a sensor producing its own energy and recording the amount of that energy reflected back after interacting with the object of interest (Kalantari et al., 2019). This technology utilizes longer wavelengths than other types of sensors, allowing it to see through clouds, which is a major advantage for data collection (Verhoest et al., 2008).

The availability of remote sensing data allows for past, present, and near-real time monitoring, supporting analyses of the state and evolution of land at different spatial scales (Dubovyk, 2017). The objectivity and repetitiveness of satellite imagery make it important information for supporting soil erosion assessments. Over the past few years, cohesive approaches including remote sensing data with geographic information system (GIS) analysis and modeling techniques have received increasing attention for soil erosion studies at large scales (e.g., Eniyew et al., 2021). Remote sensing enables cost-effective evaluations over extensive areas and allows monitoring of soil erosion at different temporal and spatial scales (Dubovyk, 2017).

Remote sensing applications provide relevant information for assessing soil erosion by (i) allowing soil erosion mapping, including its extent and severity (Panahi et al., 2022), and (ii) supporting analysis of its driving factors at different spatial/temporal scales or assessment of the impact of rehabilitation activities (e.g., Panahi et al., 2022). Within soil erosion mapping, one form of satellite image data that has been used successfully is that from advanced very high-resolution radiometry (AVHRR) (Dubovyk, 2017). One of the most widely used remote sensing methods for soil erosion studies is based on the analysis of LULC changes, namely, through vegetation cover dynamics. Differences in vegetation cover are important for assessing soil erosion since vegetation provides relevant protection against erosive agents (as discussed in the previous sections) and responds to environmental (e.g., soil, topography, climate) and anthropogenic factors (e.g., deforestation). Some studies have focused on distinguishing between bare or vegetated areas using different optical sensors, such as ASTER, Landsat, Ikonos, SPOT, and Quickbird (King et al., 2005). Time series of satellite images have been used to calculate vegetation-related parameters (e.g., phenometrics) and then analyze their spatiotemporal dynamics over a specific period (e.g., Bai et al.,

2021). Another possibility is to use image classification for comparison of multitemporal LULC maps, considering seasonal variability (Alem, 2022) and identifying changes in soil erosion (Yiran et al., 2011). However, these change maps can suffer from mapping inaccuracies caused by errors in the classification results (Dubovyk, 2017). Validation of remote sensing results based on site observations is an important aspect to assure the quality of the results. Recently, the European Space Agency's Climate Change Initiative (ESACCI) and the Copernicus Climate Change Service climate data store (C3S-CDS) released annual global land cover maps with high spatial resolution (300 m) for the periods 1992–2015 and 2016–18, respectively, allowing detailed analyses of global land cover changes (Hu et al., 2021). However, processing high-resolution global datasets is highly computationally demanding and time-consuming.

Remote sensing can also be used to describe spatiotemporal variations in other soil surface characteristics relevant to understanding rainfall-runoff and erosion processes. For example, it can be used to characterize the soil cover due to crop residues and even discriminate between crop residues using spectral indices in the mid-infrared or near-infrared domains or by using mixed spectral analysis (Arsenault and Bonn, 2000). Empirical equations derived from experimental data can then be used to relate soil surface roughness to water infiltration (King et al., 2005). Thus, a remote sensing approach can be useful for inferring soil surface roughness and identifying the main sources of runoff and the connectivity between these sources and the catchment outlet (Auzet et al., 1990).

Depending on the type of sensor employed, remote sensing can be useful for providing data on topographic characteristics, such as slope gradient and curvature. Digital elevation models (DEMs) have a long history of using remote sensing data, such as radar images (e.g., Radarsat, ERS), SAR interferometry and optical images (e.g., SPOT 5, ASTER, Quickbird or Ikonos), with vertical resolutions of up to 10 m, a few meters, and 1 m, respectively (King et al., 2005). Topographical morphology, including different forms of erosion (e.g., rills and gullies), can be identified from aerial photos or data from very high spatial resolution (VHSR) sensors. Google Earth has frequently been used to identify geomorphological features such as landslides and gully erosion (Borrelli et al., 2022).

Remote sensing can thus provide relevant data for several empirical models that use input data on soil characteristics, topography, land use, and precipitation to quantify soil erosion. For example, Bhadur and Praveen (2022) used remote sensing and GIS techniques to estimate different soil loss parameters for use with the Revised Universal Soil Loss Equation (RUSLE) to map soil erosion in different years. Specifically, they used bands four and five of Landsat 8 images to calculate the normalized difference vegetation index (NDVI) and the corresponding values of the crop management factor (C). Acharki et al. (2022) also combined remote sensing with GIS to quantify some parameters of the RUSLE model, using multitemporal Landsat data to provide land use information in different time periods. In that study, these images were geometrically corrected, a specific algorithm was applied to reduce atmospheric effects in the images, and the corrected bands were clipped using the boundaries of the study site. The authors also used machine learning algorithms for image classification and to identify LULC categories, which were validated by visual interpretation, Google Earth images and field trips through several accuracy metrics (Acharki et al., 2022). Each land cover class was then assigned a C-factor for RUSLE, and ASTER Global Digital Elevation Map satellite information was used to estimate elevation, slope and slope length, and steepness factor (LS) in RUSLE (Acharki et al., 2022). A similar study using RUSLE, remote sensing, and GIS, but applied to a different area, was performed by Eniyew et al. (2021), who used the Shuttle Radar Topography Mission (SRTM) of NASA to prepare the DEM and Landsat 8 images to produce the LULC map and calculate the C-factor after classification, preprocessing, and geometric and radiometric corrections of the image data. Despite the advantages, inappropriate integration between empirical models and remote sensing tools can cause oversimplification and, as a consequence, distortion and discrepancies between the results and reality (Brychta et al., 2022).

One of the current challenges when using remote sensing for soil erosion assessments is the mismatch between the temporal and spatial resolution of available satellite images and the scale of the environmental and socioeconomic drivers of erosion. For example, high spatial resolution imagery (e.g., from the Sentinel program) is not always available over the long periods required for tendency analysis. Thus, there is no optimal remote sensing method for assessing soil erosion, and the best method must be selected for each specific case based on the location of the study area, the scale of the study, and the most appropriate data availability and quality. Aggregation of data based on different spatial scales is also an aspect to consider in some studies. For example, Graw et al. (2016) used a multiscale approach for mapping bush encroachment at the local level based on the use of aerial photographs, MODIS images (250 m spatial resolution), and Landsat images (30 m spatial resolution). Multiscale datasets may be relevant for mapping soil erosion at different scales in a specific region, especially for linking with complementary data relevant to understanding the drivers of soil erosion but collected using distinct methodologies, which are often available at different scales from those provided by remote sensing (Dubovyk, 2017).

13.6 Final considerations

Soil is a fundamental resource providing several goods and services relevant for life on Earth, but inappropriate land use and management and climate change negatively affect this natural capital. Erosion is one of the most relevant soil degradation processes and is responsible for several on-site and off-site environmental, social, and economic impacts, which threaten food and water security. Several sustainable management practices (e.g., reduced tillage, cover crops) and restoration measures (e.g., afforestation) have been implemented to mitigate soil erosion, but their effectiveness is not uncertain due to numerous factors linked, for example, with the soil type, topography, and climate and requires further investigation.

Over the last decades, advances in remote sensing and integration with GIS have received increasing attention to support soil erosion assessments at different scales, particularly at regional and global scales, and in remote areas of difficult access. Remote sensing data with distinct spatial and temporal resolutions, determined by the type of sensor, have been used to describe soil surface characteristics, understand soil and water transfer functions, and describe hydrological and geomorphological processes, which are relevant for modeling soil erosion. Integration of distinct remote sensing data (e.g., optical and radar) has been relevant to support, for example, the assessment of hydrological variables that allow prediction of runoff fluxes. Remote sensing has been revealed to be useful for mapping soil erosion hotspots and identifying and prioritizing erosion-prone areas for conservation efforts. A better understanding of the soil erosion problem, including the current status, spatial extent and trend over time, is of utmost importance for policy and decision makers and land users to make more informed decisions that are relevant to support the achievement of land degradation neutrality.

Acknowledgments

This work was supported by Portuguese Science and Technology Foundation through the project PTDC/EEI-ROB/2459/2021.

References

Acharki, S., Qorchi, F., Arjdal, Y., Amharref, M., Bernoussi, A., Aissa, H., 2022. Soil erosion assessment in Northwestern Morocco. Remote Sens. Appl. Soc. Environ. 25, 100663.

Adams, B., Gashaw, T., Adem, A., Worqlul, A., Dile, Y., Molla, E., 2022. Identification of soil erosion hot-spot areas for prioritisation of conservation measures using the SWAT model in Ribb watershed, Ethiopia. Resour. Environ. Sustain. 8, 100059.

Alem, B., 2022. The nexus between land use land cover dynamics and soil erosion hotspot area of Girana Watershed, Awash Basin, Ethiopia. Heliyon 8, e08916.

Amiri, M., Pourghasemi, H., Ghanbarian, G., Afzali, S., 2019. Assessment of the importance of gully erosion effective factors using Boruta algorithm and its spatial modelling and mapping using three machine learning algorithms. Geoderma 340, 55–69.

Arif, M.S., Riaz, M., Shahzad, S.M., Yasmeen, T., Ashraf, M., Siddique, M., Mubarik, M.S., Bragazza, L., Buttler, A., 2018. Fresh and composted industrial sludge restore soil functions in surface soil of degraded agricultural land. Sci. Total Environ. 619–620, 517–527.

Arsenault, E., Bonn, F., 2000. Final Report of the FLOODGEN Project Canada CSA-ST-CR-2000-0013 Viasat. U. Sherbrooke.

Auzet, A.V., Boiffin, J., Papy, F., Maucorps, J., Ouvry, J.F., 1990. An approach to the assessment of erosion forms and erosion risk on agricultural land in the Northern Paris Basin, France. In: Boardman, J., Foerster, D.L., Dearing, J.A. (Eds.), Soil Erosion on Agricultural Land. John Wiley and Sons, pp. 383–400.

Bai, Z., Caspari, T., Gonzalez, M.R., Batjes, N.H., Mader, P., Bunemann, E.K., Goede, R., Brussaard, L., Xu, M., Ferreira, C.S.S., Reintam, E., Fan, H., Mihelic, R., Glavan, M., Tóth, Z., 2018. Effects of agricultural management practices on soil quality: a review of long-term experiments for Europe and China. Agric. Ecosyst. Environ. 265, 1–7.

Bai, X., Zhao, W., Wang, J., Ferreira, C.S.S., 2021. Precipitation drives the floristic composition and diversity of temperate grasslands in China. Glob. Ecol. Conserv. 32, e01933.

Barão, L., Alaoui, A., Ferreira, C.S.S., Basch, G., Schwilch, G., Geissen, V., Sukkel, W., Lemesle, J., Garcia-Orenes, F., Morugán-Coronado, A., Mataix-Solera, J., Kosmas, C., Glavan, M., Pintar, M., Tóth, B., Hermann, T., Vizitiu, O.P., Lipiec, J., Reintam, E., Xu, M., Di, Fan, H., Wang, F., 2019. Assessment of promising agricultural management practices. Sci. Total Environ. 649, 610–619.

Bezak, N., Borrelli, P., Panagos, P., 2021. A first assessment of rainfall erosivity synchrony scale at pan-European scale. Catena 198, 105060.

Bezerra, M.O., Baker, M., Palmer, M.A., Filoso, S., 2020. Gully formation in headwater catchments under sugarcane agriculture in Brazil. J. Environ. Manag. 270, 110271.

Bhadur, R., Praveen, K., 2022. Study of soil erosion by using remote sensing and GIS techniques in Sone command area in Bihar, India. Mater. Today: Proc. 62, 1664–1670.

Borrelli, P., Robinson, D.A., Fleischer, L.R., Lugato, E., Ballabio, C., Alewell, C., Meusburger, K., Modugno, S., Schutt, B., Ferro, V., Bagarello, V., Oost, K.V., Montanarella, L., Panagos, P., 2017. An assessment of the global impact of 21st century land use change on soil erosion. Nat. Commun. 8 (1), 2013.

Borrelli, P., Robinson, D.A., Panagos, P., Lugato, E., Yang, J.E., Alewell, C., Ballabio, C., 2020. In: Proceedings of the National Academy of Sciences, pp. 21994–22001.

Borrelli, P., Poesen, J., Vanmaercke, M., Ballabio, C., Hervas, J., Maerker, M., Scarpa, S., Panagos, P., 2022. Monitoring gully erosion in the European Union: a novel approach based on the Land Use/Cover Area frame survey (LUCAS). Int. Soil Water Conserv. Res. 10, 17–28.

Boulet, A.K., Alarcão, C., Ferreira, C., Kalantari, Z., Veiga, A., Campos, L., Ferreira, A., Hessel, R., 2021. Agro-ecological services delivered by legume cover crops grown in succession with grain corn crops in the Mediterranean region. Open Agric. 6, 609–626.

Bouma, J., Batjes, N.H., 2000. Trends of worldwide soil degradation. Hohenh. Umwelttagung 3, 3–43.

Brychta, J., Podhrazska, J., Stastna, M., 2022. Review of methods of spatiotemporal evaluation of rainfall erosivity and their correct application. Catena 217, 106454.

Castro, G.P., Malinovski, J.R., Nutto, L., Malinovski, R.A., 2016. Machinery and equipment in harvesting. In: Pancel, L., Köhl, M. (Eds.), Tropical Forestry Handbook. Springer, Berlin, Heidelberg.

Chevigny, E., Quiquerez, A., Petit, C., Curmi, P., 2014. Lithology, landscape structure and management practice changes: key factors patterning vineyard soil erosion at metre-scale spatial resolution. Catena 121, 354–364.

Cunha, E., Santos, C., Silva, R., Panachuki, E., Oliveira, P., Oliveira, N., Falcão, K., 2022. Assessment of current and future land use/cover changes in soil erosion in the Rio da Prata basin (Brazil). Sci. Total Environ. 818, 151811.

Djuma, H., Bruggeman, A., Camera, C., Zoumides, C., 2017. Combining qualitative and quantitative methods for soil erosion assessments: an application in a sloping Mediterranean watershed, Cyprus. Land Degrad. Dev. 28 (1), 243–254.

Dubovyk, O., 2017. The role of Remote Sensing in land degradation assessments: opportunities and challenges. Eur. J. Remote Sens. 50 (1), 601–613.

Duan, J., Liu, Y.J., Yang, J., Tang, C.J., Shi, Z.H., 2020. Role of groundcover management in controlling soil erosion under extreme rainfall in citrus orchards of southern China. J. Hydrol. 582, 124290.

EEA (European Environment Agency), 2006. Urban Sprawl in Europe: The Ignored Challenge. Copenhagen, 10/2006.

EEA (European Environment Agency), 2017. Landscapes in Transition - an Account of 25 Years of Land Cover Change in Europe. European Environment Agency. Report No 10/2017.

EPA (Environmental Protection Agency), 1993. Guidance Specifying Management Measures for Sources of Nonpoint Pollution in Coastal Waters - Chapter 4: Management Measures for Urban Areas. United States Environmental Protection Agency. EPA-840-B-92-002.

Eniyew, S., Teshome, M., Sisay, E., Bezabih, T., 2021. Integrating RUSLE model with remote sensing and GIS for evaluation soil erosion in Telkwonz Watershed, Northwestern Ethiopia. Remote Sens. Appl. Soc. Environ. 24, 100623.

Esteves, T.C.J., Kirkby, M.J., Shakesby, R.A., Ferreira, A.J.D., Soares, J.A.A., Irvine, B.J., Ferreira, C.S.S., Coelho, C.O.A., Bento, C.P.M., Carreira, M.A., 2012. Mitigating land degradation caused by wildfire: application of the PESERA model to fire-affected sites in central Portugal. Geoderma 191, 40–50.

Eswaran, H., Lal, R., Reich, P.F., 2001. Land degradation: an overview. In: Bridges, E.M., Hannam, I.D., Oldeman, L.R., Pening de Vries, F.W.T., Scherr, S.J., Sompatpanit, S. (Eds.), Responses to Land Degradation. Proceedings of 2nd International Conference on Land Degradation and Desertification, Khon Kaen, thailand. Oxford Press, New Delhi, India.

Evans, D., Janes-Bassett, V., Borrelli, P., Chenu, C., Ferreira, C.S.S., Griffiths, R., Kalantari, Z., Keesstra, S., Lal, R., Panagos, P., Robinson, D., Seifollahi-Aghmiuni, S., Smith, P., Steenhuis, T., Thomas, A., Visser, S., 2021. Sustainable futures over the next decade are rooted in soil science. Eur. J. Soil Sci. 1–16.

FAO (Food and Agricultural Organisation), 2016. Global Forest Resources Assessment 2015. How Are the World Forests Changing? Rome, p. 45.

FAO (Food and Agricultural Organisation), 2020. Global Forest Resources Assessment 2020 – Key Findings. Rome. https://doi.org/10.4060/ca8753en.

FAO, ITPS, 2015. Status of the World's Soil Resources (SWSR). Main Report. Food and Agriculture Organisation of the United Nations and Intergovernmental Technical Panel on Soils. Rome, Italy.

Ferguson, B.K., Deak, T., 1994. Role of urban storm-flow volume in local drainage problems. J. Water Resour. Plann. Manag. 120, 523–530.

Ferreira, C.S.S., Ferreira, A.J.D., Pato, R.L., Magalhães, M.C., Coelho, C.O., Santos, C., 2012. Rainfall-runoff-erosion relationships study for different land uses, in a suburban area. Z. Geomorphol. 56 (X), 005–020.

Ferreira, C.S.S., Walsh, R.P.D., Costa, M.L., Coelho, C.O.A., Ferreira, A.J.D., 2016a. Dynamics of surface water quality driven by distinct urbanisation patterns and storms in a Portuguese peri-urban catchment. J. Soils Sediments 16, 2606–2621.

Ferreira, C.S.S., Walsh, R.P.D., Kalantari, Z., Ferreira, A.J.D., 2020. Impact of land-use changes on spatiotemporal suspended sediment dynamics within a peri-urban catchment. Water 12, 665.

Ferreira, C.S.S., Walsh, R.P.D., Shakesby, R.A., Keizer, J.J., Soares, D., González-Pelayo, O., Coelho, C.O.A., Ferreira, A.J.D., 2016b. Differences in overland flow, hydrophobicity and soil moisture dynamics between Mediterranean woodland types in a peri-urban catchment in Portugal. J. Hydrol. 533, 473–485.

Ferreira, A.J.D., Soares, D., Serrano, L.M.V., Walsh, R.P.D., Ferreira, C.M.D., Ferreira, C.S.S., 2016c. Roads as sources of heavy metals in urban areas. The Covões Catchment experiment, Coimbra, Portugal. J. Soils Sediments 16, 2622–2639.

Ferreira, C.S.S., Walsh, R.P.D., Blake, W.H., Kikuchi, R., Ferreira, A.J.D., 2017. Temporal dynamics of sediment sources in an urbanising Mediterranean catchment. Land Degrad. Dev. 28 (8), 2354–2369.

Ferreira, C.S.S., Keizer, J.J., Santos, L.M.B., Serpa, D., Silva, V., Cerqueira, M., Ferreira, A.J.D., Abrantes, N., 2018a. Runoff, sediment and nutrient exports from a Mediterranean vineyard under integrated production: an experiment at plot scale. Agric. Ecosyst. Environ. 256, 184–193.

Ferreira, C.S.S., Walsh, R.P.D., Ferreira, A.J.D., 2018b. Degradation in urban areas. Curr. Opin. Environ. Sci. Health 5, 19–25.

Ferreira, C.S.S., Kalantari, Z., Salvati, L., Canfora, L., Zambon, I., Walsh, R.P.D., 2019. Urban areas. In: Pereira, P. (Ed.), Advances in Chemical Pollution, Environmental Management and Protection", vol 4. Elsevier, ISBN 9780128164150, pp. 207–249 (chapter 6).

Ferreira, C.S.S., Kalantari, Z., Seifollahi-Aghmiuni, S., Ghajarnia, N., Rahmati, O., Solomun, M.K., 2021. Rainfall-runoff-erosion processes in urban areas. In: Rodrigo-Comino, J. (Ed.), Precipitation: Earth Surface Responses and Processes". Elsevier, pp. 481–498 (chapter 21).

Ferreira, C.S.S., Seifollahi-Aghmiuni, S., Destouni, G., Ghajarnia, N., Kalantari, Z., 2022. Soil degradation in the European Mediterranean region: processes, status and consequences. Sci. Total Environ. 805, 150106.

Forrester, D.I., Smith, R.G.B., 2012. Faster growth of Eucalyptus grandis and Eucalyptus pilularis in mixed-species stands than monocultures. For. Ecol. Manag. 286, 81–86.

Gharibreza, M., Zaman, M., Porto, P., Fulajtar, E., Parsaei, Eisaei, H., 2021. Assessment of deforestation impact on soil erosion in loess formation using 137Cs method (case study: Golestan Province, Iran). Int. Soil Water Conserv. Res. 8, 393–405.

Gil, E., Kijowska-Strugala, M., Demczuk, P., 2021. Soil erosion dynamics on a cultivated slope in the Western Polish Carpathians based on over 30 years of plot studies. Catena 207, 105682.

Gonzalez-Pelayo, O., Gimeno-García, E., Ferreira, C.S.S., Ferreira, A.J.D., Keizer, J.J., Andreu, V., Rubio, J.L., 2015. Water Repellency of air-dried and sieved samples from limestone soils in Central Portugal collected before and after prescribed fire. Plant Soil 394, 199–214.

Gong, C., Tan, Q., Liu, G., Xu, M., 2022. Impacts of mixed forests on controlling soil erosion in China. Catena 213, 106147.

Graw, V., Oldenburg, C., Dubovyk, O., 2016. Bush Encroachment Mapping for Africa: Multiscale Analysis with Remote Sensing and GIS. ZEF - Center for Development Research University of Bonn. Discussion Paper No 218.

Hu, X., Naess, J., Iordan, C., Huang, B., Zhao, W.,F., 2021. Recent global land cover dynamics and implications for soil erosion and carbon losses from deforestation. Anthropocene 34, 100291.

International Atomic Energy Agency, 2014. Guidelines for Using Fallout Radionuclides to Assess Erosion and Effectiveness of Soil Conservation Strategies. IAEA publication, Vienna, Austria, p. 213. IAEA-TECDOC-1741.

Kalantari, Z., Ferreira, C.S.S., Koutsouris, A.J., Ahmer, A.K., Cerdà, A., Destouni, G., 2019. Assessing flood probability for transportation infrastructure based on catchment characteristics, sediment connectivity and remotely sensed soil moisture. Sci. Total Environ. 661, 393–406.

Kanter, D.R., Musumba, M., Wood, W.L.R., Palm, C., Antle, J., Balvanera, P., Dale, V.H., Havlik, P., Kline, K.L., Scholes, R.J., Thornton, P., Tittonell, P., Andelman, S., 2018. Evaluating agricultural trade-offs in the age of sustainable development. Agric. Syst. 163, 73–88.

Keesstra, S.D., Bouma, J., Wallinga, J., Tittonell, P., Smith, P., Cerdà, A., Montanarella, L., Quinton, J.N., Pachepsky, Y., van der Putten, W.H., Bardgett, R.D., Moolenaar, S., Mol, G., Jansen, B., Fresco, L.O., 2016. The significance of soils and soil science towards realisation of the United Nations Sustainable Development Goals. Soils 2, 111–128.

King, C., Baghdadi, N., Lecomte, V., Cerdan, O., 2005. The application of remote-sensing data to monitoring and modelling of soil erosion. Catena 62, 79–93.

Kirkby, M.J., Irvine, B.J., Jones, R.J.A., Govers, G., Boer, M., Cerdan, O., Daroussin, J., Gobin, A., Grimm, M., Le Bissonnais, Y., Kosmas, C., Mantel, S., van Puigdefabregas, J., Lynden, G., 2008. The PESERA coarse scale erosion model for Europe. I. — model rationale and implementation. Eur. J. Soil Sci. 59, 1293–1306.

Kuhwald, M., Busche, F., Saggau, P., Duttmann, R., 2022. Is soil loss due to crop harvesting the most disregarded soil erosion process? A review of harvest erosion. Soil Tillage Res. 215, 105213.

Lal, R., 2003. Soil erosion and the global carbon budget. Environ. Int. 29 (4), 437–450.

Lal, R., 2017. Soil erosion by wind and water: problems and prospects. Soil Erosion Research Methods. Routledge, pp. 1–10.

Li, Z.Y., Fang, H.Y., 2016. Impacts of climate change on water erosion: a review. Earth Sci. Rev. 163, 94–117.

Li, H.Y., Zhu, N.Y., Wang, S.C., Gao, M.N., Xia, L.Z., Kerr, P.G., Wu, Y.H., 2020. Dual benefits of long-term ecological agricultural engineering: mitigation of nutrient losses and improvement of soil quality. Sci. Total Environ. 721, 137848. https://doi.org/10.1016/j.scitotenv.2020.137848.

Li, R., Li, Q., Pan, L., 2021. Review of organic mulching effects on soil and water loss. Arch. Agron Soil Sci. 67 (1), 136–151.

Line, D.E., White, N.M., Osmond, D.L., Jennings, G.D., Mojonnier, C.B., 2002. Pollutant export from various land uses in the upper Neuse River Basin. Water Environmental Resources 74, 100e8.

Liu, Y., Zhang, J., Qin, Y., 2020. How global warming alters future maize yield and water use efficiency in China. Technol. Forecast. Soc. Change 160, 120229.

Ma, X., Zhao, C., Zhu, J., 2021. Aggravated risk of soil erosion with global warming – a global meta-analysis. Catena 200, 105129.

Martins, M.A.S., Machado, A.I., Serpa, D., Prats, S.A., Faria, S.R., Varela, M.E.T., Gonzalez-Pelayo, O., Keizer, J.J., 2013. Runoff and interrill erosion in a maritime pine and a eucalypt plantation following wildfire and terracing in north-central Portugal. J. Hydrol. Hydromechanics 61 (4), 261–268. https://doi.org/10.2478/johh-2013-0033.

Matthews, F., Panagos, P., Verstraeten, G., 2022. Simulating event-scale rainfall erosivity across European climatic regions. Catena 213, 106157.

Miranda, E.D., Attorre, F., Azevedo, J., Belen, I., Alcalde, E.E., Freitas, H., Garavaglia, V., Hódar, J.A., Iritas, O., Karaaslan, Y., Khater, C., Koutsias, N., Makinson, D., Mansour, S., Pettenella, D., Picard, N., Pino, J., Vieira, J., Viale, M., 2018. Drivers of degradation and other threats. In: FAO and Plan Bleu. 2018. State of Mediterranean Forests 2018. Food and Agriculture Organisation of the United Nations, Rome and Plan Bleu, Marseille, pp. 2–15 (Chapter 1).

Mushi, C.A., Ndomba, P.M., Trigg, M.A., Tshimanga, R.M., Mtalo, F., 2019. Assessment of basin-scale soil erosion within the Congo River Basin: a review. Catena 178, 64–76.

Niu, Y., Wang, L., Wan, X., Peng, Q., Huang, Q., Shi, Z., 2021. A systematic review of soil erosion in citrus orchards worldwide. Catena 206, 105558.

Oldeman, L.R., Hakkeling, R., Sombroek, W.G., 2017. World Map of the Status of Human-Induced Soil Degradation: An Explanatory Note. International Soil Reference and Information Centre.

Panagos, P., Ballabio, C., Lugato, E., Jones, A., Borrelli, P., Scarpa, S., Orgiazzi, A., Montanarella, L., 2018. Potential sources of anthropogenic copper inputs to European agricultural soils. Sustainability 10, 2380.

Panagos, P., Borrelli, P., Poesen, J., 2019. Soil loss due to crop harvesting in the European Union: a first estimation of an underrated geomorphic process. Sci. Total Environ. 664, 487–498.

Panagos, P., Ballabio, C., Poesen, J., Lugato, E., Scarpa, L., Montanarella, L., Borrelli, P., 2020. A soil erosion indicator for supporting agricultural, environmental and climate policies in the European Union. Rem. Sens. 12, 1365.

Panagos, P., Ballabio, C., Himics, M., Scarpa, X., Matthews, F., Bogonos, M., Poesen, J., Borrelli, P., 2021. Projections of soil loss by water erosion in Europe by 2050. Environ. Sci. Pol. 124 (2021), 380–392.

Pancel, L., 2016. Basic outline of tree plantations in the tropics. In: Pancel, L., Köhl, M. (Eds.), Tropical Forestry Handbook. Springer, Berlin, Heidelberg.

Panahi, M., Rahmati, O., Kalantari, Z., Darabi, H., Rezaie, F., Moghaddam, D.D., Ferreira, C.S.S., Foody, G., Lee, C.W., Lee, S., 2022. Large-scale dynamic flood monitoring in an arid-zone floodplain using SAR data and hybrid machine learning models. J. Hydrol. 128001.

Parlak, M., Palta, Ç., Yokuş, S., Blanco-Canqui, H., Çarkacı, D.A., 2016. Soil losses due to carrot harvesting in south central Turkey. Catena 140, 24–30.

Pimentel, D., Burgess, M., 2013. Soil erosion threatens food production. Agriculture 3, 443–463.

Rakshit, A., Abhilash, P.C., Singh, H.B., Ghosh, S., 2017. Managing soil resources for sustainable development Adaptive soil management: from theory to practices. J. Clean. Prod. 174, 199–200.

Russel, K.L., Vietz, G.J., Fletcher, T.D., 2017. Global sediment yields from urban and urbanising watersheds. Earth Sci. Rev. 168, 73e80.

Ruysschaert, G., Poesen, J., Verstraeten, G., Govers, G., 2004. Soil loss due to crop harvesting: significance and determining factors. Prog. Phys. Geogr. 28, 467–501. https://doi.org/10.1191/0309133304pp421oa.

Shakesby, R.A., 2011. Postwildfire soil erosion in the Mediterranean: review and future research directions. Earth Sci. Rev. 105, 71–100.

Shakesby, R.A., Bento, C.P., Ferreira, C.S., Ferreira, A.J., Stoof, C.R., Urbanek, E., Walsh, R.P., 2015. Impacts of prescribed fire on soil loss and soil quality: an assessment based on an experimentally burned catchment in central Portugal. Catena 128, 278–293.

Smith, P., 2016. Soil carbon sequestration and biochar as negative emission technologies. Global Change Biol. 22, 1315–1324.

Solomun, M.K., Ferreira, C.S.S., Eremija, S., Tošić, R., Lazović, N., Češljar, G., 2021. Long-term fire effects on vegetation and topsoil properties in beech forests of Manjaca Mountain (Western of Bosnia and Herzegovina). Int. J. Wildland Fire WF20111.

Stolte, J., Tesfai, M., Øygarden, L., Kværnø, S., Keizer, J., Verheijen, F., Panagos, P., Ballabio, C., Hessel, R., 2016. Soil Threats in Europe: Status, Methods, Drivers and Effects on Ecosystem Services. A Review Report, Deliverable 2.1 of the RECARE Project 2016. Available at: https://esdac.jrc.ec.europa.eu/public_path/shared_folder/doc_pub/EUR27607.pdf.

Stoof, C.R., Ferreira, A.J.D., Mol, W., den Berg, J.V., Kort, A., Drooger, S., Slingerland, E.C., Mansholt, A.U., Ferreira, C.S.S., Ritsema, C.J., 2015. Soil surface changes increase runoff and erosion risk after a low-moderate severity fire. Geoderma 239–240, 58–67.

Telak, L.J., Pereira, P., Ferreira, C.S.S., Filipovic, V., Filipovic, L., Bogunovic, I., 2020. Short-Term impact of tillage on soil and the hydrological response within a fig (Ficus carica) Orchard in Croatia. Water 12, 3295.

Teng, H., Liang, Z., Chen, S., Liu, Y., Viscarra, R.A., Chappell, A., Yu, W., Shi, Z., 2018. Current and future assessments of soil erosion by water on the Tibetan plateau based on RUSLE and CMIP5 climate models, 635, 673–686.

The environmental food crisis – the environment's role in averting future food crises. In: Nellemann, C., MacDevette, M., Manders, T., Eickhout, B., Svihus, B., Prins, A.G., Kaltenborn, B.P. (Eds.), 2009. A UNEP Rapid Response Assessment. United Nations Environment Programme, GRID-Arendal.

Thomaz, E., Marcatto, F., Antoneli, V., 2022. Soil erosion on the Brazilian sugarcane cropping system: an overview. Geogr. Sustaina. 3, 129–138.

UNEP (United Nations Environment Programme), 2012. Global Environmental Outlook: Environment for the Future We Want. Available at: http://www.unep.org/geo/sites/unep.org.geo/files/documents/geo5frontmatter.pdf.

Vanmaercke, M., Panagos, P., Vanwalleghem, T., Jarihani, B., Poesen, J., 2022. Measuring, modelling and managing gully erosion at large scales: a state of the art. Earth Sci. Rev. 218, 103637.

Vaze, J., Chiew, F.H.S., 2002. Experimental study of pollutant accumulation on an urban road surface. Urban Water 4, 379–389.

Vercruysse, K., Grabowski, R.C., Rickson, R.J., 2017. Suspended sediment transport dynamics in rivers: multiscale drivers of temporal variation. Earth Sci. Rev. 166, 38–52.

Verheijen, F., Jones, R., Rickson, R., Smith, C., 2009. Tolerable versus actual soil erosion rates in Europe. Earth Sci. Rev. 94, 23–38.

Verhoest, N., Lievens, H., Wagner, W., Álvarez-Mozos, J., Moran, M., Mattia, F., 2008. On the soil roughness parameterisation problem in soil moisture retrieval of bare surfaces from synthetic aperture radar. Sensors 8 (7), 4213.

Waggenbrenner, J.W., McDonald, L.H., Coats, R.N., Robichaud, P.R., Brown, R.E., 2015. Effects of postfire salvage logging and skid trail treatment on ground cover, soils and sediment production in the interior western United States. For. Ecol. Manag. 335, 176–193.

WHO/UNICEF, 2017. Progress on Drinking Water, Sanitation and Hygiene: 2017 Update and SDG Baselines. World Health Organisation (WHO) and the United Nations Children's Fund (UNICEF), Geneva, 2017. Licence: CC BY-NC-SA 3.0 IGO.

Xia, L.Z., Liu, G.H., Wu, Y.H., Li, M., Li, Y.D., 2015. Protection methods to reduce nitrogen and phosphorus losses from sloping citrus land in the Three Gorges Area of China. Pedosphere 25, 478–488.

Yiran, G.A.B., Kusimi, J.M., Kufogbe, S.K., 2011. A synthesis of remote sensing and local knowledge approaches in land degradation assessment in the Bawku East District, Ghana. Int. J. Appl. Earth Obs. Geoinf. 14 (1), 204–213.

Yu, S., Wu, Q., Li, Q., Gao, J., Lin, Q., Xu, Q., Wu, S., 2014. Anthropogenic land uses elevate metal levels in streamwater in an urbanising watershed. Sci. Total Environ. 488–489, 61–69.

Zhang, G.H., Xie, C.B., Pi, X.Y., Zuo, C.Q., 2015. Effectiveness of soil conservation methods in preventing red soil erosion in Southern China. J. Mt. Sci. 12, 1281–1291. https://doi.org/10.1007/s11629-014-3152-8.

Zhao, J., Wang, Z., Dong, Y., Yang, Z., Govers, G., 2022. How soil erosion and runoff are related to land use, topography and annual precipitation: insights from a meta-analysis of erosion plots in China. Sci. Total Environ. 802, 149665.

CHAPTER

14

Road-side slope erosion using MLS and remote sensing

Saleh Yousefi[1], *Somayeh Mirzaee*[2] *and Christopher Gomez*[3]

[1]**Soil Conservation and Watershed Management Research Institute, Chaharmahal and Bakhtiari Agricultural and Natural Resources Research and Education Center, AREEO, Shahrekord, Iran;** [2]**Department of Natural Resources, Shahrekord University, Shahrekord, Iran;** [3]**LofHazs Laboratory (Sabo), Graduate School of Oceaonology, Kobe University, Kobe, Japan**

14.1 Soil erosion and forest roads

Soil is one of the most important sources on the Earth, and losing soil is a major threat to many functions the Earth system provides. As a key factor of the Earth system, soil controls geochemical and hydrological processes (Keesstra et al., 2016; Smith et al., 2015). Forestry operations can have significant impacts on soil erosion processes in forested landscapes. Roads are often constructed to provide access to timber harvesting areas or other activities such as recreation, fire suppression, and research. These roads increase surface runoff and accelerate sediment transport by providing pathways for water and wind to move across the landscape. This paper reviews current knowledge of how forest road construction affects soil erosion rates and discusses potential strategies to mitigate these effects. We also explore the implications of our findings for managing soils and reducing environmental degradation associated with forestry operations. Worldwide, soil erosion is an important geo-environmental hazard with serious socioeconomic and ecological implications (Yousefi et al., 2016). Surface soil compaction, removing vegetation cover and disturbing the forest floor combination during constructive projects are the main reasons on increasing soil erosion in forest roads (Sidle et al., 2006). During the 20th century by increasing in the technology development and employment of machines, human impacts on our environment increased and natural resource (water, soil, air etc.) have been destroyed by mismanagement spatially in undeveloped countries. Roads are considered as vital arteries of a society. So, if they do not develop and expand, the life and economic and social development of the

Remote Sensing of Soil and Land Surface Processes
https://doi.org/10.1016/B978-0-443-15341-9.00019-8

society will be disturbed (Cheng et al., 2015; Navarro-Hevia et al., 2015; Rose et al., 2014; Safari et al., 2016). But on the other hand, road construction is considered as one of the factors of destruction of natural resources. In such a way that today this issue is considered as one of the basic problems of natural areas (Mahdavi, 2012; Saito et al., 2013). Roads are essential in term of fire management, disease control and timber harvesting in forest lands as well as they are important in recreational and conservation activities (Barber et al., 2014).

One of the major problems caused by road construction is the creation and increase of various forms of soil erosion around the road network and the increase in the amount of sedimentation due to the destruction of vegetation, the increase in the speed of runoff, and the instability of the slopes (Vahabzadeh et al., 2015). Due to the fact that forest basins are very important from environmental, economic, and ecological point of view, it is necessary to have information such as the impact of roads on the sedimentation of their flow, as well as more fundamental management of these roads (Haigh et al., 1995).

14.2 Road erosion measurement

Measuring roadway erosion is important for understanding its long-term effect on natural environments. Roadway erosion has been linked to increased sediment loading in rivers and streams, which can lead to habitat destruction and reduced water quality. Erosion measurements involve both direct observation and indirect methods using remote sensing techniques. In this review, we will focus primarily on traditional field-based approaches used to measure roadway erosion. We will cover common tools and techniques used to assess erosional changes over time and discuss their advantages and disadvantages. Finally, we will consider some of the challenges involved in measuring roadway erosion and suggest possible solutions. By better understanding the dynamics of roadway erosion, it may be possible to develop more effective management practices aimed at preserving natural resources. There are two main approaches for road condition monitoring and data collection in forest area (Pires et al., 2021). In many countries, they use manual data collection that is slow and mainly provide poor data in many countries (Baker et al., 2015). The geodetic method remains the most widely used approach for land measurement and surveying. This involves utilizing a Total Station and Theodolite Surveyor Geodesy Measuring, in conjunction with the Global Navigation Satellite System (GNSS) and remote sensing techniques (Kweon et al., 2020). In connection with the use of geodetic approaches, the price of equipment and cost of projects are typically too high (Shataee et al., 2011). On the other hand, GNSS and RS methods can handle projects with low cost and save more time and money (Kweon et al., 2020; Pires et al., 2021). Recently, remote sensing methods have also gained popularity in many countries and projects over the last 15 years. The obtained 3D data are mainly utilized for digital terrain (DT) and surface topographic modeling. Accurate information about roads and their surroundings have become increasingly important due to the growing number of applications, such as traffic safety, noise modeling, driver assistance, road maintenance and repair, and car and pedestrian navigation (Alexiou et al., 2021; Balado et al., 2019; Cao et al., 2021; Jaakkola et al., 2017).

14.2.1 Remote sensing for road erosion

14.2.1.1 Light detection and ranging LiDAR

During the last few decades, with the development of new technologies, the aforementioned automatic data collection and remote sensing approaches have become increasingly popular (Craven and Wing, 2014; Thompson and Croke, 2013). In addition, the modeling technology to obtaining three-dimensional data can be classified in mapping method used.

Light Detection and Ranging (LiDAR) is a highly sophisticated technique used for investigating forest road networks worldwide. This method involves distance measurement by calculating the speed of the reflected laser beam pulse emitted from the scanned object (Craven and Wing, 2014). At the beginning, the LiDAR method was used for the creation of a digital terrain model (DTM) and prepares the topographic data (Saito et al., 2013).

In addition, the LiDAR technique is mostly used as airborne laser scanning (ALS) in forest lands; in this method, the scanner is mounted on an aircraft machine. ALS is worldwide used for civil modeling, forest inventory, the generation of DTMs, or for archeological studies (Jaakkola et al., 2017). Soil erosion is one of the most significant environmental issues facing humanity today. It affects land productivity, air and water pollution, biodiversity, and climate change. LiDAR technology provides a powerful tool for studying soil erosion processes and quantifying the impacts of different factors such as rainfall intensity, slope angle, vegetation cover, etc. This paper reviews current research utilizing LiDAR data to study soil erosion and outlines potential applications of the technique. The main benefits of using LiDAR are discussed along with existing limitations associated with the use of the technology. Potential future directions for research involving LiDAR and soil erosion are also suggested. Overall, the application of LiDAR offers great promise in advancing our knowledge of soil erosion processes and developing strategies to mitigate them. The airborne scanner (AS) used in LiDAR remote sensing typically operates from a distance of several hundred meters to several kilometers from the scanned location on the Earth's surface. This is the optimal distance range for acquiring high-resolution data of the terrain and surface features. By using laser beams to measure distances and calculate speeds based on reflected pulses, LiDAR is able to provide accurate and detailed information about forest road networks and other features on the Earth's surface, making it an essential tool for digital terrain and surface topographic modeling (Yu et al., 2015).

The combination of TLS and photogrammetry allows for using two distinct measuring strategies. The first one relies on the direct distance measurements from individual points, while the second utilizes the aerial image analysis to calculate distances between (Cao et al., 2021; Jaakkola et al., 2017). Terrestrial laser scanning (TLS) by distance measurements can be used for large-scale point cloud acquisition at mid-range distances; on the other hand, image-based surface reconstruction techniques allow more flexibility with higher accuracy at close ranges. State-of-the-art TLS systems come equipped with an internal camera or offer the capability to attach an external one (Eltner and Baumgart, 2015; Jaakkola et al., 2017). This allows the capture of photos, which are then aligned to the obtained 3D point cloud data. Nevertheless, this setup has its drawbacks as it restricts the movement of both devices.

14.2.1.2 Mobile laser scanning (MLS)

In data collection approaches to TLS, mobile laser scanning (MLS) and ALS can be used (Jaakkola et al., 2017). These approaches allow for more detailed and accurate 3D models since they can collect large amounts of data in a short amount of time. Additionally, TLS is becoming increasingly popular among researchers and professionals due to its accuracy and efficiency. TLS uses multiple laser beams to scan objects from different angles and create highly detailed 3D maps. This technique also provides high point cloud densities, with some studies reporting over 10,000 points per square meter (Pires et al., 2022). Geomorphological mapping of large areas can be time consuming and laborious when done manually with traditional survey techniques. MLS, however, has been developed as a fast and efficient alternative for collecting geomorphic data in remote or difficult to access locations. MLS systems use high-precision lasers mounted on vehicles such as cars, boats, or planes to scan the terrain below at very short intervals. The resulting point clouds are then used to create digital elevation models, which provide detailed information regarding surface topography and other features such as vegetation cover and landforms. This paper presents an overview of MLS technology and its application to geomorphologic research. It discusses how this technique can help improve accuracy and reduce costs associated with traditional surveying methods while providing more comprehensive coverage of larger areas. Finally, it outlines some potential future applications of MLS within the field of geomorphology.

Currently, TLS can be considered in various fields: topographical mapping, landslides investigation, forest inventory mapping, actual state of buildings documentation, documentation of cultural heritage, industrial sites mapping, monitoring and mapping of ground surface soil erosion, and many others (Balado et al., 2019; Jaakkola et al., 2017; Pires et al., 2022; Yu et al., 2015).

Using TLS technology over a short period allows for the acquisition of vast amounts of current, high-resolution data which can then be used to model environments or areas accurately. Laser scanners utilize several strategies for data gathering including triangulation, time of flight (TOF), and phase shifting. These technologies are particularly useful when capturing fast moving targets or small features within larger scenes. Furthermore, their ability to capture millions of individual measurements makes them ideal for creating extremely precise 3D models. To measure the 3D morphology of a surface and to detect any change over time, a high-resolution topographic model with a constrained error is necessary to avoid misinterpreting errors for changes in the surface over time (Kweon et al., 2020). Point cloud datasets generated from TLS have notably proven to provide accurate repeats over time. Recent research has emphasized the importance of having clearly defined tie-points and ground control points (GCPs), either obtained from a global navigation satellite system (GNSS-RTK), global navigation satellite system, post processing kinematic (GNSS-PPK), TLS or topographic total stations, to scale the SfM-MVS model and detect any smoothing of the topographic details (Balado et al., 2019; Jaakkola et al., 2017). It has been further suggested that even the high contrast of the target could help the model locally but also provide an error assessment that is overly optimistic.

14.2.2 Road-side slope erosion using low-cost MLS

In recent years, roadside erosion and road deformation measurement and monitoring have greatly benefited from the development of LiDAR technology, especially when mounted on vehicles on roads and rails (Lei et al., 2021; White et al., 2010). However, these vehicles and associated setups are best adapted for highways and roads, which offer good height and width clearances. Forestry roads and mountain roads are thus out of reach for those technologies, and although LiDAR has become a widespread technology, it has only occurred in a handful of countries (Eltner and Baumgart, 2015; Saito et al., 2013), and traditional airborne LiDAR still has difficulties penetrating the dense canopy to offer sufficient ground information. Thus, forested roadside erosion remains a challenge to measure in most regions of the world. MLS is well suited for surveying roads due to its accuracy and speed. It has been successfully employed by many organizations prior to rebuilding existing roads or constructing entirely new ones. By collecting detailed information about the terrain and any obstructions on site, engineers have access to critical data they need to make informed decisions regarding the best course of action. Additionally, MLS provides real-time feedback so changes can be made quickly if necessary. This helps ensure projects stay on track while minimizing costs associated with delays or errors. Finally, since it is a contactless process, there is no risk of damaging infrastructure during the survey. All this adds up to faster completion times and higher quality results (Jaakkola et al., 2017). Due to its important advantages, especially high-accuracy and precision data collection, mobile laser scanner method can be applied in the road construction field to monitor the road surface wear. To maintain safe driving conditions on forest roads, regular maintenance and repair projects must be conducted in many developed countries. Additionally, MLS techniques are often employed to quickly scan and evaluate pavement projects in urban areas. The scanning process using MLS systems is quick and efficient, providing highly detailed point cloud data which accurately represents the project's reality. Thus, this technique has become popular among engineers when extracting road objects or designing transportation plans. Furthermore, these technologies are now being utilized to monitor changes in natural and geomorphic features over time. In conclusion, MLS technology is becoming increasingly valuable in various engineering applications related to road construction and management. In summary, this paper has presented the application of airborne LiDAR and MLS data to estimate forest roads erosion in mountainous areas. The results showed that both technologies can be used together with high accuracy to map and monitor changes in terrain due to road construction or other human activities. Furthermore, the combination of these two techniques provides more detailed information than traditional field surveys and aerial photography alone. Therefore, it could provide valuable insights into understanding how landscapes are changing over time as well as providing useful information on soil stability and potential risks associated with road maintenance operations. Finally, further research should focus on developing algorithms to automate the process of estimating road erosion from LiDAR and mobile laser scan data. This would enable faster and easier analysis of large datasets and help improve our ability to accurately measure and predict future changes in landscape morphology caused by human activity.

Acknowledgment

This study was funded by the Agricultural Research, Education and Extension Organization, AREEO, as part of Project No. 0-09-09-002-000095.

References

Alexiou, S., Deligiannakis, G., Pallikarakis, A., Papanikolaou, I., Psomiadis, E., Reicherter, K., 2021. Comparing high accuracy t-lidar and uav-sfm derived point clouds for geomorphological change detection. ISPRS Int. J. Geo-Inf. 10, 367. https://doi.org/10.3390/ijgi10060367.

Baker, E., Price, B.W., Rycroft, S., Villet, M.H., 2015. Global cicada sound collection I: recordings from South Africa and Malawi by BW price & MH villet and harvesting of BioAcoustica data by GBIF. Biodivers. Data J. 3. https://doi.org/10.3897/BDJ.3.e5792.

Balado, J., Martínez-Sánchez, J., Arias, P., Novo, A., 2019. Road environment semantic segmentation with deep learning from mls point cloud data. Sensors 19, 3466. https://doi.org/10.3390/s19163466.

Barber, C.P., Cochrane, M.a., Souza, C.M., Laurance, W.F., 2014. Roads, deforestation, and the mitigating effect of protected areas in the Amazon. Biol. Conserv. 177, 203–209. https://doi.org/10.1016/j.biocon.2014.07.004.

Cao, L., Wang, Y., Liu, C., 2021. Study of unpaved road surface erosion based on terrestrial laser scanning. Catena 199, 105091. https://doi.org/10.1016/j.catena.2020.105091.

Cheng, B., Lv, Y., Zhan, Y., Su, D., Cao, S., 2015. Constructing China's roads as works of art: a case study of "esthetic Greenway" construction in the shennongjia region of China. Land Degrad. Dev. 26, 324–330. https://doi.org/10.1002/ldr.2210.

Craven, M., Wing, M.G., 2014. Applying airborne LiDAR for forested road geomatics. Scand. J. For. Res. 29, 174–182. https://doi.org/10.1080/02827581.2014.881546.

Eltner, A., Baumgart, P., 2015. Accuracy constraints of terrestrial Lidar data for soil erosion measurement: application to a Mediterranean field plot. Geomorphology 245, 243–254. https://doi.org/10.1016/j.geomorph.2015.06.008.

Haigh, M.J., Rawat, J.S., Rawat, M.S., Bartarya, S.K., Rai, S.P., 1995. Interactions between forest and landslide activity along new highways int he Kumaun Himalaya. For. Ecol. Manage. 78, 173–189. https://doi.org/10.1016/0378-1127(95)03584-5.

Jaakkola, A., Hyyppä, J., Yu, X., Kukko, A., Kaartinen, H., Liang, X., Hyyppä, H., Wang, Y., 2017. Autonomous collection of forest field reference—the outlook and a first step with UAV laser scanning. Rem. Sens. 9, 785. https://doi.org/10.3390/rs9080785.

Keesstra, S.D., Bouma, J., Wallinga, J., Tittonell, P., Smith, P., Cerdà, A., Montanarella, L., Quinton, J.N., Pachepsky, Y., Van Der Putten, W.H., Bardgett, R.D., Moolenaar, S., Mol, G., Jansen, B., Fresco, L.O., 2016. The significance of soils and soil science towards realization of the United Nations sustainable development goals. Soils 2, 111–128. https://doi.org/10.5194/soil-2-111-2016.

Kweon, H., Seo, J.I., Lee, J.W., 2020. Assessing the applicability of mobile laser scanning for mapping forest roads in the Republic of Korea. Rem. Sens. 12, 1502. https://doi.org/10.3390/RS12091502.

Lei, G., Yao, R., Zhao, Y., Zheng, Y., 2021. Detection and modeling of unstructured roads in forest areas based on visual-2D lidar data fusion. Forests 12, 820. https://doi.org/10.3390/f12070820.

Mahdavi, A., 2012. Forests and rangelands? wildfire risk zoning using GIS and AHP techniques. Casp. J. Environ. Sci. 10, 43–52.

Navarro-Hevia, J., Lima-Farias, T.R., de Araújo, J.C., Osorio-Peláez, C., Pando, V., 2015. Soil erosion in steep road cut slopes in Palencia (Spain). Land Degrad. Dev. 27, 190–199. https://doi.org/10.1002/ldr.2459.

Pires, R. de P., Holmgren, J., Olofsson, K., Lindberg, E., Persson, H.J., 2021. Influence of Distance to the Sensor on Stem Detection with Car-Mounted Mobile Laser Scanner, 104. Geowissenschaftliche Mitteilungen, pp. 154–156.

Pires, R. de P., Olofsson, K., Persson, H.J., Lindberg, E., Holmgren, J., 2022. Individual tree detection and estimation of stem attributes with mobile laser scanning along boreal forest roads. ISPRS J. Photogrammetry Remote Sens. 187, 211–224. https://doi.org/10.1016/j.isprsjprs.2022.03.004.

Rose, C.W., Yu, B., Ward, D.P., Saxton, N.E., Olley, J.M., Tews, E.K., 2014. The erosive growth of hillside gullies. Earth Surf. Process. Landforms 39, 1989–2001. https://doi.org/10.1002/esp.3593.

Safari, A., Kavian, A., Parsakhoo, A., Saleh, I., Jordán, A., 2016. Impact of different parts of skid trails on runoff and soil erosion in the Hyrcanian forest (northern Iran). Geoderma 263, 161–167.

Saito, M., Goshima, M., Aruga, K., Matsue, K., Shuin, Y., Tasaka, T., 2013. Study of automatic forest road design model considering shallow landslides with lidar data of funyu experimental forest. Croat. J. For. Eng. 34, 1–16.

Shataee, S., Weinaker, H., Babanejad, M., 2011. Plot-level forest volume estimation using airborne laser scanner and TM data, comparison of boosting and random forest tree regression algorithms. Procedia Environ. Sci. 7, 68–73. https://doi.org/10.1016/j.proenv.2011.07.013.

Sidle, R.C., Ziegler, A.D., Negishi, J.N., Rahim, A., Siew, R., Turkelboom, F., 2006. Erosion processes in steep terrain—truths, myths, and uncertainties related to forest management in. Southeast Asia 224, 199–225. https://doi.org/10.1016/j.foreco.2005.12.019.

Smith, P., Cotrufo, M.F., Rumpel, C., Paustian, K., Kuikman, P.J., Elliott, J.A., McDowell, R., Griffiths, R.I., Asakawa, S., Bustamante, M., House, J.I., Sobocká, J., Harper, R., Pan, G., West, P.C., Gerber, J.S., Clark, J.M., Adhya, T., Scholes, R.J., Scholes, M.C., 2015. Biogeochemical cycles and biodiversity as key drivers of ecosystem services provided by soils. Soils 1, 665–685. https://doi.org/10.5194/soil-1-665-2015.

Thompson, C., Croke, J., 2013. Geomorphic effects, flood power, and channel competence of a catastrophic flood in confined and unconfined reaches of the upper Lockyer valley, southeast. Geomorphology 197, 156–169. https://doi.org/10.1016/j.geomorph.2013.05.006.

Vahabzadeh, G., Safari, A., Farhoudi, M.H., Abdollahi, H.R., Fathizad, H., Khosravi, G.R., 2015. Assessment of erosion value and sediment delivery ratio in the unsealed and forest roads. JWSS-Isfahan Univ. Technol. 18, 295–313.

White, R.A., Dietterick, B.C., Mastin, T., Strohman, R., 2010. Forest roads mapped using LiDAR in steep forested terrain. Rem. Sens. 2, 1120–1141. https://doi.org/10.3390/rs2041120.

Yousefi, S., Moradi, H., Boll, J., Schönbrodt-Stitt, S., 2016. Effects of road construction on soil degradation and nutrient transport in Caspian Hyrcanian mixed forests. Geoderma 284, 103–112. https://doi.org/10.1016/j.geoderma.2016.09.002.

Yu, Y., Li, J., Guan, H., Wang, C., 2015. Automated extraction of urban road facilities using mobile laser scanning data. IEEE Trans. Intell. Transport. Syst. 16, 2167–2181. https://doi.org/10.1109/TITS.2015.2399492.

CHAPTER

15

Suspended sediment load prediction and tree-based algorithms

Salim Heddam[1], *Amir Naghibi*[2], *Khabat Khosravi*[3] *and Shailesh K. Singh*[4]

[1]**Faculty of Science, Agronomy Department, University 20 Août 1955, Skikda, Algeria;** [2]**Department of Water Resources Engineering & Center for Advanced Middle Eastern Studies, Lund University, Lund, Sweden;** [3]**Department of Earth and Environment, Florida International University, Miami, FL, United States;** [4]**National Institute of Water and Atmospheric Research, Christchurch, New Zealand**

15.1 Introduction

Knowledge of suspended sediment load (Q_{ssl}) is crucial in preserving and managing water supplies and the environment. Nonpoint pollution sources, for example, different kinds of soil erosion processes, nutrient loading (mainly nitrogen and phosphorus), and corresponding pollutants form a strict risk to the ecosystem and the environment (Melesse et al., 2011). Q_{ssl} in streams is a complicated phenomenon and is treated as one of the most compelling issues water resources engineers are currently confronting (Gericke and Venohre, 2012). Q_{ssl} impacts water quality, dam reservoirs, and ecological conditions and owns several environmental influences. Q_{ssl} is a part of the sediment load, which is uplifted because of the fluid's flow and normally constitutes the key fraction of sediment loads in streams in contrast to bed load and accordingly is noted as a greatly vital variable in the field of geological, hydrological, and hydraulic modeling. For approximating Q_{ssl}, there are four major methods: (i) numerical modeling (Pan and Huang, 2010), physical models (Kuti and Ewemoje, 2021), empirical equations (Mimikou, 1982; Pektas and Cigizoglu, 2017), and artificial intelligence (AI) and machine learning (ML) algorithms (Kisi et al., 2012; AlDahoul et al., 2021; Huang et al., 2021).

Because of the uncertainties associated with Q_{ssl} transportation and its nonuniform patterns in open channels regarding the hydrologic/hydraulic driving variables, "empirical or linear formulas such as sediment rating curves (Warrick, 2015) or autoregressive integrated

https://doi.org/10.1016/B978-0-443-15341-9.00013-7

moving average (ARIMA) (Melesse et al., 2011) are restricted in generating highly accurate estimates of Q_{ssl}. "Numerical/physical" models such as Soil and Water Assessment Tool (SWAT)" demand an enormous extent of high spatiotemporal resolution data, for example, "soil texture, geophysical, hydro climatological, and ecological information" to attain high estimation accuracy; nevertheless, this type of input is not often accessible, particularly in "developing countries", for example, the Middle East region and Africa. Furthermore, generating such "initial conditions data" to train "empirical/physical models" for larger basins requires significant time and expense (Nhu et al., 2020). Lately, ML algorithms have been employed to estimate Q_{ssl}. MLs offer several advantages, for example, the proficiency to be trained with data at different scales, identify connections among driving and target variables for more complicated phenomena, and higher estimation effectiveness (Yaseen et al., 2020). Another attribute of ML algorithms is their "robustness in accommodating missing values in time series" modeling by adapting several "data imputation methods", which is a regular issue in less developed countries, for example, the Middle East.

Different ML algorithms have been executed to model Q_{ssl} in recent years. Artificial neural networks (ANNs) have been applied for modeling sediment (Melesse et al., 2011; Rezaei et al., 2021; Zeyneb et al., 2022). As reported in a study, Melesse et al. (2011) trained an ANN algorithm to simulate the Q_{ssl} at different temporal scales, that is, daily and weekly, and demonstrated that ANN generated superior predictions in contrast with "multiple linear regressions (MLR) and ARIMA." The authors claimed that the daily simulation of Q_{ssl} was more accurate than the weekly estimates. Asadiani Yekta et al. (2010) applied ANFIS with ANN for Q_{ssl} estimation and depicted the higher efficiency of the ANFIS. Kakaei Lafdani et al. (2013) modeled Q_{ssl} by "Broyden, Fletcher, Goldfarb, and Shanno (BFGS-ANN) with Support Vector Machine (SVM)" and depicted that the BFGS-ANN led to the greatest accuracy. One drawback of ANNs is that they demand enormous training sets, and their execution time is slower. Asadiani Additionally, some researchers have utilized metaheuristic approaches hybridized with ANFIS to amplify its efficiency in modeling Q_{ssl} "(Fadaee et al., 2022; Zounemat-Kermani et al., 2020; Babanezhad et al., 2021; Mehri et al., 2021; Adnan et al., 2022; Zeyneb et al., 2022)." In addition, two researchers implemented genetic programming (GP) for the simulation of Q_{ssl}, contrasted its efficiency to other algorithms, and reported that GP generated improved estimates (Aytek and Kisi, 2008; Kisi et al., 2012).

Recently, new ML algorithms have been applied to surpass the drawbacks of classical algorithms. As an example, Khosravi et al. (2018) implemented an ensembled ML algorithm constituted of "reduced error pruning tree (REPT), M5P and instance-based learning (IBK)" with Bagging algorithm to predict Q_{ssl} in a case study. They claimed that the "bagging-M5P hybrid method" yielded higher and more reliable results than other developed algorithms. In another study, Sharafati et al. (2020) implemented several ML-based models including RF, gradient boost regression (GBR), and AdaBoost (ABR), to estimate Q_{ssl} at the Mississippi River, and finally, they declared that the RF algorithm generated higher efficacy than the other developed models.

All in all, decision tree-based algorithms are perceived as one of the compelling and convincing methods in hydrology for predicting Q_{ssl}. Thus, this research spotlights the application of three decision tree-based algorithms, that is, RT, RF, and M5P, to predict Q_{ssl}. For this, the models are validated to analyze their efficiencies. Additionally, multiple combinations of input factors, including rainfall, flow discharge, and previous Q_{ssl} time steps, were incorporated. Additionally, lag times are examined to select the ideal set. M5P is applied for the first time in predicting Q_{ssl}, and it is compared with benchmark algorithms, that is, RF and RT. Moreover, the outcomes were compared with other tree-based algorithms developed by Khosravi et al. (2022).

15.2 Tree-based models

15.2.1 Random forest

Random forest (RF) is a boosted and new version of the original algorithm, called classification and regression tree (CART), proposed by Breiman (2001), and it is considered a member of the bootstrap aggregating approach "bagging" (Fig. 15.1). Building an RF model can be achieved in three steps (Fig. 15.1): (i) from the original dataset, an ensemble of a subset (*L*) is constructed, (*ii*) for each prepared subset "*L*" a decision tree (DT) model is developed, which is unpruned, and (*iii*) the final response of the RF model is calculated as the average of all DT models. It was demonstrated from several applications that the generalization ability of the RF is very strong in response to the unbiased estimation of the generalization error (Li et al., 2018). According to Breiman (2001), the performances of RF could be significantly improved by growing several trees, and at each node, a randomization procedure is adopted for the election of an ensemble of input variables for splitting and bootstrapping for the original calibration dataset. Furthermore, the nonchosen cases are called "*out-of-bag: OOB.*" Finally, the training process of the RF should take into account three kinds of parameters: (*i*) the ratio of data used for training, (*ii*) the "*L*" value, and (*iii*) the number of variables chosen randomly at each node.

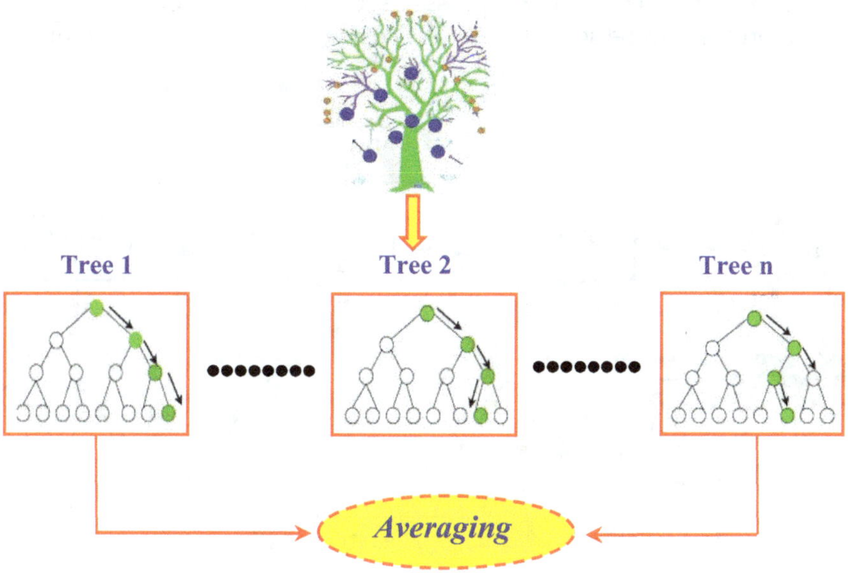

FIGURE 15.1 Random forest architecture.

15.2.2 M5 prime

The M5P algorithm is tree-based (Quinlan 1992; Xu et al., 2005). One of the most particularities of the M5P model is the presence of the multivariate linear regression (MVLR) method at the leaves of the trees. The M5P proceeds by dividing the training space into an ensemble of subspaces and combines the individual solutions provided at each subspace (Yaseen et al., 2020; Sihag et al., 2022). The M5P regression model can be achieved in two steps: (*i*) splitting the data and creating the DT based on each sub-dataset and (*ii*) choosing the separation criteria as the standard deviation ratio (SDR) for which research on reducing the variation between the subset at each branch is then activated. The SDR is calculated as follows:

$$\mathrm{SDR} = \mathrm{SD}(T) - \sum \frac{|T_i|}{|T|} \times \mathrm{StD}(T_i) \tag{15.1}$$

where T is an ensemble of examples reaching the nodes, T_i is the example that possesses the ith aftermath of the potential dataset, and StD refers to the standard deviation (Yaseen et al., 2020; Sihag et al., 2022).

15.2.3 Random trees

The random tree (RT) is a machine learning algorithm largely used for solving regression and classification problems (Geurts et al., 2006; Breiman et al., 2017) (Fig. 15.2). The structures of the RT are simply an ensemble of trees called a forest. It is important to note that the training dataset is divided into a set of trees for which all are trained using the same parameters; however, the training subsets are different and are prepared using the bootstrap procedure. The regression or the classification error is calculated using the so-called *"OOB."* According to Fig. 15.2, the RT possesses two kinds of nodes: split node and leaf node; hence, the problem to be solved is divided using a recursive method into an ensemble

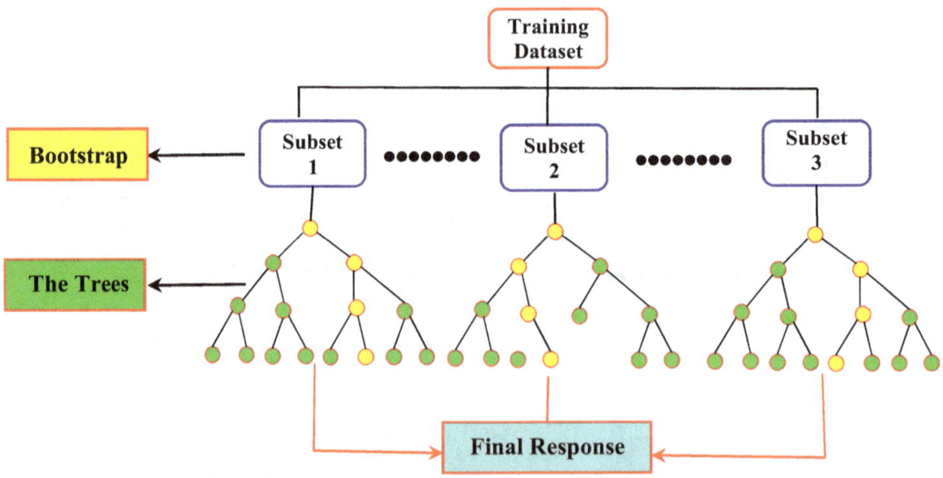

FIGURE 15.2 Decision tree algorithm.

of a lower problem with low complexity. The splitting strategy is performed at the split node, while the leaf nodes are the final node having a model for the problem to be solved (Markuš et al., 2014).

15.3 Example

15.3.1 Study area

The Talar watershed as an example case study located in northern Iran lies between the latitude of 52° 35′ 18′ and 53° 23′ 35″ E and longitude of 35° 44′ 19″ and 36° 19′ 13″ N (Fig. 15.3). The Talar River which fed by Sorkh-Abad, and Shor-Ab-mainly in Savadkooh-, after 180 km pour to Caspian Sea. The mean annual rainfall, and mean minimum and maximum temperature of the Talar basin with an area of 2100 km^2 are about 552.7 mm and 7.7 and 21.1 °C, respectively (Sahin, 2012). The main source of Q_{ssl} in the Talar watershed is landslide occurrences, while in many watersheds, mainly a complex interaction between sparse vegetation cover, land use change, and intense rainfall are the main reason (Emamgholizadeh and Demneh, 2019).

15.3.2 Data

Time-series monthly Q_w, Q_{ssl}, and rainfall data from January 1, 2004 to September 22, 2019, were used. The time series dataset was separated into three periods: 2004–2013, 2014–16, and from 2017 to 2019, for training, calibrating, and validation of developed models.

15.3.3 Model input

Twelve input scenarios were prepared according to the input variable effectiveness and finally evaluated to obtain the optimum input scenario (Table 15.1). All input scenarios are built according to the Pearson correlation coefficient (r) between inputs and output parameters as well as the principal component analysis (PCA) approach (Input No.12) in a time-series dataset. For more information, please see Khosravi et al. (2022).

15.3.4 Model evaluation

In the present study, quantitative indicators of the coefficient of determination (R^2) and time-series observational plots are applied to evaluate the performance of the different models.

15.4 Results and discussion

The result of time-series Q_{ssl} prediction revealed that although the M5P model was not able to capture peak data, the overall estimation of the M5P model is better than that of RF and RT. The RT model is not able to capture peak values or other values accurately

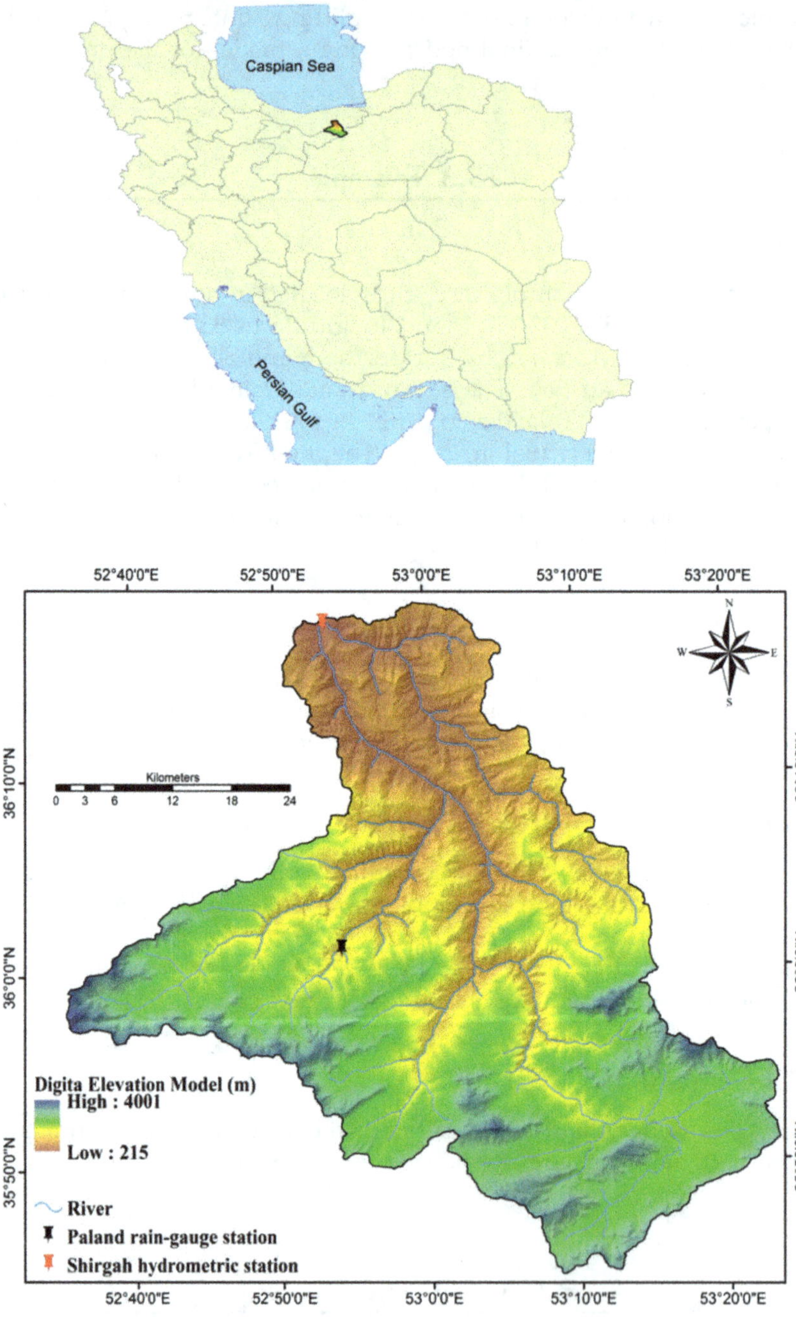

FIGURE 15.3 The position of the Talar watershed in Iran (Khosravi et al., 2022).

TABLE 15.1 Different input combinations (Khosravi et al., 2022).

Input No.	Input scenarios
1	Q_w
2	Q_w, R
3	Q_w, R_1
4	Q_w, R_{-1}, R_{-2}
5	R, R_{-1}, R_{-2}
6	Q_w, R_{-1}, R_{-2}, R
7	Q_w, Q_{w-1}, Q_{w-2}
8	Q_w, R_{-1}, R_{-2}, R, Q_{w-1}
9	Q_w, R_{-1}, R_{-2}, R, Q_{w-1}, Q_{s-1}
10	Q_w, R_{-1}, R_{-2}, R, Q_{w-1}, Q_{s-1}
11	Q_w, R_{-1}, R_{-2}, R, Q_{w-1}, Q_{s-1}, Q_{w-2}
12	Q_w, R_{-1}, R_{-2}, R, Q_{w-1}, Q_{s-1}, Q_{w-2}, Q_{s-2}

(Fig. 15.4). The RF model has a good ability to capture peak values (i.e., peak No. 1 and 3) but is not able to predict all other values accurately. All developed models were not able to capture the peak Q_{ssl} No. 2 accurately. Khosravi et al. (2022) stated that landslide occurrences, which are common in the Talar watershed, may have occurred, destroying the trends between input and output in historical data, and none of the applied modes that work based on historical data were able to capture this peak value.

Based on the scatter plot and R^2 values, the M5P model has the highest R^2 value (0.83), followed by RF (0.78) and RT (0.35). The lowest prediction of the RT model was obvious from the time-series plot described in the previous section (Fig. 15.5).

Khosravi et al. (2022) developed the AM tree, REP tree, and DPC tree, as well as four ensemble-based models that integrate a standalone model with the RF model for Q_{ssl} prediction with the same data. They stated that hybridization boosted the modeling implementation, as standalone models of AM Tree, DPC Tree, and REP Tree have R^2 values of 0.82, 0.55, and 0.79, while their hybridized versions have R^2 values of 0.835, 0.80, and 0.82, respectively. Based on our achievement, the standalone M5P model with an R^2 of 0.832 has a slightly lower performance than a hybrid model of the AM Tree-ROF model as the best model among the tree-based methods applied by Khosravi et al. (2022).

15.5 Q_{ssl}, ML, and remote sensing technique

Due to the costly Q_{ssl} measuring and collecting dataset, the remote sensing (RS) technique is one of the most practical approaches in the geoscience field of study, especially for soil

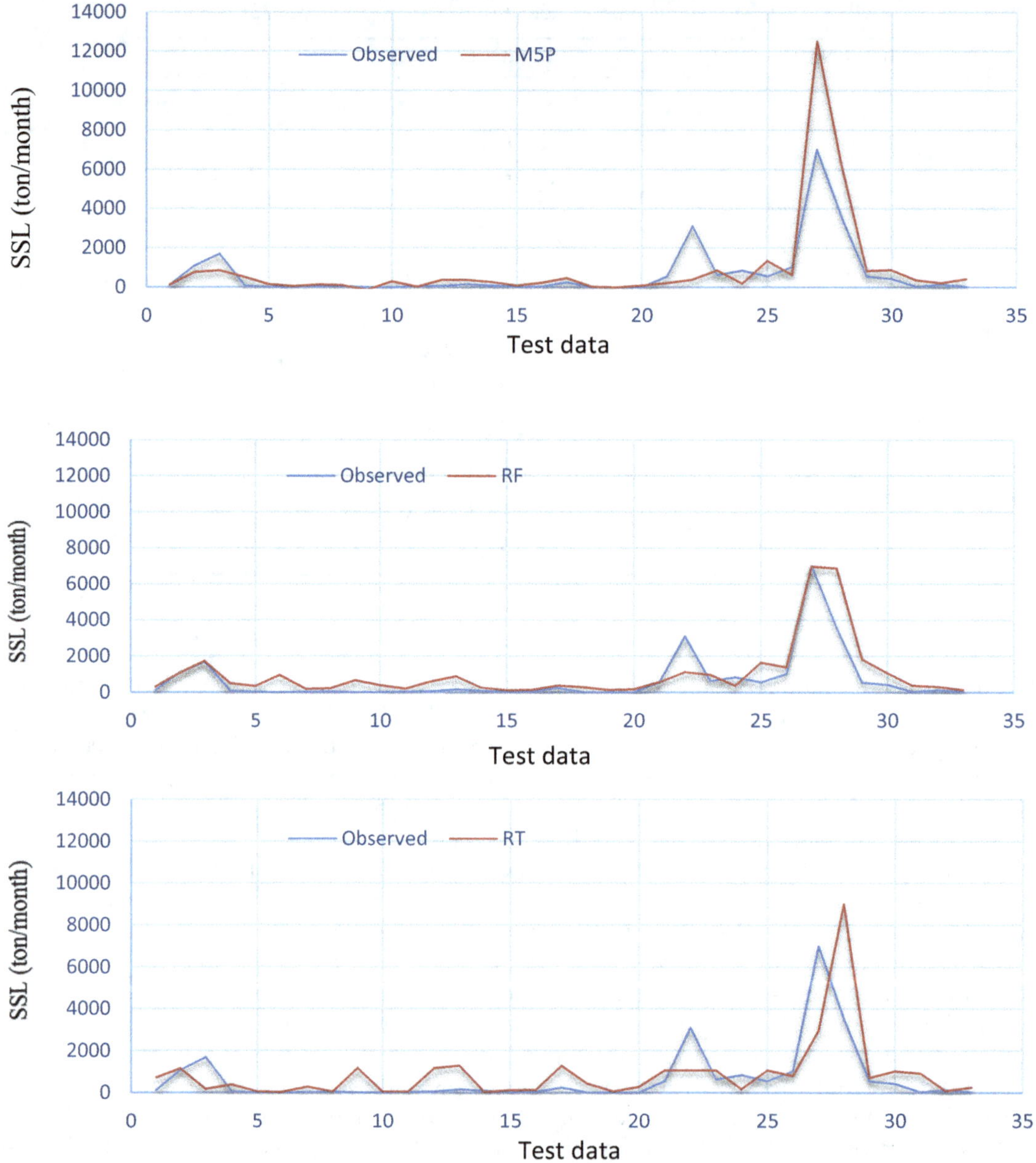

FIGURE 15.4 Q_{ssl} time series plot applied for model evaluation.

erosion and sediment yield modeling. In addition, integration of the RS technique with ML models will boost the models' prediction accuracy. The r-value between suspended sediment concentration (SSC) and reflectance of satellite-based images implemented using several approaches such as the semi-empirical approach (Duan et al., 2010; Mohsen et al., 2020;

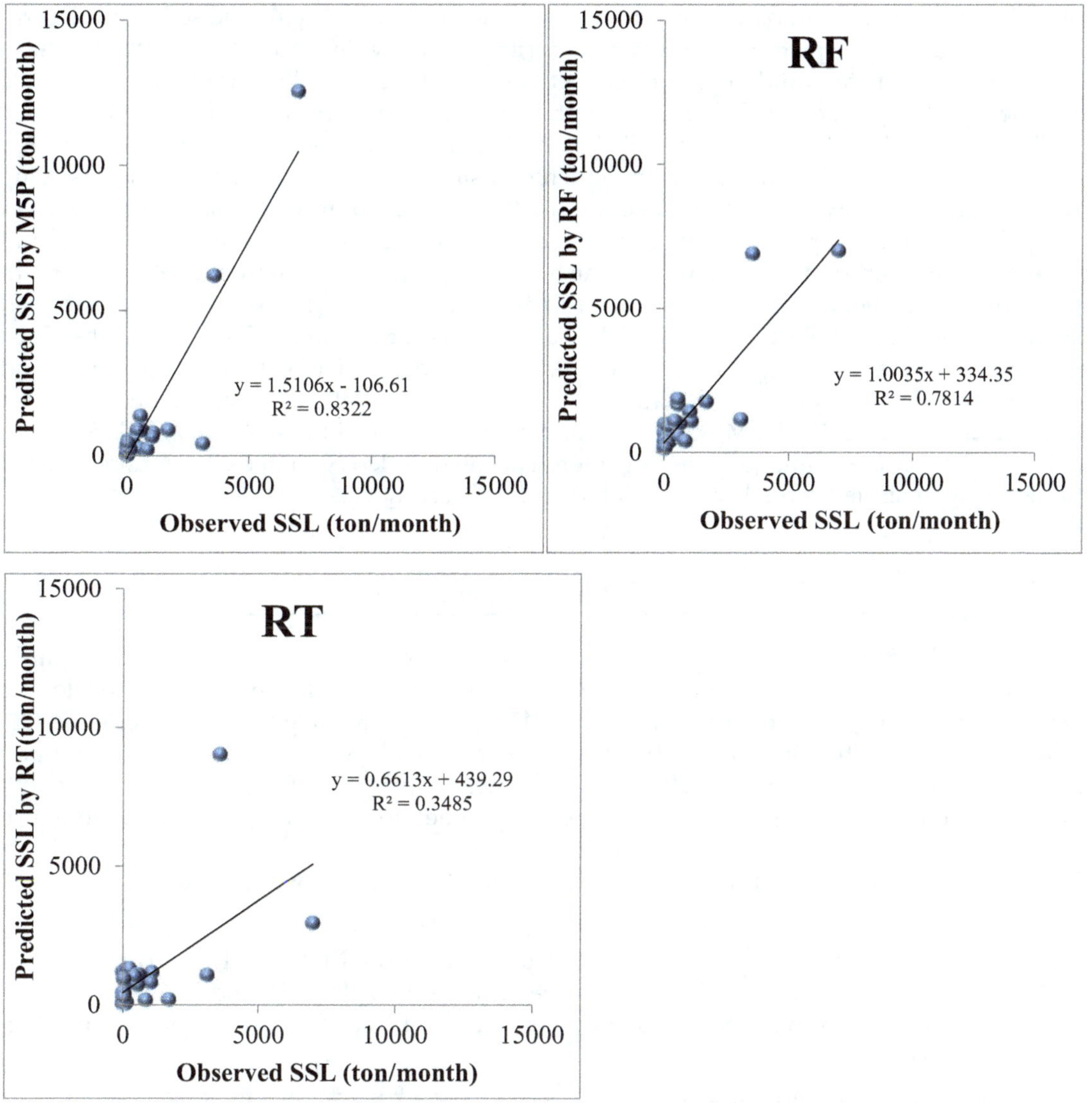

FIGURE 15.5 Scatter plot applied for model evaluation. *SSL*, suspended sediment load; *RT*, random tree.

Yu et al., 2020), bio-optical modeling approach (Ammenberg et al., 2002; Manuel et al., 2020), or ML algorithms (Mohamed and Shah, 2018; Silveira Kupssinskü et al., 2020). Although some studies predict the SSC (Martinez et al., 2009; Mouyen et al., 2018; Flores et al., 2020), in most cases, the observed SSC data are combined with satellite-based data (Stumpf and Goldschmidt, 1992; Martinez et al., 2009). Ahmed et al. (2019) combined the satellite RS and unsupervised ML techniques with appropriate validation metrics for sediment deposition at a river network. They reported that near infrared (NIR) and other bands outside the visible electospectrum band are determinant features in identifying the sediment depositional regions. They depicted that learning algorithms mostly depend on image cluster interpretability than cluster validation criteria. Dethier et al. (2020) implemented the global scale approach for the prediction of SSC using optical satellite imagery of Landsat 5 and 7. They finally calibrated by 134,697 observed SSC dataset. Variability of each river in the connection between water properties and SSC is implemented using two main approaches, (1) classifying rivers using well-known K-means clustering (KMC) or (2) calibration simulates for each river using three main criteria including observed SSC, suspended-sediment grain size, and percent organic carbon. They reported that five site-specific observed data integrated with our algorithm further decrease the average relative error to 49%. Mohsen et al. (2022) applied Sentinel-2A and 2B satellite images and ML algorithms for sediment discharge at the Tisza and Maros Rivers in Hungary. The river section was divided into several segments, and the average reflectance of Sentinel-2 images was correlated to calculated suspended sediment concentration (SSC) using SVM, RF, ANN, and combined algorithms. Bands 2, 3, 4, 5, 6, 7, 8, and 8a of the Sentinel-2 images were considered as independent variables. They finally stated that RF and combined models had a high degree of accuracy.

15.6 Conclusion

Suspended sediment load is affected by various climatic, hydrogeological, and topographical factors; however, detailed data on these factors are not available, especially in developing countries. Therefore, we aimed to implement M5P as a new ML algorithm in modeling Q_{ssl} and compared it with RF and RT. To make the framework reproducible, only simple and easy-to-provide data such as rainfall, flow discharge, and previous Q_{ssl} with different month lag times were used. Furthermore, the proper lag times for each input variable were also determined.

The main conclusions of this research can be stated in the following bullet points:

- M5P has the highest performance in estimating Q_{ssl}, followed by the RF algorithm.
- RT did not produce satisfactory results compared with the M5P and RF algorithms.
- RF is more accurate in capturing peak values compared with M5P and RT
- An effective input combination is one in which all input variables with and without lag time are involved.

Regarding the acceptable performance of the M5P and RF models, this framework can be implemented in regions with a lack of data, especially in developing countries and regions such as the Middle East. These models can be fed by and calibrated to the input factors in other regions. Therefore, it is not necessary to find areas similar to the one under study in this research.

Acknowledgments

This research is supported by the MECW research program and the Centre for Advanced Middle Eastern Studies, Lund University.

References

Adnan, R.M., Yaseen, Z.M., Heddam, S., Shahid, S., Niaraki, A., Kisi, O., 2022. Predictability performance enhancement for suspended sediment in rivers: inspection of newly developed hybrid adaptive neuro-fuzzy system model. Int. J. Sediment Res. 37 (3), 383–398.

Ahmed, N., Mahmud, S., Elahi, M.L., Ahmed, S., Sujauddin, M., 2019. Forecasting river sediment deposition through satellite image driven unsupervised machine learning techniques. Remote Sens. Appl.: Soc. Environ. 13, 435–444.

AlDahoul, N., Essam, Y., Kumar, P., et al., 2021. Suspended sediment load prediction using long short-term memory neural network. Sci. Rep. 11, 7826. https://doi.org/10.1038/s41598-021-87415-4.

Ammenberg, P., Flink, P., Lindell, T., Pierson, D., Strombeck, N., 2002. Bio-optical modelling combined with remote sensing to assess water quality. Int. J. Remote Sens. 23, 1621–1638.

Asadiani Yekta, A.H., Marsooli, R., Soltani, F. 2010. Suspended sediment estimation of Ekbatan Reservoir Sub Basin using Adaptive Neuro-Fuzzy Inference Systems (ANFIS), Artificial Neural Networks (ANN), and Sediment Rating Curves (SRC). River Flow 2010. Dittrich, Koll, Aberle & Geisenhainer (eds.) 2010 Bundesanstalt für Wasserbau. ISBN 978-3-939230-00-7.

Aytek, A., Kisi, Ö., 2008. A genetic programming approach to suspended sediment modelling. J. Hydrol. 351 (3–4), 288–298. https://doi.org/10.1016/j.jhydrol.2007.12.005.

Babanezhad, M., Behroyan, I., Marjani, A., et al., 2021. Artificial intelligence simulation of suspended sediment load with different membership functions of ANFIS. Neural Comput. Appl. 33, 6819–6833. https://doi.org/10.1007/s00521-020-05458-6.

Breiman, L., 2001. Random forests. Mach. Learn. 45 (1), 5–32. https://doi.org/10.1023/A:1010933404324.

Breiman, L., Friedman, J.H., Olshen, R.A., Stone, C.J., 2017. Classification and Regression Trees. Routledge.

Dethier, E.N., Renshaw, C.E., Magilligan, F.J., 2020. Toward improved accuracy of remote sensing approaches for quantifying suspended sediment: Implications for suspended-sediment monitoring. J. Geophys. Res. Earth Surf. 125 (7) e2019JF005033.

Duan, H., Ma, R., Xu, J., Zhang, Y., Zhang, B., 2010. Comparison of different semi-empirical algorithms to estimate chlorophyll-a concentration in inland lake water. Environ. Monit. Assess. 170, 231–244.

Emamgholizadeh, S., Demneh, R.K., 2019. A comparison of artificial intelligence models for the estimation of daily suspended sediment load: a case study on the Telar and Kasilian rivers in Iran. Water Supply 19 (1), 165–178. https://doi.org/10.2166/ws.2018.062.

Fadaee, M., Mahdavi-Meymand, A., Zounemat-Kermani, M., 2022. Suspended sediment prediction using integrative soft computing models: on the analogy between the butterfly optimization and genetic algorithms. Geocarto Int. 37 (4), 961–977. https://doi.org/10.1080/10106049.2020.1753821.

Flores, J., Wu, J., Stöckle, C., Ewing, R., Yang, X., 2020. Estimating river sediment discharge in the Upper Mississippi river using Landsat imagery. Remote Sens. 12, 2370.

Gericke, A., Venohr, M., 2012. Improving the estimation of erosion-related suspended solid yields in mountainous, non-alpine river catchments. Environ. Model. Software 37, 30–40. https://doi.org/10.1016/j.envsoft.2012.04.008.

Geurts, P., Ernst, D., Wehenkel, L., 2006. Extremely randomized trees. Mach. Learn. 63, 3–42. https://doi.org/10.1007/s10994-006-6226-1.

Huang, C.C., Chang, M.J., Lin, G.F., Wu, M.C., Wang, P.H., 2021. Real-time forecasting of suspended sediment concentrations reservoirs by the optimal integration of multiple machine learning techniques. J. Hydrol. Reg. Stud. 34, 100804. https://doi.org/10.1016/j.ejrh.2021.100804.

Khosravi, K., Golkarian, A., Melesse, A., Deo, R., 2022. Suspended sediment load modeling using advanced hybrid rotation forest based elastic network approach. J. Hydrol. 610 (2022), 127963.

Khosravi, K., Mao, L., Kisi, O., Yaseen, Z.M., Shahid, S., 2018. Quantifying hourly suspended sediment load using data mining models: case study of a glacierized Andean catchment in Chile. J. Hydrol. 567, 165–179. https://doi.org/10.1016/j.jhydrol.2018.10.015.

Kisi, O., Dailr, A.H., Cimen, M., Shiri, J., 2012. Suspended sediment modeling using genetic programming and soft computing techniques. J. Hydrol. 450, 48–58. https://doi.org/10.1016/j.jhydrol.2012.05.031.

Kuti, I.A., Ewemoje, T.A., 2021. Modelling of sediment yield using the soil and water assessment tool (SWAT) model: a case study of the Chanchaga Watersheds, Nigeria. Sci. Afr. 13, e00936. https://doi.org/10.1016/j.sciaf.2021.e00936.

Lafdani, E.K., Nia, A.M., Ahmadi, A., 2013. Daily suspended sediment load prediction using artificial neural networks and support vector machines. J. Hydrol. 478, 50–62. https://doi.org/10.1016/j.jhydrol.2012.11.048.

Li, C., Tao, Y., Ao, W., Yang, S., Bai, Y., 2018. Improving forecasting accuracy of daily enterprise electricity consumption using a random forest based on ensemble empirical mode decomposition. Energy 165, 1220–1227. https://doi.org/10.1016/j.energy.2018.10.113.

Manuel, A., Blanco, A.C., Tamondong, A.M., Jalbuena, R., Cabrera, O., Gege, P., 2020. Optmization of bio-optical model parameters for turbid lake water quality estimation using Landsat 8 and wasi-2D. Int. Arch. Photogram. Remote Sens. Spatial Inf. Sci. 42, 67–72.

Markuš, N., Frljak, M., Pandžić, I.S., Ahlberg, J., Forchheimer, R., 2014. Eye pupil localization with an ensemble of randomized trees. Pattern Recogn. 47 (2), 578–587. https://doi.org/10.1016/j.patcog.2013.08.008.

Martinez, J.-M., Guyot, J.-L., Filizola, N., Sondag, F., 2009. Increase in suspended sediment discharge of the Amazon River assessed by monitoring network and satellite data. Catena 79, 257–264.

Mehri, Y., Nasrabadi, M., Omid, M.H., 2021. Prediction of suspended sediment distributions using data mining algorithms. Ain Shams Eng. J. 12 (4), 3439–3450. https://doi.org/10.1016/j.asej.2021.02.034.

Melesse, A.M., Ahmad, S., McClain, M.E., Wang, X., Lim, Y.H., 2011. Suspended sediment load prediction of river systems: an artificial neural network approach. Agric. Water Manag. 98 (5), 855–866. https://doi.org/10.1016/j.agwat.2010.12.012.

MIMIKOU, M., 1982. An investigation of suspended sediment rating curves in western and northern Greece. Hydrol. Sci. J. 27 (3), 369–383. https://doi.org/10.1080/02626668209491116.

Mohamed, I., Shah, I., 2018. Suspended sediment concentration modeling using conventional and machine learning approaches in the Thames River, London Ontario. J. Water Manag. Model. 26, 1–12.

Mohsen, A., Elshemy, M., Zeidan, B., 2020. Water quality monitoring of Lake Burullus (Egypt) using Landsat satellite imageries. Environ. Sci. Pollut. Res. 28, 1–14.

Mohsen, A., Kovács, F., Kiss, T., 2022. Remote sensing of sediment discharge in rivers using sentinel-2 images and machine-learning algorithms. Hydrol 9 (5), 88.

Mouyen, M., Longuevergne, L., Steer, P., Crave, A., Lemoine, J.-M., Save, H., Robin, C., 2018. Assessing modern river sediment discharge to the ocean using satellite gravimetry. Nat. Commun. 9, 1–9.

Nhu, V.H., Khosravi, K., Cooper, J.R., Karimi, M., Kisi, O., Pham, B.T., Lyu, Z., 2020. Monthly suspended sediment load prediction using artificial intelligence: testing of a new random subspace method. Hydrol. Sci. J. 65 (12), 2116–2127. https://doi.org/10.1080/02626667.2020.1754419.

Pan, C., Huang, W., 2010. Numerical modeling of suspended sediment transport affected by tidal bore in Qiantang Estuary. J. Coast Res. 26 (6), 1123–1132. https://doi.org/10.2112/JCOASTRES-D-09-00024.1.

Pektas, A.O., Cigizzoglu, H.K., 2022. Long-range forecasting of suspended sediment. Hydrol. Sci. J. 62 (14). https://www.tandfonline.com/doi/full/10.1080/02626667.2017.1383607.

Quinlan, J.R., 1992. Learning with continuous classes. In: 5th Australian Joint Conference on Artificial Intelligence, vol 92, pp. 343–348. https://doi.org/10.1142/9789814536271.

Rezaei, K., Pradhan, B., Vadiati, M., et al., 2021. Suspended sediment load prediction using artificial intelligence techniques: comparison between four state-of-the-art artificial neural network techniques. Arabian J. Geosci. 14, 215. https://doi.org/10.1007/s12517-020-06408-1.

Sahin, S., 2012. An aridity index defined by precipitation and specific humidity. J. Hydrol. 444, 199–208. https://doi.org/10.1016/j.jhydrol.2012.04.019.

Sharafati, A., Haji Seyed Asadollah, S.B., Motta, D., Yaseen, Z.M., 2020. Application of newly developed ensemble machine learning models for daily suspended sediment load prediction and related uncertainty analysis. Hydrol. Sci. J. 65 (12), 2022–2042. https://doi.org/10.1080/02626667.2020.1786571.

Sihag, P., Pandhiani, S.M., Sangwan, V., Kumar, M., Angelaki, A., 2022. Estimation of ground-level O3 using soft computing techniques: case study of Amritsar, Punjab State, India. Int. J. Environ. Sci. Technol. 19 (6), 5563–5570. https://doi.org/10.1007/s13762-021-03514-9.

Silveira Kupssinskü, L., Thomassim Guimarães, T., Menezes de Souza, E., Zanotta, D.C., Roberto Veronez, M., Gonzaga Jr., L., Mauad, F.F., 2020. A method for chlorophyll-a and suspended solids prediction through remote sensing and machine learning. Sensors 20, 2125.

Stumpf, R.P., Goldschmidt, P.M., 1992. Remote sensing of suspended sediment discharge into the western Gulf of Maine during the April 1987 100-year flood. J. Coast Res. 8, 218–225.

Warrick, J.A., 2015. Trend analyses with river sediment rating curves. Hydrol. Process. 29 (6), 936–949.

Xu, M., Watanachaturaporn, P., Varshney, P.K., Arora, M.K., 2005. Decision tree regression for soft classification of remote sensing data. Remote Sens. Environ. 97 (3), 322–336. https://doi.org/10.1016/j.rse.2005.05.008.

Yaseen, Z.M., Naghshara, S., Salih, S.Q., Kim, S., Malik, A., Ghorbani, M.A., 2020. Lake water level modeling using newly developed hybrid data intelligence model. Theor. Appl. Climatol. 141, 1285–1300. https://doi.org/10.1007/s00704-020-03263-8.

Yu, Y., Chen, S., Qin, W., Lu, T., Li, J., Cao, Y., 2020. A semi-empirical chlorophyll-a retrieval algorithm considering the effects of Sun glint, bottom reflectance, and non-algal particles in the optically Shallow water zones of Sanya Bay using SPOT6 data. Remote Sens. 12, 2765.

Zeynab, T., Nadir, M., Boualem, R., 2022. Modeling of suspended sediment concentrations by artificial neural network and adaptive neuro fuzzy interference system method—study of five largest basins in Eastern Algeria. Water Pract. Technol. 17 (5), 1058–1081.

Zounemat-Kermani, M., Mahdavi-Meymand, A., Alizamir, M., Adarsh, S., Yaseen, Z.M., 2020. On the complexities of sediment load modeling using integrative machine learning: application of the great river of Loíza in Puerto Rico. J. Hydrol. 585, 124759. https://doi.org/10.1016/j.jhydrol.2020.124759.

CHAPTER

16

Soil erosion and sediment change detection using UAV technology

Somayeh Mirzaee[1], *Christopher Gomez*[2], *Mehdi Pajouhesh*[1] *and Khodayar Abdollahi*[1]

[1]Faculty of Earth Science, Shahrekord University, Shahrekord, Iran; [2]Graduate School of Oceanology, Laboratory of Sediment Hazards and Disaster Risk, Kobe University, Kobe, Japan

16.1 Introduction

Soil is an essential resource for Earth's system, which is why soil loss is a significant threat to many of Earth's functionalities. Hence, understanding the complicated erosion process is helpful for both soil erosion and land degradation, especially for our farming land (Eltner et al., 2013). On the other hand, this key component of the earth system controls hydrological and geomorphological processes such as water erosion (Keesstra et al., 2018; YORULMAZ and GÜMÜS 2022). Worldwide, soil erosion is a major environmental problem with severe ecological and socioeconomic consequences (Yousefi et al., 2016). As a result, monitoring water erosion is considered one of the most significant issues in land management. There are two current approaches for collecting data in soil erosion monitoring (Pires et al., 2022). The first approach is based on manual data collection, a method commonly used in wide range of countries that is dull and produces insufficient details. The other technique relies on higher technologies. The geodetic method remains the most common measurement method that makes use of the combination of total stations, collected data from the Global Navigation Satellite System (GNSS) and remote sensing (RS) (methods (Kweon et al., 2020). GNSS and RS methods can carry out projects at low costs, as they save money and time (Kweon et al., 2020; Pires et al., 2022). RS technologies have been commonly used over the past 15 years. Their derived 3D data are systematically used for digital territory and surface simulation. With the recent development of new technologies, automated data collection and RS methods have been increasingly expanded (Thompson and Croke, 2013; Craven and Wing, 2014). Due to the attractiveness of 3D data, modeling technology to obtain this information is gradually becoming more mature. This has led to less misinterpretation of surface change over time. To represent the 3D morphology of a surface and/or to detect any changes over time,

Remote Sensing of Soil and Land Surface Processes
https://doi.org/10.1016/B978-0-443-15341-9.00025-3

high-resolution topographic models with bounded error are required (Shataee et al., 2011). Recent devices such as terrestrial light detection and ranging (tLiDAR) methods have appeared for more than a decade (Gulyaev and Buckeridge, 2004; Wawrzyniec et al., 2007; Hodge et al., 2009; Resop and Hession, 2010; O'Neal and Pizzuto, 2011; Colomina and Molina, 2014; Khan et al., 2018). In addition, rapidly developing unmanned aerial vehicle (UAV)-based photogrammetry facilities (Marteau et al., 2017; Alexiou et al., 2021) have been developed. Their high-precision digital terrain models are widely used in emphasizing change detection or even jungle canopy framework. The structure-from-motion (SfM) algorithm is a broad approach that can be used for such purposes. Drone terraces with cameras can be used for wide-scale photogrammetric studies, particularly in inaccessible locations; the SfM technique is an inactive surface restoration method. SfM photogrammetry through UAV imagery has proven to be a useful instrument in collection of topographic information, particularly in the case of geomorphological applications (Lucieer et al., 2014; Cucchiaro et al., 2018). Several investigators have revealed the connotation of SfM techniques for generating good-quality digital terrain models (DTMs), point clouds, and surface reconstructions from drone imagery (Niethammer et al., 2012; Eltner et al., 2013; Azizi et al., 2014). A large number of studies deal with SfM processes and compare DTMs and their potential to be used effectively for postflood research (Diakakis et al., 2019; Andreadakis et al., 2020), change detection (Tamminga et al., 2015; Cook 2017; Langhammer et al., 2017), or even sand identification (Mirijovský and Langhammer 2015; Woodget and Austrums, 2017). The SfM approach uses standard characteristics/objectives in consecutive images from various locations. The drone can be used in places that are not easily available, such as harsh and variable slides. Successive pneumatic examinations provide multifold DTMs that can be used for comparing and estimating measurable yield surface changes. For this reason, software that follows a multiview stereoscopic progress using image detail detection and detail matching is commonly applied (Lucieer et al., 2014). Point cloud datasets generated from UAVs have proven to provide accurate replications over time. Sequences of images can be relied upon, where most of the time, the scale invariant feature transform (SIFT) algorithm (Lowe, 2004) or one of its derivatives is used to match different image features (Gomez, 2022). The error is determined through the Random and Consensus Sampling (RANSAC) algorithm (Satkin et al., 2015). These two algorithms are among the main developing procedures for Structure from Motion and Multiview Stereophotogrammetry (SfM-MVS), which allows the automation of feature matching and the photogrammetry process (James and Robson, 2012). These results state that SfM-MVS produces comparable outcomes to UAV imagery. These high-resolution data could be used in applications such as modeling elevation changes due to rainfall-runoff processes. A fusion of unmanned aerial vehicle (UAV) image information and ground control data (GCPs) is particularly suitable for noninvasive (nondamage) field measurements.

16.2 Remote sensing for earth surface studies

The RS approach refers to a series of methods of analyzing and investigating the structure of objects and complications without direct contact with them. Recent progress in science and technology has allowed us to improve the quality of RS products over time. In regard to earth sciences applications, RS provides a great advantage because valuable information from

TABLE 16.1 Summary of the differences between RS and ground-based methods.

Remote sensing data	Ground-based data
High accuracy	Low accuracy
Low cost	High cost
Collecting data from places with no access (for example: Extreme topography)	Lack of access to extreme topography
Wide range of information	Low range of information
Remote sensing data is digital	Mostly are analog
Periodicity and repetition data	Frequent data collection
Good coverage of the region	Insufficient station data

objects can be gathered from a distance (Table 16.1). Electromagnetic waves are the key elements that are generally used. The reflection of the waves from different objects permits us to obtain further information about target objects. The structure of the object of interest varies according to its type, pattern, and texture. Reflected and emitted radiation is considered the key element for monitoring the physical characteristics of the Earth's surface that can be used for military, soil condition monitoring, agriculture, mining exploration, etc.

16.3 RS-based earth surface change detection for sediment and soil erosion modeling

16.3.1 LiDAR

Light detection and ranging (LiDAR) is a kind of laser imaging technology for generating high-resolution maps (Thompson and Croke, 2013; Craven and Wing, 2014). LiDAR is a remote sensing technology that measures distance by targeting an object with a laser and analyzing the reflected light (Craven and Wing, 2014). Since LiDAR functions very similarly to radar, it is sometimes called laser radar. The main difference between LiDAR and radar is the radiation wavelengths used. Radar uses wavelengths in the radio range, while LiDAR utilizes laser wavelengths.

The time difference between sending the electromagnetic pulse and receiving the reflected pulse is a common way to determine the distance from the current location to an object or surface. This method uses laser pulses such as radar technology or radio waves to determine the distance to the desired object and primarily creates a digital terrain model (DTM) (Saito et al., 2013). The LiDAR technique is mainly applied in airborne laser scanning (ALS), where the optical scanner is scaled on a small airplane.

ALS is progressively meant for purposes such as town modeling, generating digital terrain models, studying archeological features, or undertaking forest inventories (Jaakkola et al., 2017). The length of the airborne scanner from the scanned position on the ground is few hundred meters or kilometers (Liu et al., 2014). Complementing terrestrial laser scanning (TLS) with auxiliary photogrammetric acquisition techniques makes it possible to exploit two different measurement principles (Jaakkola et al., 2017; Gao et al., 2021). Distance-measured TLS can obtain large-scale point clouds at medium-range distances, while image-based surface reconstruction methods enable high accuracy at short distances. Advanced TLS systems have an integrated camera or external attachment option (Stöcker et al., 2015; Cook, 2017). It permits the collection of images automatically registered in the resulting 3D point cloud. However, the relative position stabilization between the two sensors in practice leads to a significant limitation.

16.3.2 Mobile laser scanning (MLS)

Mobile laser scanning (MLS) system denotes another active 3D data acquisition using one or few light scanners climbed on an ambulatory platform. The aim of MLS is to record three-dimensional information from the textures. To conduct this technique in a better manner, the following requirements must be taken into account: time-effective information gain in amplified mark areas, automatic record of three-dimensional information in a general coordinate system, high separability, and high correctness of the recorded information (Yu et al., 2015; Pires et al., 2022). Furthermore, airborne laser scanning, terrestrial laser scanning (TLS) and mobile laser scanning (MLS) can be applied (Yu et al., 2015). These methods provide higher point clouds with densities up to thousands of points per square meter (Pires et al., 2022). The TLS method is a well-developed method for 3D data collection. In general, an MLS generally includes minimum one laser scanner, supplying a two-dimensional path scanner, and inertial measurement units/global positioning system, that estimate the position/orientation of the movable platform in a projected coordinate system. The distinct global positioning system consists of a stationary base station and a so-called rover on the movable terrace. It also utilizes software aimed at absorbing the geometric section information (laser scan data) with the position/orientation data of the scanning terrace. A rigid and shock-absorbing mobile terrace and selective synchronized digital photo camera(s) were boosted on the identical platform (Niethammer et al., 2012; Westoby et al., 2012).

Using such technique can supply a huge rate of modern high-separability information within a concise term of time. This allows us to achieve more accurate modeling of the environment. Laser scanners have different types of applications including triangulation, time-of-flight (TOF), and step change. To measure the 3D morphology of a desired surface or to detect its changes over time and to avoid misinterpretation errors for surface change over time, accessibility of a high-resolution topographic model with bounded error is necessary (Shataee et al., 2011). Point cloud datasets generated from TLS have been proven to provide accurate repeats over time. Recent studies have emphasized the importance of having clearly defined anchor points and ground control points (GCPs), either from a global navigation satellite system (GNSS-RTK) or global navigation satellite system, postkinematic processing (GNSS-PPK), or from TLS or topographic total stations, to scale the SfM-MVS model and detect any smoothing of topographic details (Jaakkola et al., 2017; Balado et al., 2019).

16.3.3 Change detection using MLS

A laser scanner can investigate surface microtopography changes caused by erosion and sedimentation processes. This technique can reveal small details that remain hidden from the eyes of any expert or any experimental and physical model. In recent years, the measurement and monitoring of land erosion and deformation have dramatically benefited from the development of LiDAR technology.

TABLE 16.2 Summary of LiDAR productions.

Dataset	Resolution	Time scale	Period	References
UAV SfM-MVM and LiDAR	1 m	Annual	2012–14	Gomez (2022)
Satellite, UAV and video camera	3 m, 10 cm and 1 m	Weekly, daily, and minutes	2019	Angnuureng et al. (2022)
UAV	2 cm	Monthly	2018–19	Meinen and Robinson (2021)
UAV-SfM	2.5 cm	Monthly	2018–19	de Haas et al. (2020)
UAV and TLS	3 and 4 mm	Seasonal	2012–13	Eltner et al. (2013)
UAV	15 mm	Annual	2010–11	D'Oleire-Oltmanns et al. (2012)
LiDAR	1 m	Single record	2011	Azizi et al. (2014)
MLS	1 cm	–	2017–18	Kweon et al. (2020)
UAV	4 cm	Single record	2011	Lucieer et al. (2014)

The point cloud dataset generated from TLS has proven to provide accurate replications over time (Gao et al., 2021), and sequences of images (Yao et al., 2019) can be relied upon, where most of the time, the scale invariant feature transform (SIFT) algorithm (Shim, 2014). Additionally, one of its derivatives may be used to match the different features of the images (Gomez, 2022). On the one hand, the error might be determined by the RANSAC (Random Sampling and Consensus) algorithm (Satkin et al., 2015). These two algorithms have been the key to developing the Structure from Motion and Multiview Stereo photogrammetry (SfM-MVS) method, which allows the automation of feature matching (James and Robson, 2012). Various research results state that SfM-MVS produces comparable results to TLS (Terrestrial Laser Scanner) (James and Robson, 2012). A summary of laser scanning production is presented in Table 16.2.

16.4 Conclusion

With some differences in terms of coverage area and spectral, spatial and temporal resolution, satellite sensors, and drone-based technologies are among the standard remote sensing methods. Today, it has been proven that satellite and aerial remote sensing are robust and extensive tools in many applications of earth science. In recent years, drones have been considered the most widely used mapping tools in earth science. Recent attention to using drones can be due to the following reasons:

1. Flight efficiency: drones can fly in the ideal range at different altitudes (100 m–30,000 m) with a high grade of independence (1–48 h). Utilization of drones or tiny unmanned aircraft provides the possibility of a fast set of aerial images at low and medium heights from both vertical and diagonal corners.

2. Compatibility with all types of missions and
3. being cheap technology: UAVs with changeable scales and relatively low gravity can perform operations at low cost. This is the case when other normal flight devices have a great cost.

To identify changes over time, in applications such as map-based geomorphological studies, time series data are needed (daily, seasonal and yearly). Some high-accuracy uses of land mapping tools, such as Total Station and GPS, are expensive and time-consuming, cannot be used in all geomorphic landforms, or it is difficult to prepare data using them. Therefore, the use of drones provides better conditions for data preparation. In addition to two-dimensional images, using drones and the SFM4 technique, three-dimensional characteristics of complications can be obtained from the sequence of two-dimensional images. This photogrammetric technique provides a high-resolution topographic (less than 0.1 m), which is very important in geomorphological studies.

Numerical measurement of soil loss with high-resolution UAV and TLS knowledge can be performed in large subhectare plots. Applicable software packages such as PhotoScan and Pix4-UAV provide fast and automatic image editing. The digital earth models extracted from both systems agree with the local steady reference system. The combined use of digital land models derived from TLS and UAV can be beneficial to investigate soil erosion, and further revealing the change requires more detailed investigation. For example, weight point assignment can be taken as a function of impact tilt, surface variety, or roughness (Eltner et al., 2013).

The UAV system from other map preparation methods is an automatic image processing, which significantly increases the speed of map preparation (Kariminejad et al., 2019). The UAV has made it even easier to obtain details with greater accuracy (Colomina and Molina, 2014). As a result, a new insight can be seen under the research of land evolution. Collecting flexible information at a lower cost is one of the advantages of using these systems (Xu et al., 2019). Advances in computer vision and digital aerial mapping have intensified this issue. This has caused the very fast and easy production of surface digital models with great accuracy and spatial resolution power that has led to the sharp use of drones as a suitable tool for surveying and field monitoring (Cucchiaro et al., 2018). To date, several researchers have used a kind of surface digital model with centimeter accuracy and high spatial resolution (extracted from UAV images) for purposes such as estimating the volume, measuring the erosion rate, analyzing the morphology, and monitoring ditches in addition to photography systems (Marzolff et al., 2011; Peter et al., 2014; Hosseinalizadeh et al., 2019; Kariminejad et al., 2019). Therefore, the use of these methods focused on erosion (tunnel, headcut, gully, etc.) to develop and apply appropriate methods to monitor and better understand the creation of this type of erosion and examine their change at different temporal and spatial scales is considered essential (Castillo et al., 2014; Xu et al., 2019). Therefore, this type of research along with advanced tools expands our knowledge about change detection, erosion evaluation, and erosion computation.

References

Alexiou, S., Deligiannakis, G., Pallikarakis, A., et al., 2021. Comparing high accuracy t-lidar and uav-sfm derived point clouds for geomorphological change detection. ISPRS Int. J. Geo-Inf. 10, 367. https://doi.org/10.3390/ijgi10060367.

Andreadakis, E., Diakakis, M., Vassilakis, E., et al., 2020. Unmanned aerial systems-aided post-flood peak discharge estimation in ephemeral streams. Remote Sens. 12, 1–27. https://doi.org/10.3390/rs12244183.

Angnuureng, D.B., Brempong, K.E., Jayson-Quashigah, P.N., et al., 2022. Satellite, drone and video camera multi-platform monitoring of coastal erosion at an engineered pocket beach: a showcase for coastal management at Elmina Bay, Ghana (West Africa). Reg. Stud. Mar. Sci. 53, 102437. https://doi.org/10.1016/j.rsma.2022.102437.

Azizi, Z., Najafi, A., Sadeghian, S., 2014. Forest road detection using LiDAR data. J. Res. 25, 975–980. https://doi.org/10.1007/s11676-014-0544-0.

Balado, J., Martínez-Sánchez, J., Arias, P., Novo, A., 2019. Road environment semantic segmentation with deep learning from mls point cloud data. Sensors 19, 3466. https://doi.org/10.3390/s19163466.

Castillo, C., Taguas, E.V., Zarco-Tejada, P., et al., 2014. The normalized topographic method: an automated procedure for gully mapping using GIS. Earth Surf. Process. Landforms 39, 2002–2015. https://doi.org/10.1002/esp.3595.

Colomina, I., Molina, P., 2014. Unmanned aerial systems for photogrammetry and remote sensing: a review. ISPRS J. Photogrammetry Remote Sens. 92, 79–97. https://doi.org/10.1016/j.isprsjprs.2014.02.013.

Cook, K.L., 2017. An evaluation of the effectiveness of low-cost UAVs and structure from motion for geomorphic change detection. Geomorphology 278, 195–208. https://doi.org/10.1016/j.geomorph.2016.11.009.

Craven, M., Wing, M.G., 2014. Applying airborne LiDAR for forested road geomatics. Scand. J. For. Res. 29, 174–182. https://doi.org/10.1080/02827581.2014.881546.

Cucchiaro, S., Cavalli, M., Vericat, D., et al., 2018. Monitoring topographic changes through 4D-structure-from-motion photogrammetry: application to a debris-flow channel. Environ. Earth Sci. 77, 1–21. https://doi.org/10.1007/s12665-018-7817-4.

D'Oleire-Oltmanns, S., Marzolff, I., Peter, K.D., Ries, J.B., 2012. Unmanned aerial vehicle (UAV) for monitoring soil erosion in Morocco. Remote Sens. 4, 3390–3416. https://doi.org/10.3390/rs4113390.

de Haas, T., Nijland, W., McArdell, B., Kalthof, M., 2020. Short communication: optimizing UAV-SfM based topographic change detection with survey co-alignment. Earth Surf. Dyn. Discuss 1–17.

Diakakis, M., Andreadakis, E., Nikolopoulos, E.I., et al., 2019. An integrated approach of ground and aerial observations in flash flood disaster investigations. The case of the 2017 Mandra flash flood in Greece. Int. J. Disaster Risk Reduc. 33, 290–309. https://doi.org/10.1016/j.ijdrr.2018.10.015.

Eltner, A., Mulsow, C., Maas, H.-G., 2013. Quantitative measurement of soil erosion from Tls and uav data. Int. Arch. Photogramm. Remote Sens. Spat. Inf. Sci. XL-1/W2:119–124. https://doi.org/10.5194/isprsarchives-xl-1-w2-119-2013.

Gao, C., Li, P., Hu, J., et al., 2021. Development of gully erosion processes: a 3D investigation based on field scouring experiments and laser scanning. Remote Sens. Environ. 265, 112683. https://doi.org/10.1016/j.rse.2021.112683.

Gomez, C., 2022. Point-cloud technology for coastal and floodplain geomorphology. In: Point Cloud Technologies for Geomorphologists: From Data Acquisition to Processing. Springer, pp. 53–81.

Gulyaev, S.A., Buckeridge, J.S., 2004. Terrestrial methods for monitoring cliff erosion in an urban environment. J. Coast Res. 20, 871–878. https://doi.org/10.2112/1551-5036(2004)20[871:tmfmce]2.0.co;2.

Hodge, R., Brasington, J., Richards, K., 2009. Analysing laser-scanned digital terrain models of gravel bed surfaces: linking morphology to sediment transport processes and hydraulics. Sedimentology 56, 2024–2043. https://doi.org/10.1111/j.1365-3091.2009.01068.x.

Hosseinalizadeh, M., Kariminejad, N., Chen, W., et al., 2019. Gully headcut susceptibility modeling using functional trees, naïve Bayes tree, and random forest models. Geoderma 342, 1–11. https://doi.org/10.1016/j.geoderma.2019.01.050.

Jaakkola, A., Hyyppä, J., Yu, X., et al., 2017. Autonomous collection of forest field reference—the outlook and a first step with UAV laser scanning. Remote Sens. 9, 785. https://doi.org/10.3390/rs9080785.

James, M.R., Robson, S., 2012. Straightforward reconstruction of 3D surfaces and topography with a camera: accuracy and geoscience application. J. Geophys. Res. Earth Surf. 117. https://doi.org/10.1029/2011JF002289.

Kariminejad, N., Hosseinalizadeh, M., Pourghasemi, H.R., et al., 2019. Evaluation of factors affecting gully headcut location using summary statistics and the maximum entropy model: golestan Province, NE Iran. Sci. Total Environ. 677, 281–298. https://doi.org/10.1016/j.scitotenv.2019.04.306.

Keesstra, S., Mol, G., de Leeuw, J., et al., 2018. Soil-related sustainable development goals: four concepts to make land degradation neutrality and restoration work. Land 7, 133. https://doi.org/10.3390/land7040133.

Khan, J., Wang, P., Xie, Y., et al., 2018. Mapping MODIS LST NDVI imagery for drought monitoring in Punjab Pakistan. IEEE Access 6, 19898–19911.

Kweon, H., Seo, J.I., Lee, J.W., 2020. Assessing the applicability of mobile laser scanning for mapping forest roads in the Republic of Korea. Remote Sens. 12, 1502. https://doi.org/10.3390/RS12091502.

Langhammer, J., Lendzioch, T., Miřijovský, J., Hartvich, F., 2017. UAV-based optical granulometry as tool for detecting changes in structure of flood depositions. Remote Sens. 9, 240. https://doi.org/10.3390/rs9030240.

Liu, Y.-J., Wang, T.-W., Cai, C.-F., et al., 2014. Effects of vegetation on runoff generation, sediment yield and soil shear strength on road-side slopes under a simulation rainfall test in the three gorges reservoir area, China. Sci. Total Environ. 485, 93–102.

Lowe, D.G., 2004. Distinctive image features from scale-invariant keypoints. Int. J. Comput. Vis. 60, 91–110. https://doi.org/10.1023/B:VISI.0000029664.99615.94.

Lucieer, A., Turner, D., King, D.H., Robinson, S.A., 2014. Using an unmanned aerial vehicle (UAV) to capture microtopography of antarctic moss beds. Int. J. Appl. Earth Obs. Geoinf. 27, 53–62. https://doi.org/10.1016/j.jag.2013.05.011.

Marteau, B., Vericat, D., Gibbins, C., et al., 2017. Application of Structure-from-Motion photogrammetry to river restoration. Earth Surf. Process. Landforms 42, 503–515. https://doi.org/10.1002/esp.4086.

Marzolff, I., Ries, J.B., Poesen, J., 2011. Short-term versus medium-term monitoring for detecting gully-erosion variability in a Mediterranean environment. Earth Surf. Process. Landforms 36, 1604–1623. https://doi.org/10.1002/esp.2172.

Meinen, B.U., Robinson, D.T., 2021. Agricultural erosion modelling: evaluating USLE and WEPP field-scale erosion estimates using UAV time-series data. Environ. Model. Software 137, 104962. https://doi.org/10.1016/j.envsoft.2021.104962.

Mirijovský, J., Langhammer, J., 2015. Multitemporal monitoring of the morphodynamics of a mid-mountain stream using UAS photogrammetry. Remote Sens. 7, 8586–8609. https://doi.org/10.3390/rs70708586.

Niethammer, U., James, M.R., Rothmund, S., et al., 2012. UAV-based remote sensing of the Super-Sauze landslide: evaluation and results. Eng. Geol. 128, 2–11. https://doi.org/10.1016/j.enggeo.2011.03.012.

O'Neal, M.A., Pizzuto, J.E., 2011. The rates and spatial patterns of annual riverbank erosion revealed through terrestrial laser-scanner surveys of the South River, Virginia. Earth Surf. Process. Landforms 36, 695–701. https://doi.org/10.1002/esp.2098.

Peter, K.D., d'Oleire-Oltmanns, S., Ries, J.B., et al., 2014. Soil erosion in gully catchments affected by land-levelling measures in the Souss Basin, Morocco, analysed by rainfall simulation and UAV remote sensing data. Catena 113, 24–40. https://doi.org/10.1016/j.catena.2013.09.004.

Pires, R. de P., Olofsson, K., Persson, H.J., et al., 2022. Individual tree detection and estimation of stem attributes with mobile laser scanning along boreal forest roads. ISPRS J. Photogrammetry Remote Sens. 187, 211–224. https://doi.org/10.1016/j.isprsjprs.2022.03.004.

Resop, J.P., Hession, W.C., 2010. Terrestrial laser scanning for monitoring streambank retreat: comparison with traditional surveying techniques. J. Hydraul. Eng. 136, 794–798. https://doi.org/10.1061/(asce)hy.1943-7900.0000233.

Saito, M., Goshima, M., Aruga, K., et al., 2013. Study of automatic forest road design model considering shallow landslides with lidar data of funyu experimental forest | Istraživanje automatskog projektiranja šumskih cesta s obzirom na plitka klizišta pomoću lidar-a na pokusnom objektu funyu. Croat. J. For. Eng. 34, 1–16.

Satkin, S., Rashid, M., Lin, J., Hebert, M., 2015. 3DNN: 3D nearest neighbor: data-driven geometric scene understanding using 3D models. Int. J. Comput. Vis. 111, 69–97. https://doi.org/10.1007/s11263-014-0734-4.

Shataee, S., Weinaker, H., Babanejad, M., 2011. Plot-level forest volume estimation using airborne laser scanner and TM data, comparison of boosting and random forest tree regression algorithms. Procedia Environ. Sci. 7, 68–73. https://doi.org/10.1016/j.proenv.2011.07.013.

Shim, D., 2014. Remote sensing place: satellite images as visual spatial imaginaries. Geoforum 51, 152–160.

Stöcker, C., Eltner, A., Karrasch, P., 2015. Measuring gullies by synergetic application of UAV and close range photogrammetry - a case study from Andalusia, Spain. Catena 132, 1–11. https://doi.org/10.1016/j.catena.2015.04.004.

Tamminga, A.D., Eaton, B.C., Hugenholtz, C.H., 2015. UAS-based remote sensing of fluvial change following an extreme flood event. Earth Surf. Process. Landforms 40, 1464–1476. https://doi.org/10.1002/esp.3728.

Thompson, C., Croke, J., 2013. Geomorphic effects, flood power, and channel competence of a catastrophic flood in confined and unconfined reaches of the upper Lockyer valley, southeast. Geomorphology 197, 156–169. https://doi.org/10.1016/j.geomorph.2013.05.006.

Wawrzyniec, T.F., McFadden, L.D., Ellwein, A., et al., 2007. Chronotopographic analysis directly from point-cloud data: a method for detecting small, seasonal hillslope change, Black Mesa Escarpment, NE Arizona. Geosphere 3, 550–567. https://doi.org/10.1130/GES00110.1.

Westoby, M.J., Brasington, J., Glasser, N.F., et al., 2012. "Structure-from-Motion" photogrammetry: a low-cost, effective tool for geoscience applications. Geomorphology 179, 300–314. https://doi.org/10.1016/j.geomorph.2012.08.021.

Woodget, A.S., Austrums, R., 2017. Subaerial gravel size measurement using topographic data derived from a UAV-SfM approach. Earth Surf. Process. Landforms 42, 1434–1443. https://doi.org/10.1002/esp.4139.

Xu, Q., Kou, P., Wang, C., et al., 2019. Evaluation of gully head retreat and fill rates based on high-resolution satellite images in the loess region of China. Environ. Earth Sci. 78, 1–15. https://doi.org/10.1007/s12665-019-8483-x.

Yao, Y., Li, J., Duan, P., Cheng, F., 2019. Research on three-dimensional model reconstruction of slope erosion based on sequence images. Opt Laser. Technol. 110, 209–218. https://doi.org/10.1016/j.optlastec.2017.12.006.

YORULMAZ, G., GÜMÜS, S., 2022. Comparison of Wepp:road batch model and ABAG model results for estimating the amount of top soil erosion in forest road fill slopes. Eur J For Eng 8, 26–34. https://doi.org/10.33904/ejfe.1105553.

Yousefi, S., Moradi, H., Boll, J., Schönbrodt-Stitt, S., 2016. Effects of road construction on soil degradation and nutrient transport in Caspian Hyrcanian mixed forests. Geoderma 284, 103–112. https://doi.org/10.1016/j.geoderma.2016.09.002.

Yu, Y., Li, J., Guan, H., Wang, C., 2015. Automated extraction of urban road facilities using mobile laser scanning data. IEEE Trans. Intell. Transport. Syst. 16, 2167–2181. https://doi.org/10.1109/TITS.2015.2399492.

CHAPTER

17

Monitoring and detection of land subsidence

Shamsollah Asgari[1], *Kourosh Shirani*[2] *and Freidoon Soleimani*[1]

[1]Soil Conservation and Watershed Management Research Institute, Ilam Agricultural and Natural Resources Research and Education Center, Agricultural Research, Education and Extension Organization (AREEO), Ilam, Iran; [2]Soil Conservation and Watershed Management Research Institute, Agricultural Research, Education and Extension Organization (AREEO), Tehran, Iran

17.1 Introduction

Land subsidence is one of the main problems in geology worldwide that causes compression of underground layers. The cause of this phenomenon can be attributed to human activities (such as extraction of water from underground water) and natural activities (such as earthquakes) (Du et al., 2016; Ge et al., 2014, 2016).

This phenomenon is also one of the main hazards of regional geology and is considered to cause serious damage to buildings, infrastructure, roads, and bridges (Poland et al., 1982; Yang et al., 2015; Guang et al., 2016). According to the definition of the US Geological Survey, the phenomenon of ground subsidence, including collapse or downward settlement, is the surface of the Earth that can have a small displacement vector (Brambati et al., 2003).

Land subsidence is globally recognized as a disastrous geological phenomenon, especially in fast-growing urban regions. It destroys urban infrastructure and the environment, leads to the loss of resources, poses a serious threat to human lives and routine activities, and results in economic crises and constraints to progress (Pacheco et al., 2006).

Analyzing and monitoring subsidence in large cities allows meaningful insights for mitigating the probable loss of property and life due to increasing development rates in many regions around the globe (Khorrami et al., 2020). A range of variables that could produce urban subsidence includes human and natural reasons such as building loads (Chen et al., 2018; Yang et al., 2018).

Remote Sensing of Soil and Land Surface Processes
https://doi.org/10.1016/B978-0-443-15341-9.00021-6

Rapid urbanization and the construction of megacities play a key role in the occurrence of land subsidence. A huge increase in population demands excessive groundwater extraction, which leads to the abrupt sinking of the surface due to the decline in the groundwater level. It is a globally established reality that groundwater extraction leads to land subsidence in larger cities (Budhu and Adiyaman, 2010; Kanwal et al., 2018; Cian et al., 2019). Furthermore, very large-scale infrastructure development and excessive transportation pressure also contribute to the sinking of the surface, particularly on soft soils. Cities located on soft unstratified soils composed of silt, sand, and clay are more prone to subsidence, and these areas are mainly located onshores and riverbanks. Groundwater extraction causes large-scale subsidence, while the infrastructure load is the source of localized and small-scale subsidence (Zeitoun and Wakshal, 2013). Subsidence is a continuous and widespread phenomenon, although its manifestation is usually not prominent enough to alert stakeholders before a hazard takes place; therefore, continuous monitoring is very important.

Conventionally, a large number of techniques and methods are available to detect subsidence at small and large area levels. These techniques include Global Navigation Satellite System (GNSS)-based monitoring, groundwater monitoring, photo geological analysis, repeated optical leveling, tape extensometers, and ground reconnaissance. These methods are suitable for evaluating land subsidence, but they are not suitable for evaluating subsidence in large areas because they require a lot of time and money. One of the methods for identifying and measuring land subsidence is the use of radar remote sensing. Interferometry Synthetic Aperture Radar (InSAR) can be used as a suitable alternative method. Investigating land subsidence in the delta Nile using 1-Sentinel and GPS satellite images. In this research, using the radar interferometry method, 144 studies used images from 2015 to 2019, and the results showed that, contrary to the results of previous studies, changes in the shape of areas that are mainly controlled by natural processes indicate local subsidence and that dominant human control is changing the shape of the Earth. They demanded that flood models related to the rise of the sea level in the Nile Delta be reviewed (Rateb and Abu Talib, 2020). Land subsidence of the Isfahan metropolis has been evaluated using the Sentinel1 image series and PS method in the period of 2014-2019. The results of this research showed that subsidence occurred between 5 and 100 years ago, with an increasing trend from south to north, northeast and east (Gourabi et al., 2020). Investigating the deformation of the ground surface in the French salt mine using 1-Sentinel images in the period from 2015 to 2017 and radar interferometry technique and comparing SAR images with data combined the groundwater level (Liora et al., 2021). Extreme subsidence in a populated city is (Mashhad) detected by PSInSAR considering groundwater withdrawal and geotechnical properties (Khorrami et al., 2020). Therefore, in this research, land subsidence has been described using remote sensing and InSAR.

17.2 Review and definitions

17.2.1 Definition of land subsidence

Subsidence is a general term for downward vertical movement of the Earth's surface, which can be caused by both natural processes and human activities. Subsidence induced by groundwater withdrawals has become a significant problem and considered as anthropogenic subsidence (Brambati et al., 2003).

According to UNESCO's definition, subsidence means collapse or subsidence of the surface of the Earth that occurs for different reasons on a large scale. Usually, this term refers to the vertical downward movement of the surface (Allaby et al. 2013; Chen et al., 2018).

17.2.2 Processes that lead to subsidence

These processes include the release of lava from the hard rock substrate, the discharge of underground water, the dissolution of carbonate sinks underground, the release of oil and gas and mines, tectonic forces, and folding of the Earth's crust, which are known as natural and human processes (Jackson et al., 1997).

17.2.3 Dissolution of limestone

Limestone and gypsum that are inside the ground dissolve if they are exposed to water and form caves, and the surface of the earth sinks in these areas. This subsidence creates sinkholes that may be hundreds of meters deep (Waltham et al., 2005).

17.2.4 Mining

Several types of sub-surface mining, specifically methods that intentionally cause the extracted void to collapse (such as pillar extraction, long wall mining, and any metal leprous mining method that uses "caving" such as "block caving" or "sublevel caving") will result in surface subsidence. Mining-induced subsidence is relatively predictable in magnitude, manifestation, and extent, except where a sudden pillar or near-surface tunnel collapse occurs (usually very old workings (Herrera et al., 2007).

17.2.5 Extraction of petroleum and natural gas

Extraction from oil and gas reservoirs will result in a drop in the pore pressure of these reservoirs, and as a result, the amount of effective pressure due to the overburden weight of the reservoir rock (which, before extraction, will be some of it by the pressure of the fluid inside the reservoir and some of it by the rock the tank is tolerated) increases and causes its condensation. If the reservoir rock density exceeds a certain limit, the overburden rocks will start to settle due to their own weight. This subsidence will be proportional to the tank pressure drop (Maurice and Jackson, 2014).

17.2.6 Groundwater-related subsidence

Excessive use of underground water in different parts of the world has caused land subsidence (Chitsazan et al., 2022). Extraction of underground water and drop of underground water causes land subsidence, which creates unconsolidated sediments in the aquifer (Bhattarai et al., 2017).

In China, due to excessive use of underground water, it caused subsidence in some cities of this country, such as Beijing in the 1950s and Shanghai (Bai et al., 2022). The subsidence was about 10 cm per year in the city of Beijing, which reached its peak in the city of Shanghai in 1965 (Gao et al., 2018). For example, in Mexico City, groundwater has been extracted too much, and now parts of the city are sinking 30 cm a year. Now the city is caught in a vicious cycle. (Cigna et al., 2021 the city of San Joaquin, California, America, whose main industry is agriculture, faced a 60-centimeter land subsidence per year, which is the fastest subsidence in the world (Miller et al., 2020), Land subsidence in Vietnam (Braun et al., 2020), and Gorgan in Iran (Rezaei et al., 2019).

17.2.7 How to use remote sensing in detecting land subsidence

Interferometry Synthetic Aperture Radar (InSAR) is one of the most accurate and economical methods possible for the detection and diagnosis of height differences created in providing the surface of the earth in the shortest time and for large areas (Zare et al., 2016; Aslan et al., 2018; Hooper et al., 2012). Previous studies have used multitemporal InSAR in different fields, including subsidence resulting from harvesting and using a large water underground (Béjar-Pizarro et al., 2017; Bell et al., 2008; Caló et al., 2017; Chaussard et al., 2017; Chen et al., 2016), changing the shape of urban surfaces (Aslan et al., 2018; Bonì et al., 2018; Erten et al., 2019; Hooper et al., 2012), subsidence resulting from mining, spatial modeling of land subsidence in the south of the watershed (Abdikan et al., 2014; Carlà et al., 2018; Czikhardt et al., 2017), landslides (Bayer et al., 2018; Béjar-Pizarro et al., 2017; Imamoglu et al., 2018), pits and pits (Caló et al., 2017; Kim et al., 2016; Massonnet et al., 1998), earthquakes (Hussain et al., 2018; Liu et al., 2018), determination of snow height (Zare et al., 2016) and the movements of salt domes (Mehrabi, 2017). In all these fields, this method has proven its ability, and the results and the obtained accuracy are measured.

Over the past 2 decades, InSAR techniques have revolutionized the study of land subsidence induced by a wide variety of subsurface processes, including natural phenomena, such as faulting, folding, volcanism and dissolution in karst environments, or anthropogenic processes such as fluid extraction, mining activities, and other processes involving the removal of a supporting volume from beneath an area.

17.3 Material and methods

The study area, with an area of 97.82 $km^{2,}$ is located at 51°43′–52°3′E and 33°9′–32°19′N on the Mahyar Plain, with an approximate area of 558.2 km^2. The lowest and highest elevations of the study area are 1501 and 2343 m, respectively. This plain is limited to Isfahan city in the north and the east, to Shahreza city in the south and Mobarakeh town in the west (Fig. 17.1).

17.3.1 Research methodology

The employed data in this research are composed of data from radar satellite sensors, accessory data such as optical remote sensing images to extract land use, a 1:50,000

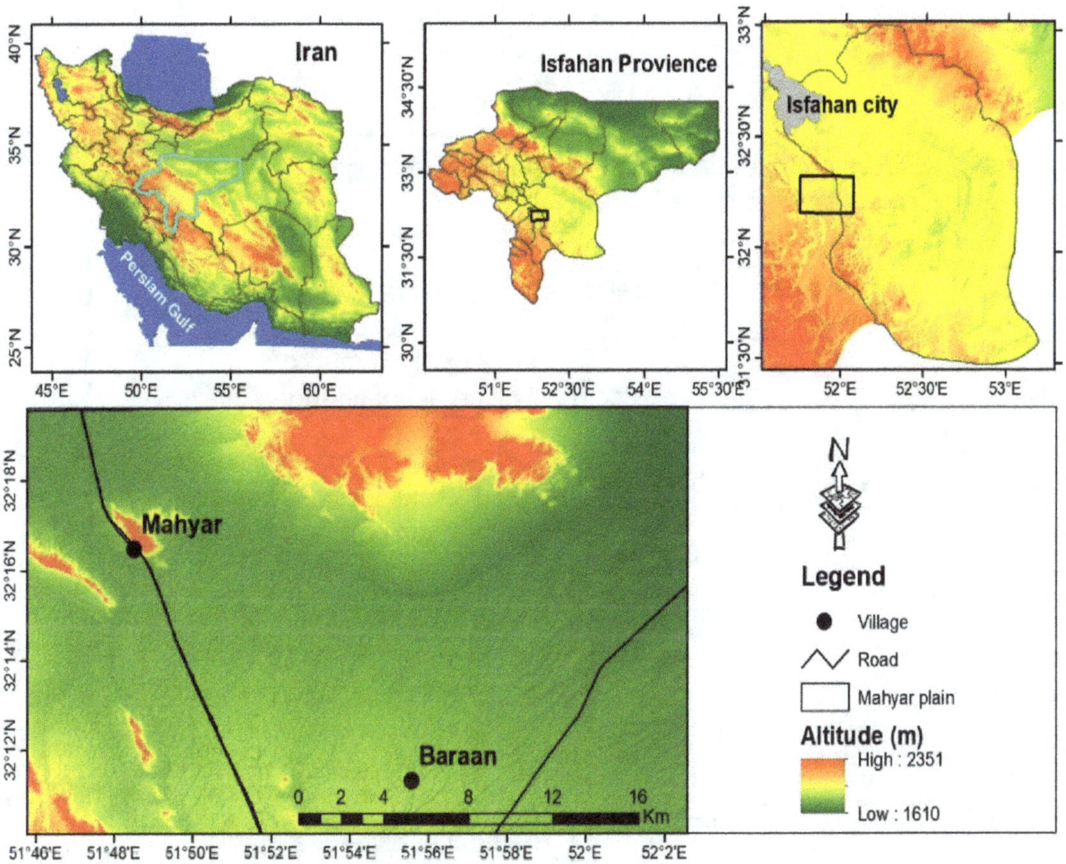

FIGURE 17.1 Map of the study area.

topographic map, 1:100,000 geological maps, data from observation wells, and data from field surveying with GPS. The research is of laboratory-surveying type. The laboratory method was initially performed to process images, and then surveying data were gathered according to the results. Eventually, the laboratory results were verified by means of the surveying data. In the processing stage of images, the time series of radar data acquired by the C-band ASAR sensor of the ENVISAT satellite were employed. The regions subjected to land subsidence and the amount of deformation were determined through satellite radar interferometry (SAR) by means of SARSCAPE software on the ENVI platform. In the second stage, features resulting from land subsidence in the study area were detected and recorded through field surveying in Fig. 17.2.

17.3.2 Radar interferometry

The InSAR method makes it possible to generate digital elevation models (DEMs) with an optimum altitudinal resolution for C-band data with a 5.6 cm wavelength of approximately

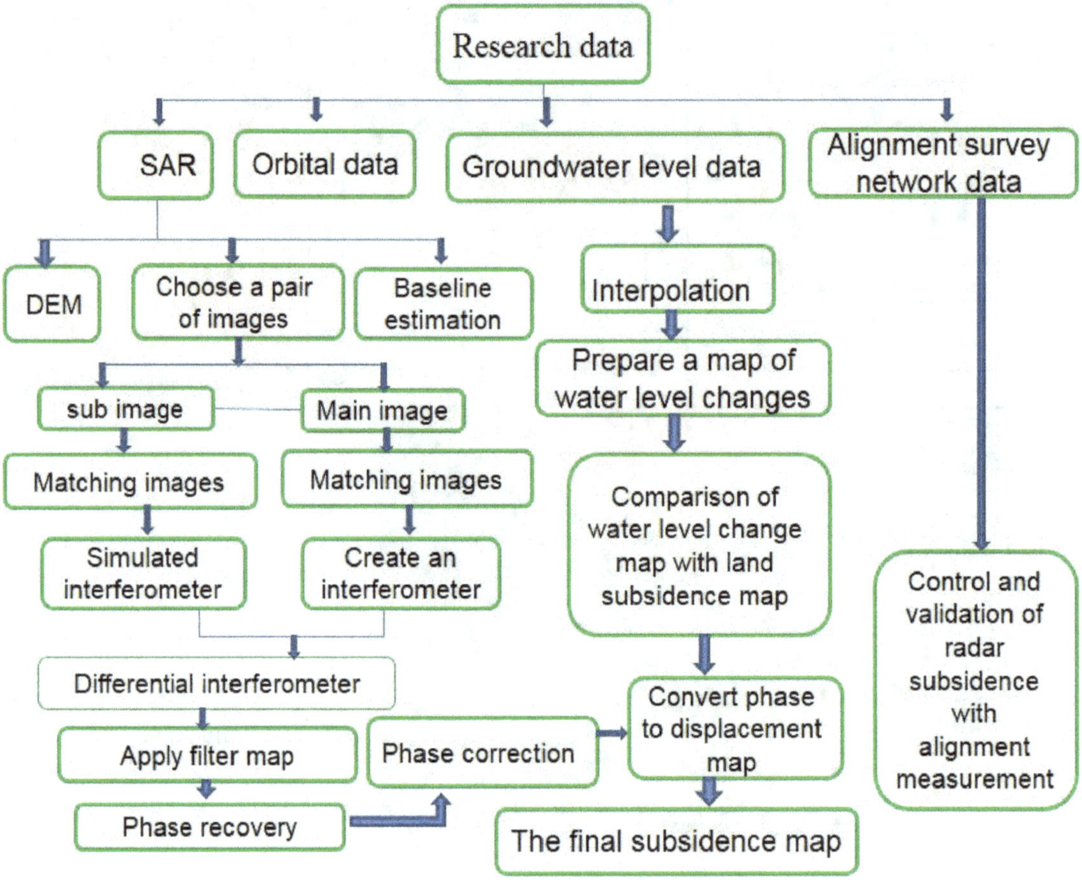

FIGURE 17.2 Conceptual model of the research stages.

5 m (Sharifikia, 2009). This methodology was developed by Pacheco et al. (2006) for the first time. This method can detect ground deformations by means of at least 3 (two images + DEM) or more radar images in different intervals and millimeter resolution. The fundamental principles of ground surface deformation measurement by SAR are illustrated in Fig. 17.3. In the figure, P is the specified indicator on the surface projected. The amount of subsidence is the distance (D_t) between P and P1 that occurred during a specified time. For measurement of this value, the sensor measures phase (φ_s) for the second image (slave) acquired in time t and a geometry completely such as the first image. Differential Interferometry Synthetic Aperture Radar (DInSAR) method shows the distance between φs and φm in the form of phase interferometer ($\Delta\varphi_{\text{int}}$). SP-MP is the distance between two images, which is shown in formula 17.1.

$$\Delta\Phi_{\text{int}} = 4\pi\frac{\text{SP} - \text{MP}}{\lambda} \tag{17.1}$$

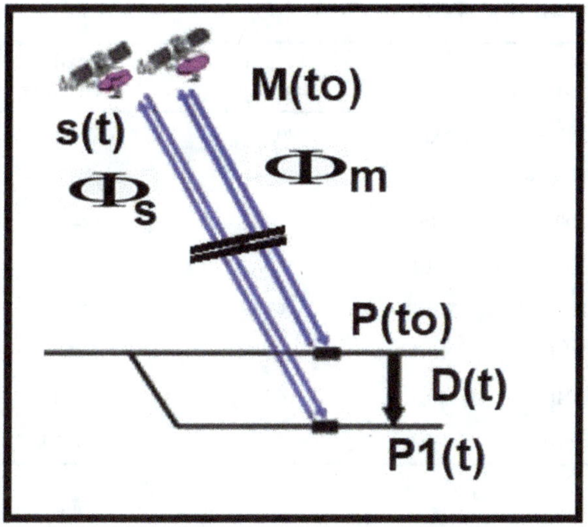

FIGURE 17.3 The Differential Interferometry Synthetic Aperture Radar (DInSAR) method for the calculation of land subsidence.

For measurement of subsidence, the surface is imagined as unstable as the surface declines from P to P1. The determination of subsidence in the time interval (Dt) between two images is a function of the phase difference between two images ($\Delta\Phi_{int}$) associated with the phase resulting from topography (Φ_{Top}), the phase resulting from surface displacement (Φ_{Mov}), and the phase resulting from the atmosphere (Φ_{Atm}) (Zebker et al., 1994).

$$\Delta\Phi_{int} = 4\pi\frac{PS - MP}{\lambda} = \Phi_{Top} + \Phi_{Mov} + \Phi_{Atm} \tag{17.2}$$

In this method, if a third image is not available, a synthetic interferogram is generated by means of DEM and conversion of elevation to phase. In this way, the effect of the phase resulting from topography is calculated by employing the inverse of DEM information. Eventually ignoring the atmospheric effect on high displacements (few centimeters) or its removal by the help of optical images, the phase difference between two interferograms only indicates surface displacement (land subsidence).

17.4 Results

Repetitive and nonrepetitive two-pass DInSAR is the method employed in this research to determine land subsidence. The suitable selection of radar pair images is the most fundamental stage in SAR processing, in which factors such as the frequency of sensors, spatial baseline, temporal baseline, and spatial overlap in the direction of sensor movement are effective in the selection of pair images. In this research, 8 ASAR pair images were used. The

TABLE 17.1 Selected ASAR pair images.

Abbreviated name	Date (year-month-day)	Temporal baseline (day)	Spatial baseline (m)	Threshold baseline (m)
A_1	24/08/2004–07/11/2006	803	79.5	2160.2
A_2	20/06/2006–09/06/2009	1071	148.6	2155.8
A_3	20/02/2007–09/06/2009	839	124.6	2159.8
A_4	27/03/2007–20/04/2010	1118	54.05	2158.3
A_5	14/08/2007–29/06/2010	1045	41.20	2161.7
A_6	31/03/2009–29/06/2010	454	117.5	2157.9
A_7	20/04/2010–29/06/2010	69	221.03	2158.8
A_8	20/04/2010–07/09/2010	137	309.13	2158.8

selected pair images were converted to SLC format in the SARScape software, and calculations related to the perpendicular baseline and the other mentioned factors effective in the selection of pair images were performed (Table 17.1).

After conformity between two images, the phase resulting from two SLC images was subtracted, and an interferogram was generated (Fig. 17.4).

SLC is composed of a complex value including amplitude and phase. This phase is not absolutely specified, but it is measured based on the 2π value. The phase difference received from the reflections resulting from two close orbits changes slowly. The phase difference calculated for every pixel is called the interferogram. The changes pertaining to the distance

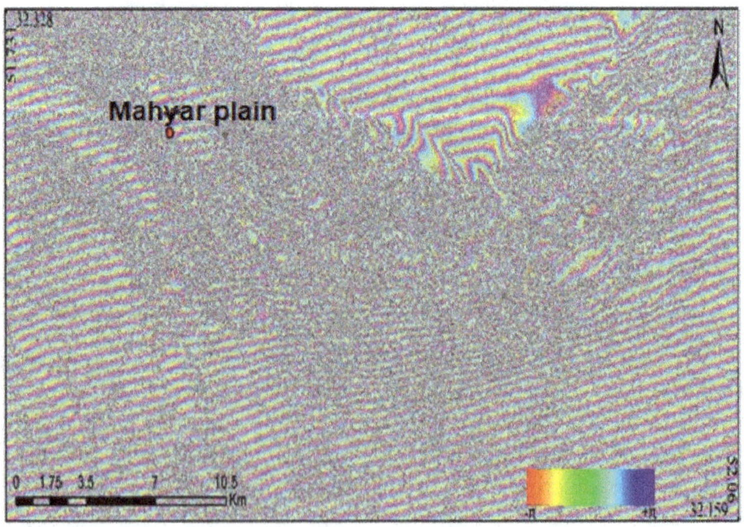

FIGURE 17.4 The interferogram resulting from 2004/08/24–2006/11/07 pair images.

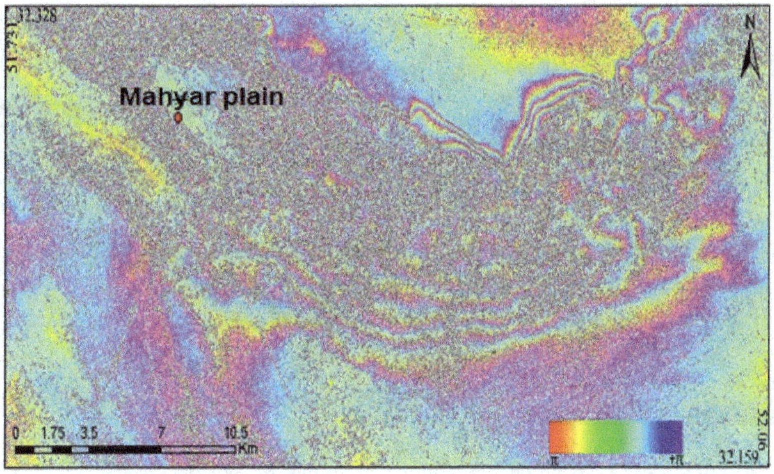

FIGURE 17.5 Flattened interferogram resulting from 2004/08/24–2006/11/07 pair images.

of the feature to the radar sensor exist in the interferogram. These changes are related to the satellite orbital error ($\Delta\varphi\Gamma$), the topography effect ($\Delta\varphi T$), and the feature displacement ($\Delta\varphi D$). All other effects should be removed to extract changes in surface displacement ($\Delta\varphi\Delta$).

$$\varphi - \varphi = \Delta\varphi\Gamma + \Delta\varphi T + \Delta\varphi A\tau\mu + \Delta\varphi\Delta + \Delta\varphi\nu o\iota\sigma \tag{17.3}$$

The simplest method for removal of the topography effect is minimizing the perpendicular baseline of the selected pair images. In this way, most of the topography effect is removed from the phase difference. To more equalize the topography effect, the DEM acquired by shuttle (SRTM) was used (Fig. 17.5).

The DInSAR resulting from the removal of topographic effects includes noise created by changes in the environmental conditions of the region, and conform filters can be employed to omit the noise and subsequently recover the interferogram. The Goldstein filter was used in this research to recover the interferogram (Fig. 17.6). This filter acts as conform and dynamic, and it performs filtering as local. This filter is modified by the help of the bandwidth variable obtained directly from the correlation of the spectral resolution of the differential interferogram. In this way, it operates in regions with high correlation and low bandwidth or limited pixels and in regions with low correlation and high bandwidth and more pixels. According to the lower coherency of eight employed pair images than the critical limit (0.5) (Table 17.2), all the images were used to determine the amount of subsidence.

One of the most completed and important stages in the processing of DInSAR is the phase correction stage. As an interferogram can only change as 2π in a phase radiance phase, its changes may be more than the mentioned value, and the correction process can cause computation of real values of the interferogram phase, and the ambiguity of 2π can be corrected in this way. For this purpose, various and complicated algorithms have been developed that have advantages according to the type of data and the study area. The general methods employed for this purpose include the algorithms called regional growth and flow with

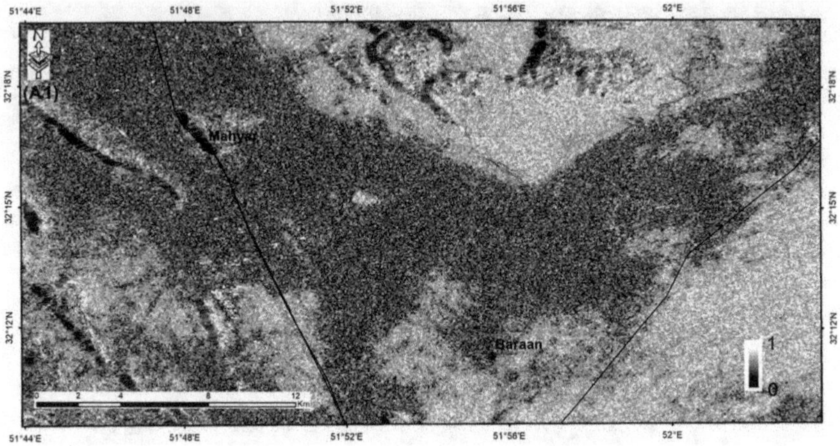

FIGURE 17.6 Coherency image for pair images of 2004/08/24–2006/11/07.

TABLE 17.2 Correlation values of pair radar images in coherent images.

Pair images	Temporal baseline (day)	The minimum value	The average	The maximum value	Standard deviation
A_1	803	0.00	0.33	0.99	0.16
A_2	1071	0.00	0.26	0.98	0.13
A_3	839	0.00	0.20	0.98	0.11
A_4	1118	0.00	0.20	0.99	0.13
A_5	1045	0.00	0.30	0.99	0.15
A_6	454	0.00	0.23	0.98	0.13
A_7	69	0.00	0.43	0.98	0.16
A_8	137	0.00	0.39	0.98	0.16

the lowest cost. The first algorithm is more adaptable to regions with relatively high correlation in coherent images. In this research, according to the correlation values in the resulting interferogram, the first method was employed for the interferogram resulting from the ASAR sensors (Fig. 17.7). The threshold values of this algorithm were also calculated and modified according to the average correlation in the coherent image.

To correctly convert the modified phase to elevation values and calculate the displacement amount of the ground surface, the monitoring stage of reflating is needed. Performing this stage leads to removal of the other probable disturbing phases, such as the remaining phases of topographic effects and the remaining phases of orbital displacements that occurred in the

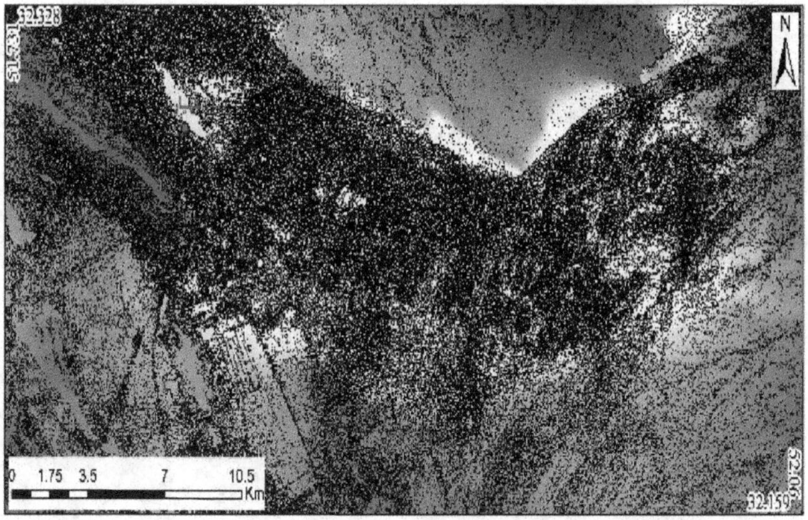

FIGURE 17.7 The image resulting from fuzzy modification of the pair interferogram of 2004/08/24–2006/11/0.

ground during the acquisition interval of the two radar images. This operation is necessary for interferograms showing clues of platform shake in ramp form. The logical phase is achieved by the phase difference resulting in the phase correction stage caused by topography and orbital errors. This logical phase is related to displacements occurring in the ground during acquisition of the two radar images. The mentioned process has been independently implemented on every pair of interferograms resulting from the ASAR sensor. After purification and double correction of the phase, the resulting absolute phase can be converted to the displacement map and georeferenced. This process can be performed by geodetic conversion and by the help of the Doppler amplitude equation. The methodology is similar to the conventional methods of converting image systems to international reference systems. The only difference is the simultaneous implementation of the equation of Doppler amplitude on both the receiver and creator of the SAR image in addition to the geodetic conversions. In this way, the displacement and subsequently the Cartesian coordinates of every pixel will be obtained. The output map indicates the intensity of displacement in every pixel in the metric unit. In the obtained maps, the positive values indicate displacement pertaining to ground surface movement toward the line of sight (uplifting), and the negative values indicate getting away of ground surface from the sensor in the line of sight (land subsidence).

The results of the displacement of the pair images and the maximum displacement in every image are presented in Fig. 17.8 and Table 17.3, respectively.

Every interferogram individually contains surface variations during a time interval. Surface variations over time can be evaluated by applying a few interferograms associated with time series analysis. Time series analysis by means of SAR is a suitable method for estimating ground surface variations with high spatial resolution. The displacement of the ground surface for each point and every data point can be calculated with access to a significant number of interferograms and using the least square method. In this study, eight interferograms of the ASAR images were employed in the time series analysis. The amount of

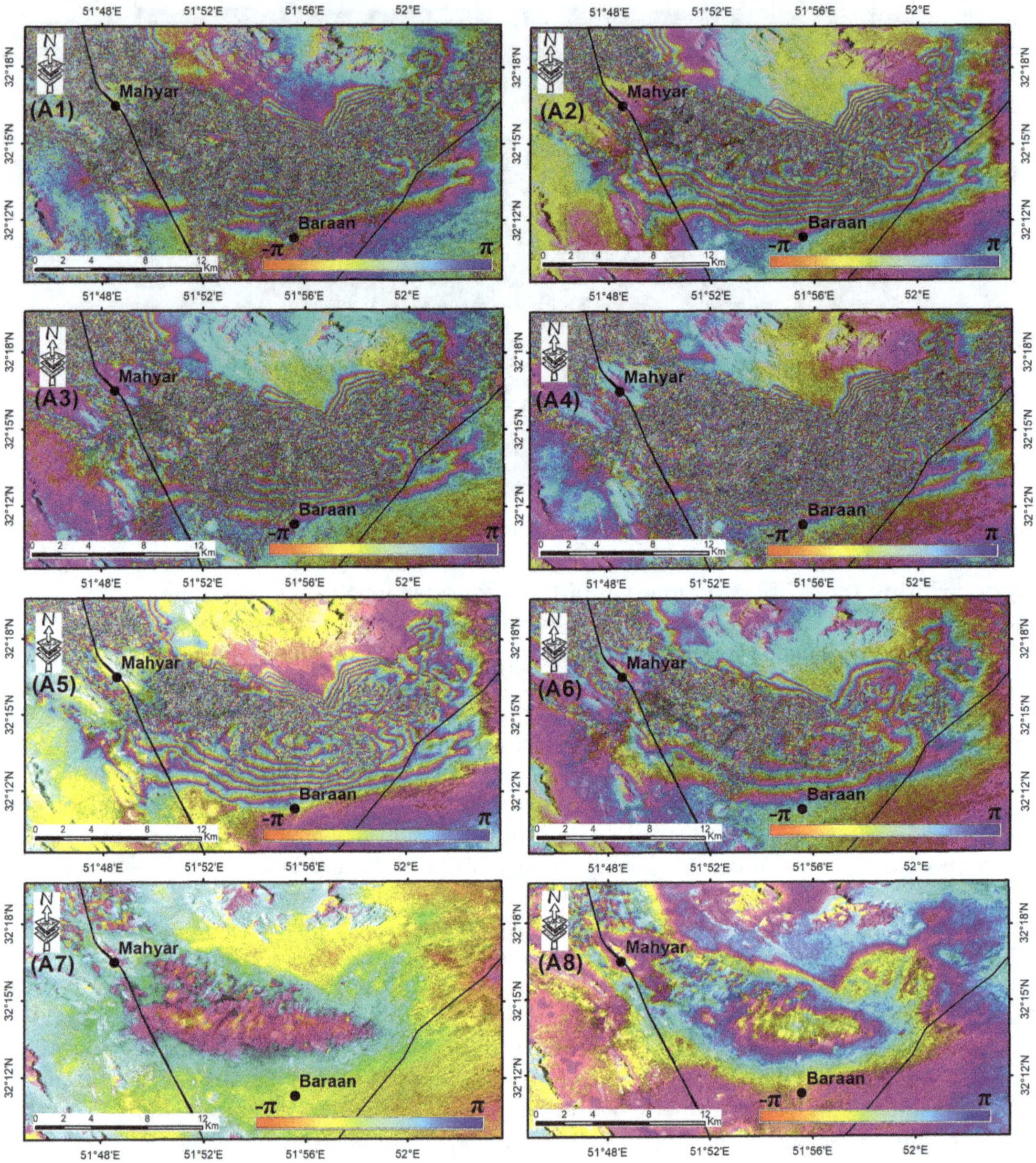

FIGURE 17.8 Maps of land subsidence resulting from processing of the paired images.

TABLE 17.3 The maximum land subsidence values extracted from paired images of the ASAR sensor.

The interferogram	The time interval (day)	The maximum land subsidence (cm)
A_1	803	−13.90
A_2	1071	−12.90
A_3	839	−8.60
A_4	1118	−10.40
A_5	1045	−14.06
A_6	454	12.80
A_7	69	−1.30
A_8	137	−6.10

TABLE 17.4 Land subsidence classes in the study area.

Percentage	Area (hectare)	Land subsidence classes
48.51	28,232.75	Low
29.73	17,300.53	Medium
16.73	9735.4	High
5.02	2922.28	Very high

land subsidence is indicated for every interferogram in Table 17.4. The final map of the average rate of land subsidence resulting from time series analysis of the total area exposed to land subsidence has been illustrated as classified into four classes in Fig. 17.9. The maximum land subsidence rate in the area has been estimated to be approximately 6.4 cm/year.

17.4.1 Verification

In this research, the rate and amplitude of land subsidence extracted by the DInSAR technique were verified by surveying and observation data. For this purpose, the research findings were controlled and verified by employing lithology, land use, and field surveying. In the first step, a surveying operation was performed to validate the findings of the employed technique and gather spatial documents from the occurrence of land subsidence hazards associated with surveying outcrops and simultaneously detecting and observing the reasons for land subsidence. The scattering of visited points is illustrated in Fig. 17.10. The field research indicates few morphological outcrops caused by land subsidence in the study area.

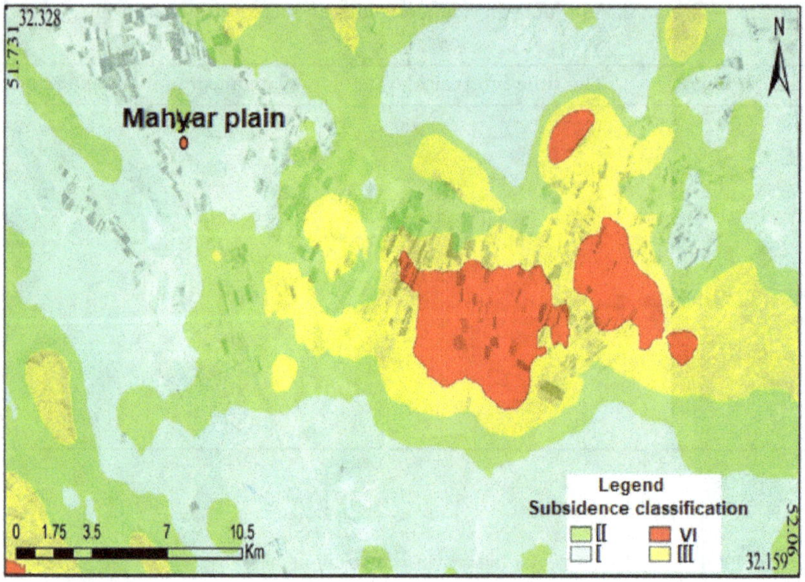

FIGURE 17.9 Classification of land subsidence in the study area.

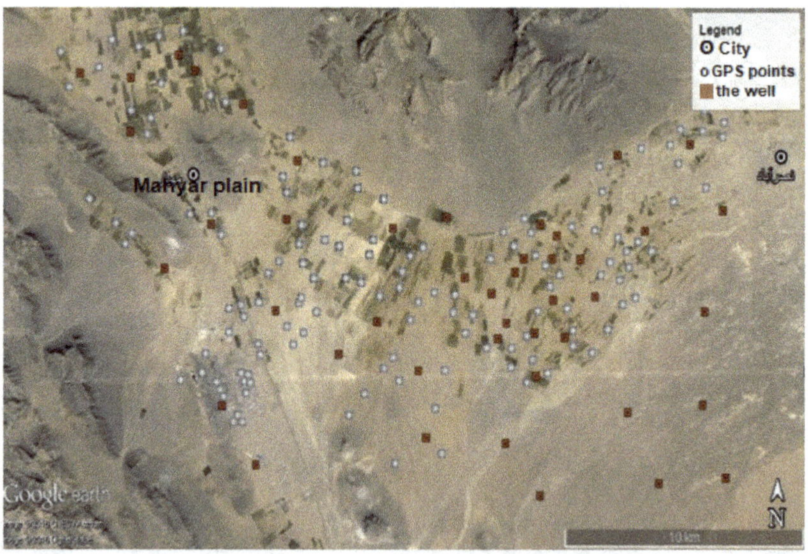

FIGURE 17.10 The scattering of surveying points associated with piezometric wells in the study area.

The results of land use (Fig. 17.11) indicate that the maximum land subsidence is related to agricultural land use and water irrigation in the study area. This problem can mainly be attributed to groundwater extraction for agriculture/drinking and probably other influencing factors, such as geology (Fig. 17.12) and tectonics. The concentration of the maximum rate of

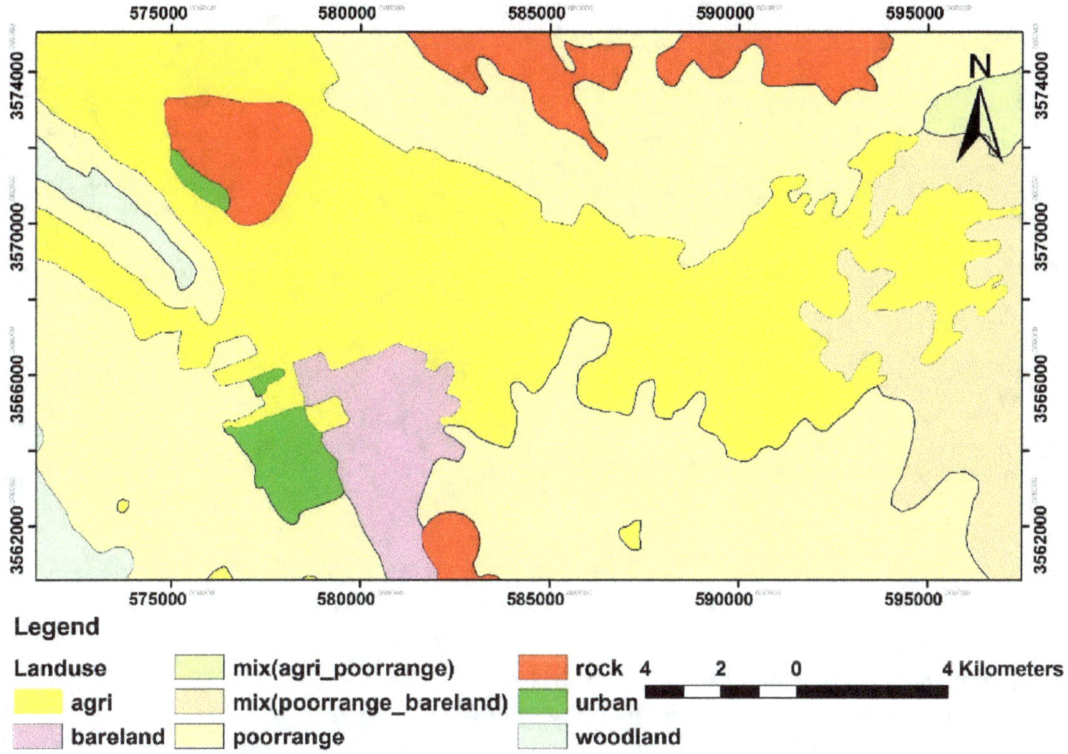

FIGURE 17.11 The land use map.

land subsidence in residential areas has created relatively unsafe environmental conditions for residents and has increased the risk of geohazards (Figs. 17.13–17.16). Analysis of groundwater fluctuations plays an important role in studies of land subsidence controlling factors. The fluctuations in groundwater level observed from 2005 to 2009 in 40 observation wells located in the Mahyar Plain were employed to verify the results. The groundwater level dropped 2 m (from 1519 to 1517 m) from 2005 to 2009. Therefore, the annual average drawdown of groundwater is 33.33 cm/year in a 6-year period.

17.5 Discussion

This research focuses on the analysis of the land subsidence problem in the Mahyar Plain by means of the DInSAR method to estimate the rate and amplitude of this phenomenon. For this purpose, time series of 12 ascending pair images of the C-band AAR sensor were used. Forty-eight interferograms were prepared by the data acquired from 2004 to 2010. Eight pair images were selected out of 12 pair images according to the temporal and spatial baselines and the coherency value, and they were employed to prepare the

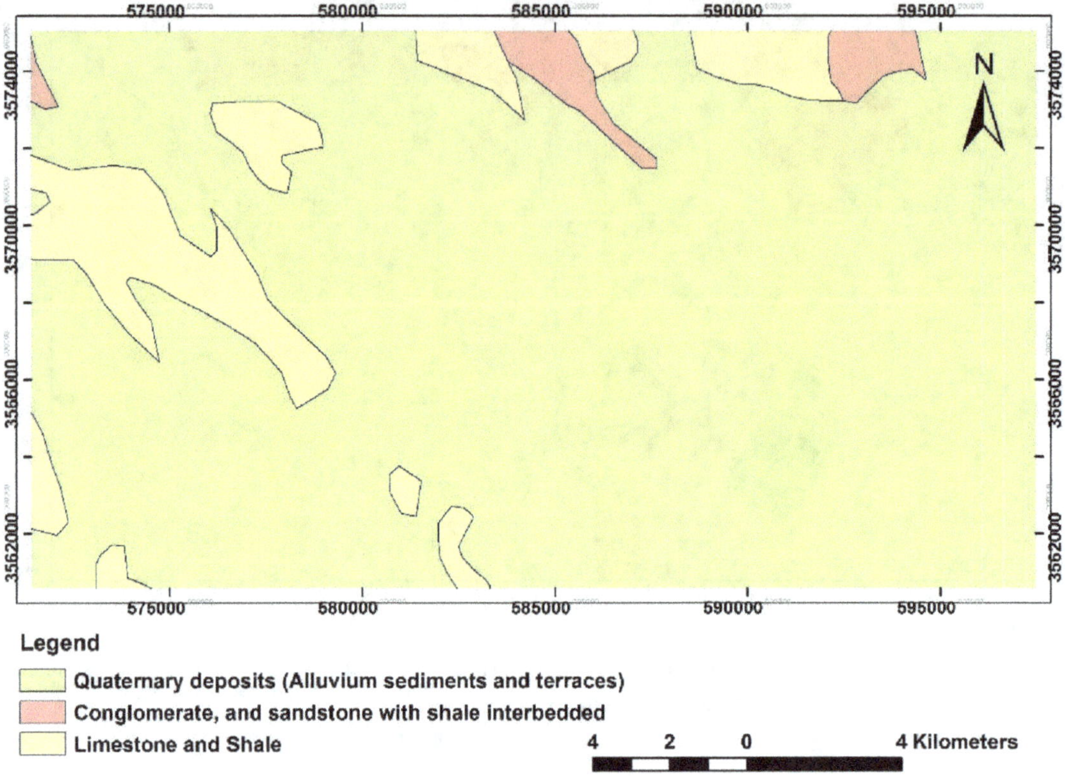

FIGURE 17.12 The lithological map.

FIGURE 17.13 Samples of the observed features in the study area.

land subsidence of the study area by means of the least squares method in the MATLAB environment. The results indicate that the maximum rate of land subsidence in the study area is approximately 6.4 cm/year during the 6-year time period. Salehi et al. (2007) also analyzed the land subsidence phenomenon in the Mahyar Plain in the 2003 to 2006 and 2004 to 2010 time intervals, and they estimated the annual maximum land subsidence of the Mahyar Plain to be approximately 8.2 and 7 cm, respectively. The ASAR data significantly outperform the PALSAR data in the detection of even low land subsidence during short intervals due to their shorter wavelength (56 mm) than the wavelength of the PALSAR data (23 cm). (Shirani et al., 2014) have compared the resolution of ASAR data with PALSAR data for the estimation of the vertical displacement of the ground surface. Their results are similar to those of the present research. To analyze the compatibility of the results of interferometry, other complementary data of field surveying, land use, geological maps, and piezometric wells were used. Adaptation of the digital geological map with the distribution pattern of the regions subjected to land subsidence confirms the possibility of land subsidence of the ground surfaces due to the underlying lithological structures. For this purpose, the effect of tectonic controlling factors on land subsidence was also analyzed. Adaptation of land subsidence maps with the land use layers also confirms the maximum occurrence of subsidence in water irrigation land use. Adaptation of the final map of the amount and extent of land subsidence in this plain with the focal points of residential areas revealed that these areas are located in this district of land subsidence. Therefore, it has threatened significantly for densely populated focal points. Field visits from the plain also confirm the existence of many morphological outcrops (such as cracks and fractures in buildings and structures) caused by land subsidence in the study area (Figs. 17.14–17.16). Analysis of the existing observation wells in the study area indicates that the groundwater level has dropped 2 m, which has been effective in land subsidence. Because of the persistence of the effective conditions on the land subsidence of this plain, continuation of remote sensing monitoring and installation of periodical measurement stations (GPS) can provide the possibility of monitoring the amount and extent of land subsidence and verifying the results.

17.6 Conclusion

Land subsidence phenomenon is one of the natural disasters, which natural factors and human factors reduce underground resources such as underground water, oil, etc. Several methods are used in the field of identifying and determining the amount of land subsidence. Studying and measuring the amount of land subsidence in the early years was often estimated based on local observations and their characteristics, which was done in a traditional way. This method is very expensive and time-consuming, and on the other hand, updating them requires re-reviewing the basic data. With the widespread use of GPS tools, the possibility of monitoring areas subject to land subsidence with the aim of determining its amount was considered. This method was accepted and confirmed in terms of accuracy, but due to issues such as the high cost of installing and establishing permanent stations, the difficulty of determining the scope and extent of subsidence, and the disorder in the monitoring period due to changes in budgets and credits. It is not a successful year. Based on this, the use of data and techniques that can have sufficient precision and accuracy while eliminating and reducing the mentioned problems has been given more attention. Remote sensing as a new science in the 19th century created a new era in earth science studies. Remote sensing technology, which is based on the use of earth observation satellites to collect data, caused a huge revolution in the field of data collection process and the production of various information and maps. Differential radar interferometric technique is the most important and best method of using radar data and remote sensing to evaluate the phenomenon of land subsidence. This method can estimate the ground subsidence phenomenon very accurately with less cost and less time and on a large scale. One of the advantages of D-InSAR is its ability to prepare a comprehensive map of deformation with ground justification and with a denser number of sampling points than detailed mapping. Although in the mentioned method, it is possible to image the amount of the resulting displacement along the line of sight of the satellite in all the north, south, east, west, and height directions, but since the sensitivity of the imaging sensors to the displacement in the vertical direction is much higher, the resulting images are suitable for checking subsidence. Of course, many research records confirm the use of radar interferometric technique and the use of remote sensing in evaluating and investigating the subsidence phenomenon of this method. In this chapter of the book, how to investigate the phenomenon of subsidence using remote sensing method and differential radar interferometric technique was explained, and we hope that this chapter and this book will be placed in the direction of human progress and perfection.

Appendix

Pictures of the subsidence of the Mahyar Plain in Isfahan Province, Iran (Figs. 17.14–17.16).

FIGURE 17.14 The early stage of a sinkhole.

FIGURE 17.15 Parallel cracks (fissures) at the junction of plain and mountain landform.

FIGURE 17.16 A long crack (fissure) in the middle part of Mahyar Plain in the south of Isfahan.

References

Abdikan, S., Arıkan, M., Sanli, F.B., Cakir, Z., 2014. Monitoring of coal mining subsidence in periurban area of Zonguldak city (NW Turkey) with persistent scattered interferometry using ALOSPALSAR. Environ. Earth Sci. 71 (9), 4081–4089.

Allaby, M., 2013. Subsidence. A Dictionary of Geology and Earth Sciences, fourth ed. Oxford University Press, Oxford, ISBN 9780199653065.

Aslan, G., Cakır, Z., Ergintav, S., Lasserre, C., Renard, F., 2018. Analysis of secular ground motions in istanbul from a long-term insar time-series (1992–2017). Remote Sens. 10 (3), 408–419.

Bai, Z., Wang, Y., Balz, T., 2022. Beijing, land subsidence revealed using PS-InSAR with long time series TerraSAR-X SAR data. Remote Sens. 14 (11), 2529.

Bayer, B., Simoni, A., Mulas, M., Corsini, A., Schmidt, D., 2018. Deformation responses of slow moving landslides to seasonal rainfall in the Northern Apennines, measured by InSAR. Geomorphology 308, 293–306.

Béjar-Pizarro, M., Notti, D., Mateos, R.M., Ezquerro, P., Centolanza, G., Herrera, G., Bru, G., Sanabria, M., Solari, L., Duro, J., 2017. Mapping vulnerable urban areas affected by slow-moving landslides using 32 Sentinel-1 InSAR data. Remote Sens. 9 (9), 876–889.

Bell, J.W., Amelung, F., Ferretti, A., Bianchi, M., Novali, F., 2008. Permanent caterer InSAR reveals seasonal and long-term aquifer-system response to groundwater pumping and artificial recharge. Water Resour. Res. 44 (2), 1–18.

Bhattarai, R., Alifu, H., Maitiniyazi, A., Kondoh, A., 2017. Detection of land subsidence in Kathmandu Valley, Nepal, using DInSAR technique. Land 6 (2), 39.

Bonì, R., Bosino, A., Meisina, C., Novellino, A., Bateson, L., McCormack, H., 2018. A methodology to detect and characterize uplift phenomena in urban areas using Sentinel-1 data. Remote Sens. 10 (4), 607–620.

Brambati, A., Carbognin, L., Quaia, T., Teatini, P., Tosi, L., 2003. The Lagoon of Venice: geological setting, evolution and land subsidence. Episodes 26 (3), 264–268.

Braun, A., Hochschild, V., Pham, G.T., Nguyen, L.H.K., Bachofer, F., 2020. Linking land subsidence to soil types within Hue city in Central Vietnam. J. Viet. Env. 12 (7), 1–6.

Budhu, M., Adiyaman, I.B., 2010. Mechanics of land subsidence due to groundwater pumping. Int. J. Numer. Anal. Methods Aeromech. 34 (14), 1459–1478.

Caló, F., Notti, D., Galve, J., Abdikan, S., Görüm, T., Pepe, A., BalikSanli, F., 2017. Dinsar-based detection of land subsidence and correlation with groundwater depletion in konya plain. Turk. Remote Sens. 9 (1), 83–95.

Carlà, T., Farina, P., Intrieri, E., Ketizmen, H., Casagli, N., 2018. Integration of ground-based radar and satellite InSAR data for the analysis of an unexpected slope failure in an open-pit mine. Eng. Geol. 23 (5), 39–52.

Chaussard, E., Milillo, P., Bürgmann, R., Perissin, D., Fielding, E.J., Baker, B., 2017. Remote sensing of ground deformation for monitoring groundwater management practices: application to the Santa Clara Valley during the 2012–2015 California drought. J. Geophys. Res. Solid Earth 122 (10), 8566–8582.

Chen, G., 2018. Detection of land subsidence associated with land creation and rapid urbanization in the Chinese loess plateau using time series insar: a case study of Lanzhou new district. Remote Sens. 10 (8), 270–289.

Chen, M., Tomás, R., Li, Z., Motagh, M., Li, T., Hu, L., Gong, H., Li, X., Yu, J., Gong, X., 2016. Imaging land subsidence induced by groundwater extraction in Beijing (China) using satellite radar interferometry. Remote Sens. 8 (6), 468–479.

Chitsazan, M., Rahmani, G., Ghafoury, H., 2022. Investigation of subsidence phenomenon and impact of groundwater level drop on alluvial aquifer, case study: Damaneh-Daran plain in west of Isfahan province, Iran Model. Adv. Appl. Geol. 12 (2), 259–275.

Cian, F., Blasco, J., Carrera, L., 2019. Sentinel-1 for monitoring land subsidence of coastal cities in Africa using PSInSAR: a methodology based on the integration of SNAP and Stamps. Geosciences 9 (3), 124. https://doi.org/10.3390/geosciences9030124.

Cigna, F., Tapete, D., 2021. Present-day land subsidence rates, surface faulting hazard and risk in Mexico City with 2014–2020 Sentinel-1 IW InSAR. Remote Sens. 253, 112–161.

Czikhardt, R., Papco, J., Bakon, M., Liscak, P., Ondrejka, P., Zlocha, M., 2017. Ground stability monitoring of undermined and landslide prone areas by means of sentinel-1 multitemporal InSAR, case study from Slovakia. Geosciences 7 (3), 87–101.

Du, Z., Ge, L., Li, X., Ng, A., 2016. Subsidence monitoring over the Southern Coalfield, Australia using both L-Band and C-Band SAR time series analysis. Remote Sens. 8 (7), 543–556.

Erten, E., Rossi, C., 2019. The worsening impacts of land reclamation assessed with Sentinel-1: the Rize (Turkey) test case. Int. J. Appl. Earth Obs. Geoinf. 74, 57–64.

Gao, M., Gong, H., Chen, B., Li, X., Zhou, C., Shi, M., Yuan, S., Zheng, C., Duan, G., 2018. Regional land subsidence analysis in eastern Beijing plain by insar time series and wavelet transforms. Remote Sens. 10, 365.

Ge, L., Ng, A.H.M., Li, X., Abidin, H.Z., Gumilar, I., 2014. Land subsidence characteristics of Bandung Basin as revealed by ENVISAT ASAR and ALOS PALSAR interferometry. Remote Sens. Environ. 154, 46–60.

Goorabi, A., Karimi, M., Yamani, M., Perissin, D., 2020. Land subsidence in Isfahan metropolitan and its relationship with geological and geomorphological settings revealed by Sentinel-1A InSAR observations. J. Arid Environ. 181, 104238.

GuangYao, D., HuiLi, G., Huanhuan, L., Youquan, Z., BeiBei, C., KunChao, L., 2016. Monitoring and analysis of land subsidence along Beijing-Tianjin Intercity railway. J. Indian Soc. Remote Sens. 44 (6), 915–931.

Guo, J., Zhou, L., Yao, C., Hu, J., 2016. Surface subsidence analysis by multitemporal insar and grace: a case study in Beijing. Sensors 16 (9), 1495–1503.

Herrera, G., Tomás, R., López-Sánchez, J.M., Delgado, J., Mallorquí, J., Duque, S., Mulas, J., 2007. Advanced DInSAR analysis on mining areas: La Union case study (Murcia, SE Spain). Eng. Geol. 90, 148–159.

Hooper, A., Bekaert, D., Spaans, K., Arıkan, M., 2012. Recent advances in SAR interferometry time series analysis for measuring crustal deformation. Tectonophysics 514, 1–13.

Hussain, E., Wright, T.J., Walters, R.J., Bekaert, D.P., Lloyd, R., Hooper, A., 2018. Constant strain accumulation rate between major earthquakes on the North Anatolian Fault. Nat. Commun. 9 (1), 1392–1404.

Imamoglu, M., Kahraman, F., Abdikan, S., 2018. Preliminary results of temporal deformation analysis in Istanbul using multitemporal InSAR with Sentinel-1 SAR data. In: IGARSS 2018.

Jackson, A., Julia, e.d., 1997. Subsidence. In: Glossary of Geology, fourth ed. American Geological Institute, Alexandria, Virginia, ISBN 0922152349.

Kanwal, S., Ding, X.L., Zhang, L., 2018. Measurement of vertical deformation in Karachi using multitemporal insar. IEEE International Geoscience and Remote Sensing Symposium 22 (27), 1395–1398.

Khorrami, M., Abrishami, S., Maghsoudi, Y., Alizadeh, B., Perissin, D., 2020. Extreme subsidence in a populated city (Mashhad) detected by PSInSAR considering groundwater withdrawal and geotechnical properties. Sci. Rep. 10, 11357.

Kim, J.W., Lu, Z., Degrandpre, K., 2016. Ongoing deformation of sinkholes in Wink, Texas, observed by time-series Sentinel-1A SAR interferometry (preliminary results). Remote Sens. 8 (4), 313.

Liora, F.S., Doucet, S., Vernant, P., Champollion, C., Carme, J., 2021. Monitoring surface deformation of deep salt mining in Vauvert (France), combining InSAR and leveling data for multisource inversion. Solid Earth 12, 15–34.

Liu, C., Ji, L., Zhu, L., Zhao, C., 2018. InSARconstrained interseismic deformation and potential seism genic asperities on the Altyn Tagh Fault at 95 E, Northern Tibetan Plateau. Remote Sens. 10 (6), 943–960.

Massonnet, D., Feigl, K.L., 1998. Radar interferometry and its application to changes in the Earth's surface. Rev. Geophys. 36 (4), 441–500.

Maurice, B.D., Richard, J, 2014. Seepage pathway assessment for natural gas to shallow groundwater during well stimulation, in production, and after abandonment. Environ. Geosci. 21 (3), 107–126. https://doi.org/10.1306/eg.04231414004.

Mehrabi, A., 2017. Identifying evidence of the existence of a salt dome Buried and new in the Zagros region using interference method Radar measurement of Sentinel 1 and Isar. J. Senzadvar Geogr. Inf. Syst. Nat. Resour. 4 (9), 90–101.

Miller, M.M., Jones, C.E., Sangha, S.S., Bekaert, D., 2020. Rapid drought-induced land subsidence and its impact on the California aqueduct. Remote Sens. Environ. 251, 112063.

Pacheco, J., Arzate, J., Rojas, E., Arroyo, M., Yutsis, V., Ochoa, G., 2006. Delimitation of ground failure zones due to land subsidence using gravity data. Eng. Geol. 84 (40636), 143–160.

Poland, J.F., 1982. The occurrence and control of land subsidence due to ground-water withdrawal with special reference to the san Joaquin and Santa Clara Valleys, California. U.S. Geological Survey, Federal Center, Box 25425, Denver, CO 80225 19 (14), 497.

Rateb, A., Abotalib, Z., 2020. Inferencing the land subsidence in the Nile Delta using Sentinel1 satellites and GPS between 2015 and 2019. Sci. Total Environ. 729, 1–10.

Rezaei, A., Mousavi, Z., Hydrol, J., 2019. Characterization of land deformation, hydraulic head, and aquifer properties of the Gorgan confined aquifer, Iran, from InSAR observations. J. Hydrol. 579, 124196.

Salehi, R., Ghafouri, M., Lashkaripour, G., Dehghani, M., 2007. Investigating the subsidence of South Mehyar Plain using radar interferometric method. Irrig. Water Eng. 11 (3), 47–57.

Sharifi KIA, M., Afzali, A.A., Shayan, S., 2009. Extraction and evaluation of geomorphological effects caused by subsidence in Damghan Plain. Quant. Geomorphol. Stud. 14 (2), 60–78.

Shirani, k., Ebrahimi, B., Pasandi, M., 2014. Investigating land subsidence in the Najafabad plain of Isfahan using differential interferometric technique radar. J. Water Soil Sci. 25 (1), 105–127.

Waltham, T., Bell, F.G., Culshaw, M.G., 2005. Sinkholes and Subsidence. Karst and Cavernous Rocks in Engineering and Const, ISBN 978-3-540-20725-2. https://doi.org/10.1007/b138363.

Yang, Q., 2018. Multiscale analysis of the relationship between land subsidence and buildings: a case study in an eastern Beijing Urban Area using the PS-InSAR technique. Remote Sens. 10, 1006.

Yang, H., Peng, J., 2015. Monitoring urban subsidence with multi master radar interferometry based on coherent targets. J. Indian Soc. Remote Sens. 43 (3), 529–538.

Zare Kamali, M.s., Hosseini, A., Madrasi, Al, 2016. Comparison of the amount of vertical displacement of the earth using the algorithm SBAS in X and C radar bands) Case study: Lands Tehran). J. Remote Sens. Geogr. Inf. Syst. 3 (8), 104–120.

Zebker, H.A, Rosen, P.A., Goldstein, R.M., Gabriel, A., Werner, C.L., 1994. On the derivation of coseismic displacement fields using differential radar interferometry: The Landers earthquake. J. Geophys. Res. 99 (10), 19617–19634. https://doi.org/10.1029/94JB01179.

Zeitoun, D.G., Wakshal, E., 2013. Land Subsidence Analysis in Urban Areas, A Case Study: The Bangkok Plain, pp. 209–263. https://doi.org/10.1007/978-94-007-5506-2_8.

CHAPTER

18

Drought mapping, modeling, and remote sensing

Davood Mafi-Gholami[1] and Abolfazl Jaafari[2]

[1]Department of Forest Sciences, Faculty of Natural Resources and Earth Sciences, Shahrekord University, Shahrekord, Iran; [2]Research Institute of Forests and Rangelands, Agricultural Research, Education and Extension Organization (AREEO), Tehran, Iran

18.1 Introduction

Due to climate change, social—ecological systems have been adversely affected as there has been a surge in the frequency and intensity of severe weather phenomena like hurricanes and droughts, leading to detrimental consequences (Change, 2007; Mafi-Gholami et al., 2021, 2022). Therefore, understanding the adverse effects caused by the mentioned hazards is one of the most important prerequisites for developing mitigation tools and preparing for the occurrence of natural hazards, as well as formulating effective strategies against future changes (AghaKouchak et al., 2020; Birkmann et al., 2022). Among destructive weather events, drought can cause a wide range of adverse economic, social, and environmental effects with an abnormal lack of rainfall and an imbalance between water supply and demand. Some estimations suggest that the loss of the global economy due to the occurrence of droughts is approximately $8 billion per year, which is much more than other destructive weather events (Huynh et al., 2020). Although drought occurs in all climatic regions worldwide, drought hazards are the highest in arid and semiarid regions (Carrão et al., 2016) (Fig. 18.1).

Due to its wide temporal and spatial scale, drought is the main environmental factor that causes the lack and disruption of water flow, the loss of ecological capacity, and food poverty worldwide (Godfray et al., 2010; Huynh et al., 2020). Therefore, the countries of the world, especially developing countries, which have sensitive and unstable environments and weak economic and technological facilities, are looking for tools and solutions to decrease the adverse effects of drought, achieve sustainable development, and even maintain the existing situation (Aghakouchak et al., 2015). To this end, monitoring and modeling the temporal and spatial changes in drought in different regions using various tools play an important role

https://doi.org/10.1016/B978-0-443-15341-9.00005-8

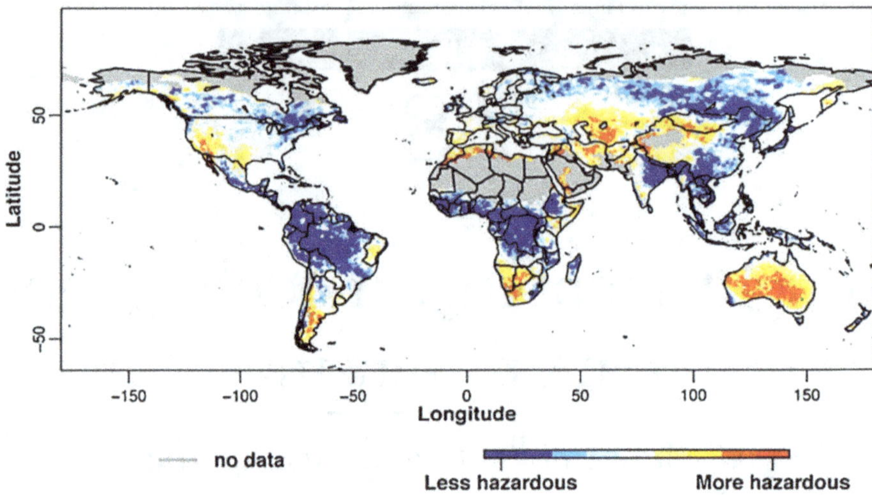

FIGURE 18.1 Global drought map for the period spanning from 1901 to 2010 (Carrão et al., 2016).

in obtaining vital information for assessing drought risk and mitigating the destructive effects of this climatic phenomenon on coupled social—ecological systems (Mafi-Gholami et al., 2021, 2022). To assess and spatially map the occurrence of drought, ground-based point observations (Aghakouchak et al., 2014) or satellite observations (Krajewski et al., 2006) are used. Due to the lack of meteorological stations and the lack of climatic data or the inappropriate distribution of stations in many regions of the world, satellite observations are used due to their characteristics, such as global, consistent data records, improved spatial resolution, and near-real-time observations. Remote sensing datasets, including total water storage, soil moisture, precipitation, snow, evaporation, land surface temperature, and vegetation and land cover, are utilized to monitor drought in different regions of the world from diverse perspectives (e.g., ecological, agricultural, meteorological, and hydrological).

18.2 Modeling temporal and spatial changes in drought

Drought modeling is a vital process used to simulate the hydrological system's behavior under different drought scenarios. Mathematical equations are employed in these models to predict how much water will be available in a specific region during a dry weather spell. This information can then be used for planning purposes, like determining the amount of water that can be safely allocated for various uses.

To effectively monitor and model the spatio-temporal variations of meteorological drought, various indices are employed. These indices rely on either ground-based point observations or satellite data. The most commonly used indices include the Standardized Precipitation-Evapotranspiration Index (SPEI), Palmer Drought Severity Index (PDSI), and Standard Precipitation Index (SPI) (Palmer, 1965; McKee et al., 1993; Vicente-Serrano et al., 2010). Among these three indices, SPI holds prominence globally for assessing drought events and related risk management due to its ease of application, spatial compatibility, and temporal flexibility (Mafi-Gholami et al., 2017, 2019, 2021, 2022).

The SPI is used to evaluate drought conditions based on long-term rainfall data at various time scales ranging from 1 to 48 months. It accomplishes this by contrasting the total cumulative precipitation in a specific area during a given period with the average precipitation during the same duration, hence enabling a temporal analysis of dry and wet periods (McKee et al., 1993). SPI values corresponding to various time scales provide information about the effects of drought duration on water resources. SPI values corresponding to shorter time scales (e.g., 30–90 days) indicate alterations in soil moisture, which are critical to the agricultural sector, while longer time scale SPI values (e.g., 6 months, 12 months, or more) signify prolonged variations in surface and groundwater levels. As such, these longer time scale SPI values are essential when analyzing the impact of precipitation changes on natural ecosystems and managing water resources in human settlements.

The calculation of SPI involves fitting the long-term rainfall data from each station with the gamma distribution function and the extended function to ascertain the cumulative probability of rainfall for that station, given a specified month and time scale (McKee et al., 1993). As a result of the fitting procedure, an SPI value is generated with a standard deviation of one and an average of zero. Consequently, the SPI indicates the number of standard deviations by which a given event differs from normal conditions. Positive SPI values imply above-average rainfall, while negative values indicate below-average precipitation (McKee et al., 1993). Drought severity and likelihood are determined by categorizing different SPI values into various classes (Table 18.1) (McKee et al., 1993).

Table 18.1 indicates that the threshold of drought occurrence is SPI < 0, and the drought duration is computed as all months when the value of SPI is less than zero. Therefore, the probability of occurrence of each class of drought is computed by dividing the number of months with negative SPI values for that class by all the months of the statistical period at each station. The average of the total values of the probability of occurrence of each drought class in the long term at all meteorological stations of a region is considered the probability of drought occurrence for that class.

TABLE 18.1 Drought classes defined for the Standard Precipitation Index (SPI) values.

Range of SPI values	Precipitation regime
>2	Extremely wet
1.99–1.5	Very wet
1.49–1	Moderately wet
0.99 to −0.99	Normal precipitation
−0.99 to −1.49	Moderately dry
−1.5 to −1.99	Very dry
< −2	Extremely dry

After calculating the drought severity and probability of occurrence at every weather station, mapping and modeling of temporal and spatial changes in drought are performed in different areas. Hydrological investigations in a large area require the integration of information related to different locations with specific geographical locations (Shahid and Nath, 2002; Shahid and Behrawan, 2008). Geographic information systems (GISs) can be effectively used to analyze spatio-temporal changes in patterns of precipitation and droughts (Mafi-Gholami et al., 2022). Studies show that among the different interpolation methods, the kriging method has higher efficiency and predictive power than other methods, such as inverse distance weighted (IDW), Thyssen, or Laplace (Oliver and Webster, 1990; van Beers and Kleijnen, 2004; Zimmerman and Holland, 2005). However, the kriging accuracy depends on the selected variogram model, and since drawing a suitable variogram model depends on the number of observations and their distance from each other (Van Meirvenne and Hofman, 1989), accordingly, the small number and inappropriate spatial distribution of meteorological stations do not make it possible to calculate an optimal variogram model. Meul and Van Meirvenne (2003) considered the kriging method inappropriate for a small number of samples with large spatial changes. Therefore, Burrough and McDonnell (1998a,b) and Patel et al. (2007) introduced the IDW method as a suitable method for interpolation using sample points with a small number and inappropriate spatial distribution. Based on the number and distribution of meteorological stations and choosing the appropriate interpolation method, the interpolation of drought intensity in different classes is performed based on SPI values in the study area. Finally, the interpolated map of each of the drought classes is multiplied by the probability of its occurrence, and by merging the obtained maps, the final map that shows the risk of drought occurrence is provided in the study region (Shahid and Behrawan, 2008).

In addition to the mentioned methods, the computation of the drought magnitude at each meteorological station is another method for assessing the risk of drought (Mafi-Gholami et al., 2019, 2021, 2022). To calculate the drought magnitude at each meteorological station, all negative SPI values during the study period are added together using the following equation (Mafi-Gholami et al., 2019, 2021, 2022).

$$DM = -\left(\sum_{j=1}^{x} SPI_j\right) \tag{18.1}$$

Where DM is drought magnitude and SPI_j gives the value of negative SPI during period j.

Using the drought magnitude values calculated for each of the stations and GIS functions, a drought risk map can be prepared throughout the region of interest.

18.3 Monitoring temporal changes in drought

Undoubtedly, obtaining more information about changes in the trend of destructive weather events (e.g., drought occurrences) over time is crucial for analyzing and designing coping strategies or mitigating the adverse effects caused by these events. Numerous statistical methods and techniques exist for identifying trends in data series, each having its own advantages and limitations. The World Meteorological Organization developed the Mann-

Kendall (MK) test, which has become one of the most widely used techniques. The MK test is represented by:

$$S = \sum_{i=1}^{n-1} \sum_{j=i+1}^{n} \operatorname{sgn}(x_j - x_i) \tag{18.2}$$

where

$$(x_j - x_i) = \begin{cases} +1. & x_j > x_i \\ 0\,. & x_j = x_i \\ -1. & x_j < x_i \end{cases}$$

and n is equal to the sample size. The S statistic has an almost normal distribution when $n \geq 8$. In this case, the mean and variance can be given by:

$$E(s) = 0$$

$$V\,(S) = \frac{n\,(n-1)(2n+5) - \sum_{i=1}^{n} t_i i(i-1)(2i+5)}{18} \tag{18.3}$$

where t_i is the number of nodes in chain i. The standard statistic (Z) for the one-tailed test is formulated as the following relationship:

$$Z = \begin{cases} \dfrac{S-1}{\sqrt{\mathrm{Var}\,(S)}} & S > 0 \\ 0 & S = 0 \\ \dfrac{S+1}{\sqrt{\mathrm{Var}\,(S)}} & S < 0 \end{cases} \tag{18.4}$$

During a two-tailed test aimed at identifying trends in data series, rejecting the null hypothesis (which assumes no trend) in favor of the one hypothesis (stating the presence of a trend) occurs when Eq. (18.5) is confirmed to be true.

$$|Z| \geq Z\,\alpha/2 \tag{18.5}$$

where α represents the significance level and $Z\alpha$ is the standard normal distribution statistic calculated at this significance level. The value of $\alpha/2$ is used to consider both tails of the distribution. When the Z statistic is positive, it specifies a rising trend in the data series. Conversely, a negative Z statistic suggests a descending trend in the data series. The MK test at 95% and 99% confidence levels is used to investigate the trend in the series of SPI values of selected stations.

18.4 Change detection in drought intensity time series

Once changes in the drought occurrence trend have been detected, identifying breakpoints is a crucial step in obtaining an improved interpretation and more precise examination of the effects of droughts on ecosystems (Chu et al. al., 2012; Mafi-Gholami et al., 2017). A breakpoint is typically understood as a change occurring at a specific time in the frequency distribution function of a variable (Paturel et al., 1997). To detect such changes in data series, breakpoint tests are applied, and some of the most effective methods for this purpose include Pettitt (1979), Buishand (1984), Lee and Heghinian (1977), and Hubert et al. (1989). The Pettitt-Mann-Whitney test is widely considered the most commonly used method (Pettitt, 1979; Kiely, 1999) for detecting breakpoints. This method is a modified version of the Mann–Whitney test and is particularly sensitive to breakpoints in continuous data series compared to other methods. It divides the time series into two subgroups before and after the breakpoint and determines the probability of the change point across different time intervals such as annual, six-monthly, quarterly, or monthly periods (Maftei and Barbulescu, 2012). The Pettitt-Mann-Whitney method involves treating each time series, represented by data series $x_1, x_2 \ldots, x_T$ with a length of T, as two distinct samples: $x_1, x_2 \ldots x_t$ and $x_{t+1}, x_{t+2} \ldots, x_T$. The $V(t)$ and $U(t)$ indices are then computed using the following equations:

$$V_{t.T} = \sum_{j=1}^{T} \mathrm{sgn}(x_t - x_j) \tag{18.6}$$

$$U_{t.T} = U_{t-1.T} + V_{t.T}\, t = 2.T \tag{18.7}$$

and

$$U_{1.T} = V_{1.T} \tag{18.8}$$

where

$$\mathrm{sgn}(x) = \begin{cases} 1. & \text{for} \quad x > 0 \\ 0. & \text{for} \quad x = 0 \\ -1. & \text{for} \quad x < 0 \end{cases}$$

The approximate probability $P(t)$ for the change point is equal to:

$$P(t) = 1 - \exp\left(- 6U_{t.T}^2 \,/\, T^3 + T^2 \right) \tag{18.9}$$

The point of greatest change occurs when either the absolute value of $|U_{t.T}|$ or $P(t)$ achieves its highest value.

$$K_T = \max|U_{t.T}| \tag{18.10}$$

Finally, the probability of trend change for each year of the time series is calculated by considering the average annual rainfall values using the nonparametric Pettitt-Mann-

Whitney test ($\alpha = 0.05$). In order to accurately determine the probability of a change in the trend of drought occurrence during a specific period, it is important to pinpoint the most significant breakpoint or change point within that timeframe. The cumulative sum (CUSUM) method is a highly effective technique for identifying the most significant change points in time series data. Initially developed for industrial control purposes (Barratt et al., 2007), it has since become a valuable tool in environmental monitoring programs as well (Carslaw et al., 2006; Andersen et al., 2009). The simplicity and effectiveness of the CUSUM method make it an ideal choice for detecting breakpoints or change points in a wide variety of data types, including climate variables (Li et al., 2008; Maftei and Barbulescu, 2012). Therefore, it is commonly used in studies seeking to identify such breakpoints in time series data. Overall, by utilizing the CUSUM method in conjunction with the Pettitt-Mann-Whitney method, researchers can more accurately identify key changes in drought occurrence trends over a given period of time. The value of CUSUM is calculated by Eq. (18.11).

$$S_i = \sum_{i=1}^{i} (x_i - k) \tag{18.11}$$

where k represents the average value of the time series. The maximum value of S_i indicates the highest probability of a change occurring.

Once the most critical year of change (i.e., breakpoint) has been determined in the time series of average annual rainfall, the significance of the difference between the mean annual rainfall values before and after the breakpoint is investigated using a t-test.

18.5 Remotely sensed monitoring of changes in ecosystems in response to drought

Due to the close relationship between rainfall and ecosystems, variations in rainfall characteristics and drought events will significantly influence the structure and function of ecosystems (Ellison, 2000). For example, Eslami-Andargoli et al. (2009) showed a decrease in the production potential and extent of mangroves in the face of decreasing rainfall and drought occurrence in recent decades. Satellite imagery has been widely used to model the historical changes that occurred in the structure and function of ecosystems in the face of drought (Fromard et al., 2004; Giri et al., 2007, Giri et al., 2011). To model the response of ecosystems to changes in drought severity over time, by conducting extensive field surveys and recording vegetation characteristics in sample plots, the best equation is established between spectral reflectance in satellite images and vegetation characteristics related to ecosystems in sample plots. Then, by implementing the developed equation on the time series of satellite images in the past decades, maps of changes in the vegetation characteristics are prepared over time. Finally, statistical analyses are performed between the changes in drought occurrence and vegetation characteristics in the past decades, and based on this, the response of the ecosystems to the changes in the amount of rainfall and the occurrence of drought is modeled (Fig. 18.2).

Previous studies have indicated a rise in the severity of climate change, particularly with regard to drought periods (Chiang et al., 2021), and this trend is especially pronounced in low latitudes (Etemadi et al., 2016). In some areas, for example, the Middle East, these changes

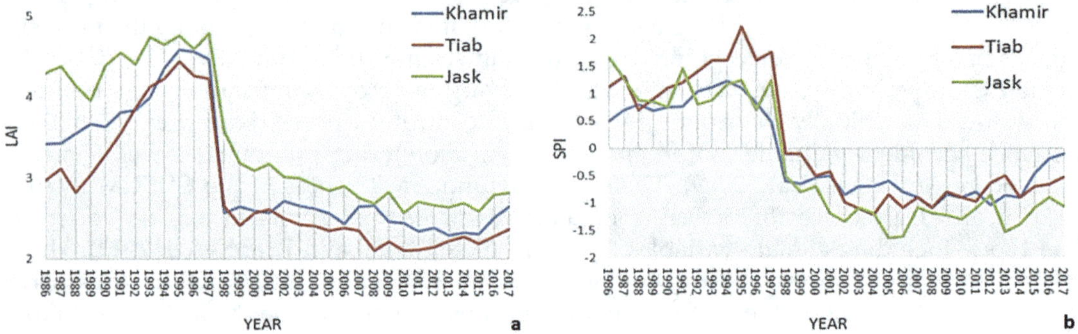

FIGURE 18.2 (A) Changes in the predicted leaf area index (LAI) of mangroves of Iran, which were produced from satellite data and (B) the observed Standard Precipitation Index (SPI) computed based on monthly rainfall data from 1986 to 2017 (Mafi-Gholami et al., 2019).

align with the most pessimistic climate scenario (RCP 8.5) (Mafi-Gholami et al., 2019, 2022), which will increase the vulnerability of environments to multiple disasters in forthcoming decades. Given this, predicting the adverse impact of future climate changes on ecosystems is a crucial prerequisite for preparing climate change adaptation programs and reducing their vulnerability (Ellison, 2015). To achieve this objective, researchers have endeavored to simulate the response of ecosystems to future climate changes (Mafi-Gholami et al., 2019). By developing regression equations that model the relationship between the changes in the function of environments during recent decades in response to drought, researchers have been able to project the response of ecosystems to various climate change scenarios in the future (as depicted in Fig. 18.3) (Mafi-Gholami et al., 2019).

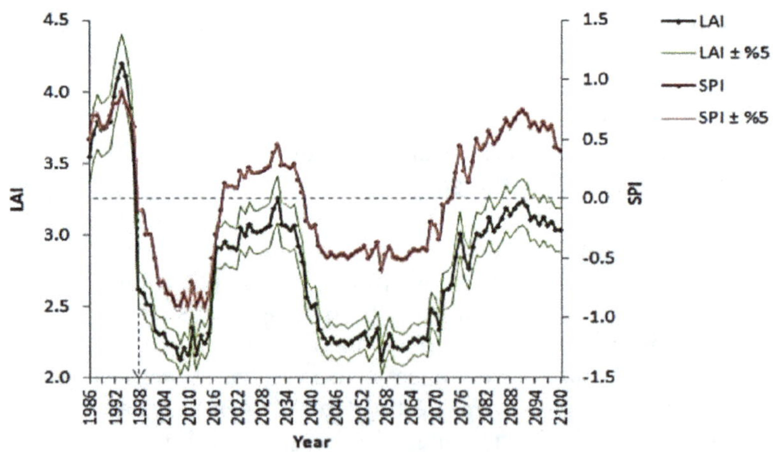

FIGURE 18.3 Projections for Standard Precipitation Index (SPI) and leaf area index (LAI) of mangroves of Iran from 1986 to 2100. To project the SPI values, the RCP8.5 climate change scenario was employed, whereas regression equations estimated from the observed SPI values derived from rainfall and the LAI values obtained from NDVI data between 1986 and 2017 were utilized to predict the LAI values (Mafi-Gholami et al., 2019).

18.6 Conclusions

The occurrence of droughts and changes in rainfall patterns can have severe consequences on the coupled social–ecological systems worldwide. Therefore, it is vital to monitor and model such events to provide valuable information for safeguarding these systems against potential harm. To achieve this goal, climate data recorded in ground-based point observations or satellite observations are used to calculate drought monitoring indicators. Because of the temporal and spatial limitations of ground-based point observation data, satellite observation data that have features such as enhanced resolution, consistent data records, and near-real-time observations are currently used for drought monitoring and modeling. Based on the climatic data (such as rainfall) obtained from the mentioned sources, suitable indicators for drought monitoring are calculated, and while using appropriate statistical methods, the trend of changes in the occurrence of drought and the existence of change points (break-points) in this trend is detected, and the possibility of better interpretation and more detailed examination of the influences of drought events and changes in precipitation patterns on environments is provided for managers. For drought risk mapping throughout the study area, the values of drought probability are multiplied in its intensity interpolated map, and drought risk mapping is performed using the appropriate functions in ArcGIS. By modeling the relationship between variations in the function and performance of ecosystems (inferred from satellite images) and variations in drought intensity over time, it becomes possible to predict how ecosystems will respond to changes in drought intensity across diverse climate scenarios during forthcoming decades. Such projections offer valuable insights that can help managers and planners enhance the efficiency and success of climate change adaptation programs, design better coping strategies, and mitigate the undesirable influences of climate change phenomena on environments and ecosystems. This is crucial for protecting these complex systems and ensuring their long-term sustainability.

References

AghaKouchak, A., Cheng, L., Mazdiyasni, O., Farahmand, A., 2014. Global warming and changes in risk of concurrent climate extremes: insights from the 2014 California drought. Geophys. Res. Lett. 41 (24), 8847–8852.

AghaKouchak, A., Chiang, F., Huning, L.S., Love, C.A., Mallakpour, I., Mazdiyasni, O., et al., 2020. Climate extremes and compound hazards in a warming world. Annu. Rev. Earth Planet Sci. 48, 519–548.

AghaKouchak, A., Farahmand, A., Melton, F.S., Teixeira, J., Anderson, M.C., Wardlow, B.D., Hain, C.R., 2015. Remote sensing of drought: progress, challenges and opportunities. Rev. Geophys. 53 (2), 452–480.

Andersen, T., Carstensen, J., Hernandez-Garcia, E., Duarte, C.M., 2009. Ecological thresholds and regime shifts: approaches to identification. Trends Ecol. Evol. 24 (1), 49–57.

Barratt, B., Atkinson, R., Anderson, H.R., Beevers, S., Kelly, F., Mudway, I., Wilkinson, P., 2007. Investigation into the use of the CUSUM technique in identifying changes in mean air pollution levels following introduction of a traffic management scheme. Atmos. Environ. 41 (8), 1784–1791.

Birkmann, J., Jamshed, A., McMillan, J.M., Feldmeyer, D., Totin, E., Solecki, W., et al., 2022. Understanding human vulnerability to climate change: a global perspective on index validation for adaptation planning. Sci. Total Environ. 803, 150065.

Buishand, T.A., 1984. Tests for detecting a shift in the mean of hydrological time series. J. Hydrol. 73 (1–2), 51–69.

Burrough, P.A., McDonnell, R.A., 1998a. Principles of Geographical Information Systems. Oxford University Press, Oxford, p. 330.

Burrough, P.A., McDonnell, R.A., 1998b. Creating continuous surfaces from point data. In: Burrough, P.A., Goodchild, M.F., McDonnell, R.A., Switzer, P., Worboys, M. (Eds.), Principles of Geographic Information Systems. Oxford University Press, Oxford, UK. Chaplot, V., Walter, C.

Carrão, H., Naumann, G., Barbosa, P., 2016. Mapping global patterns of drought risk: an empirical framework based on subnational estimates of hazard, exposure and vulnerability. Global Environ. Change 39, 108−124.

Carslaw, D.C., Ropkins, K., Bell, M.C., 2006. Change-point detection of gaseous and particulate traffic-related pollutants at a roadside location. Environ. Sci. Technol. 40 (22), 6912−6918.

Change, C., 2007. IPCC Fourth Assessment Report. The Physical Science Basis, vol. 2, pp. 580−595.

Chiang, F., Mazdiyasni, O., AghaKouchak, A., 2021. Evidence of anthropogenic impacts on global drought frequency, duration, and intensity. Nat. Commun. 12 (1), 2754.

Chu, H.J., Pan, T.Y., Liou, J.J., 2012. Change-point detection of long-duration extreme precipitation and the effect on hydrologic design: a case study of south Taiwan. Stoch. Environ. Res. Risk Assess. 26, 1123−1130.

Ellison, A.M., 2000. Mangrove restoration: do we know enough? Restor. Ecol. 8 (3), 219−229.

Ellison, J.C., 2015. Vulnerability assessment of mangroves to climate change and sea-level rise impacts. Wetl. Ecol. Manag. 23, 115−137.

Eslami-Andargoli, L., Dale, P.E.R., Sipe, N., Chaseling, J., 2009. Mangrove expansion and rainfall patterns in Moreton Bay, southeast Queensland, Australia. Estuar. Coast Shelf Sci. 85 (2), 292−298.

Etemadi, H., Samadi, S.Z., Sharifikia, M., Smoak, J.M., 2016. Assessment of climate change downscaling and nonstationarity on the spatial pattern of a mangrove ecosystem in an arid coastal region of southern Iran. Theor. Appl. Climatol. 126, 35−49.

Fromard, F., Vega, C., Proisy, C., 2004. Half a century of dynamic coastal change affecting mangrove shorelines of French Guiana. A case study based on remote sensing data analyses and field surveys. Mar. Geol. 208 (2–4), 265−280.

Giri, C., Ochieng, E., Tieszen, L.L., Zhu, Z., Singh, A., Loveland, T., et al., 2011. Status and distribution of mangrove forests of the world using earth observation satellite data. Global Ecol. Biogeogr. 20 (1), 154−159.

Giri, C., Pengra, B., Zhu, Z., Singh, A., Tieszen, L.L., 2007. Monitoring mangrove forest dynamics of the Sundarbans in Bangladesh and India using multitemporal satellite data from 1973 to 2000. Estuar. Coast. Shelf Sci. 73 (1–2), 91−100.

Godfray, H.C.J., Beddington, J.R., Crute, I.R., Haddad, L., Lawrence, D., Muir, J.F., et al., 2010. Food security: the challenge of feeding 9 billion people. Science 327 (5967), 812−818.

Hubert, P., Carbonnel, J.P., Chaouche, A., 1989. Segmentation des séries hydrométéorologiques—application à des séries de précipitations et de débits de l'Afrique de l'ouest. J. Hydrol. 110 (3–4), 349−367.

Huynh, T.D., Nguyen, T.H., Truong, C., 2020. Climate risk: the price of drought. J. Corp. Finance 65, 101750.

Kiely, G., 1999. Climate change in Ireland from precipitation and streamflow observations. Adv. Water Resour. 23 (2), 141−151.

Krajewski, W.F., Anderson, M.C., Eichinger, W.E., Entekhabi, D., Hornbuckle, B.K., Houser, P.R., et al., 2006. A remote sensing observatory for hydrologic sciences: a genesis for scaling to continental hydrology. Water Resour. Res. 42 (7), 1−13.

Lee, A.F., Heghinian, S.M., 1977. A shift of the mean level in a sequence of independent normal random variables—a Bayesian approach. Technometrics 19 (4), 503−506.

Li, Z.L., Xu, Z.X., Li, J.Y., Li, Z.J., 2008. Shift trend and step changes for runoff time series in the Shiyang River basin, northwest China. Hydrol. Process.: Int. J. 22 (23), 4639−4646.

Mafi-Gholami, D., Mahmoudi, B., Zenner, E.K., 2017. An analysis of the relationship between drought events and mangrove changes along the northern coasts of the Persian Gulf and Oman Sea. Estuar. Coast Shelf Sci. 199, 141−151.

Mafi-Gholami, D., Pirasteh, S., Ellison, J.C., Jaafari, A., 2021. Fuzzy-based vulnerability assessment of coupled social-ecological systems to multiple environmental hazards and climate change. J. Environ. Manag. 299, 113573.

Mafi-Gholami, D., Zenner, E.K., Jaafari, A., 2022. Mapping recent (1997−2017) and future (2030) county-level social vulnerability to socio-economic conditions and natural hazards throughout Iran. J. Clean. Prod. 355, 131841.

Mafi-Gholami, D., Zenner, E.K., Jaafari, A., Ward, R.D., 2019. Modeling multi-decadal mangrove leaf area index in response to drought along the semi-arid southern coasts of Iran. Sci. Total Environ. 656, 1326−1336.

Maftei, C., Barbulescu, A., 2012. Statistical analysis of precipitation time series in Dobrudja region. Mausam 63 (4), 553−564.

McKee, T.B., Doesken, N.J., Kleist, J., 1993. The relationship of drought frequency and duration to time scales. In: Proceedings of the 8th Conference on Applied Climatology, pp. 179–183.

Meul, M., Van Meirvenne, M., 2003. Kriging soil texture under different types of nonstationarity. Geoderma 112 (3–4), 217–233.

Oliver, M.A., Webster, R., 1990. Kriging: a method of interpolation for geographical information systems. Int. J. Geogr. Inf. Syst. 4 (3), 313–332.

Palmer, W.C., 1965. Meteorological Drought. Department of Commerce, Weather Bureau – Climatology 30, 58.

Patel, N.R., Chopra, P., Dadhwal, V.K., 2007. Analyzing spatial patterns of meteorological drought using standardized precipitation index. Meteorol. Appl. J. Forecast. Pract. Appl. Train. Tech. Model. 14 (4), 329–336.

Paturel, J.E., Servat, E., Kouamé, B., Lubes, H., Ouedraogo, M., Masson, J.M., 1997. Climatic variability in humid Africa along the Gulf of Guinea Part II: an integrated regional approach. J. Hydrol. 191 (1–4), 16–36.

Pettitt, A.N., 1979. A non-parametric approach to the change-point problem. J. Roy. Stat. Soc.: Ser. C Appl. Stat. 28 (2), 126–135.

Shahid, S., Behrawan, H., 2008. Drought risk assessment in the western part of Bangladesh. Nat. Hazards 46, 391–413.

Shahid, S., Nath, S.K., 2002. GIS integration of remote sensing and electrical sounding data for hydrogeological exploration. J. Spatial Hydrol. 2 (1), 1–12.

Van Beers, W.C., Kleijnen, J.P., 2004. Kriging interpolation in simulation: a survey. In: Proceedings of the 2004 Winter Simulation Conference, 2004, vol. 1. IEEE.

Van Meirvenne, M., Hofman, G., 1989. Spatial variability of soil nitrate nitrogen after potatoes and its change during winter. Plant Soil 120, 103–110.

Vicente-Serrano, S.M., Beguería, S., López-Moreno, J.I., 2010. A multiscalar drought index sensitive to global warming: the standardized precipitation evapotranspiration index. J. Clim. 23 (7), 1696–1718.

Zimmerman, D.L., Holland, D.M., 2005. Complementary co-kriging: spatial prediction using data combined from several environmental monitoring networks. Environmetrics: The official Journal of the International Environmetrics Society 16 (3), 219–234.

Further reading

Chelani, A.B., 2011. Change detection using CUSUM and modified CUSUM method in air pollutant concentrations at traffic site in Delhi. Stoch. Environ. Res. Risk Assess. 25, 827–834.

CHAPTER

19

Predictive pedometric mapping of soil texture in small catchments: Application of the integrated computer-assisted digital maps, machine learning, and limited soil data

Sedigheh Maleki[1,2] *and Aiding Kornejady*[3]

[1]Department of Plant Production, Faculty of Agriculture, University of Torbat Heydarieh, Torbat Heydarieh, Iran; [2]Saffron Institute, University of Torbat Heydarieh, Torbat Heydarieh, Iran; [3]Research Department of Natural Resources, Golestan Agricultural and Natural Resources Research and Education Center, Agricultural Research, Education and Extension Organization (AREEO), Gorgan, Iran

19.1 Introduction

Soil particle size is an effective factor in soil erosion and consequently, reduction in soil fertility and soil degradation (Makabe et al., 2009; Warrington et al., 2009; Pahlavan-Rad and Akbarimoghaddam, 2018; Kaya et al., 2022). High-resolution soil maps are essential for land-use planning and other activities related to forestry, agriculture, and environmental protection (Stenberg et al., 2010). Since one of the most important management challenges in the catchments of Iran can be attributed to the lack of appropriate detailed studies and research; therefore, it was decided to select some representative basins across Iran to conduct a detailed study and install advanced monitoring environmental sensors. The representative basin and the paired catchments within were established in 1998 in 12 provinces of Iran. In this regard, soil studies are considered one of the pivotal parts of detailed

https://doi.org/10.1016/B978-0-443-15341-9.00018-6

studies and action plans. Despite the great importance of soil properties, we are faced with a lack of relevant data and soil maps at the required scales for environmental modeling (Adhikari et al., 2013).

The main measurement and monitoring limitations of soil texture in arid and semi-arid catchments and sub-catchments, in combination with the destructive impacts of erosion, amplifies the need for better soil mapping methods (Andrade et al., 2020; Kaya et al., 2022). Remote sensing (RS) and machine/deep learning have recently been partaken in predicting the spatial distributions of soil attributes (McBratney et al., 2003; Minasny and Hartemink, 2011). Digital soil mapping (DSM) is a relatively new method for preparing soil maps (Taghizadeh-Mehrjardi et al., 2016; Forkuor et al., 2017; Khaledian and Miller, 2020). DSM techniques such as SCORPAN model are efficient and can be used to overcome some of the limitations of conventional soil-mapping methods through assessing the reciprocal relationships between environmental predictors and soil properties (McBratney et al., 2003; Malone et al., 2017; Zeraatpisheh et al., 2019; Lamichhane et al., 2019; Khaledian and Miller, 2020; Maleki et al., 2020; Taghizadeh-Mehrjardi et al., 2021). DSM has been widely utilized to produce soil texture maps (Pahlavan-Rad and Akbarimoghaddam, 2018; Amirian-Chakan et al., 2019; Andrade et al., 2020; Taghizadeh-Mehrjardi et al., 2020; Žížala et al., 2022; Kaya et al., 2022; Purushothaman et al., 2022).

DSM can adopt machine learning algorithms, such as random forest (RF), to improve its accuracy (Wiesmeier et al., 2011; Nussbaum et al., 2018; Pahlavan-Rad and Akbarimoghaddam, 2018; Kaya et al., 2022), hence was also applied in this study. The optimal use of DSM to apply more accurate auxiliary variables and use of experts' knowledge in the field of soil sciences should be carried out to facilitate soil studies and the selection of uniform map units, which consequently leads to the reduction of soil sampling and laboratory analysis in areas with no soil legacy data (Maleki et al., 2020). In the meantime, the digital elevation model (DEM) and orthophotos with high spatial resolution acquired by unmanned aerial vehicles (UAV) have proved of tremendous value (Maleki et al., 2020; Kariminejad et al., 2022a, b), which can be used in the preparation of different maps with high accuracy as input variables of erosion models in different catchments.

Studies have shown that measuring soil texture is a valuable index for investigating issues such as runoff and erosion in soils of arid and semi-arid regions. Therefore, considering the importance of the particle size distribution, for the first time in the studies of paired catchments, a soil texture map was prepared in three sub-catchments of Iran, that is, Goorband (Sistan and Baluchestan province), Kakhk (Khorasan Razavi province), and Dehgin (Hormozgan province) using DSM method with a limited soil data and environmental variables extracted from the UAV, which is different from previous studies in this field and has innovation in applicable soil science studies. For this purpose, the conditional Latin Hypercube sampling (cLHS) method was used to select sampling sites of soils.

19.2 Case studies across Iran

This study was conducted in three paired catchments across Iran, including Kakhk (Khorasan Razavi province), Goorband (Sistan and Baluchestan province), and Dehgin (Hormozgan province) (Fig. 19.1), which are further described in Table 19.1.

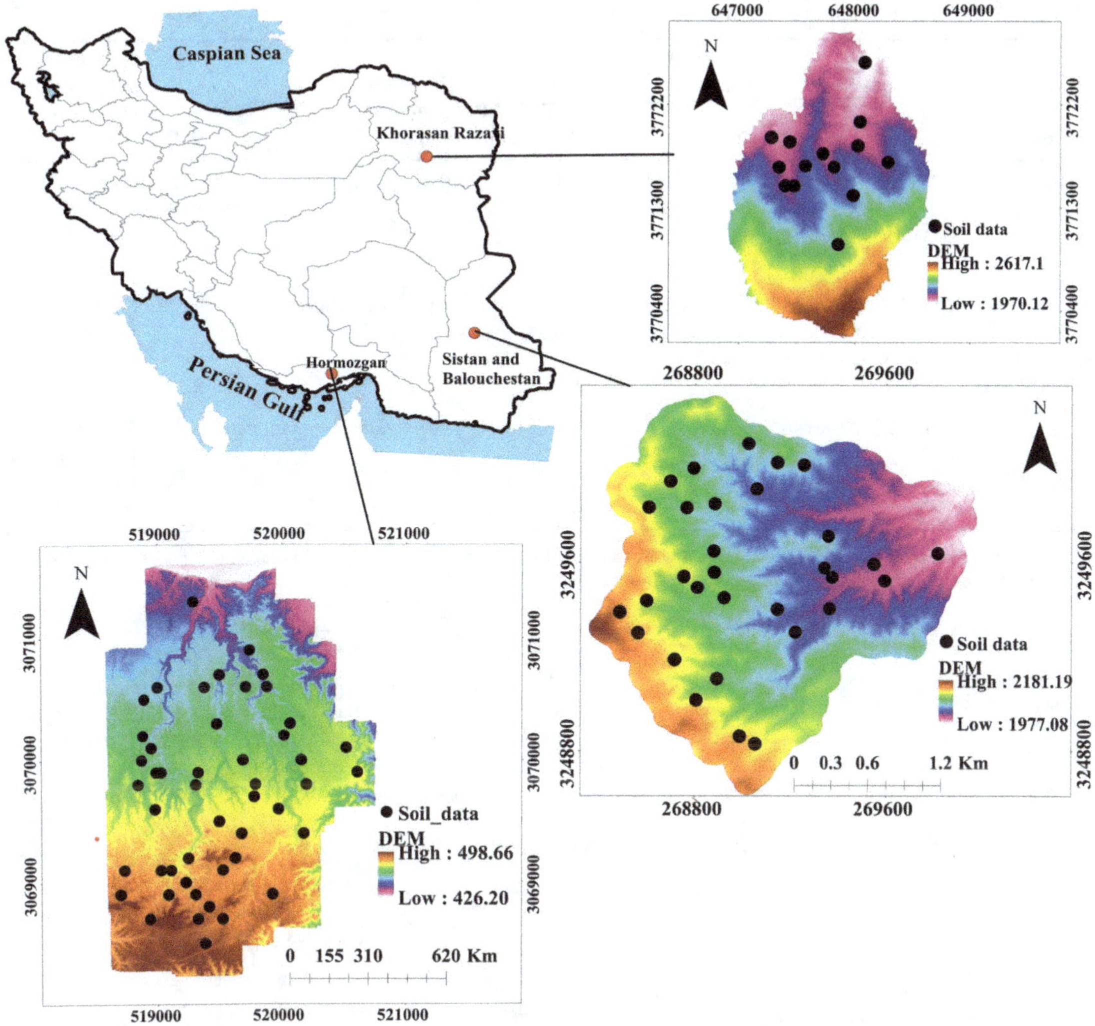

FIGURE 19.1 The location of the study area and soil samples distribution on digital elevation model (DEM) obtained from an unmanned aerial vehicle (UAV). *(Maps generated by authors).*

19.3 A rigorous predictive approach for pedometric mapping of soil texture

The methodology (Fig. 19.2) includes five phases: selection of the study areas; data preparation composed of environmental covariates with UAV, field investigation, and soil sampling; prediction of soil texture class maps using the RF algorithm and the identification and ranking of environmental covariates; validation of predictive maps; and the examination of the spatial variability of texture classes to determine strategies for management.

TABLE 19.1 Characteristics of paired study basins.

Name of paired catchments	Location		Area of sub-catchment (ha)		MAT (°C)	MAP (mm)	Elevation ranges (m.a.s.l.)	STR	SMR	Soil subgroup (Soil Survey Staff, 2014)
	Long.	Lat.	Control	Sample						
Kakhk, Khorasan Razavi province, Iran	58°35′37″ to 58°36′41″	34°03′46″ to 34°05′06″	110.6	106.5	14.3	243	1970–2617	Mesic	Dry Xeric	Lithic Torriorthents, Typic Torriorthents
Goorband, Sistan and Baluchestan province, Iran	60°36′52″ to 60°37′53″	29°20′45″ to 29°21′39″	58	83	15.4	179	2152–2180	Thermic	Aridic	Lithic Torriorthents
Dehgin, Hormozgan province, Iran	57°11′22″ to 57°12′36″	27°46′48″ to 27°44′16″			25.8	186.1	426–499	Hyperthermic	Aridic	Typic Torriorthents; Typic Haplocalcids

m a.s.l, meters above sea level; *MAT*, mean annual temperature, *MAP*, mean annual precipitation; *SMR*, soil moisture regime; *STR*, soil temperature regime; *Long*, longitude; *Lat*, latitude.

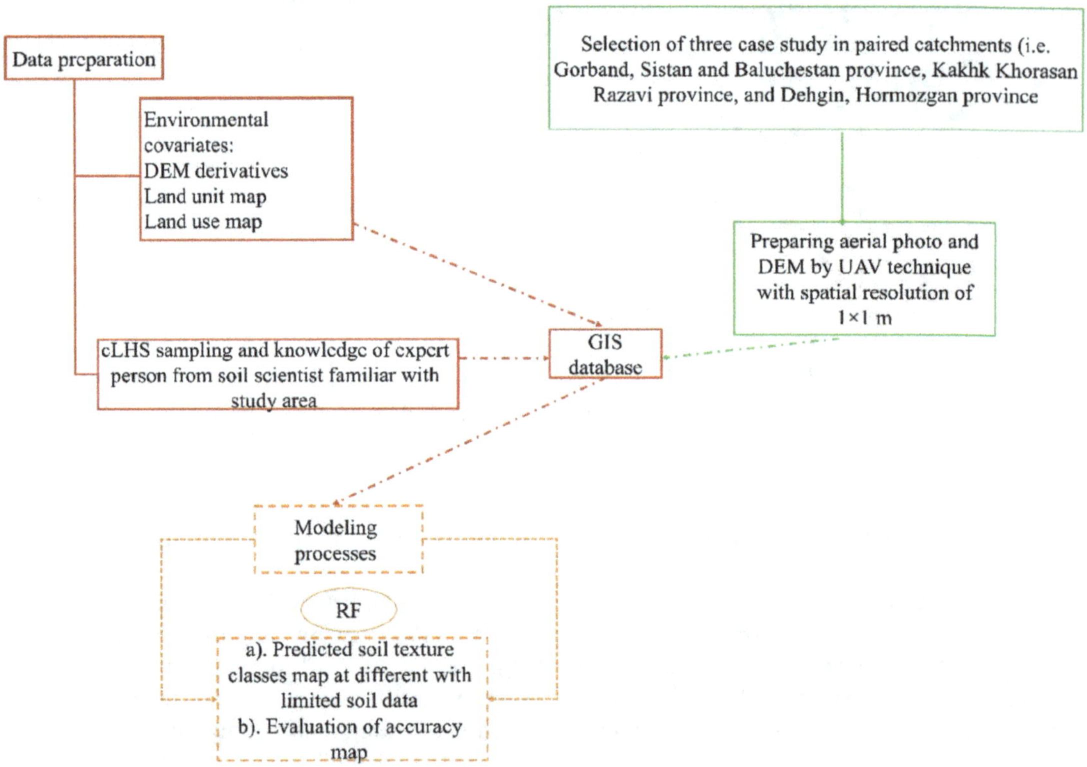

FIGURE 19.2 Methodology, approach, and its components.

19.3.1 Aerial images obtained from unmanned aerial vehicles and environmental covariates

After conducting thorough inspections across case studies including wind conditions and the roughness of topography, detecting remote and inaccessible locations to build benchmarks, a multirouter UAV, Phantom 4 pro v2, was adopted as the most suitable option, configuring a 70%–80% longitudinal and transverse overlap, supplied with DJI FC6310 camera with 5472 × 3648 resolution and 8.8 mm focal length. The UAV images were processed using different photogrammetry software, ContextCapture, Agisoft Metashape, and UAS Master, by which different photogrammetry produced such as 3D model, dense point cloud, orthophoto, and most importantly Digital Surface Model (DSM) and Digital Terrain Model (DTM) (Fig. 19.3). Previous soil studies have also admired the application of UAV-derived products for DSM (Maleki et al., 2020).

19.3.2 Sampling design and soil analysis

In soil analysis studies, an efficient sampling design can be borrowed from the intrinsic changes of the study area. Such a strategic design can be adopted in large areas where the cost of soil study can be dramatically high. The conditioned Latin Hypercube Sampling (cLHS) is an intuitive approach to circumvent such a predicament. First expounded by Minasny and McBratney (2007), the cLHS method is now provided as an R package (R Development Core Team, 2015). The cLHS method, in a nutshell, sieves through and randomly stratifies the environmental factors representing a given area into the desired sample classes defined by the user and further selects the sampling locations based on the availed prior information of the environmental variables. According to the literature, the cLHS method has been proven promising for soil studies (Zeraatpisheh et al., 2017, 2019; Maleki et al., 2020; Fathizad et al., 2020).

The UAV-derived DEM with a 1 × 1 m resolution was used to retrieve morphometric indices in SAGA software (Conrad et al., 2015) and then used along with other soil/vegetation factors as inputs in the cLHS R package to identify the soil samples for each case study (Table 19.2). Fig. 19.1 depicts the spatial distribution of cLHS-generated soil samples across the three case studies, which is considered the first attempt at using a strategic soil sampling technique for producing precise soil maps in catchment detailed studies, especially areas with limited environmental data. As with the rationale of cLHS which is underpinned by retrieving the most unique information from environmental variables, the reality attested that cLHS-generated soil samples align perfectly with the discernible and somewhat coincidental changes in vegetation cover across case studies. It also uncovers the pivotal role of morphometric indices and in general DEM-derivatives in unraveling the most unique soil attributes. Ultimately, we collected 21, 31, and 55 soil samples, respectively, in Kakhk, Goorband, Dehgin study areas.

All soil samples were passed through a 2 mm sieve after air-drying. The soil texture was determined based on the hydrometer method (Gee and Bauder, 1986). It is noteworthy that by field-related mission and choosing the best points to describe the components of the land unit, two soil pedons were dug up to a depth of 100 cm in each of the study areas. Excavated soil pedons were described, classified, and sampled from genetic horizons based on the keys to Soil Taxonomy (Soil Survey Staff, 2014) classification. All the excavated soil pedons were

FIGURE 19.3 Examples of unmanned aerial vehicles (UAVs) photogrammetry products for Goorband case study ((A) 3D model; (B) Phantom 4 Pro V2.0.; (C) Orthophotomosaic image of the region obtained from UAV; (D): DTM; (E) streams digitized using 3D stereo glasses overlaid on a hillshade map). *(UAV products generated by Spatial Science Innovators Co., Photogrammetry Group).*

TABLE 19.2 The environmental covariates used as a predictor for the conditioned Latin Hypercube Sampling (cLHS) method and digital soil texture classes mapping in the three study area.

Category	Covariate name (abbreviation)	Definition	References
Topographic indices	Elevation (DEM)	Height over sea level	Wilson and Gallant (2000)
	Slope (Slope)	Gradient of line is changing of elevation in the direction and steepness	Gallant and Dowling (2003)
	Aspect (ASP)	Slope direction	
	Curvature (Curvature)	The degree to which a curve deviates from a straight line	
	Plan curvature (Plancurv)	Perpendicular line to the direction of maximum slope	
	Profile curvature (Profcurv)	Direction of maximum slope	
	Convergence (Convergence)	Highlights the convergent areas as channels and divergent areas as ridges.	
	Analytical hillshade	Angle between the surface and the incoming light beams	
	Channel network	The interpolated channel network base level elevations	Conrad et al. (2015)
	Vertical distance to channel network (VDTCN)	Calculates the vertical distance to a channel network base level	Olaya (2004)
	Topographic roughness index (TRI)	Measures terrain ruggedness	Moore et al. (1993)
	Stream power index (SPI)	Connotation of sediment transport	
	Topographic wetness index (TWI)	A measure of the topographic control on soil wetness or the ratio	
	Multi-resolution valley bottom flatness index (MRVBF)	Flat areas in low altitudes and valleys	Gallant and Dowling (2003)
	Multi-resolution of ridge top flatness index (MRRTF)	Flat areas in high altitudes and ridges	
	Topographic position index (TPI)	Elevation difference between a particular cell and the average elevation of the surrounding cells	Olaya (2004)
	Mass balance index (MBI)	Soil erosion and deposition balance	
	Position Openness	Represents areas visible from a line of sight	
	Wind effect	Dimensionless index indicating areas exposed to wind	Boehner and Antonić (2009)
	Valley depth	The vertical distance to a channel network base level	

(Continued)

TABLE 19.2 The environmental covariates used as a predictor for the conditioned Latin Hypercube Sampling (cLHS) method and digital soil texture classes mapping in the three study area.—cont'd

Category	Covariate name (abbreviation)	Definition	References
	Relative slope position (RSP)	The position of one point relative	Olaya (2004)
	Terrain surface texture (Texure)	The geometrical irregularities present at a surface or the nested-means terrain classification	
Organism		Land use map	Prepared using orthophoto of unmanned aerial vehicles (UAVs)
—			
Soil, vegetation, topography	Curve Number (CN)	Prediction of direct runoff or amount of infiltration resulting from excess precipitation	Natural resources protection organization (USDA)
Parent material, topography, soil	Land unit map	—	FAO (1976)

classified to the level of soil family. For each pedon, information about the location of the soil in the landscape, including landform unit, land components, parent material, vegetation diversity, slope, elevation, erosion, and the number of surface pebbles was recorded.

19.3.3 Mapping spatial variability of soil classes using random forest model

A random forest is a decision-tree approach (Breiman, 2001; Lamichhane et al., 2019). Following a bootstrap approach, samples are randomly selected to train each decision tree (Lamichhane et al., 2019). The ntree and mtry are two crucial parameters of the random forest model, which determine the number of trees and influential predictors (i.e., variables) in each random subset (Breiman, 2001; Lamichhane et al., 2019). Also, response curves, one of the valuable outputs of the random forest model, show the most significant contributing factors. Based on a rule of thumb, factors with less than a 15% contribution percentage can be considered less significant and can be omitted from the modeling procedure (Zhi et al., 2017). The rest of the environmental variables are used to train the model. The spatial modeling of the soil properties used the *caret* package of R3.3.1 (R Development Core Team 2015).

19.3.4 Evaluation of model

To evaluate the accuracy of the used models, the overall accuracy and the Kappa index have been used. Kappa index measures (Eq. 19.1) the degree of agreement between two variables or judgments. Kappa compares the probability of agreement with the expected value, while the ratings are considered independent.

$$K = \frac{\mathrm{OA} - \mathrm{CA}}{1 - \mathrm{CA}} \tag{19.1}$$

The kappa index calculates the accuracy level, unlike the overall accuracy method, based on all the pixels that are correctly and incorrectly classified (OA) and the pixels that are based on chance and randomness (CA).

Overall accuracy computes the weighted average of sensitivity and specificity. Sensitivity is weighted by prevalence, and specificity is weighted by the complement of prevalence.

$$\mathrm{OA} = \sum_{u=1}^{U} \frac{A_{\mathrm{uu}}}{A} \tag{19.2}$$

where U is the number of classes, A_{uu} is the correctly classified surface unit u and A is the whole surface.

19.4 Soil development in different paired catchments

As natural land features, the Kakhk catchment is mostly characterized by mountains and hills where poor soil development exhibits sporadically across the area (Fig. 19.4A and B). Moreover, artificial terraces and arable soil for rainfed farming are discernible in small portions of low-hill land units with gentle slopes, which is evident from the freshly planted Almond trees (Fig. 19.4B). The latter accounts for 4.18% of the entire Kakhk catchment. Also, a stabilized and balanced erosion-deposition mechanism is evident in these areas. On the other hand, most investigated soil profiles in hill land units were identified as the subgroup of Lithic Torriorthents. In general, soil developments in these areas rely highly on relief and topography, most notably slope, and the degrees to which the land is subjected to erosional processes. South-facing areas (i.e., receiving more precipitation and hence contains higher moisture content) with steep slopes and sparse vegetation contain less gravel soil cover are exposed to consecutive erosion/deposition cycles and exhibit a lesser thickness of A-horizon (topsoil); hence, the soil is poorly developed. Antithetically, north-facing areas with gentle slopes and dense vegetation can withstand erosion/deposition processes, and hence, soil is highly developed. Our finding is perfectly in line with Teka et al. (2015) and Lozano-García et al. (2016), who reported that the presence of a dense drainage network and higher moisture content in north-facing slopes can signify highly developed soils.

According to the field visits and desk research based on UAV-derived orthophoto and similar base maps, two land units were identified in Hormozgan's paired catchment, Dehgin, namely hills and plateau/upper terraces. Most notably, soil profiles observed in hill land units are categorized in Typic Torriorthents. Similar to the soil study conducted in Goorband, soil development in Dehgin can be attributed to relief/slope and the level of erosion, where dense vegetation in some areas decelerates the erosion-deposition cycle. Soils in plateaus/terraces land units, similar to hills, are poorly developed due mainly to the pervasive gully erosion and concentrated stream networks across the area which have left a thin A-horizon with a thickness of about 0–5 cm (Fig. 19.4C and D).

Goorband catchment is situated in the arid climate of the southeast of Iran, Sistan and Baluchestan province, yet the soil development in this area is incomparable to other arid

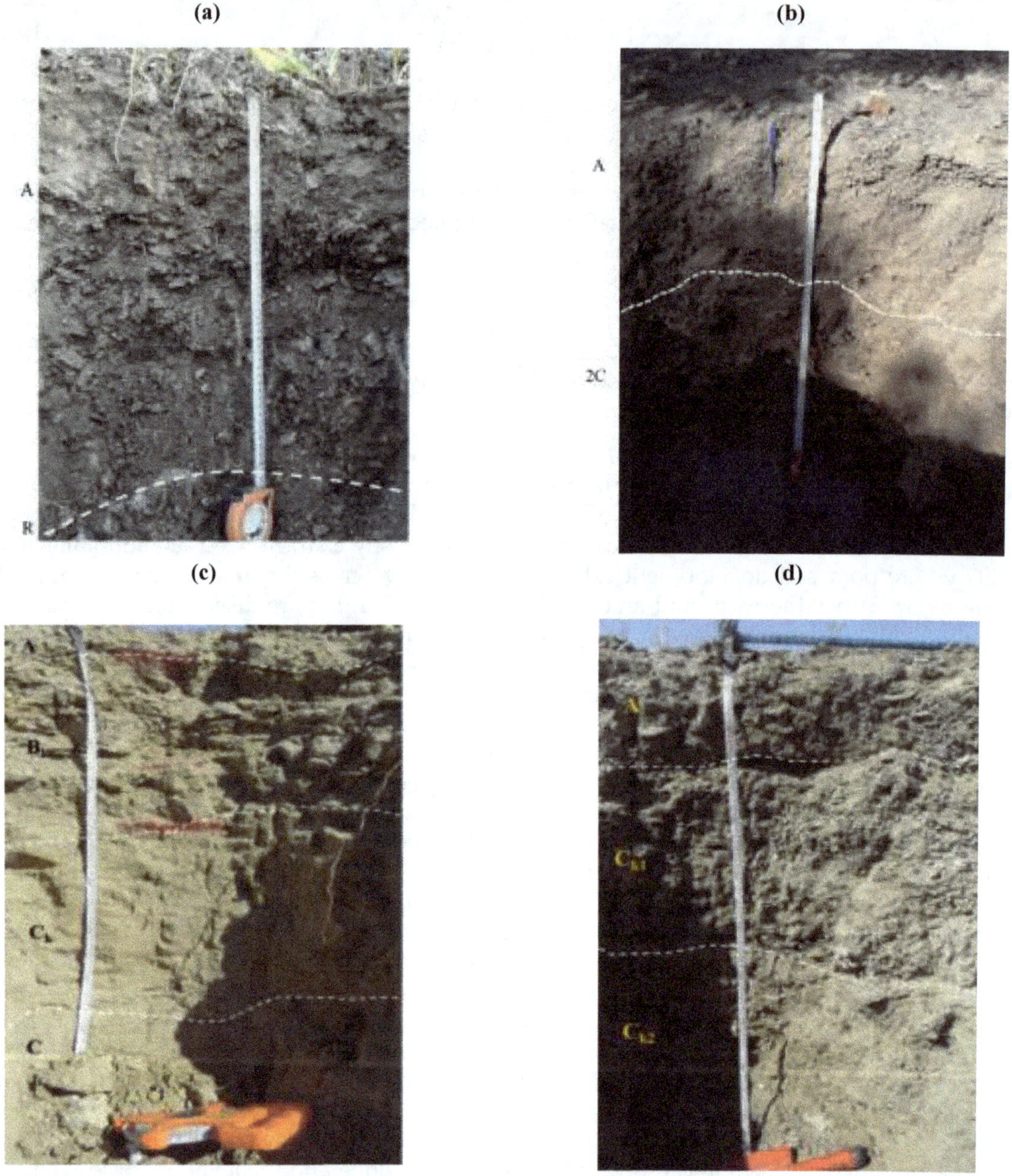

FIGURE 19.4 Examples of the studied pedons in; (A) Loamy-skeletal, mixed, mesic, Lithic Torriorthents located on hill land type in Kakhk catchment, Khorasan Razavi, Iran located on the land type of hills, (B), Sandy-skeletal, mixed, mesic, Typic Torriorthents located on the land type of hills (conservation terraces) in Kakhk catchment, Khorasan Razavi, Iran, (C) Sandy, mixed, calcareous, hyperthermic, Typic Torripsamments located on the land type of plateau and upper terraces in Dehgin, Hormozgan province, Iran, (D) Coarse-loamy, mixed, hyperthermic, Typic Haplocalcids. Located on the land type of plateau and upper terraces in Dehgin, Hormozgan province, Iran. *photos taken by authors.*

and semi-arid zones. It seems that soil parent material in Goorband is mainly supplied by erosion-deposition processes in the form of colluviums and alluviums. The large portion of the catchment is characterized by hill land units, 15.22% of which is covered by rocky outcrops, and poorly developed soil cover is sporadically scattered across the area (Fig. 19.5). The latter attests that, according to the American classification (WRB, 2015),[1] thin layers of young soil across the area can be categorized as Lithic Torriorthents. The eroded materials from the highlands in upstream are gradually sorted out—based on granule size distribution—toward the lowlands in downstream.

19.5 Prediction map of soil texture classes

Textural analysis indicates that Kakhk, Goorband, and Dehgin case studies are all identified with a high proportion of sand content (Table 19.3). The lower proportion of clay stems from the influence of parent material on soil formation and poorly developed soils in these areas. It is noteworthy that no particular outcrop was identified in the Dehgin case study.

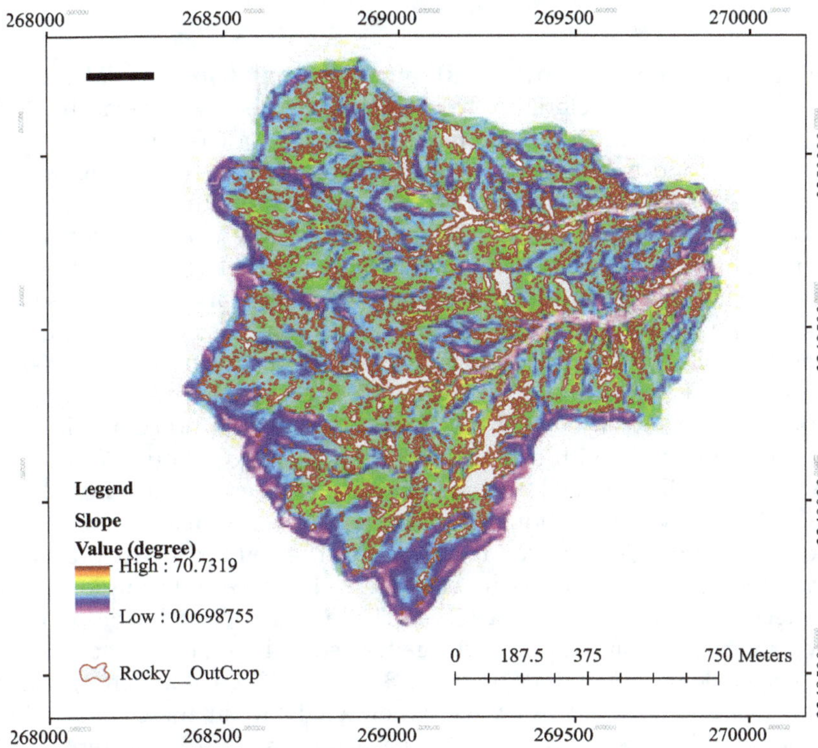

FIGURE 19.5 Rocky outcrop pattern overlaid on the slope map of Goorband paired catchment, Sistan and Baluchestan province, Iran. *maps generated by authors.*

[1] World Reference Base for Soil Resources.

TABLE 19.3 Average particle size distribution in the studied areas.

Case study	Sub-catchment	Area of outcrops (ha)	Sand	Silt	Clay
				(%)	
Kakhk	Control	8.84 (42.39%)	49.02	30.60	20.37
	Treatment	12.01 (57.60%)	47.90	34.00	18.09
	Total	20.85 (9.51%)	48.33	32.70	18.96
Goorband	Control	8.1 (13.96%)	54.09	32.90	13
	Treatment	13.37 (16.10%)	54.09	28.95	12.19
	Total	21.47 (15.22%)	55.83	31.29	12.87
Dehgin	Control	—	64.04	29.41	6.56
	Treatment	—	73.72	20.96	5.48
	Total	—	68.70	25.41	5.98

Soil texture map was generated by the RF algorithm following the rules of DSM. As a result, Kakhk is mostly characterized by loamy and loamy-clay-sandy soil textures (Fig. 19.6A). Most notably, loamy soils are mostly present in areas with gentle slopes where the erosion level is low and the loss of silts rarely occurs. Similarly, soil texture in flat areas concentrated by stream networks was classified as loamy soils with the respective proportion of silt and sand to be 40% and below 50%. Dehgin case study is mostly covered by sandy-loamy and loamy-sandy soils (Fig. 19.6B). Most particularly, the high proportion of sand content is evident in the north and central parts of the catchment marked by intense propagation of gullies, which has a connotation of high erosion level and loss of silt and clay materials. The high proportion of silt materials alongside the stream network (about 50%) signifies the deposition process of fine materials. Loamy soils in the Goorband case study are mostly evident in flat areas with lower levels of the erosion-deposition process, while streams are covered by sandy-loamy soils with a high proportion of sand (about 83%) and an almost negligible proportion of silt (i.e., 6%), which signifies severe erosional process.

In this work, DEM-derived morphometric indices proved promising due to their valuable contribution to generating a detailed soil texture map. The soil texture map of the Goorband catchment across the stream network is generated only by three soil samples, which surprisingly aligns perfectly with what can be observed in the UAV-derived orthophoto (Fig. 19.6C and D). The overall accuracy and kappa of the generated soil texture map for Kakhk, Dehgin, and Goorbad, respectively, are 50%-0.36, 46%-0.35, and 75.5%-0.42. Hence, using morphometric indices directly extracted from UAV-generated DEM in the realm of DSM merits more studies. Moreover, this work is in line with Maleki et al. (2020) that reported the significant performance of the RF algorithm in soil texture mapping with limited soil samples across the loess plateau of Golestan province.

Amirian-Chakan et al. (2019) reported that DSM techniques can produce informative maps of soil texture for a 4600-ha area in Khuzestan province, southwestern Iran. The

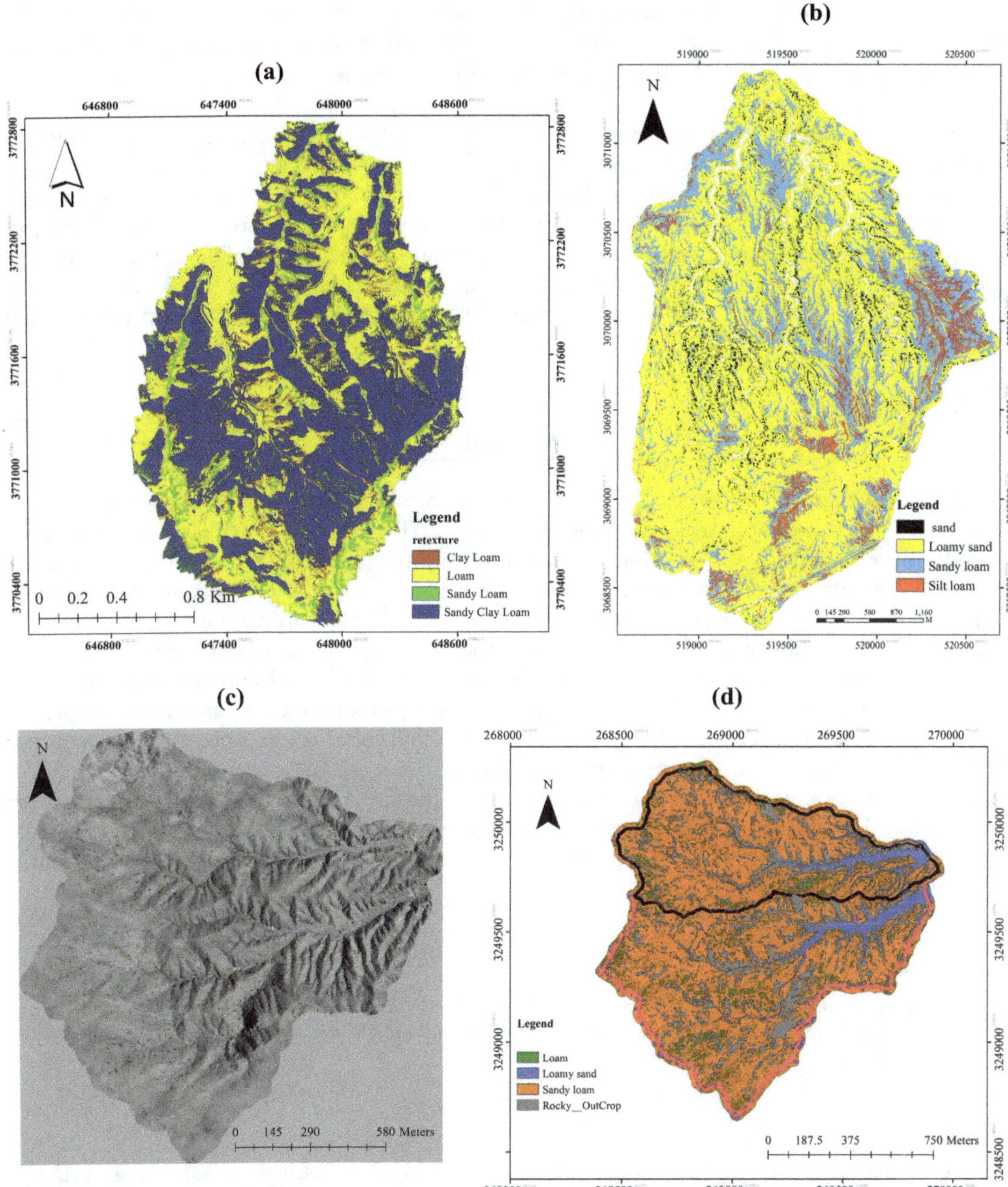

FIGURE 19.6 The predicted soil texture classes map of the paired catchment (A) Kakhk, Khorasan Razavi, Iran; (B) Dehgin, Hormozgan, Iran; (C) Orthophoto of Goorband, Sistan and Baluchestan, Iran; and (D) Goorband, Sistan and Baluchestan, Iran. *Maps generated by authors.*

feasibility of the prediction and mapping of soil particle size distribution through the RF algorithm in an area with varying parent materials, soil classes, and land use are also reported by Andrade et al. (2020) and Žížala et al. (2022), respectively, in Brazil and the Czech Republic.

Our findings indicate that the abundance of soil samples and selecting the most relevant morphometric and environmental indices can directly alter DSM model's accuracy. Also, the resolution of DEM can directly influence the accuracy of the terrain-based syntheses, compensating for the most economic costs and the time spent on soil and water conservation by providing the most varied and valuable information for decision-makers, authorities, and experts. Large areas with limited soil data and high labor expenses would dictate more efficient and cost-effective methods, which is what UAVs and photogrammetry products with suitable flight scales would offer.

19.6 Conclusions

Since many parts of the study areas are formed of steep slopes, it is very important to consider the impact of soil erosion in the entire area. Therefore, it is advised that the maps of erodibility coefficient and erosion classification should be prepared for the study areas and become available to the relevant planners. These maps can rather complement the information on the region's soil and vegetation status, as well as morphometric characteristics. This simple evaluation can decide which products can be used at the local scale and if more accurate DSM products are required. The proportion of sand content and the loss of silt materials in the treatment sub-catchment gives a notion that management strategies may have not been successful. In this regard, the ways of dealing with these limitations should be investigated. For example, it is possible to increase organic matter and carry out the correct methods of water distribution and generally create a suitable cultivation pattern in the region through appropriate terracing. Based on the success of creating accurate digital soil maps in the case studies, the application of UAV-derived products merits more studies and should be further investigated.

Acknowledgments

We would like to thank Abbas Goli Jirandeh, Hadi Alizadeh, Amir Alizadeh, and Mehdi Karimi of *Spatial Science Innovators Co.* for putting meticulous effort for the UAV flight missions and providing us with high-resolution photogrammetry products.

References

Adhikari, K., Kheir, R.B., Greve, M.B., Bøcher, P.K., Malone, B.P., Minasny, B., McBratney, A.B., Greve, M.H., 2013. High-resolution 3-D mapping of soil texture in Denmark. Soil Sci. Soc. Am. J. 77 (3), 860–876.

Amirian-Chakan, A., Minasny, B., Taghizadeh-Mehrjardi, R., Akbarifazli, R., Darvishpasand, Z., Khordehbin, S., 2019. Some practical aspects of predicting texture data in digital soil mapping. Soil Tillage Res. 194, 104289.

Andrade, R., Silva, S.H.G., Faria, W.M., Poggere, G.C., Barbosa, J.Z., Guilherme, L.R.G., Curi, N., 2020. Proximal sensing applied to soil texture prediction and mapping in Brazil. Geoderma Reg. 23, e00321.

Boehner, J., Antonić, O., 2009. Land-surface parameters specific to topo-climatology. Geomorphometry– Concepts, Software, Applications. Developments in Soil Science, 33. Elsevier, pp. 195–226.

Breiman, L., 2001. Random forests. Mach. Learn. 45, 5–32.

Conrad, O., Bechtel, B., Bock, M., Dietrich, H., Fischer, E., Gerlitz, L., Wehberg, J., Wichmann, V., Böhner, J., 2015. System for automated geoscientific analyses (SAGA) v. 2.1. 4. Geosci. Model Dev. (GMD) 8 (7), 1991–2007.

FAO, 1976. A framework for land evaluation. FAO Soils Bulletin, No 32. FAO, Rome, 71 pp.

Fathizad, H., Ardakani, M.A.H., Sodaiezadeh, H., Kerry, R., Taghizadeh-Mehrjardi, R., 2020. Investigation of the spatial and temporal variation of soil salinity using random forests in the central desert of Iran. Geoderma 365, 114233.

Forkuor, G., Hounkpatin, O.K., Welp, G., Thiel, M., 2017. High resolution mapping of soil properties using remote sensing variables in south-western Burkina Faso: a comparison of machine learning and multiple linear regression models. PLoS One 12 (1) e0170478.

Gallant, J.C., Dowling, T.I., 2003. A multiresolution index of valley bottom flatness for mapping depositional areas. Water Resour. Res. 39 (12), 1347. https://doi.org/10.1029/2002WR001426.

Gee, G.W., Bauder, J.W., 1986. Particle size analysis. In: Klute, A. (Ed.), Methods of Soil Analysis, Part 1. American Society of Agronomy, Madison, WI, USA, pp. 383–411.

Kariminejad, N., Pourghasemi, H.R., Maleki, S., Hosseinalizadeh, M., 2022b. Digital soil mapping and modeling in Loess-derived soils of Iranian Loess Plateau. Geocarto Int. 37 (26), 11633–11651.

Kariminejad, N., Shariati, M., Hosseinalizadeh, M., Pourghasemi, H.R., Maleki, S., 2022a. Investigating geometrical characteristics of collapsed pipes and the changing role of driving factors. J. Environ. Manag. 312, 114910.

Kaya, F., Basayiğit, L., Keshavarzi, A., Francaviglia, R., 2022. Digital mapping for soil texture class prediction in northwestern Türkiye by different machine learning algorithms. Geoderma Reg. 31, e00584.

Khaledian, Y., Miller, B.A., 2020. Selecting appropriate machine learning methods for digital soil mapping. Appl. Math. Model. 81, 401–418.

Lamichhane, S., Kumar, L., Wilson, B., 2019. Digital soil mapping algorithms and covariates for soil organic carbon mapping and their implications: a review. Geoderma 352, 395–413.

Lozano-García, B., Parras-Alcántara, L., Brevik, E.C., 2016. Impact of topographic aspect and vegetation (native and reforested areas) on soil organic carbon and nitrogen budgets in Mediterranean natural areas. Sci. Total Environ. 544, 963–970.

Makabe, S., Kakuda, K.I., Sasaki, Y., Ando, T., Fujii, H., Ando, H., 2009. Relationship between mineral composition or soil texture and available silicon in alluvial paddy soils on the Shounai Plain, Japan. Soil Sci. Plant Nutr. 55 (2), 300–308.

Maleki, S., Khormali, F., Mohammadi, J., Bogaert, P., Bodaghabadi, M.B., 2020. Effect of the accuracy of topographic data on improving digital soil mapping predictions with limited soil data: an application to the Iranian loess plateau. Catena 195, 104810.

Malone, B.P., Minasny, B., McBratney, A.B., 2017. Using R for Digital Soil Mapping, vol. 35. Springer International Publishing, Cham, Switzerland.

McBratney, A.B., Santos, M.M., Minasny, B., 2003. On digital soil mapping. Geoderma 117 (1–2), 3–52.

Minasny, B., Hartemink, A.E., 2011. Predicting soil properties in the tropics. Earth Sci. Rev. 106 (1–2), 52–62.

Minasny, B., McBratney, A.B., 2007. Incorporating taxonomic distance into spatial prediction and digital mapping of soil classes. Geoderma 142 (3–4), 285–293.

Moore, I.D., Gessler, P.E., Nielsen, G.A.E., Peterson, G.A., 1993. Soil attribute prediction using terrain analysis. Soil Sci. Soc. Am. J. 57 (2), 443–452.

Nussbaum, M., Spiess, K., Baltensweiler, A., Grob, U., Keller, A., Greiner, L., Schaepman, M.E., Papritz, A., 2018. Evaluation of digital soil mapping approaches with large sets of environmental covariates. Soils 4 (1), 1–22.

Olaya, V.F., 2004. A Gentle Introduction to Saga GIS. The SAGA User Group eV, Göttingen, Germany.

Pahlavan-Rad, M.R., Akbarimoghaddam, A., 2018. Spatial variability of soil texture fractions and pH in a flood plain (case study from eastern Iran). Catena 160, 275–281.

Purushothaman, N.K., Reddy, N.N., Das, B.S., 2022. National-scale maps for soil aggregate size distribution parameters using pedotransfer functions and digital soil mapping data products. Geoderma 424, 116006.

R Development Core Team, 2015. R: A Language and Environment for Statistical Computing. R Foundation for Statistical Computing, Viena, Austria.

Soil Survey Staff, 2014. Keys to Soil Taxonomy. 12th USDA-Natural Resources Conservation Service, Washington, DC. Available at: https://www.nrcs.usda.gov/wps/PA_NRCSConsumption/download?cid=stelprdb1252094&ext=pdf. (Accessed 3 January 2018).

Stenberg, B., Rossel, R.A.V., Mouazen, A.M., Wetterlind, J., 2010. Visible and near infrared spectroscopy in soil science. Adv. Agron. 107, 163–215.

Taghizadeh-Mehrjardi, R., Mahdianpari, M., Mohammadimanesh, F., Behrens, T., Toomanian, N., Scholten, T., Schmidt, K., 2020. Multi-task convolutional neural networks outperformed random forest for mapping soil particle size fractions in central Iran. Geoderma 376, 114552.

Taghizadeh-Mehrjardi, R., Nabiollahi, K., Kerry, R., 2016. Digital mapping of soil organic carbon at multiple depths using different data mining techniques in Baneh region, Iran. Geoderma 266, 98–110.

Taghizadeh-Mehrjardi, R., Schmidt, K., Toomanian, N., Heung, B., Behrens, T., Mosavi, A., Band, S.S., Amirian-Chakan, A., Fathabadi, A., Scholten, T., 2021. Improving the spatial prediction of soil salinity in arid regions using wavelet transformation and support vector regression models. Geoderma 383, 114793.

Teka, K., Nyssen, J., Teha, N., Haile, M., Deckers, J., 2015. Soil, land use and landform relationship in the Precambrian lowlands of northern Ethiopia. Catena 131, 84–91.

Warrington, D.N., Mamedov, A.I., Bhardwaj, A.K., Levy, G.J., 2009. Primary particle size distribution of eroded material affected by degree of aggregate slaking and seal development. Eur. J. Soil Sci. 60 (1), 84–93.

Wiesmeier, M., Barthold, F., Blank, B., Kögel-Knabner, I., 2011. Digital mapping of soil organic matter stocks using Random Forest modeling in a semi-arid steppe ecosystem. Plant Soil 340, 7–24.

Wilson, J.P., Gallant, J.C., 2000. Terrain Analysis: Principles and Applications. Wiley & Sons, New York.

WRB-World Reference Base for Soil Resources, 2015. Food and agriculture organization of the United Nations. Update 201 (5), 204.

Zeraatpisheh, M., Ayoubi, S., Jafari, A., Finke, P., 2017. Comparing the efficiency of digital and conventional soil mapping to predict soil types in a semi-arid region in Iran. Geomorphology 285, 186–204.

Zeraatpisheh, M., Ayoubi, S., Jafari, A., Tajik, S., Finke, P., 2019. Digital mapping of soil properties using multiple machine learning in a semi-arid region, central Iran. Geoderma 338, 445–452.

Zhi, J., Zhang, G., Yang, F., Yang, R., Liu, F., Song, X., Zhao, Y., Li, D., 2017. Predicting mattic epipedons in the northeastern Qinghai-Tibetan plateau using random forest. Geoderma Reg. 10, 1–10.

Žížala, D., Minařík, R., Skála, J., Beitlerová, H., Juřicová, A., Rojas, J.R., Penížek, V., Zádorová, T., 2022. High-resolution agriculture soil property maps from digital soil mapping methods, Czech Republic. Catena 212, 106024.

CHAPTER

20

Object-based image analysis approach for gully erosion detection

Ramyar Aliramaee[1], *Omid Rahmati*[2], *Farnoush Mohammadi*[3] *and Seyed Masoud Soleimanpour*[4]

[1]Department of Watershed Management Engineering, Faculty of Natural Resources, Tarbiat Modares University, Noor, Iran; [2]Soil Conservation and Watershed Management Research Institute, Agricultural Research, Education and Extension Organization (AREEO), Kurdistan Agricultural and Natural Resources Research and Education Center, Sanandaj, Iran; [3]Department of Reclamation of Arid and Mountainous Regions, Faculty of Natural Resources, University of Tehran, Karaj, Iran; [4]Soil Conservation and Watershed Management Research Institute, Fars Agricultural and Natural Resources Research and Education Center, Agricultural Research, Education and Extension Organization (AREEO), Shiraz, Iran

20.1 Introduction

The vulnerability of humans to natural and anthropogenic hazards continues to increase. Natural hazards are prevalent in our world today, including, droughts, desertification, floods, fires, earthquakes, and landslides, and they, directly and indirectly, cause land degradation. We will likely experience more destruction shortly due to these disasters, which will be commonplace as time goes by. Due to the rapid growth of the global population along with globalization, this situation began with increased vulnerability, and catastrophic outcomes will only increase unless improved methods are developed for mitigating the effects, such as forecasting and warning, and strengthening community resilience and preparedness. A wide range of natural resources is affected by natural hazards. These resources include water, air, and soil, which are crucial for human beings.

The soil serves as a source of nourishment for human activity and is an integral part of the human environment in agriculture and forestry. Therefore, the soil needs to be given much more attention, and greater control needs to be exercised concerning its rational use, protection, and improvement. Degradation of soil is one of the challenges that soil faces.

Remote Sensing of Soil and Land Surface Processes
https://doi.org/10.1016/B978-0-443-15341-9.00009-5

The degradation of soil is a biophysical process that is further exacerbated by socioeconomic and political factors. A soil degradation process refers to a long-term reduction in soil productivity as well as the ability of soils to moderate the environment (Lal, 1994). The soil degradation is aggravated by poverty caused by population growth and industrialization, and it can cause more serious problems on subsistence farms without external input than on commercial farms. So, the soil has to be conserved carefully. Further industrialization and urbanization are also being caused by rapid technological progress that results in deterioration, destruction, intoxication, and contamination of the environment.

20.2 Soil erosion as a natural hazard

Soil erosion has been known as one of the primary drivers of land degradation. Soil erosion is a serious problem in semi-arid or sub-humid regions (Lal, 2001). Since sedimentation is a leading pollutant of water resources, watershed planners must understand the mechanism of sedimentation, which is induced by erosion.

The process of soil erosion is the detachment of surface materials by wind and water. As a result of this process, specially under intensive rainfall, the structure of soils deteriorates. The erosion process involves three stages: detachment, transport, and deposition of soil. During erosion, energy is supplied by agents of erosion, and the type of erosion is determined by the source of energy. In general, there are four types of energy sources: physical such as wind and water, gravity, chemical reactions, and anthropogenic perturbations such as tillage.

20.3 Types of soil erosion

Researchers, conservationists, and policymakers have been paying considerable attention to erosion by surface runoff. Water erosion, which damages productive soil and agricultural grounds, is a key factor in land degradation in many regions of the world. Due to its socioeconomic effects, loss of organic matter and nutrients, reduction in water availability, and decrease in water retention, water erosion is a serious issue (Morgan, 1995; Poesen et al., 1996; Kirkby, 2002).

There are three types of water erosion: rill, surface/sheet, stream bank, and gully. The highest amount of soil loss occurs as a result of gully erosion, which can be both temporary and permanent (Ma and Zhao, 1994; Poesen et al., 2003). Despite this, studies of soil erosion have mainly concentrated on rill and sheet erosion, while the study of gully erosion has not been conducted a lot (Valentin et al., 2005). Gully initiation forms deep and long trenches in localized areas, leading to much soil erosion (Fig. 20.1). In contrast to sheet erosion, which affects only the surface layers of the profile, it impacts a wider area, and sheet erosion occurs over a wider area than gully erosion, which is more of a localized catastrophe.

FIGURE 20.1 A field photograph of a gully in Talwar watershed, Iran. *Taken by Omid Rahmati.*

20.4 The effects of soil erosion

The hazard of soil erosion has traditionally been associated with agriculture, and it has significant long-term implications for soil productivity and land sustainability. And as a result of soil erosion, soil degradation is exacerbated and vice versa. However, it is a problem that also affects forestry, transportation, and recreation areas. In addition to sedimentation and pollution, erosion also contributes to increased flooding and environmental damage.

There are offsite and onsite consequences associated with soil erosion. Onsite effects are those which occur at the site where the soil is removed, and the main onsite effect is the loss of soil productivity, pollution, siltation, and blocking of channels and waterways are examples of offsite consequences. In addition to increased productivity on downstream lands, sedimentation of reservoir dams and eutrophication of water bodies and streams are also important offsite influences.

It is a general consequence of soil erosion that the area of fertile land suited for agriculture is diminishing, the ratio of cultivated land is declining at an even faster rate, the proportion of deteriorated soil caused by human activity is on the rise, and the concentrations of exogenous, often toxic substances are increasing in the soil, reducing its ability to perform its biotic functions in the ecosphere.

The sedimentation behind dams can reduce the lifespan of reservoirs. Carrying nutrients by sediments contributes to the eutrophication of water bodies (Rabalais et al., 2010). In streams and rivers, benthic habitats can be degraded by sediment and biodiversity can be affected (Bilotta and Brazier, 2008). There is often a greater economic cost associated with the movement of sediment and its deposition in the landscape than the cost associated with the long-term loss of soil associated with eroding fields.

20.4.1 Gully erosion and identification

An important issue in natural resource management is gully erosion, which can have severe economic, environmental, and socioeconomic consequences. As defined by Poesen et al. (1996), a gully forms a channel that conventional tillage techniques cannot fill it because it has a cross-sectional area of more than 1 ft^2. In gullies, sediment deposits or bedrock are carved into ephemeral drainage channels.

Gullies occur when runoff water accumulates due to land surface conditions (Morgan, 2009; Blanco and Lal, 2008). Tillage operations can remove ephemeral gullies, though they tend to reappear in the same location regularly (Casalí et al., 1999). Wider gullies cannot be eliminated with tillage.

Gully erosion, which makes slopes useless for farming and supplies sediment to a broad range of habitats, is one of the main signs of desertification and land degradation (Poesen et al., 2003). Although these features show serious land degradation, their size cannot be clearly attributed to recent rainfall (Seeger et al., 2009) or runoff on the surface of the earth (Marzolff and Ries, 2007), and in contrast to their appearance of freshness and activity, they may actually be very old and stable (Valentin, 2004). So, it is often difficult to determine when they formed and when they changed.

It has also been noted that little is identified about gullying process and its importance at different geographic dimensions and that the majority of study has concentrated on sheet erosion (inter-rill) and rill erosion (Poesen et al., 2006; Vrieling et al., 2007). This is because gullies, particularly in semi-arid areas, can persist unaltered for long periods after they have developed.

20.4.2 The condition occurring and factors affecting

The process of gully's incision and development usually does not require a specific timeframe, but it does need suitable conditions. In general, semi-arid and sub-humid areas where rainfall amount is low with considerable intensity are susceptible to gully formation. It generally does not take a precise amount of time for a gully to be cut and developed, but favorable circumstances are required. Gully growth typically happens in semi-arid areas with infrequent but extreme rainfall. In addition, gully erosion is generally associated with unsustainable developments, population growth, overgrazing, deforestation, and other geo-environmental factors.

Geomorphological factors play a major role in gully occurrence. Selected geomorphological conditioning factors are as follows: topography (Tang et al., 2022), lithology (Golestani et al., 2014), drainage density (Arabameri and Pourghasemi, 2019), soil texture (Wang et al., 2021), physical properties of the surface soil (Rahmati et al., 2022), slope degree (Zabihi et al., 2018), slope aspect (Rahmati et al., 2017a,b), plan curvature (pal et al., 2020), distance from roads and rivers (Dube et al., 2014), erodibility of soil (Torri and Borselli, 2003), and altitude (Mokarram and Zarei, 2021). Each of them has a specific effect on gully erosion.

In addition, the environment has a significant influence on the expansion of gullies. Several factors contribute to gully growth, including precipitation (Anderson et al., 2021), indices of surface runoff (Grissinger, 1996), antecedent precipitation (Bocco, 1991), and soil moisture (Sadeghi et al., 2020). Additionally, the gully's appearance and development can be strongly influenced by land cover and land use. Generally, barren lands and sparsely vegetated areas are more prone to gully erosion than rangeland and forests (Shit et al., 2015; Debanshi and Pal, 2020).

20.5 Effects of gully erosion on land degradation

Throughout history, gully erosion has contributed to unsustainable land management practices (Zaimes and Schultz, 2012) and has been recognized as a primary cause of land degradation (Pal et al., 2022). It affects soil productivity (Tang et al., 2013), limits land use, and causes socioeconomic problems around the world (Jahantigh and Pessarakli, 2011). Furthermore, gully erosion causes damage to roads and structures (Junior et al., 2010), and advancing gully heads can result in the loss of residential and recreational areas (Burkard and Kostaschuk, 1997). Moreover, gully erosion is a serious concern when it comes to the management of natural resources, such as soil and water (Joseph and Westhuizen, 2021).

20.6 The effects of gully erosion on sediment production

The gully erosion is a substantial source of sediment for agricultural areas (Poesen et al., 1996) and, thereby, a major threat to their sustainability. It has been reported that 90% of the sediment production of a watershed in some parts of the world is due to gully erosion (Vandekerckhove et al., 2000). As a result of soil erosion and sedimentation, water quality, and reservoir capacity are drastically reduced (Kidane and Alemu, 2015). As well, sediments can transport both nutrients and pollutants (Wijesiri et al., 2019). The human activities have led to accelerated erosion, although quantitative information is lacking about its contribution to sediment production.

20.7 The importance of monitoring gully erosion

Understanding the environment and its processes is essential for sustainable land management, which calls for an effective method of comprehension, surveying, and tracking. Gully sites must be carefully monitored and mapped because they are one of the major contributors to soil loss in the watershed landscape. Also, identifying gullies and adopting suitable land use planning are primary principles in soil conservation. Additionally, planning for soil protection and watershed management both take into account the severity and geographic spread of soil erosion.

Resource managers and residential planners need to relate the expansion rates of gullies to other measurable variables to explain and estimate future gully morphology and behavior. Most analyses of gully growth are empirical in approach. In other words, gully development over time is generally investigated using aerial photographs. Using traditional methods could be time-consuming and hard, so there is a substantial need for using remote sensing (RS) methods for the identification of gully erosion.

20.7.1 Investigating the geomorphological processes of the Earth's surface

Researchers are interested in investigating the geomorphological features of the Earth's surface. In this regard, two methods are available; field-based methods and RS methods. Before the advent of aerial photographs and later satellite imagery, field-based methods were used for the interpretation of images and processing of images. It should be noted, however, that field investigation methods are expensive, time-consuming, and labor-intensive, making them unsuitable for large-scale investigations (Karami et al., 2015).

Researchers have used RS to study soil erosion since the 1940s (Smith, 1943). In recent years, spaceborne data have become increasingly available with suitable spatial and spectral information, allowing erratic instances of land degradation to be detected, thus reducing the relevance of airborne data. In 1972, the scientific community gained access to satellite imagery with the launch of Landsat-1(Wulder et al., 2019). So, Landsat imagery was successfully delineated by visual analysis after aerial photo interpretation (Singh, 1977).

A variety of types of soil erosion have been investigated using RS (Lu et al., 2003; Tagore et al., 2012; Khosrokhani and Pradhan, 2014). It has been relatively unchanged for decades to assess gully erosion using aerial photos and photogrammetric techniques (Langran, 1983; Stromquist et al., 1985). Photo interpretations were restricted to visual interpretations of photos, which were frequently obtained, typically stored in files that were difficult to access and organize, and lacked multispectral data. These images were also typically taken for a narrower area of interest. Despite the obvious advantages of visual analysis, it is time-consuming, laborious, and does not meet the needs of the continuous land-surface monitoring (Bogrekci and Godwin, 2007; Mararakanye and Nethengwe, 2012). Hence, semi-automatic or automatic techniques are required to extract gully formations (Shruthi et al., 2011).

Given the variety of gully sizes, forms, occurrences, and the dynamic nature of gullying process, an RS-based analysis is the only feasible method for recording gullies (Knight et al., 2007). Since RS images have become more readily available, and geographic information systems (GIS) have been developed, image-based interpretation methods have become more prevalent. These techniques can be categorized into two categories: visual interpretation and automatic extraction. The visual interpretation approach depends on expert knowledge and experiences about gully formation and image characteristics (Liu et al., 2018; Wang et al., 2020). However, the subjectivity of manual work remains limitations.

Understanding the dynamics of gully growth, particularly, the rise in the area and a number of gully systems are crucial in addition to recording characteristics like gully dimensions and locations (Poesen, 2011). The same unbiased technique must be used to evaluate all images in order to detect changes in pre-existing gullies, particularly in big areas with widespread gullying (Shruthi et al., 2015).

20.7.2 Pixel-based methods

Researchers now have new chances to study gully erosion, thanks to developments in RS techniques, including the availability of digital surface models and improved temporal and spatial precision, which need little ground-based and in situ measured data (Martha et al., 2010), and the expansion and development of digital imaging technology and GIS. Some studies have tried to improve automatic tools that can map gullies with a high degree of efficiency and repeatability. The first approach to spectral similarity analysis in pictures was pixel-based, but this approach ran into the inescapable issue of false positives because of spectrally similar pixels (Rahmati et al., 2017a,b). Different thresholds and geometric indicators including vegetation (shrubs, hedgerows, and agricultural lands), shadow, small-scale features, and bright areas are used to remove false positives. It also has another negativity that pixel-based methods only use the surface reflectance values contained in pixels (Shruti et al., 2011).

It has been the main scheme for extracting gullies for many years to analyze pixel-based images, using only surface reflectance values (Bocco, 1991; Metternicht and Zinck, 1998). In order to map eroded areas thematically and quantitatively, different pixel-based categorization techniques can be used (Lillesand et al., 2015). However, it is essential to have a thorough understanding of the study area in order to select appropriate training pixels for a successful classification (Laliberte and Rango, 2009).

Additionally, factors like moisture, biological matter, mineral composition, and shadows have an impact on how spectral heterogeneity is distributed throughout an area. Multispectral picture classifications are negatively impacted as a consequence. Furthermore, according to Vrieling and Rodrigues (2005), the geographic resolutions of the first platforms and satellites, for example, SPOT, Landsat, IRS, ASTER, and ENVISAT, were inadequate to detect gully forms (Vrieling and Rodrigues, 2005).

A high-resolution RS image (HRI) requires a significant reduction of pixel size, a considerable increase of information to be processed, as well as a higher degree of spectral ambiguity (Shruthi et al., 2011). As a result, a technique that only uses spectral data is unable to catch the spectral variations linked to gully characteristics like vegetation or shadow and moisture effects. With HRI becoming more widely accessible, a transition from conventional pixel-based techniques has become imperative. It is more appropriate to analyze objects or features rather than individual pixels to address the heterogeneity, as well as for a knowledge-based analysis similar to the interpretation of visual images.

20.7.3 Object-based methods

Object-based image analysis (OBIA), also known as object-oriented analysis (OOA), has been demonstrated to be superior to pixel-based methods for identifying and describing gullies (Shruthi et al., 2014). As compared to pixel-based approaches, object-based analysis has improved the identification of gully features and determination of gully extents (Johansen, 2010; Knight et al., 2007).

In comparison to pixel-based methods, OBIA approach can consider more data in terms of conceptual, geometry and textures, as well as an optimized feature space (Blaschke and Strobl, 2001; Karami et al., 2015). The use of OBIA for mapping gully erosion by HRI, along with ground information and expert knowledge, has been reported by several researchers (d'Oleire-Oltmanns et al., 2014; Shruthi et al., 2015). Accurate gully identification requires extra data or specialized expertise (Bocco and Valenzuela, 1993). It should be noted, however, that the determination of extraction rules, including the variety of thresholds and feature types, can only be carried out semi-automatically, which implies that these procedures keep a definite element of subjectivity.

In order to more precisely identify the gully system and define its limits, the OBIA technique demonstrated its efficacy in recording finer gully tracks. Because of the constraints of ocular picture analysis and digitization, several mistakes in the OBIA processing can actually be ascribed to them (Shruthi et al., 2011). Further, the OBIA method of gully mapping remains objective and practical as compared to the traditional method (visually interpreting images or using pixels). In this case, it is more objective because it relies on the same evaluation criteria for all participants, thus avoiding the effects of human subjectivity. As the region to be monitored becomes larger, the procedure will certainly become more advantageous as the effort required to analyze manually would increase with the number of gully forms, but the work required to process OBIA would not. Thus, it is better suited to supporting the decision-making processes of land managers and erosion researchers. The time required to record gully dimensions throughout the test site using the same strategy/rule-set was significantly less than the time needed to manually map gully features after limits for the test region were established.

20.7.4 Environmental criteria used in object-oriented analysis to identify gullies

With their shape (d'Oleire-Oltmanns et al., 2014), length (Liu et al., 2017), topological entities (Karami et al., 2015), and textural characteristics (Shruthi et al., 2014), gully erosion features can be considered spatial objects that can be distinguished by their geometric properties as well as their spatial association with neighboring features. Several factors influence the spectral heterogeneity of gullies, such as bare soil, shadow, vegetation, and moisture-associated brightness changes (Taruvinga, 2009). Due to their spectral heterogeneity, gullies tend to produce "speckled" or "salt-and-pepper" image classification results. Some previous studies have demonstrated the inability to separate water erosion features at an acceptable level of accuracy through pixel-based classification techniques including support vector machine (SVM) and maximum likelihood classification (MLC) algorithms due to their spectral similarity to nonerosion features (Torkashvand, 2022; Shruthi et al., 2011).

20.8 Digital surface model generation

First, we need a digital surface model that characterizes land surface, man-made features and vegetation above ground. A digital terrain model (DTM), which only represents the earth's surface and excludes above-ground characteristics, would be generated from the digital surface model (DSM) to measure topographic factors. It should be noticed that parameter accuracy is primarily determined by DTM accuracy (Dragut and Blaschkeb, 2006).

Corrective measures are taken to minimize large errors in DTM derivatives using the Leica Photogrammetric Suite (LPS) (Martha et al., 2010). Utilizing 3D break-lines, the vertex or elevation values of specific regions, such as streams, rivers, and steep gullies, are typically carefully rectified. Finally, SAGA will be used to correct the DTM hydrologically

(Böhner and McCloy, 2006) so that slope, flow direction, and sub-watersheds can be extracted. These types of information along with the terrain model will be used as input data for OBIA approach.

20.9 Gully extraction

A semi-automatic approach for feature extraction is used in eCognition Developer (Baatz, 2000). The eCognition Developer is a software tool for analyzing and interpreting images based on objects. The software extracts information from digital imagery, such as aerial photographs, satellite images, and other RS data. Land cover changes can be detected, land cover types can be classified, and features like gullies, roads, buildings, and vegetation can be mapped. Over time, new features and capabilities have been added to the software since its initial release.

20.9.1 Texture measure based on flow direction

There is no doubt that texture metrics, for instance, those obtained from the gray level co-occurrence matrix (GLCM) (Haralick et al., 1973) and can greatly improve image classification systems' efficiency (Laliberte and Rango, 2009). Gully features can be distinguished from nongully features based on their direction. The direction of flow in gullies is generally along the slope, whereas freshly plowed land, which has textural and spectral characteristics similar to gullies, usually runs parallel to the slope. In order to differ them from gully forms, rotation-variant GLCM metrics are used. By resampling the original DTM to lower resolution, experts will be able to recognize directionality in the texture by creating a simulated flow direction boundary that includes an adequate number of PAN cells. The flow direction (FD) will then be calculated and classified into four main classes: N–S (0°), NE–SW (45°), E–W (90°), and NW–SE (135°). Following this, the normalized GLCMs will be calculated in the flow direction (Pi,j(FD)) and perpendicular to the flow (Pi,j(FD-90°)) within each flow direction segment.

20.9.2 Spatial classification: background removal

For HRI results to be used, a large amount of information must be handled. To achieve better results, it is advantageous to eliminate the majority of noninterested or background segments. A relatively low specific catchment area, an area with a flat to gentle slope and a texture that is primarily perpendicular to flow direction (quasi-linear features like plow lines), as well as vegetation and shadow, is classified as background and will not be included in the analysis. Drăgut et al. (2010) proposed the estimation of scale parameter (ESP) tool to estimate the ideal scale factor for conducting a segmentation, a critical step in the image classification. To remove most segments with relatively low SCA and areas with flat to gentle slopes, a multiresolution segmentation will be applied on the slope and terrain model using a homogeneity principle derived from compactness and shape elements determined by a scale factor of 72. Following that, the unclassified area is further considered for segmentation process with multiresolution using the P-B-G-R-NIR bands and an S parameter of 92, determined using ESP. False positives will be removed later when segments cannot be classified as background.

20.9.3 Extraction of gully candidates

Once the background has been removed, the unclassified segments are segmented using chessboard segmentation with a specific object size (1) to refound the original components of images by dividing them or selected areas into equal squares of a given size. After this, a contrast filter is applied to improve characteristics in dark boundaries and shadowed areas. Using this method, potential gully segments are identified and then merged. As a second phase, an edge detection filter named Lee-Sigma will be utilized to the PAN information to identify dark borders and sites. As a result of this process, the outlines of objects in the image, as well as the boundaries between background and objects are detected. This extraction strategy was used to detect all potential and actual gullies. Among potential gully candidates, false positives will be removed later.

20.9.4 Removal of false positives

Finally, false positives will be eliminated from the gully feature extraction process, which included other quasi-linear features which were mistakenly classified as gullies, as well as segments identified based on locally erroneous DSM derivatives. In the elevation data, slopes along vegetation boundaries are represented as small humps, cattle tracks, and plow lines that are orthogonal to bright patches and slope probably represent outcrops can be observed.

20.9.5 Accuracy assessment using ground-truth data

Although accuracy calculation is typically focused to comparing ground-truth data (reference data) and results, and application of the evidences can also be important. Therefore, it is not meant to match gully features individually. This is due to the challenges associated with producing complete reference data (e.g., by digitizing), as well as the partial limitations in its use from the perspective of information users. As gullies limit the use of land, their extent is the most significant parameter. Visual image interpretation and field observations will be used to generate reference data. Since direct comparison of gully polygons with digitized features is affected by digitizing process, polylines will be digitized and converted to a raster format. Therefore, instead of digitizing single gullies, polygons presenting gullies can be digitized by linking the gully incision points identified in classified (AOOA) gully systems and reference (AREF) by the convex hull method proposed by Melkman (1987).

References

Anderson, R.L., Rowntree, K.M., Le Roux, J.J., 2021. An interrogation of research on the influence of rainfall on gully erosion. Catena 206, 105482.

Arabameri, A., Pourghasemi, H.R., 2019. Spatial modeling of gully erosion using linear and quadratic discriminant analyses in GIS and R. Spat. Model. GIS R Earth Environ. Sci. 299–321.

Baatz, M., 2000. Multiresolution segmentation: an optimization approach for high quality multi-scale image segmentation. Angewandte geographische informationsverarbeitung 12–23.

Bilotta, G.S., Brazier, R.E., 2008. Understanding the influence of suspended solids on water quality and aquatic biota. Water Res. 42 (12), 2849–2861.

Blanco, H., Lal, R., 2008. Principles of Soil Conservation and Management, vol. 167169. Springer, New York.

Blaschke, T., Strobl, J., 2001. What's wrong with pixels? Some recent developments interfacing remote sensing and GIS. Zeitschrift für Geoinformationssysteme 12–17.

Bocco, G., Valenzuela, C.R., 1993. Integrating satellite-remote sensing and geographic information systems technologies in gully erosion research. Rem. Sens. Rev. 7 (3–4), 233–240.

Bocco, G., 1991. Gully erosion: processes and models. Prog. Phys. Geogr. 15 (4), 392–406.

Bogrekci, I., Godwin, R.J., 2007. Development of an image-processing technique for soil tilth sensing. Biosyst. Eng. 97 (3), 323–331.

Böhner, J., McCloy, K.R., 2006. SAGA-analysis and modelling applications. Collect. Göttinger geographische Abhandlungen 115.

Burkard, M.B., Kostaschuk, R.A., 1997. Patterns and controls of gully growth along the shoreline of Lake Huron. Earth Surf. Process. Landforms: J. Brit. Geomorphol. Group 22 (10), 901–911.

Casalı, J., López, J.J., Giráldez, J.V., 1999. Ephemeral gully erosion in southern Navarra (Spain). Catena 36 (1–2), 65–84.

Debanshi, S., Pal, S., 2020. Assessing gully erosion susceptibility in Mayurakshi river basin of eastern India. Environ. Dev. Sustain. 22, 883–914.

d'Oleire-Oltmanns, S., Marzolff, I., Tiede, D., Blaschke, T., 2014. Detection of gully-affected areas by applying object-based image analysis (OBIA) in the region of Taroudannt, Morocco. Rem. Sens. 6 (9), 8287–8309.

Drăgut, L., Blaschke, T., 2006. Automated classification of landform elements using object-based image analysis. Geomorphology 81 (3–4), 330–344.

Drăgut, L., Tiede, D., Levick, S.R., 2010. ESP: a tool to estimate scale parameter for multiresolution image segmentation of remotely sensed data. Int. J. Geogr. Inf. Sci. 24 (6), 859–871. https://doi.org/10.1080/13658810903174803. In this issue.

Dube, F., Nhapi, I., Murwira, A., Gumindoga, W., Goldin, J., Mashauri, D.A., 2014. Potential of weight of evidence modelling for gully erosion hazard assessment in Mbire District–Zimbabwe. Phys. Chem. Earth, Parts A/B/C 67, 145–152.

Golestani, G., Issazadeh, L., Serajamani, R., 2014. Lithology effects on gully erosion in Ghoori chay Watershed using RS and GIS. Int. J. Biosci. 4 (2), 71–76.

Grissinger, E., 1996. Rill and gullies erosion. Soil Eros., Conservat. Rehabilit. 153–167.

Haralick, R.M., Shanmugam, K., Dinstein, I.H., 1973. Textural features for image classification. IEEE Transact. Syst., Man, Cybernet. (6), 610–621.

Jahantigh, M., Pessarakli, M., 2011. Causes and effects of gully erosion on agricultural lands and the environment. Commun. Soil Sci. Plant Anal. 42 (18), 2250–2255.

Johansen, K., 2010. Object-based mapping of gullies from SPOT-5 imagery and ancillary data over catchment extents. ISPRS 38 (4), C7.

Joseph, L.F., Van der Westhuizen, H.C., 2021. Low-cost soil conservation technique for controlling gully erosion in the semi-arid area of the Free State province. S. Afr. J. Agric. Ext. 49 (3), 136–142.

Junior, O.C., Guimaraes, R., Freitas, L., Gomes-Loebmann, D., Gomes, R.A., Martins, E., Montgomery, D.R., 2010. Urbanization impacts upon catchment hydrology and gully development using mutli-temporal digital elevation data analysis. Earth Surf. Process. Landforms: J. British Geomorphol. Res. Group 35 (5), 611–617.

Karami, A., Khoorani, A., Noohegar, A., Shamsi, S.R.F., Moosavi, V., 2015. Gully erosion mapping using object-based and pixel-based image classification MethodsGully erosion mapping. Environ. Eng. Geosci. 21 (2), 101–110.

Khosrokhani, M., Pradhan, B., 2014. Spatio-temporal assessment of soil erosion at Kuala Lumpur metropolitan city using remote sensing data and GIS. Geomatics, Nat. Hazards Risk 5 (3), 252–270.

Kidane, D., Alemu, B., 2015. The effect of upstream land use practices on soil erosion and sedimentation in the Upper Blue Nile Basin, Ethiopia. Res. J. Agric. Environ. Manage. 4 (2), 55–68.

Kirkby, M., 2002. Modelling the interactions between soil surface properties and water erosion. Catena 46 (2–3), 89–102.

Knight, J., Spencer, J., Brooks, A., Phinn, S., 2007. May. Large-area, high-resolution remote sensing based mapping of alluvial gully erosion in Australia's tropical rivers. In: Proceedings of the 5th Australian Stream Management Conference. Charles Sturt University, pp. 199–204.

Lal, R., 1994. Soil Erosion Research Methods. CRC Press.

Lal, R.A.T.T.A.N., 2001. Soil degradation by erosion. Land Degrad. Dev. 12 (6), 519–539.

Laliberte, A.S., Rango, A., 2009. Texture and scale in object-based analysis of subdecimeter resolution unmanned aerial vehicle (UAV) imagery. IEEE Trans. Geosci. Rem. Sens. 47 (3), 761–770.

Langran, K.J., 1983. January. Potential for monitoring soil erosion features and soil erosion modeling components from remotely sensed data. In: 1983 International Geoscience and Remote Sensing Symposium (IGARSS"83).

Lillesand, T., Kiefer, R.W., Chipman, J., 2015. Remote Sensing and Image Interpretation. John Wiley and Sons.

Liu, K., Ding, H., Tang, G., Zhu, A.X., Yang, X., Jiang, S., Cao, J., 2017. An object-based approach for two-level gully feature mapping using high-resolution DEM and imagery: a case study on hilly loess plateau region, China. Chin. Geogr. Sci. 27, 415–430.

Liu, K., Ding, H., Tang, G., Song, C., Liu, Y., Jiang, L., Zhao, B., Gao, Y., Ma, R., 2018. Large-scale mapping of gully-affected areas: an approach integrating Google Earth images and terrain skeleton information. Geomorphology 314, 13–26.

Lu, H., Prosser, I.P., Moran, C.J., Gallant, J.C., Priestley, G., Stevenson, J.G., 2003. Predicting sheetwash and rill erosion over the Australian continent. Soil Res. 41 (6), 1037–1062.

Ma, H., Zhao, H., 1994. United Nations: convention to combat desertification in those countries experiencing serious drought and/or desertification, particularly in Africa. Int. Leg. Mater. 33, 1328–1382.

Mararakanye, N., Nethengwe, N.S., 2012. Gully features extraction using remote sensing techniques. S. Afr. J. Geol. 1 (2), 109–118.

Martha, T.R., Kerle, N., Jetten, V., Van Westen, C.J., Kumar, K.V., 2010. Landslide volumetric analysis using Cartosat-1-derived DEMs. Geosci. Rem. Sens. Lett. IEEE 7 (3), 582–586.

Marzolff, I., Ries, J.B., 2007. Gully erosion monitoring in semi-arid landscapes. Z. Geomorphol. 405–425.

Melkman, A.A., 1987. On-line construction of the convex hull of a simple polyline. Inf. Process. Lett. 25 (1), 11–12.

Metternicht, G.I., Zinck, J.A., 1998. Evaluating the information content of JERS-1 SAR and Landsat TM data for discrimination of soil erosion features. ISPRS J. Photogrammetry Remote Sens. 53 (3), 143–153.

Mokarram, M., Zarei, A.R., 2021. Determining prone areas to gully erosion and the impact of land use change on it by using multiple-criteria decision-making algorithm in arid and semi-arid regions. Geoderma 403, 115379.

Morgan, R.P.C., 1995. Soil Erosion and Conservation Longman (Harlow, UK).

Morgan, R.P.C., 2009. Soil Erosion and Conservation. John Wiley and Sons.

Pal, S.C., Arabameri, A., Blaschke, T., Chowdhuri, I., Saha, A., Chakrabortty, R., Lee, S., Band, S.S., 2020. Ensemble of machine-learning methods for predicting gully erosion susceptibility. Rem. Sens. 12 (22), 3675.

Pal, S.C., Chakrabortty, R., Arabameri, A., Santosh, M., Saha, A., Chowdhuri, I., Roy, P., Shit, M., 2022. Chemical weathering and gully erosion causing land degradation in a complex river basin of Eastern India: an integrated field, analytical and artificial intelligence approach. Nat. Hazards 110 (2), 847–879.

Poesen, J.W., Vandaele, K., Van Wesemael, B., 1996. Contribution of gully erosion to sediment production on cultivated lands and rangelands. IAHS Publicat.-SeriesProceed. Repor.-Intern Assoc. Hydrol.Sci. 236, 251–266.

Poesen, J., Nachtergaele, J., Verstraeten, G., Valentin, C., 2003. Gully erosion and environmental change: importance and research needs. Catena 50 (2–4), 91–133.

Poesen, J., Vanwalleghem, T., De Vente, J., Knapen, A., Verstraeten, G., Martínez-Casasnovas, J.A., 2006. Gully erosion in Europe. Soil erosion in Europe 515–536.

Poesen, J., 2011. Challenges in gully erosion research. Landform analysis 17, 5–9.

Rabalais, N.N., Diaz, R.J., Levin, L.A., Turner, R.E., Gilbert, D., Zhang, J., 2010. Dynamics and distribution of natural and human-caused hypoxia. Biogeosciences 7 (2), 585–619.

Rahmati, O., Tahmasebipour, N., Haghizadeh, A., Pourghasemi, H.R., Feizizadeh, B., 2017a. Evaluation of different machine learning models for predicting and mapping the susceptibility of gully erosion. Geomorphology 298, 118–137.

Rahmati, O., Tahmasebipour, N., Haghizadeh, A., Pourghasemi, H.R., Feizizadeh, B., 2017b. Evaluating the influence of geo-environmental factors on gully erosion in a semi-arid region of Iran: an integrated framework. Sci. Total Environ. 579, 913–927.

Rahmati, O., Kalantari, Z., Ferreira, C.S., Chen, W., Soleimanpour, S.M., Kapović-Solomun, M., Seifollahi-Aghmiuni, S., Ghajarnia, N., Kazemabady, N.K., 2022. Contribution of physical and anthropogenic factors to gully erosion initiation. Catena 210, 105925.

Sadeghi, S.H., Ghaffari, G.A., Rangavar, A., Hazbavi, Z., Singh, V.P., 2020. Spatiotemporal distribution of soil moisture in gully facies. Int. Soil and Water Conservat. Res. 8 (1), 15–25.

Seeger, K.M., Marzolff, I., Ries, J.B., 2009. Identification of gully-development processes in semi-arid NE-Spain. Z. Geomorphol. 53 (4), 417–431.

Shit, P.K., Paira, R., Bhunia, G., Maiti, R., 2015. Modeling of potential gully erosion hazard using geo-spatial technology at Garbheta block, West Bengal in India. Model. Earth Syst. Environ. 1, 1–16.

Shruthi, R.B., Kerle, N., Jetten, V., 2011. Object-based gully feature extraction using high spatial resolution imagery. Geomorphology 134 (3–4), 260–268.

Shruthi, R.B., Kerle, N., Jetten, V., Stein, A., 2014. Object-based gully system prediction from medium resolution imagery using Random Forests. Geomorphology 216, 283–294.

Shruthi, R.B., Kerle, N., Jetten, V., Abdellah, L., Machmach, I., 2015. Quantifying temporal changes in gully erosion areas with object oriented analysis. Catena 128, 262–277.

Singh, B.M., 1977. Interpretation of satellite imagery for delineation of ravines. J. Indian Soci. Photo-Interpret. 5, 31–34.

Smith, H.T.U., 1943. Aerial Photographs and Their Applications.

Stromquist, L., Lundén, B., Chakela, Q., 1985. Sediment sources, sediment transfer in a small Lesotho catchment—a pilot study of the spatial distribution of erosion features and their variation with time and climate. S. Afr. Geogr. J. 67 (1), 3–13.

Tagore, G.S., Bairagi, G.D., Sharma, N.K., Sharma, R., Bhelawe, S., Verma, P.K., 2012. Mapping of degraded lands using remote sensing and GIS techniques. J. Agric. Phys. 12 (1), 29–36.

Tang, W.J., Liu, H.H., Liu, B.Y., 2013. Effects of gully erosion and gully filling on soil degradation in the black soil region of Northeast China. J. Mt. Sci. 10, 913–922.

Tang, J., Liu, G., Xie, Y., Wu, Y., Wang, D., Gao, Y., Meng, L., 2022. Effect of topographic variations and tillage methods on gully erosion in the black soil region: a case-study from Northeast China. Land Degrad. Dev. 33 (18), 3786–3800.

Taruvinga, K., 2009. Gully Mapping Using Remote Sensing: Case Study in KwaZulu-Natal, South Africa. Master's thesis. University of Waterloo.

Torkashvand, A.M., 2022. Investigation of the possibility to prepare supervised classification map of gully erosion by RS and GIS. MMM Inter. J. Advanc. Eng., Architect. Mathemat. 1 (1).

Torri, D., Borselli, L., 2003. Equation for high-rate gully erosion. Catena 50 (2–4), 449–467.

Valentin, C., Poesen, J., Li, Y., 2005. Gully erosion: impacts, factors and control. Catena 63 (2–3), 132–153.

Valentin, C., 2004. Overland flow, erosion and associated sediment and biogeochemical transports. Vegetation, water, humans and the climate: a new perspective on an interactive system. Global Change Series–The IGBP Series 317–322. Springer Verlag, Berlin.

Vandekerckhove, L., Poesen, J., Wijdenes, D.O., Gyssels, G., Beuselinck, L., De Luna, E., 2000. Characteristics and controlling factors of bank gullies in two semi-arid Mediterranean environments. Geomorphology 33 (1–2), 37–58.

Vrieling, A., Rodrigues, S.C., 2005. Erosion assessment in the Brazilian cerrados. In: Envisat and ERS Symposium, vol. 572.

Vrieling, A., Rodrigues, S.C., Bartholomeus, H., Sterk, G., 2007. Automatic identification of erosion gullies with ASTER imagery in the Brazilian Cerrados. Int. J. Rem. Sens. 28 (12), 2723–2738.

Wang, B., Zhang, Z., Wang, X., Zhao, X., Yi, L., Hu, S., 2020. Object-based mapping of gullies using optical images: a case study in the black soil region, Northeast of China. Rem. Sens. 12 (3), 487.

Wang, R., Li, P., Li, Z., Yu, K., Han, J., Zhu, Y., Su, Y., 2021. Effects of gully head height and soil texture on gully headcut erosion in the Loess Plateau of China. Catena 207, 105674.

Wijesiri, B., Liu, A., Deilami, K., He, B., Hong, N., Yang, B., Zhao, X., Ayoko, G., Goonetilleke, A., 2019. Nutrients and metals interactions between water and sediment phases: an urban river case study. Environ. Pollut. 251, 354–362.

Wulder, M.A., Loveland, T.R., Roy, D.P., Crawford, C.J., Masek, J.G., Woodcock, C.E., Allen, R.G., Anderson, M.C., Belward, A.S., Cohen, W.B., Dwyer, J., 2019. Current status of Landsat program, science, and applications. Rem. Sens. Environ. 225, 127–147.

Zabihi, M., Mirchooli, F., Motevalli, A., Darvishan, A.K., Pourghasemi, H.R., Zakeri, M.A., Sadighi, F., 2018. Spatial modelling of gully erosion in Mazandaran Province, northern Iran. Catena 161, 1–13.

Zaimes, G.N., Schultz, R.C., 2012. Assessing riparian conservation land management practice impacts on gully erosion in Iowa. Environ. Manag. 49, 1009–1021.

CHAPTER

21

Landslide detection and monitoring using remote sensing approach

Freidoon Soleimani[1], Kourosh Shirani[2], Shamsollah Asgari[1] and Zeinab Hazbavi[3]

[1]Soil Conservation and Watershed Management Research Institute, Ilam Agricultural and Natural Resources Research and Education Center, Agricultural Research, Education and Extension Organization (AREEO), Ilam, Iran; [2]Soil Conservation and Watershed Management Research Institute, Agricultural Research, Education and Extension Organization (AREEO), Tehran, Iran; [3]Rangeland and Watershed Management Department, Faculty of Agriculture and Natural Resources, University of Mohaghegh Ardabili, Ardabil, Iran

21.1 Introduction

Landslide is the costliest geological hazards that threaten and affect many countries in the world. The impact of landslide on the environment is including loss of life, harm to land and loss of natural resources, impact on topography, rivers, vegetation cover, and habitats of wildlife (Geertsema et al., 2009). Landslide material can restrict rivers and increase the floods.

The natural disaster that has a significant impact on the world is landslide (Casagli et al., 2016; Lari et al., 2014). Compared to other disasters, landslides are one of the costlier and deadly geological hazards that threaten and affect the socioeconomic conditions of several countries (Tofani et al., 2013; Pradhan, 2010; McKean et al., 1991).

Landslides occur when layers of soil on sloping surfaces become saturated due to heavy rainfall or due to external forces and move downwards, cause to loss of life, property, environment, and economic. In the United States, landslide damages about 3.5 billion dollars and every year kills 25 to 50 people approximately (USGS, 2004). Also Nepal, in 2014, was hit by a landslide, where livestock losses were many times higher than human losses, and damages of infrastructure were expensive (Pathak, 2016).

Remote sensing widely is used in landslide researches. For scientists and decision-makers mapping of landslide, assessment of hazard and evaluation of risk are important goals. New techniques of remote sensing, based on satellite, aerial, and ground platforms, can enable

Remote Sensing of Soil and Land Surface Processes
https://doi.org/10.1016/B978-0-443-15341-9.00002-2

simple and systematic data collection over large areas and improve landslide mapping and reducing costs (Guzzetti et al., 2012; Casagli et al., 2017, 2018). Landslides play an important role in the evolution and shaping of subaerial landscapes (Brunetti et al., 2015) and are a major cause of loss of life, injuries, property damage, socio-economic disruption, and environmental degradation in many areas of the emerged continents (Alexander, 1993).

Infrared thermography (IRT) as a new tool for characterization, mapping, and surveying of landslide is used. IRT uses both ground and aerial platforms to detect and map the structural and morphological features of landslides, in an integrated way compared to other geomatics methods such as terrestrial laser scanning (TLS) and global positioning systems (GPS) (Frodella et al., 2017).

21.2 Definition of landslide types

> "The term landslide describes a wide variety of processes that result in the downward and outward movement of slope-forming materials including rock, soil, artificial fill, or a combination of these" (Varnes, 1978). "Landslide, also called landslip, is the movement downslope of a mass of rock, debris, earth, or soil (soil is a mixture of earth and debris)."

When gravity force and other shear stress types within a slope exceed the shear strength, then landslides are happened (Hungr et al., 2014). Generally, landslides by type of materials and movements are classified.

Within a single landslide, sometimes several types of movements happened, and because the spatiotemporal relationships of these movements are often complex, its analysis needs to detailed interpretation of landforms and different parts of geology or cores. **Five different types of mass movements are listed below in this chapter**.

Falls: Falls are abrupt mass movements of geological material, such as rocks that detach from cliffs or steep slopes. The movement of parts is usually fast, and the separation surfaces often remain smooth and polished. Along discontinuities such as fractures, joints, and bedding planes, detachment happens and by free-fall, bouncing, and rolling, the movement occurs. Gravity, mechanical weathering, and the moisture strongly effect on falls.

Topples: In this type of landslide, pieces of rock are separated from the main mass in the form of columns and fall. Overturning of the rock takes place against the direction of the layering slope. The dimensions of the stone columns are evaluated according to the condition of the fractures.

Slides: In this type of landslide, the rock slides perpendicularly along the layering surfaces and in the direction of the greatest slope of the layer. Rotational and translational are two main types of slides.

a. ***Rotational slide:*** *This is a slide in which the surface of rupture is curved concavely upward, and the slide movement is roughly rotational about an axis that is parallel to the ground surface and transverse across the slide.*

b. ***Translational slide:*** *In this type of slide, the landslide mass moves along a roughly planar surface with little rotation or backward tilting.*

Lateral spreads: *Lateral spreads are distinctive because they usually occur on very gentle slopes or flat terrain. The dominant mode of movement is lateral extension accompanied by shear or tensile fractures. The failure is caused by liquefaction, the process whereby saturated, loose, cohesion less sediments (usually sands and silts) are transformed from a solid into a liquefied state.*

Flows: *Five basic categories of flows differ from one another in fundamental ways*

a. *Debris flow:* *A debris flow is a form of rapid mass movement in which a combination of loose soil, rock, organic matter, air, and water mobilize as a slurry that flows downslope*

b. *Debris avalanche:* *This is a variety of very rapid to extremely rapid debris flow*

c. *Earth flow:* *Earth flows have a characteristic* "hourglass" *shape. The slope material liquefies and runs out, forming a bowl or depression at the head. The flow itself is elongated and usually occurs in fine-grained materials or clay-bearing rocks on moderate slopes and under saturated conditions. However, dry flows of granular material are also possible* (Varnes, 1978).

d. Mud flow: In special conditions, when the fine-grained soil becomes colloidal and becomes a mud, the resulting phenomenon is called mud flow. Sometimes in daily reports and news, debris flows and mudflows are known as "mudslides."

e. Creep: The extremely slow and downward movement of soil, rock cover, and rock under the influence of gravity is called creep. In general, there are three types of creep including seasonal, continuous, and progressive.

Complex: Combination of two or more of the above types is known as a complex landslide.

21.3 Identification, detection, mapping, and monitoring

The availability of advanced remote sensing technologies in the field of landslide analysis enables fast and easy data updating, mapping, and monitoring, optimizing fieldwork, and allowing the investigation of dangerous and inaccessible areas at the same time and provides operator safety (Casagli et al., 2018). Advanced ground-based remote sensing technologies are nowadays available in the field of slope instability detection, mapping, and monitoring, during short/long-term landslide management (Lillesand et al., 2014).

To study landslide processes several remote sensing data have been used, including spaceborne synthetic aperture radar (SAR) and optical remote sensing, airborne light detection and ranging (LiDAR), ground-based SAR, and terrestrial LiDAR, incorporating in situ measurements and observations of environmental factors.

Theoretical principles of ground-based remote sensing techniques for landslides mapping and monitoring including: (1) terrestrial laser scanner (TLS), (2) ground-based interferometric synthetic aperture radar (GB-InSAR), (3) infrared thermography (IRT), (4) robotic total station (RTS), and (5) digital photogrammetry workstation (DPW). Landslide monitoring includes detailed information on topography, geology, groundwater levels, material properties, and possible mass movements (Ray and Jacobs, 2007). Several types of instruments, such as extensometers, inclinometers, piezometers, strain meters, pressure cells, geophones, tilt meters, and crack meters, have been used to monitor slope movements and deformation (Mantovani et al., 1996).

The technique of radar interferometry is an efficient method for measuring the displacement of the Earth's surface. Using this technology, it is possible to monitor small movements of the earth's surface continuously and on a large scale. Today, investigating the environmental and natural hazards of the earth, including mass movements of slopes, subsidence, earthquakes, and volcanic activities, using this technology commonly has become (Abedini and Mohammadzadeh, 2022).

Radar interferometric technique with ability to work in all weather conditions throughout the day and night, with ability to cover the Earth's surface and high spatiotemporal resolution, is one of the most accurate and least expensive remote sensing techniques for detecting and monitoring the Earth surface changes, slow and unstable movements in the world (Metternich et al., 2005; Singhroy, 2002).

Object-based image analysis (OBIA) of optical satellite images can also support the landslide detection and mapping of past and present mass movements. The mechanism of the DInSAR method is to use the phase difference of the SAR radar image with a standard format, which was taken at different times from a specific area, which ultimately leads to the estimation of the displacement of the Earth along the line of sight of the satellite (Costantini et al., 2000; Massonnet and Feigl, 1998; Singhroy et al., 1998; Gabriel et al., 1989; Zebker and Goldstein, 1986).

Many different techniques to process multitemporal synthetic aperture radar data have been developed and classified to PSI and small baseline subset (SBAS) methods (Ferretti et al., 2001). One of the simple, low-cost, and accurate methods for identifying, detecting, and mapping landslides is using Google Earth imagery. Google Earth is a suitable tool to collect data for analysis the geographic information system is used. In a research (Nikandish and Hosseinizadeh, 2013) by visual interpretation of Google Earths images, landslide components such as ravines, cracks, isolated masses, landslide sediments, landslide claws, drainage networks, and other features of the landslide zone were identified, and their boundaries were determined.

Rotating the landscape to create a better 3D landscape, providing data with geographical coordinates and measuring distance and dimensions are other functions of Google Earth.

Due to the simplicity and availability of Google Earth images, at the end of this chapter, a case study of landslides in Ilam province of Iran using these images, which were carried out by the authors of this chapter, is given.

21.4 Zagros area (Iran) as a case study

Many small and large landslides have been investigated in Iran. Some of them have caused financial damage and some have caused loss of life. Investigations show that until the beginning of 1999, about 2590 landslides in the country caused the death of 162 people, the destruction of 176 houses, the financial loss of 1866 billion Rials, the destruction of 6763 ha of forest, the destruction of 170 km of communication roads, and sedimentation. The annual volume is 963,708 cubic meters (Baharvand et al., 2018). Considering the large amount of human and financial damage caused by these landslides, as well as their occurrence in Ilam province, scientific investigation of this phenomenon is necessary and necessary in order to take the necessary measures to reduce the risks and damages caused by it.

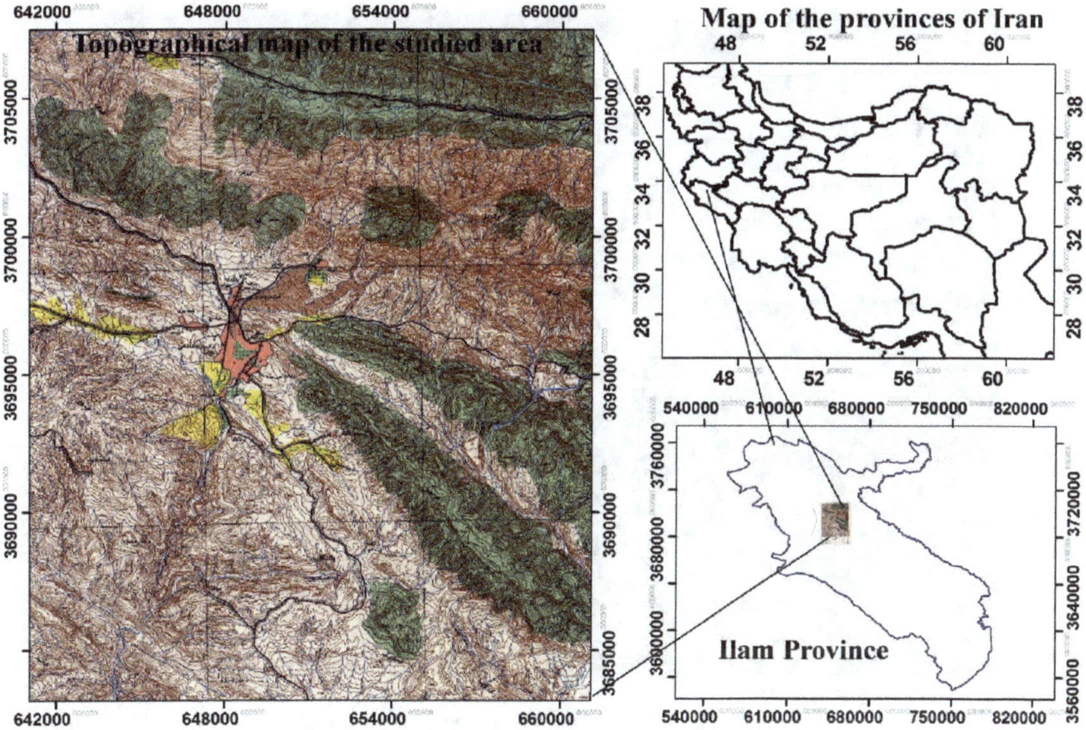

FIGURE 21.1 Geographic location map of the study area.

The study area is located in Ilam province of Iran (Fig. 21.1). This research examines the impact of landslides on the destruction of natural resources in the Arquaz-Malekshahi region of Ilam province using Google Earth satellite images.

The purpose of this research is to identify the landslides in the region and the threatened elements and the areas of forest and pasture lands that have been destroyed or are being destroyed. Google Earth images were used to identify new landslides and vegetation cover conditions of the region and aerial photographs from 1965 were used to identify old landslides and past vegetation cover conditions of the region. To investigate the state of mass movements in the studied area, the topographical map sheet with a scale of 1:50,000 of Arkuaz Malekshahi, which is significant in terms of the types of domain movements, was identified and evaluated.

To prepare a distribution map of landslides using the interpretation of aerial photographs, the relevant photographs were first classified according to the flight lines.

Aerial photos of 1:20,000 area related to the topographic map of 1:50,000 were matched based on the flight lines, according to the complications and evidence available on the photos and geographical coordinates and some important points such as villages, cities, and roads were written inside or behind the photos.

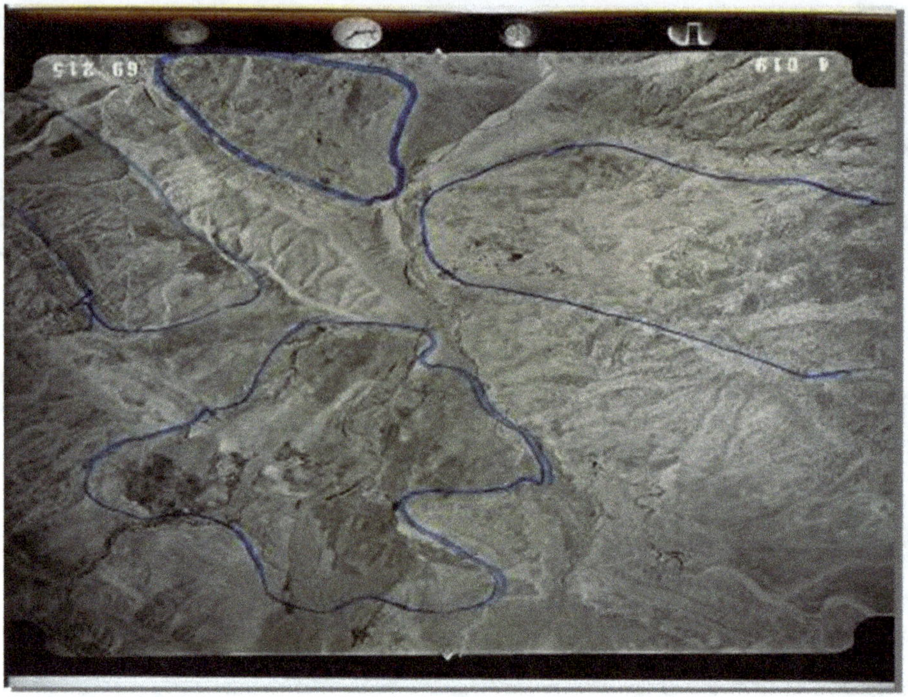

FIGURE 21.2 Some samples of the landslide area are shown in the aerial photo of Arquaz Malekshahi.

According to the stereoscopic observations of 1:20,000 aerial photographs in 1965, photo index 215, and flight line number 69–215 and field considerations, the slip zones in the studied area were identified and drawn as polygons on the aerial photograph (Fig. 21.2).

After this step, the relevant photos are scanned and transferred to ILWIS software under GIS during the process of format change, and after georeferencing each photo, all kinds of hillslope movements in the form of areas, individual and points in the form of falling, sliding, flows and complex, and old landslides were prepared based on the photographing taken in 1965.

By using Google Earth image magnification, the boundary of current landslides in the area, which are considered a serious risk, was transferred to ILWIS software under GIS, and after georeferencing, it was digitized for accuracy as much as possible landslide area was controlled by field visiting using GPS and to evaluate the information, it is shown on the 1:50,000 topographic map of Arquaz Malekshahi (Figs. 21.3 and 21.4).

By transferring the prepared images, photos, and GPS information to the GIS and superimposing the sliding zones and the vegetation map, from the total area of the 1:50,000 topography sheet of Arquaz Malekshahi, which is equivalent to 65,000 ha, 45,824 ha have forest and pasture cover, 18,324 ha are agriculture and horticulture, and 852 ha are residential and facilities. Of these values, 13,363 ha of forest and pasture lands, 3200 ha of agricultural lands, and 85 ha of residential land are under the destruction of landslides or at risk of landslides (Figs. 21.5 and 21.6).

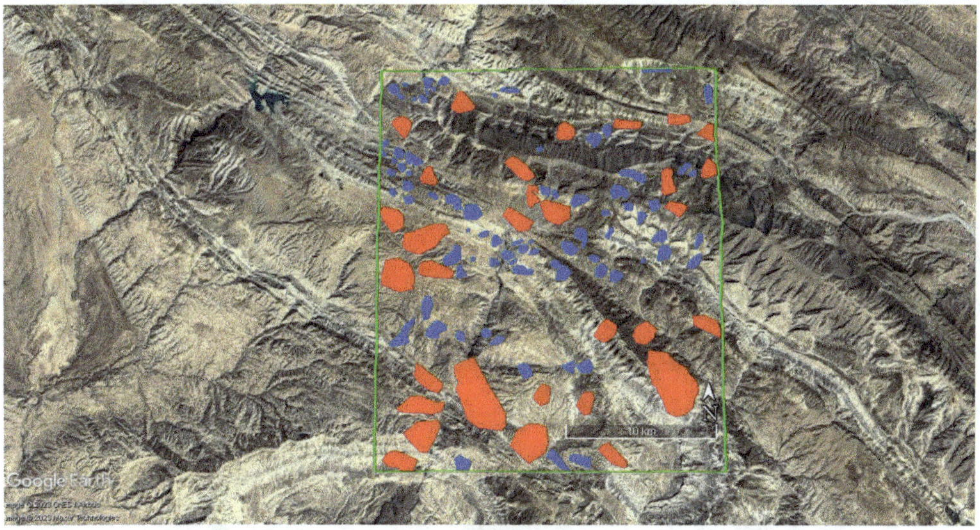

FIGURE 21.3 Landslide areas on Google Earth image.

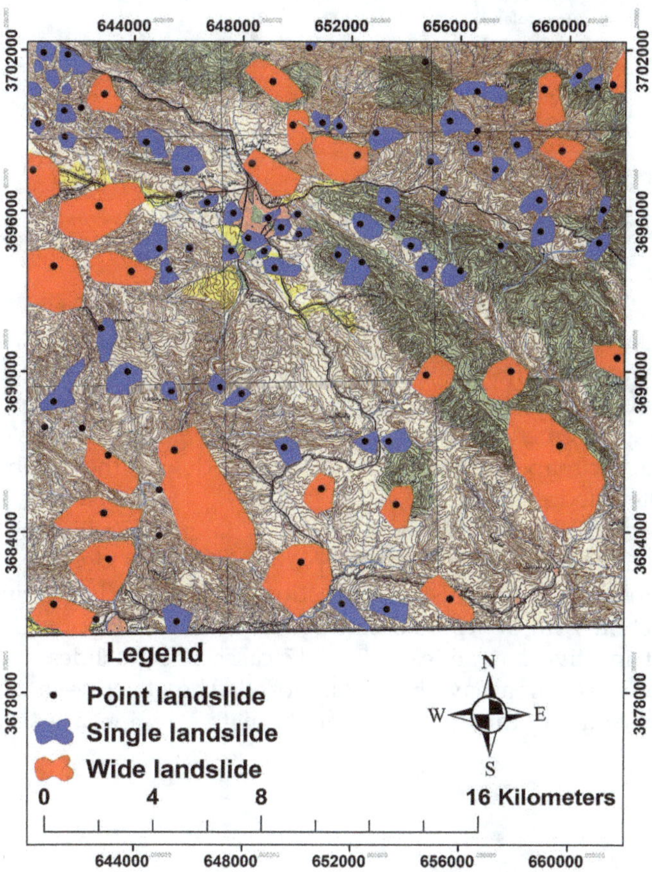

FIGURE 21.4 Matched landslides with the topographical map of Arquaz.

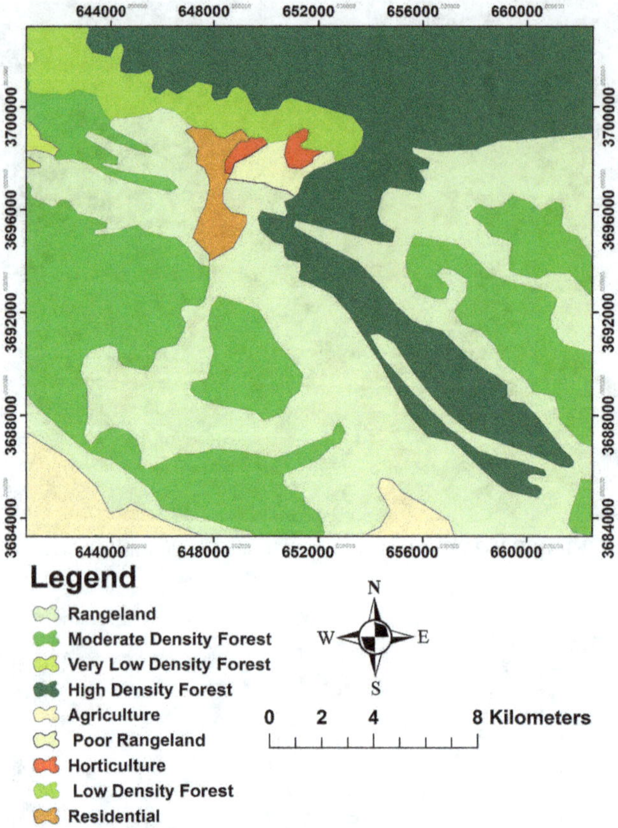

FIGURE 21.5 Vegetation cover map.

In order to evaluate the landslides, the exact location of the landslide, dimensions of the landslide, height above sea level, type of movement, dominant direction of movement, type of activity, materials involved, effective factors in causing landslides and threatened elements for any mass movement in the region using aerial photos, images Google Earth, vegetation map, topographic map, and field visit using GPS device, the number of 173 landslides in the region was detected.

According to the distribution of landslide types, 57 fallings, 50 sliding, 5 flows, 8 complex, 6 falling zones, 33 sliding zones, and 14 complex zones were identified and detected. The most direct of movements is in the south direction with 48 landslides in the first rank and the southwest direction with 37 in the second rank. The number of active landslides is 147 and the number of inactive landslides is 26. 147 cases of landslides amounting to 85% of all landslides in the active area have been diagnosed. The effective factors in causing landslides in the region and the threatened elements in this region are shown in Tables 21.1 and 21.2.

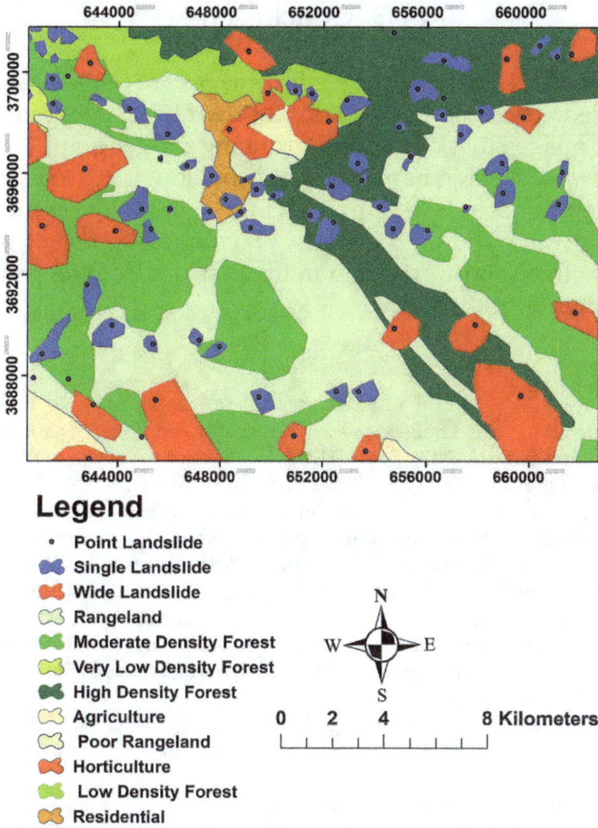

FIGURE 21.6 Combined landslides with vegetation map.

TABLE 21.1 Effective factors in causing landslides in the region.

Erosion and piping	Stratification	Fault	Fractures	Agriculture	Eliminating the support	Impact of hydraulic structure	Land use change	Others
42	23	28	47	6	23	0	3	1

TABLE 21.2 Threatened elements in this region.

Forest and Rangeland	Farmland	Road	Residential area	Rural road	Village	Hydraulic structure
13,363 ha	3200 ha	14 cases	18 cases	9 cases	3 cases	11 cases

21.5 Conclusion

What is certain is that aerial photos and satellite images are tools for scientific research that use these tools as needed. The accurate drawing of the range, polygon, or any point on the aerial photo or satellite images that can accurately show the geographical location and mathematical dimensions of a problem in nature depends on how it is used. Today, the tool that is used in precise geographic location, map drawing, and high-volume data analysis uses remote sensing, GIS software, and GPS devices. Of course, the use of this tool is related to an expert user who uses the utmost precision in the first stage of work, which is georeferencing, which is the foundation of the work.

References

Abedini, M., Mohammadzadeh, M., 2022. Landslide assessment using radar images and radar interferometry Case area: Nirchai Basin. Environ. Sci. Stud. 7 (3), 5161–5171.

Alexander, D.E., 1993. Natural Disasters. Springer Science & Business Media.

Baharvand, S., Rahnamarad, J., Soori, S., 2018. Prioritization of landslide effective factors and its hazard mapping using weighted linear combination method (case study: Vark watershed). J. Nat. Environ. Hazards 7 (16), 195–210.

Brunetti, M.T., Xiao, Z., Komatsu, G., Peruccacci, S., Guzzetti, F., 2015. Terrestrial and extraterrestrial landslide size statistics. In: European Planetary Science Congress 2015, 27 Sept–2 Oct 2015, Nantes, France. Copernicus. org/EPSC2015, Id. EPSC2015-776 (10:776).

Casagli, N., Cigna, F., Bianchini, S., Hölbling, D., 2016. Landslide mapping and monitoring by using radar and optical remote sensing: examples from the EC-FP7 project SAFER. Remote Sens. Appl.: Soc. Environ. 4, 92–108. https://doi.org/10.1016/j.rsase.2016.07.001.

Casagli, N., Morelli, S., Frodella, W., Intrieri, E., Tofani, V., 2018. TXT-Tool Ground-Based Remote Sensing Techniques for Landslides Mapping, Monitoring and Early Warning. https://www.researchgate.net/publication/322102480.

Casagli, N., Frodella, W., Morelli, S., Tofani, V., Ciampalini, A., Intrieri, E., Raspini, F., Rossi, G., Tanteri, L., Lu, P., 2017. Space borne, UAV and ground-based remote sensing techniques for landslide mapping, monitoring and early warning. Geoenviron. Disasters 4, 9.

Costantini, M., Iodice, A., Magnapane, L., Pietranera, L., 2000. Monitoring terrain movements by means of sparse SAR differential interferometric measurements. In: Proceedings of the 20th IEEE International Geoscience and Remote Sensing Symposium (IGARSS), Honolulu, Hawaii, USA, 24–28 July 2000, pp. 3225–3227.

Ferretti, A., Prati, C., Rocca, F., 2001. Permanent scatterers in SAR interferometry. IEEE Trans. Geosci. Rem. Sens. 39 (1), 8–20.

Frodella, W., Gigli, G., Morelli, S., Lombardi, L., Casagli, N., 2017. Landslide mapping and characterization through infrared thermography (IRT): suggestions for a methodological approach from some case studies. Rem. Sens. 9, 1281. https://doi.org/10.3390/rs9121281.

Gabriel, A.K., Goldstein, R.M., Zebker, H.A., 1989. Mapping small elevation changes over large areas: differential radar interferometry. J. Geophys. Res. 94, 9183–9191.

Geertsema, M., Highland, L., Vaugeouis, L., 2009. Environmental impact of landslides. In: Sassa, K., Canuti, P. (Eds.), Landslides – Disaster Risk Reduction. Springer, Berlin, Heidelberg. https://doi.org/10.1007/978-3-540-69970-5_31.

Guzzetti, F., Mondini, A.C., Cardinali, M., Fiorucci, M., Santangelo, M., Chang, K.T., 2012. Landslide inventory maps: new tools for an old problem. Earth Sci. Rev. 112, 1–25.

Hungr, O., Leroueil, S., Picarelli, L., 2014. The Varnes classification of landslide types, an update. Landslides 11, 167–194. https://doi.org/10.1007/s10346-013-0436-y.

Lari, S., Frattini, P., Crosta, G.B., 2014. A probabilistic approach for landslide hazard analysis. Eng. Geol. 182, 3–14. https://doi.org/10.1016/j.enggeo.2014.07.015.

Lillesand, T., Kiefer, R.W., Chipman, J., 2014. Remote Sensing and Image Interpretation. Wiley, New York.

Mantovani, F., Soeters, R., Westen, C.J., 1996. Remote sensing techniques for landslide studies and hazard zonation in Europe. Geomorphology 15, 213–225. https://doi.org/10.1016/0169-555X(95)00071-C.

Massonnet, D., Feigl, K.L., 1998. Radar interferometry and its application to changes in the Earth's surface. Rev. Geophys. 36, 441–500.

McKean, D., Calkins, H., Peuquet, D., 1991. Remote sensing and landslide hazard assessment. Photogramm. Eng. Rem. Sens. 57 (9), 1185–1193.

Metternicht, G., Hurni, L., Gogu, R., 2005. Remote sensing of landslides: an analysis of the potential contribution to geospatial systems for hazard assessment in mountain environments. Remote Sens. Environ. 98, 284–303.

Nikandaish, N., Hosseinizadeh Arani, S.F., 2013. Google Earth, geographic information system and four-board landslide; Simple and new method for landslide investigation. Sepehr J. 22 (87), 81–86.

Pathak, D., 2016. Remote sensing and GIS application in landslide risk assessment and management. Nepalese J. Geoinform. 15, 52–57.

Pradhan, B., 2010. Remote sensing and GIS-based landslide hazard analysis and cross-validation using multivariate logistic regression model on three test areas in Malaysia. Adv. Space Res. 45, 1244–1256. https://doi.org/10.1016/j.asr.2010.01.006.

Ray, R.L., Jacobs, J.M., 2007. Relationships among remotely sensed soil moisture, precipitation and landslide events. Nat. Hazards 43, 211–222. https://doi.org/10.1007/s11069-006-9095-9.

Singhroy, V., 2002. Landslide Hazards: CEOS, the Use of Earth Observing Satellites for Hazard Support: Assessments and Scenarios. Final report of the CEOS Disaster Management Support Group, NOAA, p. 98.

Singhroy, V., Mattar, K., Gray, A., 1998. Landslide characterization in Canada using interferometric SAR and combined SAR and TM images. Adv. Space Res. 21 (3), 465–476.

Tofani, V., Segoni, S., Agostini, A., Catani, F., Casagli, N., 2013. Technical note: use of remote sensing for landslide studies in Europe. Nat. Hazards Earth Syst. Sci. 13, 299–309. https://doi.org/10.5194/nhess-13-299-2013.

USGS (United State Geological Survey), 2004. Landslide types and processes. Available from: https://pubs.usgs.gov/fs/2004/3072/fs-2004-3072.html [Accessed: 27 April 2020].

Varnes, D.J., 1978. Slope movement types and processes. In: Schuster, R.L., Krizek, R.J. (Eds.), Special Report 176: Landslides: Analysis and Control, Transportation and Road Research Board. National Academy of Science, Washington D. C., pp. 11–33

Zebker, H.A., Goldstein, R.M., 1986. Topographic mapping from interferometric synthetic aperture radar observations. J. Geophys. Res. 91, 4993–4999.

CHAPTER

22

Classification algorithms for remotely sensed images

Neda Mohamadzadeh[1] *and Amin Ahmadisharaf*[2]

[1]Department of Geography and Geospatial Sciences, Kansas State University, Manhattan, KS, United States; [2]Kansas State University, Manhattan, KS, United States

22.1 Introduction

Geographical systems often pose a significant challenge when attempting to model and analyze, using traditional statistical methods; however, Geo computational methods offer a promising alternative. In this chapter, we will focus specifically on the application of Geo computation to geospatial data classification (DiBiase, 2006) and investigate different classification algorithms for remotely sensed images. Classification is a widely used analytical technique in the analysis of remotely sensed images (Sun et al., 2013), based on the notion that different features have distinctive spectral reflectance and remittance characteristics. So, image classification is the systematic process of assigning all pixels within remotely sensed images into a predefined set of land cover themes or classes. This classification is determined by the spectral reflectance of each pixel, which is used to identify its respective land cover classification (Al-Doski et al., 2013).

Over the last few decades, while a significant amount of basic research on GIS has focused on internal data representations and the algorithms that operate on them, several challenging problems remain in the field. One of these challenges is the development of efficient algorithms for specific purposes (Goodchild, 1992). When it comes to image classification, many algorithms are available, ranging from supervised to unsupervised, parametric to nonparametric, nonmetric, and hard or soft (fuzzy) classification. Although these approaches have been developed and are widely used, there is a significant lack of knowledge regarding the differences between the existing algorithms, making it a challenge to determine the most efficient and accurate one for producing a landcover map. To fill this gap, this research will focus on examining the functions of existing image classification algorithms and highlighting their differences.

Remote Sensing of Soil and Land Surface Processes
https://doi.org/10.1016/B978-0-443-15341-9.00001-0

The first section of this chapter will define various image classification approaches, including supervised, unsupervised, and hybrid classification, and introduce the most used classifiers, namely maximum likelihood classification (MLC) and K-means. This section will also review the advantages and disadvantages of supervised and unsupervised classification. The second section will focus on recent research that applies different image classification algorithms in a case study and compares their results. The section will also analyze the performance of these algorithms and highlight the factors that affect their accuracy. In the final section, the study will conclude the factors that affect the classification algorithms' performance, which provides insights into the key findings of the literature review.

22.2 Different image classification approaches

There are multiple methods of image classification that have been extensively employed for the creation of maps depicting land use and land cover. Supervised and unsupervised approaches are two broad types of classification procedures. Although these two can be used as alternatives, they are occasionally combined into hybrid approach using multiple methods (Al-Doski et al., 2013).

22.2.1 Supervised classification method

Supervised classification is a method in which the analyst creates on the image small regions known as training sites for each land cover/land use class. Then, the algorithm assigns the pixels on the image to each land cover/land use class based on the predictor variables measured in the sampling unit (Al-Doski et al., 2013). In supervised classification, the analyst must be knowledgeable about the study area and the spectral properties of land cover classes (Salah, 2017). The commonly used supervised classification algorithms include MLC, minimum distance classification, and parallelepiped classification.

Maximum likelihood classification is the most widely used supervised and pixel-based classification algorithm with high precision and accuracy (Al-Doski et al., 2013). This algorithm focuses on the similarity of spectral properties of pixels and the maximum probability of belonging to a specific category (Wu and Li, 2022). In other words, the probability density function for each category is taken into account while classifying unclassified pixels to determine which group they belong to (Sun et al., 2013).

Minimum distance classification is based on the smallest Euclidian distance between the pixel to be divided and the mean vector of the training sample as the classification scale (Wu and Li, 2022). In the minimum distance classifier, classes that are close to one another are grouped as a single prototype. For each class, this classifier calculates the mean vector, for each class and assigns the pixel to the closest class (Nivedita Priyadarshini et al., 2018).

Parallelepiped classifier utilizes a standard threshold to determine whether a specific pixel is a member of the specified class. This classifier is based on the standard deviation threshold from the mean of each class. The pixels are classified if they fall within the specified range of the threshold, which is defined by a low and high value. However, if the pixels fall outside of this range, they are deemed unclassified (Nivedita Priyadarshini et al., 2018).

22.2.2 Unsupervised classification method

The unsupervised classification method groups pixels from satellite images into unlabeled groupings using clustering techniques and based on their reflectance values and without any direction from the analyst. Only the quantity of clusters to create and the land cover class of each cluster are specified by the analyst. Multiple clusters often represent a single land cover class, and the analyst combines these clusters into a single class. This classification technique is widely used when there are no training samples (Salah, 2017). K-means and Iterative Self-Organizing Data Analysis Technique (ISODATA) are the two widely used clustering widely used for unsupervised classification. These methods depend on spectrally pixel-based statistics and do not integrate any knowledge of the properties of the themes being examined (Al-Doski et al., 2013).

In K-means classifier, the process of determining different classes and assigning pixels to the closest cluster means based on all available bands is the concept. The K-Means clustering algorithm begins with an initial set $C-(0)$ and performs a sequence of iterations. At each iteration t, every $c \in C$ pixel is allocated to one of the clusters based on the closest neighbor principle. The calculation of a new center for a cluster follows Eq. (22.1):

$$C_j^{(t+1)} = \frac{1}{N_j} \sum (C_i \mid C_i \in S_j^{(t)} \tag{22.1}$$

There are various factors that have effects on the results of the clustering process, such as the number of specified cluster centers, the selection of the initial cluster center, the sampling nature, the geometrical characteristics of the data, and the parameters employed in the clustering algorithm (Al-Doski et al., 2013).

ISO data unsupervised classification groups the pixels that are equally distributed and groups the remaining pixels based on a prespecified threshold. In this algorithm, the number of classes as well as the number of iterations must be set manually (Nivedita Priyadarshini et al., 2018).

22.2.3 Advantages and disadvantages of supervised and unsupervised classification

The supervised technique has some advantages. In this approach, the functional categories of information are initially distinguished. Then the spectral separability of each category is evaluated, which is an advantage over the unsupervised in which the computer determines spectrally separable class before defining the information values. In comparison, unsupervised classification is easy to apply when it automatically transforms raw data into useful information and does not require analyst-specified knowledge. Although both supervised and unsupervised classifications have advantages and drawbacks, both major methods have some limitations. For example, although unsupervised is easy to apply, the process should be repeated if new data are added. Also, achieving satisfactory results with supervised and unsupervised classification methods can be challenging when using new images with finer spectral and spatial resolution (Al-Doski et al., 2013). While being fast and simple to process is the most common advantage of classic classifiers, including ISODATA, K-means, etc.,

TABLE 22.1 Advantages and disadvantages of supervised and unsupervised classifiers.

Classifier	Advantages	Disadvantages
Supervised	_ distinct useful information categories first	_ difficult to achieve satisfactory results, utilizing fine spatial and spectral resolution data. _ being affected by the number of training samples
Unsupervised	_ simple and fast to process	_ spectrally separable classes are defined by computer before defining the information values. _ repeating the process whenever new data is added. _ considering only mean value. _ being affected by the number of clusters and clustering parameters

considering only mean value and being influenced by the number of clusters and their parameters are the most common disadvantages of these algorithms (Salah, 2017). Table 22.1 shows the advantages and disadvantages of supervised and unsupervised classifiers.

22.2.4 Hybrid classification method

The limitations of both supervised and unsupervised classification led the scientists to develop an advanced classification procedure and introduce the automated classification approach employed by Rx classification method, object-based classification, support vector machine (SVM), standardized object-oriented automatic classification technique based on fuzzy logic, artificial neural networks (ANN), decision tree classification method, and Bayesian and hybrid classifier. This method begins with an unsupervised classification of the image, and then the resulting output is interpreted using ground truth knowledge to identify any classification errors or misclassifications. Next, a supervised classification is performed on the original image, and the statistics obtained from the unsupervised classification serve as the training knowledge for the supervised classification (Al-Doski et al., 2013). Hybrid classification takes strengths of both supervised and unsupervised classification techniques. The high computation rate of large datasets is a common advantage of these classifiers; however, their most common disadvantage is that the results all depend on user-defined parameters (Salah, 2017). ANN, Fuzzy ArtMap classification, self-organized feature map (SOM), classification trees (CTs), and random forest (RF) are some of the widely used hybrid classifications.

22.3 The comparison between different image classification algorithms

Being familiar with the concept of different classification algorithms, including supervised, unsupervised, and hybrid classification, as well as specific algorithms like MLC, K-means, ISODATA, and others in the previous section, in this section, we focus on the results of each classifier to identify the most efficient one for remotely sensed data. Numerous research studies have been conducted on land use/land cover classification algorithms, where the performance of each algorithm was evaluated considering the accuracy assessment, and the most suitable and accurate classifier was recommended for the study area based on the results.

Six supervised and unsupervised classifiers, including minimum distance, maximum likelihood, parallelepiped, SVM, ISO data, and Kmeans, were applied in land use/land cover classification in an area located in India, using Sentinel 2 A images. Based on the overall accuracy and Kappa coefficient, it was concluded that the maximum likelihood classifier, with overall accuracy of 89.30% and Kappa coefficient of 0.8481, outperformed the other classifiers in this region, followed by parallelepiped classifier, SVM, minimum distance, ISO data, and K-means (Nivedita Priyadarshini et al., 2018). Another research analyzed the function of minimum distance, maximum likelihood, and RF classifiers in the land use classification of Weinan, a city located in China. The results showed that among these three classifiers, MLC showed the highest classification accuracy (85.28%), and its result was highly consistent with the actual ground objects when the Kappa coefficient was reported as 0.8206. The minimum distance method and RF stood at the second and third stages. The overall accuracy and the Kappa coefficient for these two classifiers were 72.34% and 67.75% and 0.6728 and 0.5609, respectively (Wu and Li, 2022).

The accuracy of six different machine learning algorithms, namely RF, SVM, ANN, fuzzy adaptive resonance theory-supervised predictive mapping (Fuzzy ARTMAP), spectral angle mapper (SAM), and Mahalanobis distance (MD) for land use classification were examined in another study. The results showed that although all of the classifiers can be treated as applicable, RF classifier is the most accurate algorithm, followed by ANN, SVM, Fuzzy ARTMAP, SAM, and MD (Kappa coefficients are 0.89, 0.87, 0.86, 0.85, 0.84, and 0.82 respectively) (Talukdar et al., 2020).

In another research, the Landsat TM imagery of Israel was classified using supervised, unsupervised, and hybrid classifiers. The results were compared based on both the Kappa coefficient and, Z-test. ISODATA was used as the unsupervised classifier, MLC was used as the supervised classifier, and an iterative classification approach was used as hybrid classification method where unsupervised training was utilized to create spectral signatures for each land cover classes, followed by supervised training. When the signatures from both training methods that contributed to the most accurate class assignment were combined, MLC was once more applied. Finally, the classification accuracy was assessed using 600 random points and creating a confusion matrix. The results of the Z-test showed that unsupervised classification, with an overall accuracy of 70.67%, outperformed the supervised classification (overall accuracy of 60.83%); however, if the study only focused on Kappa, the accuracy of supervised and unsupervised classification was similar. Furthermore, the accuracy improvement brought about by the hybrid classification was statistically significant compared to the supervised classification but not the unsupervised one. So, while most studies found that supervised classification performed better than unsupervised, the opposite was found in this research (Rozenstein and Karnieli, 2011).

In a study conducted by Guo et al. (2008), decision tree classification (DTC) was identified as the best classifier, while MLC was deemed as the most stable classical method among MLC, DTC, SVM, and self-organizing neural network (SONN). However, in another study, SVM was the most accurate classifier among SVM, MLC, DTC, and ANN (Pal and Mather, 2005). Moreover, comparing the classic and modern classifiers, Salah (2017) concluded that

ANN, SVMs, and CTs mostly outperformed and showed more accuracy among the modern classifiers.

Another study compared the performance of 15 distinct classification algorithms when applied to the set of Landsat images for a city in China named Guangzhou. Based on the literature, land use was classified into eight land use classes including water, residential area, natural forest, orchard, farmland, industrial/commercial, cleared land, and Idle land, and 14 subclasses. Then, the training samples were selected separately for both pixel- and object-based methods. After that, all the algorithms, including ISODATA, CBEST, MLC, K-nearest neighbor, logistic regression, C4.5, classification, regression tree, quick unbiased efficient and statistical tree algorithm, RF, SVM, radial basis function network, logistic model tree, bagging C4.5, AdaBoost C4.5, and Stochastic gradient boosting were tested, considering their specific parameters. The results indicated that unsupervised classification algorithms yield classification results that are comparable to those obtained from supervised algorithms when a very large number of clusters were used.

Furthermore, most supervised algorithms showed satisfactory classification accuracy if the training samples were sufficient, and MLC, LR, and LMT algorithms are the most proper. Moreover, within the pixel-based approach and for the 6-band case, logistic regression, MLC, neural network, SVM, and logistic model tree algorithms showed the best classification accuracy. In contrast, the CBEST and KNN showed the lowest accuracies. Also, for the 4-band case, the results show that including the two middle infrared bands would not increase the classification accuracy significantly, and the MLC produces the highest accuracy. Regarding the object-oriented classification, the results showed that the accuracy of this classification heavily depends on the segmentation, while the smallest scale, 5, led to the highest accuracy; by increasing the threshold, the accuracy was decreased. SGB and RF, C4.5, LMT, LR, and MLC showed the highest accuracy, respectively (Li et al., 2014).

Li et al. (2012) tested four different classification techniques, including MLC, as the most common parametric algorithm and CTA, Fuzzy ARTMAP, and KNN as the most common nonparametric machine learning algorithms in Altamira, Brazilian Amazon as a case study, to find out the most efficient classification algorithm using Landsat five TM image and PALSAR as datasets. It concluded that CTA performed well for Landsat TM image and PALSAR data, while for the combination of Landsat and PALSAR image, KNN was the algorithm that showed the best performance, followed by CTA. Table 22.2 compares various approaches for image classification.

22.4 Factors affect classification results

Based on the literature review, it is evident that while MLC, SVM, RF, ISODATA, and CTA are among the most frequently used classifiers with superior performance in various studies, they may not necessarily yield the most appropriate and accurate classification results in all research cases. This is because the outcomes of classification are influenced by multiple factors other than the classifier itself, such as the dataset used (Salah, 2017; Al-Doski et al., 2013), the knowledge and expertise of the analyst (Al-Doski et al., 2013; Salah, 2017; Li et al., 2012), the classification scheme (Li et al., 2014), the appropriate size of training samples and clusters'

TABLE 22.2 Contrasting various approaches for image classification.

Tested classifiers	Image dataset	The most suitable and accurate classifier	References
Minimum distance, maximum likelihood, parallelepiped, support vector machine (SVM), ISO data, and K-means	Sentinel 2 A	Maximum likelihood	Nivedita Priyadarshini et al. (2018)
Minimum distance, maximum likelihood and random forest (RF)	Landsat 8	Maximum likelihood	Wu and Li (2022)
Random forest (RF), support vector machine (SVM), artificial neural network (ANN), fuzzy adaptive resonance theory-supervised predictive mapping (fuzzy ARTMAP), spectral angle mapper (SAM), mahalanobis distance (MD)	Landsat 8	Random forest (RF)	Talukdar et al. (2020)
ISODATA, maximum likelihood, and an iterative classification approach	Landsat TM	ISODATA (based on the Z-test) ISODATA and maximum likelihood (based on Kappa)	Rozenstein and Karnieli (2011)
Maximum likelihood (MLC), decision tree (DC), support vector machine (SVM), self-organized neural network (SONN)	MODIS	Decision tree, maximum likelihood	Guo et al. (2008)
Support vector machine (SVM), maximum likelihood (MLC), decision tree (DC), artificial neural network (ANN)	Landsat 7 ETM and hyperspectral (DAIS) data	Support vector machine (SVM)	Pal and Mather (2005)
Artificial neural network (ANN), support vector machine (SVM), Classification trees (CTs), ISODATA, K-means, etc.	–	Artificial neural network (ANN), support vector machine (SVM), and classification trees (CTs)	Salah (2017)
ISODATA, CBEST, maximum likelihood, K-nearest neighbor, logistic regression, C4.5, classification and regression tree, quick unbiased efficient and statistical tree algorithm, random forest, support vector machine, etc.	Landsat TM	Maximum likelihood (MLC) logistic regression (LR) and logistic model tree (LMT) (With sufficient training samples)	Li et al. (2014)
Maximum likelihood (MLC), classification tree analysis (CTA), fuzzy adaptive resonance theory-supervised predictive mapping (fuzzy ARTMAP), and k-nearest neighbors (KNN)	Landsat TM	CTA	Li et al. (2012)
	PALSAR data	CTA, ARTMAP	
	Combination of Landsat and PALSAR	KNN, CTA	

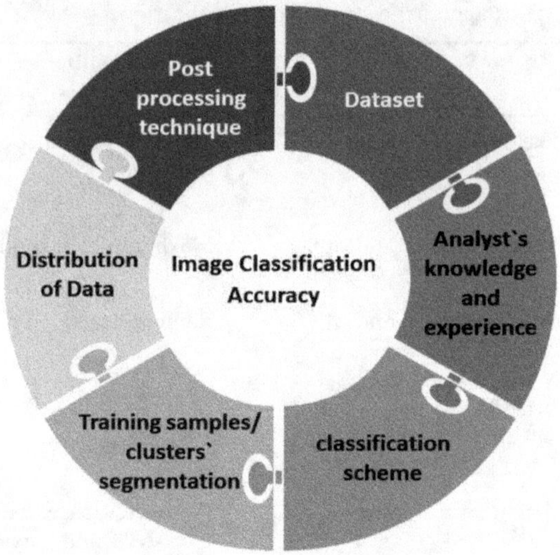

FIGURE 22.1 Influential variables in image classification accuracy, beyond the algorithm.

segmentation (Li et al., 2014; Salah, 2017), the distribution of data (Li et al., 2012; Manandhar et al., 2009), and postprocessing techniques (Li et al., 2014; Manandhar et al., 2009). Fig. 22.1 shows the influential variables in image classification accuracy, beyond the algorithm.

22.4.1 Data set

Based on recent studies, it has become evident that no algorithm can be considered the best for all datasets. For instance, CTA, followed by KNN and MLC, performed well for a Landsat TM image, while ARTAMAP produced the lowest accuracy. On the other hand, for PALSAR data, CTA and ARTMAP performed best, followed by MLC and KNN. For a combination of Landsat and PALSAR images, KNN produced the highest accuracy, followed by CTA, with fuzzy ARTMAP resulting in the lowest overall accuracy (Li et al., 2012). Among supervised classification algorithms using Land sat images as the data set, MLC is mostly the ideal land use classifier (Wu and Li, 2022), and SVM primarily performed well in comparison with other algorithms, using Landsat images, IKINOS, Quick bird, and hyperspectral data (Salah, 2017). The other important finding about the input data is that although Landsat images are still the most reliable data sources for LULC classification, combining them with PALSAR data can help improve vegetation classification performance (Li et al., 2012).

22.4.2 Knowledge and expertise of the analyst

An analyst's familiarity with the region being studied is critical in supervised classification, where selecting appropriate training samples directly impacts the accuracy of classification results (Al-Doski et al., 2013; Li et al., 2012). However, for these parametric algorithms, there is no need to be concerned about the analyst's expertise, which is necessary for nonparametric algorithms (Li et al., 2012). Although it is claimed that SVM results in higher accuracy than the traditional method, the Gamma parameter in kernel function and the penalty parameter should be wisely chosen in this technique to avoid misclassification. Since there is limited information available in the literature regarding the definition of these two parameters and also it needs a trial and error process to find out the optimal value if the parameters are not defined correctly, the results will not be reliable (Al-Doski et al., 2013).

22.4.3 Classification scheme

Recent studies have highlighted that although the accuracy assessment and Kappa coefficient are critical factors in comparing different classifiers, no single algorithm is invariably superior for all Land Use and Land Cover (LULC) classes. In fact, the overall accuracy of a classifier may not be an accurate indicator of its performance for specific LULC classes (Li et al., 2012). A closer examination of the calculated standard deviation (SD) and the coefficient of variation (CV) among the percentage share of the region in a single LULC class by various classifiers reveals that some land covers are classified more accurately than others (Talukdar et al., 2020). These findings demonstrate that the accuracy of a classifier varies depending on the LULC class being considered, and different classifiers may perform better for specific LULC classes. For example, Nivedita Priyadarshini et al. (2018) found that minimum distance, ISO data, and K-means algorithms showed misclassification among scrub land and crop land use.

Wu and Li (2022) also reported that although minimum distance, maximum likelihood, and RF performed well in classifying woodland and construction land, they were not ideal for classifying cultivated land and grassland; in contrast, MLC had the least error for these classes. In the case of water bodies, although the results of three classifiers were similar, RF produced a less clear water boundary in the land use map compared to other classifiers (Wu and Li, 2022). In another study, Talukdar et al. (2020) focused on machine learning classifiers and found that SVM and RF classifiers were more accurate in classifying built-up areas, rivers, and wetlands. The ANN)classifier proved to be the most effective for accurately classifying fallow and agricultural lands, with the Fuzzy ARTMAP and Mahalanobis distance classifiers following behind. Interestingly, all machine learning classifiers successfully classified vegetation cover and sandbar. These findings emphasize the importance of selecting appropriate classifiers for specific LULC classes to achieve accurate results.

22.4.4 Training samples and clusters' segmentation

Supervised classification accuracy heavily depends on the quality and quantity of the training samples used. A study showed that most supervised algorithms yielded satisfactory classification accuracy outcomes when an adequate number of training samples were available. MLC, LR, and LMT algorithms were the most suitable for this purpose (Li et al., 2014). Similarly, in unsupervised classification, these algorithms showed good classification results when a large number of clusters were used (Li et al., 2014).

To enhance the efficiency and accuracy of the MLC algorithm, Sun et al. (2013) implemented it in a distributed and parallel processing environment called a grid environment. This was achieved by using a sample library in the form of an XML file, which automated the entire classification process. The MLC algorithm was first implemented on a computer that acted as the master node and then deployed on six child nodes in the grid environment. The results showed that classification efficiency improved in the grid environment, especially when the number of child nodes was less than four. However, the accuracy assessment results were very similar in both environments, indicating that the training sample library's contribution played a significant role in ensuring accuracy.

Moreover, data distribution is also a critical factor in selecting an efficient algorithm. For normally distributed data, MLC is considered to be the most reliable algorithm (Li et al., 2012). The algorithm's accuracy is heavily dependent on the statistical properties of the data, making it important to ensure that the input samples conform to this requirement (Manandhar et al., 2009). These findings highlight the importance of selecting appropriate algorithms, ensuring adequate training samples, and considering the data distribution to achieve accurate results in supervised and unsupervised classification.

22.4.5 Postprocessing technique

Combining remotely sensed data with geographical data such as spatial texture, digital elevation models, geology, soil, etc. can enhance the accuracy of LULC maps. Regarding the MLC algorithm, applying postclassification correction (PCC) and utilizing ancillary data and knowledge-based logic has been shown to improve accuracy. A study applied MLC to a Landsat image and identified five land cover categories, including woodland, pasture/scrubland, vineyard, built-up, and waterbody. However, the results were unsatisfactory for built-up and vineyard categories with high commission errors and for the pasture/scrubland category with high omission errors. To improve the accuracy of the classification, a postclassification refinement was implemented utilizing ancillary data and a knowledge engineering approach based on hypothesis testing. The Singleton land use map, DEM, textural analysis, and NDVI were all integrated in the framework. As a result, the built-up and vineyard categories' commission errors were significantly decreased, while the pasture/scrubland category's omission errors also dropped. The user's accuracy in the pasture/scrubland category improved along with the user's accuracy in the built-up and vineyard categories. All three PCC maps were more accurate than 85% overall, with a Kappa statistics of over 0.8 (Manandhar et al., 2009).

22.5 Conclusion

The results of this study indicate that choosing the optimum classification method is difficult since, in remote sensing, the quality of an image classification depends on more than just the algorithm. Factors such as the availability of remotely sensed images data, the knowledge of the analyst regarding the study area, and each classifier's characteristics and required data are all critical. The dominant land use and land cover in the research region should also be taken into account when selecting a classification method. A proper classification procedure that involves training sample preparation and postclassification refinement must also be developed. To determine the most appropriate approach for a particular study area, researchers should conduct a comparative analysis of various classification methods while considering these factors.

In addition to the factors mentioned earlier, there are suggestions for improvements in future research. A comprehensive guideline and programming process can be developed to analyze these factors and identify the best classifiers for a particular study area. Future research can also focus on studying specific regions with various land cover types using supervised, unsupervised, and hybrid classifiers and different remotely sensed images to eliminate ambiguities. Additionally, it is important to study the differences in accuracy assessment statistics, such as Z-test and Kappa statistic, to identify the best statistic for image classification.

References

Al-Doski, J., Mansorl, S.B., Shafri, H.Z.M., 2013. Image classification in remote sensing. Department of Civil Engineering, Faculty of Engineering, University Putra, Malaysia 3 (10).

DiBiase, D., 2006. University consortium for geographic information science. Model curricula task force, and body of knowledge advisory board. In: Geographic Information Science and Technology Body of Knowledge, first ed. Association of American Geographers, Washington, D.C.

Goodchild, M.F., 1992. Geographical information science. Int. J. Geogr. Inf. Syst. 6, 31–45.

Guo, J., Zhang, J., Zhang, Y., Cao, Y., 2008. Study on the comparison of the land cover classification for multitemporal MODIS images. In: 2008 International Workshop on Earth Observation and Remote Sensing Applications, 1–6. IEEE, Beijing, China. https://doi.org/10.1109/EORSA.2008.4620305.

Li, C., Wang, J., Wang, L., Hu, L., Gong, P., 2014. Comparison of classification algorithms and training sample sizes in urban land classification with Landsat thematic mapper imagery. Rem. Sens. 6 (2), 964–983. https://doi.org/10.3390/rs6020964.

Li, G., Lu, D., Moran, E., Sant'Anna, S.J.S., 2012. Comparative analysis of classification algorithms and multiple sensor data for land use/land cover classification in the Brazilian Amazon. J. Appl. Remote Sens. 6 (1), 061706. https://doi.org/10.1117/1.JRS.6.061706.

Manandhar, R., Odeh, I., Ancev, T., 2009. Improving the accuracy of land use and land cover classification of Landsat data using post-classification enhancement. Rem. Sens. 1 (3), 330–344. https://doi.org/10.3390/rs1030330.

Nivedita Priyadarshini, K., Kumar, M., Rahaman, S.A., Nitheshnirmal, S., 2018. A comparative study of advanced land use/land cover classification algorithms using Sentinel-2 data. The International Archives of Photogrammetry, Remote Sensing and Spatial Information Sciences, XLII–5, pp. 665–670. https://doi.org/10.5194/isprs-archives-XLII-5-665-2018.

Pal, M., Mather, P.M., 2005. Support vector machines for classification in remote sensing. Int. J. Rem. Sens. 26 (5), 1007–1011. https://doi.org/10.1080/01431160512331314083.

Rozenstein, O., Karnieli, A., 2011. Comparison of methods for land-use classification incorporating remote sensing and GIS inputs. Appl. Geogr. 31 (2), 533–544. https://doi.org/10.1016/j.apgeog.2010.11.006.

Salah, M., 2017. A survey of modern classification techniques in remote sensing for improved image classification. J. Geomatics 11 (1), 21.

Sun, J., Yang, J., Zhang, C., Yun, W., Qu, J., 2013. Automatic remotely sensed image classification in a grid environment based on the maximum likelihood method. Math. Comput. Model. 58 (3–4), 573–581. https://doi.org/10.1016/j.mcm.2011.10.063.

Talukdar, S., Singha, P., Mahato, S., Shahfahad, Pal, S., Liou, Y.-A., Rahman, A., 2020. Land-use land-cover classification by machine learning classifiers for satellite observations—a review. Rem. Sens. 12 (7), 1135. https://doi.org/10.3390/rs12071135.

Wu, D., Li, Y., 2022. Study on land use classification of Weinan based on Landsat8. Front. Bus. Econ. Manage. 6 (2), 240–243. https://doi.org/10.54097/fbem.v6i2.3036.

CHAPTER

23

Spatial analysis of sediment connectivity and its applications

Raana Javidan[1], *Omid Rahmati*[2], *Artemi Cerdà*[3], *Francisco Escriva-Saneugenio*[3] *and Saskia D. Keesstra*[4,5]

[1]University of Environment, Karaj, Iran; [2]Soil Conservation and Watershed Management Research Institute, Agricultural Research, Education and Extension Organization (AREEO), Kurdistan Agricultural and Natural Resources Research and Education Center, Sanandaj, Iran; [3]Soil Erosion and Degradation Research Group, Departament de Geografia, Universidad de Valencia - Universitat de València, Valencia, Spain; [4]Climate-Kic Holding B.V., Amsterdam, the Netherlands; [5]University of Granada, Granada, Spain

23.1 Introduction

Connectivity is a key concept to understand the processes that shape the Earth system (Keesstra et al., 2018), and it is a key contribution to the knowledge of Geomorphology (Poeppl et al., 2017). Within Earth Sciences, connectivity is related mainly to the transport of sediments along the basins and how the materials are detached, transported, and sedimented along the fluvial system. This is initiated at the soil pedon scale (Cerdà et al., 2017; Rodrigo-Comino et al., 2018) and affects the fluvial system such as the flood research demonstrates (Kalantari et al., 2019), overland flow generation (Masselink et al., 2016), sediment delivery (Masselink et al., 2017), or both, sediments, and water (Keesstra et al., 2019), and it is applied at different scales (Rodrigo-Comino et al., 2020). Connectivity is not only applied in Earth Sciences but also by multiple disciplines and is a growing field of science (Turnbull et al., 2018) and very useful to understand the degradation thresholds and detect transitions in natural systems (Saco et al., 2020) such as forest fires (López-Vicente et al., 2021). Recent reviews on this topic show the importance of connectivity to improve the understanding of natural ecosystems and the ones man-made or man-disturbed (Parsons et al., 2015). The research on connectivity is expanding to other fields too. One of them is the perception of connectivity (Smetanová et al., 2018).

Soil erosion is a key process to understanding the landforms of the Earth planet and the geochemical cycles (Thornes, 1985; Borrelli et al., 2017). Soil erosion sculpture valleys,

https://doi.org/10.1016/B978-0-443-15341-9.00024-1

mountains, and slopes and determine the formation of alluvial terraces, deltas, and estuaries (Gerrard, 1981; Tan et al., 2022). Soil erosion is the detachment, transport, and sedimentation of material and induces the denudation of the continents and the wash of sediments, rocks, and organic matter to the oceans (Howell and Murray, 1986; Auzet et al., 2004). The research about soil erosion has been very successful for the last 100 years since the "Dustbowl" in United States triggered the research (Phillips, 1999). Most of the findings we achieved in the last century show that soil erosion is a natural process that can be enhanced by humans due to the mismanagement and misuse of the land, but also that soil erosion can be controlled using sustainable management and can contribute to the health of the planet (Dotterweich, 2008). Meanwhile, we know the process, we monitored the mechanism, and we know how to control soil erosion, but there is little knowledge about how the sediments are transferred along the watersheds and slopes (Ares et al., 2016). The understanding of what we call "connectivity" is a new approach, a holistic view, of the soil erosion process that must contribute to understanding better the process and finding more nature-based solutions to Earth management (Turley et al., 2021). This is relevant for the future of humankind as the soil is the key resource for humans, and a better knowledge of the connectivity along the landscape will bring wise knowledge to better manage the soil system (Mekonnen et al., 2017; Boardman et al., 2019).

This chapter reviews the current knowledge about connectivity and the implications of knowledge for land management under the 21st-century environmental conditions of the world. In this study, for assessing sediment connectivity, we render the SedInConnect, a stand-alone and free tool, as developed by Cavalli et al. (2013). The effects of sediment accumulation and their importance, the applications used in this field, and their functionalities will be discussed further.

23.2 Sediment connectivity

The physical linkage of sediments between sources and sinks defines sediment connectivity. One of the emerging characteristics of a morphological system is connectivity, and it can be defined as follows: a watershed facilitates the transport of materials including sediment (Heckmann et al., 2018; Wohl et al., 2019). Sediment connectivity is the extent to which a system can improve sediment transport through connections between its components, and it is a significant feature to study the redistribution of sediment in a basin (Bracken et al., 2015; Cossart and Fressard, 2017; Calsamiglia et al., 2018b).

Natural transporters such as wind, water, gravity, glaciers, and animals move sediment materials over large spatial and temporal scales, and this creates sediment connectivity between regions (Peters et al., 2008). Energy and materials that flow in an area (cascading system) and landforms (morphological system) form parts of an area (Chorley, 1971).

The features of sediment deposition and retention time are influenced by sediment binding (Preston and Schmidt, 2003). Among the factors that are considered very important in determining sediment connectivity, we can mention the spatio-temporal energy transfer pathways. According to the results of previous studies, there is a reverse relationship between sediment transfer distances and sediment size (Hoonhout and Vries, 2016). Therefore, it can be concluded that the sediments move in unrelated ways to the flows that transport them.

In the study of (Medeiros et al., 2014; Liu and Fu, 2016), some of the indices in the models based on the spatially distributed process have been investigated. In addition to conceptualizing connectivity, sediment connectivity indicators have been studied, and they have examined sediment connectivity indices and their morphological evaluation (Liu and Fu, 2016; Marchamalo et al., 2016).

Some topics are investigated in sediment studies including the resolution of digital terrain model, the size of watersheds, the types of models used, and geo-environmental influences (Persichillo et al., 2018). A wide range of resolutions from low to very high resolutions has been used to produce digital terrain models.

Soil loss increases because of the excessive use of natural resources and thus leads to an increase in the acceptance rate of soil erosion. Sediment load in streams increases due to extreme human use of natural resources, which causes difficulties inside and outside watersheds. Therefore, one of the essential approaches to managing sediment production and transport in watersheds is to identify areas with a high probability of sediment loss (Di Stefano and Ferro, 2018).

23.3 Sediment transport

Important structures and facilities in an area can be affected by significant erosion or accumulation due to the movement of sediments along the coastlines (Phillips, 2005). Coastal systems are dynamic, so to ensure the accuracy of the collected data, data must be collected both from the visible parts of the beach and the lower parts (littoral zone). Since the recent monitoring methods are limited to the visible parts of the beach, there are many limitations in the current monitoring methods. As a result, managers are not only faced with a lack of important data for planning and finding appropriate approaches to resolve or reduce coastal erosion but also due to the manual nature of the current monitoring equipment; there are also many limitations in the frequency of performing monitoring activities (Watt et al., 2019). Additionally, the accumulation of sediment in the rivers leads to the saturation of riverbeds with fine sediment (Naden et al., 2016), which is of vital importance.

As a result of the movement of sediments from different sources (the movement of those sources from a landscape or any other permanent or temporary source), the yield of fine sediment from the basin in the river channel is obtained. The movement of fine sediments is influenced by the climate, soil type, and geology of the region; however, environmental factors such as land use, vegetation cover, and topography of the region, as well as human management, reduce these effects (Wilkes et al., 2019). Considering the impact of landscape geomorphology on sediment connectivity, Fig. 23.1 illustrates a field photograph of a badland landform that causes a highly connected system.

23.4 Measuring connectivity

One of the main challenges today is the measurement of sediment connectivity. The reasons for this importance are the concept complication and directly evaluating sediment dynamic in various spatio-temporal scales (Keesstra et al., 2018). Factors that make this

FIGURE 23.1 A highly connected landscape. Badlands at El Cautivo. Almería. *photo by Artemi Cerdà.*

measurement even more difficult include the lack of universal protocols for field research, the incomparability of results across scales, and the difficulty of summarizing variables designed in a single index (Turnbull et al., 2018). Although it is difficult to measure real connectivity, attempts have been done to indirectly estimate connectivity in terms of the structure, which is essential to represent the configuration of connections (Heckmann et al., 2018).

Two concepts have been formed based on sediment connectivity, including structural and functional (Wainwright et al., 2011). The first is quantified by physical links in the landscape, and the second investigates the role of different mechanisms of a landscape in sediment transport processes (Calsamiglia et al., 2017). By examining these concepts, it is possible to reduce the uncertainties of sediment transport, and distribution, accumulation, and sediment cascades can be delineated (Hooke and Sandercock, 2012). Various techniques have been used to achieve these goals and quantify sediment connectivity, such as indices, existing models, and graph theory (Cossart and Fressard, 2017).

To depict the quantitative values of the potential of sediment delivery between basin compartments, the sediment connectivity indices method is simple and applicable. Although many studies have been conducted in this field, most of them have investigated on the physical (structural) connectivity of sediment, and functional connectivity has been less investigated. In addition, there is no consensus in different sources about the interpretation of the meaning of sediment connectivity, its methods, and indicators; as a result, the quantification and terminology of sediment binding in temporal and spatial scales are not properly investigated (Messenzehl et al., 2014; López-Vicente et al., 2017).

For sediment management and the importance of sediment in risk assessment, evaluation of sediment connectivity is of high importance, and structural measures should be designed at the basin scale. The potential movement of sediment in a given watershed is determined by the variation that occurs over time and space, related to the characteristics of the watershed system, its components, processes, and the input pressures (Poeppl et al., 2012).

The development of GIS-based tools to better understand ecosystem services, and processes associated with sediment connectivity is of great importance (Walling and Zhang, 2004; Le Roux et al., 2013). For a quick investigation of the impact of land use and management strategies on sediment flux and transport, geo-processing tools are effective (Newby and Pullar, 2009; Hamel et al., 2015).

23.5 Sediment fingerprinting

To find the sediment sources in river studies, sediment source fingerprinting should be applied. Samples are collected from the sediments carried by the river, and their physical and geochemical properties are compared with the potential resources of the catchment area. Since it is easier to identify the source of fine sediment due to the wide range of its sources, most fingerprinting studies have investigated the source of fine sediment or suspended sediment load. In addition, fine sediments should be further investigated in terms of their role in the destruction of aquatic habitats because they are considered a diffuse source pollutant. On the other hand, as coarse sediments move slower in the system and are usually moved from the channel system, their sources are much more obvious (Gellis and Walling, 2011; Mukundan et al., 2012). Potential sources can be classified in terms of source type as follows: rill erosion, sheet erosion, channel erosion, gully erosion, mass movements, and mobilization of sediment from man-made features such as unmetalled roads and urban surfaces (Walling, 2013).

23.6 Index of sediment connectivity

There are two groups to characterize sediment connectivity with the traditional approach. The first group considers the potential of upstream stream pressure along with hydraulic-related variables that affect sediment (Lu et al., 2006), and the second group considers downslope approaches that emphasize the energy expenditure of the travel pathway to get to a specific target (Novotny and Chesters, 1989). Index of connectivity (IC) is an efficient method to combine the two traditional approaches mentioned above (Borselli et al., 2008; Cavalli et al., 2016).

To compute the IC in an area that is under the effect of different types of landslide processes, a free and open-source tool named SedInConnect (Crema et al., 2015) is used. The output of this model is a topography-based index that assesses the potential link between upstream and sediment-depositing sites such as roads, retention basins, and lakes. If the potential connection for sediment to reach these features is high, the connectivity index has high values.

The components of this system include an upslope section that represents the forcing for downward routing of the sediment and a downslope section that considers the length of the flow pathway that a sediment particle moves to reach the target site (Cavalli et al., 2013).

The two weighting factors used in both components are slope and a proxy of impedance factor (W) to sediment dynamic. Several researchers consider different variables as the W factor to calculate the index of sediment connectivity. The slope index map can be an informative layer because it includes details of upslope, impedance, slope, and preferred flow paths. It also highlights a wide range of geomorphological features without a linkage to water-laden processes or debris flows.

In the work of Cavali et al. (2013), the theory and approach of this tool are presented, but the distinction between the sink and the target as well as the normalization process of the W factor are not underscored (Trevisani and Cavalli, 2016). Hence, improvements have been made including the implementation of a new algorithm developed for these purposes. The IC is easily calculated based on the procedure of Borselli et al. (2008). In this approach, IC has been defined as (Borselli et al., 2008):

$$IC = \log_{10}\frac{D_{up}}{D_{dn}} \tag{23.1}$$

where D_{up} is the upslope element and D_{dn} is the downslope element (Fig. 23.2). As the connectivity increases for larger values of IC, IC is defined in the range of $[-\infty, +\infty]$. D_{up} is the downward routing potential of the sediment produced in the high slope and is evaluated as follows:

$$D_{up} = \overline{WS}\sqrt{A} \tag{23.2}$$

where W and S are the weighting factor and the average slope gradient of the upslope element, respectively. A is the area of upslope site of the cell. The length of the flow path that a sediment moves downward to get a target is calculated using the downslope component (D_{dn}), which is explained as:

$$D_{dn} = \sum_i \frac{d_i}{W_i S_i} \tag{23.3}$$

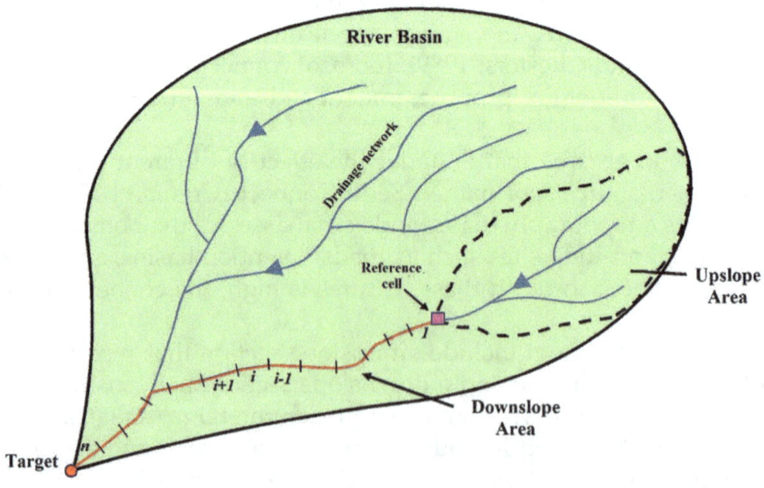

FIGURE 23.2 A schematic river basin and the main concept of the index of sediment connectivity.

where d_i is the flow path length along the ith cell according to the downslope direction, S_i and W_i are the slope gradient and the weighting factor of the ith cell, respectively (Crema and Cavalli, 2018). Fig. 23.2 indicates a schematic river basin and the main concept of the sediment connectivity index.

23.7 Applications of sediment connectivity to assess and predict impact

The sediment connectivity has been applied in various disciplines and contexts. The IC first introduced by Cavalli et al. (2013) allows us to consider the effects of different natural and anthropogenic effects and analysis their roles in sediment dynamics. To simulate sediment dynamics at the basin scale as well as within the framework of a broader approach including field observations or monitoring, evaluating, and describing sediment connectivity is an important step. In this section, we describe several areas of research in which connectivity has been used to assess or predict impact.

23.7.1 Prediction of the flood probability

In the study of Kalantari et al. (2019), a new approach is used for predictive modeling of the flood probability with an emphasis on road intersections, where sediment, water, and even debris can cause road damage. Several factors including soil type, sediment connectivity topography, precipitation, morphology, soil moisture, and land use are a set of descriptors used in this study to predict flood modeling. According to the obtained results, there are five categories of descriptors, including average elevation, soil properties, channel slope, and land use cover, and sediment connectivity index, which are important to quantify the probability of flooding and investigate flood-susceptible crossings roads (railways, etc.) (Kalantari et al., 2019).

23.7.2 Role of topographic and land use

Llena et al. (2019) have used the IC to investigate the effects of topographic land use variations and on sediment connectivity. The results demonstrate that the connectivity has been reduced due to afforestation practices. This approach can also be considered a suitable way to recognize potential risks associated with land use and morphological changes. Fig. 23.3 indicates how mountain trails can increase the connectivity of the sediments and water flows (Salesa and Cerdà, 2019, 2020). Fig. 23.4 illustrates a field photograph of a rangeland pitting and reseeding project in the Marivan Region, Iran, which restores degraded rangelands and causes a significant reduction of sediment connectivity.

Lizaga et al. (2018) have applied the IC to evaluate the impact of land cover changes in the previous years on the hydrological behavior of the watershed. According to the values obtained for the IC, agricultural land has decreased by 71%. Therefore, for land management in Mediterranean agricultural ecosystems, it is necessary to understand landscape dynamic, patterns, and interfaces of human activities (Lizaga et al., 2018). Fig. 23.5 shows a project of vegetation enhancement in rangelands of Sirvan watershed, Sanandaj, Iran, that reduces the sediment connectivity degree of the hillslope.

FIGURE 23.3 Mountain trails increase the connectivity of the sediments and water flows. *photo by Artemi Cerdà.*

FIGURE 23.4 Range pitting and reseeding project causes reduction of the sediment connectivity in Marivan Region, Iran. *photo by Omid Rahmati.*

23.7.3 Landslide effect assessment

The research of Persichillo et al. (2018) evaluates the impact of human interventions on sediment delivery. In this study, they investigated two watersheds in the Oltrepò Pavese region, with different geomorphological settings and sizes. According to the results, landscape complexity plays an effective role in coupling/decoupling upstream sediment sources. Also, the analysis of the results shows that instability phenomena that have high connectivity values can identify areas where mobilized sediment can potentially destruct critical infrastructure including road networks (Persichillo et al., 2018). Fig. 23.6 indicates how landslide occurrence produces huge sediment yield in the Kamiaran region, Iran.

FIGURE 23.5 Vegetation enhancement in rangelands of Sirvan watershed, Sanandaj, Iran. *photo by Omid Rahmati.*

FIGURE 23.6 A field photograph of landslide occurrence in the Kamiaran region, Iran. *photo by Omid Rahmati.*

In the work of Bordoni et al. (2018), a data-driven model is developed and tested to identify road networks that are prone to landslides and also threaten infrastructures. Based on the obtained results, the developed methodology can evaluate susceptible roads, and as a result, it is useful in managing the criticality of various road characteristics (Bordoni et al., 2018).

Kharismalatri et al. (2018) developed a two-section flume tributary inflow and the main channel to quantify sediment connectivity. They used a combination of different topographic factors such as main channel gradient, tributary inflow angle, and water content of sediment to determine sediment mobility. The value obtained for the IC in this study shows that there is a threshold between the occurrence of debris flow and the creation of a landslide dam (Kharismalatri et al., 2018). Broeckx et al. (2016) reported that range pitting and reseeding project causes reduction of the sediment connectivity in Marivan Region otential of a landslide susceptibility map to explain the spatial variation of sediment yield (SY) in Romania from a variety of proxies that potentially explain SY, including several landslide occurrence indicators. They used slippage. Based on the obtained results, landslide susceptibility maps (if they include all related factors) can be a very useful tool for predicting SY at regional scales (Broeckx et al., 2016).

23.7.4 Prediction of sediment yield through connectivity

In the study by Masselink et al. (2016), data from three catchments in Navarre, northern Spain, were evaluated to assess the factors influencing hydrological and sediment connectivity. Topography, biology, and soil were considered as three components of connectivity. They recommend using the model at spatial scales smaller than the watershed scale is a better approach to minimize the lumping of spatial variability in the components (Masselink et al., 2016).

23.7.5 Wildfire effect assessment and post-fire management

Martini et al. (2020) evaluated the effect of fire intensity on sediment connectivity in central Chile. This study investigated the relationships between sediment connectivity and fire intensity to recognize their association and influencing variables. According to the results, the applicability of this approach and its reproducibility can be improved by using satellite images and developing an impedance factor in IC to analyze the vegetation change due to wildfires (Martini et al., 2020).

In Calsamiglia et al.'s (2018) study, using a connectivity index based on geomorphometric concept and LiDAR data, they analyzed the patterns of the connectivity in a watershed. According to the results, 73% of the wall collapses were in the paths that showed a high index of connection values. The failure of walls enhanced runoff concentration, and consecutively, it speeds up hydrologic responses. A cost-effective approach for founding prioritized sites for terraced land management could be to identify the most linked paths in the most vulnerable areas (Calsamiglia et al., 2018a). Connectivity modeling can also help to assess the influence of postfire plans as was demonstrated by López-Vicente et al. (2021). They showed that postfire management is not always benificial to reduce soil erosion after a wildfire.

23.7.6 Road construction effects assessment

Jing et al. (2022) studied the response of sediment connectivity to the spatial pattern of river and road networks. In this study, main factors were recognized using buffer analysis, IC, geographic detector model (GDM), and random forest model (RF). Results also showed IC is affected by both road and river networks.

23.8 Effects of sediment accumulation on check dams

Gully erosion has been known as a significant source for suspended sediment in rivers, which affects land degradation and water quality (Valentin et al., 2005; Zaimes et al., 2009). Building control dams is one of the erosion control measures, and the purpose of

FIGURE 23.7 Sediment trapping behind a check dam at Khamsan Watershed, Iran. *(photo by Omid Rahmati).*

building these check dams is to stabilize the stream banks and streambed and to control sediments (Morgan, 2009; Rey and Burylo, 2014).

The total sediment trapped behind of check dams gives the amount of sediment yield using easy-to-use and simple geometric methods. Yet, these methods may not provide an accurate measurement of the remaining sediment volume (Romero-Díaz et al., 2007; Díaz et al., 2014; Ramos-Diez et al., 2016).

The amount of sediment that does not reach water streams can be estimated by more accurately evaluating the volume of sediment retained by control dams. It also provides the possibility to improve water quality for irrigation, supply, or to prevent silting of the tank (Díaz-Gutiérrez et al., 2019). Fig. 23.7 shows a field photograph of a check dam that trapped sediments and reduced sediment connectivity in the Khamsan Watershed, Iran.

23.9 Conclusions

Sediment connectivity is the extent to which a system can improve sediment transport through connections between its components, and it is a significant feature to study the redistribution of sediment in a basin. Connectivity has been accepted by the geoscience community as a useful concept for understanding and describing hydrologic function and sediment movement through watersheds. For sediment management and the importance of sediment in risk assessment, evaluation of sediment connectivity is of high importance and structural measures should be designed at the basin scale. To calculate the connectivity index in an area that is under the effect of different types of landslide processes, a free and open-source tool named SedInConnect is used. The output of this model is a topography-based index that assesses the potential connection between hillslopes and sediment-depositing features such as roads, catchment areas, retention basins, and lakes. If the potential connection for sediment to reach these features is high, the connectivity index has high values. There are several areas of research in which connectivity has been used to assess or predict impact comprising of the flood probability, impacts of topographic changes and land use on degree of sediment connectivity, landslide effect assessment, prediction of sediment yield through connectivity, wildfire effect assessment and postfire management, and road construction effects assessment. Sediment connectivity is one of the vital concepts to quantify the probability of flooding and identifying flood-prone intersections along roads. In a conclusion, continuous modeling over time is an interesting option for observing how connectivity develops over medium to long time scales. More study areas should be selected for comparison, especially areas with more important structural connectivity features such as mountainous landscapes or with more significant land use changes.

References

Ares, M.G., Bongiorno, F., Holzman, M., Chagas, C., Varni, M., Entraigas, I., 2016. Water erosion and connectivity analysis during a year with high precipitations in a watershed of Argentina. Nord. Hydrol. 47 (6), 1239–1252.

Auzet, A.V., Poesen, J., Valentin, C., 2004. Soil surface characteristics: dynamics and impacts on soil erosion. Earth Surf. Process. Landforms J. Br. Geomorphol. Res. Group 29 (9), 1063–1064.

Boardman, J., Vandaele, K., Evans, R., Foster, I.D., 2019. Off-site impacts of soil erosion and runoff: why connectivity is more important than erosion rates. Soil Use Manag. 35 (2), 245–256.

Bordoni, M., Persichillo, M.G., Meisina, C., Crema, S., Cavalli, M., Bartelletti, C., Galanti, Y., Barsanti, M., Giannecchini, R., D'amato Avanzi, G., 2018. Estimation of the susceptibility of a road network to shallow landslides with the integration of the sediment connectivity. Nat. Hazards Earth Syst. Sci. 18, 1735–1758.

Borrelli, P., Robinson, D.A., Fleischer, L.R., Lugato, E., Ballabio, C., Alewell, C., Panagos, P., 2017. An assessment of the global impact of 21st century land use change on soil erosion. Nat. Commun. 8 (1), 2013.

Borselli, L., Cassi, P., Torri, D., 2008. Prolegomena to sediment and flow connectivity in the landscape: a GIS and field numerical assessment. Catena 75, 268–277.

Bracken, L.J., Turnbull, L., Wainwright, J., Bogaart, P., 2015. Sediment connectivity: a framework for understanding sediment transfer at multiple scales. Earth Surf. Process. Landforms 40, 177–188.

Broeckx, J., Vanmaercke, M., Bălteanu, D., Chendes, V., Sima, M., Enciu, P., Poesen, J., 2016. Linking landslide susceptibility to sediment yield at regional scale: application to Romania. Geomorphology 268, 222–232.

Calsamiglia, A., Fortesa, J., García-Comendador, J., Lucas-Borja, M.E., Calvo-Cases, A., Estrany, J., 2018a. Spatial patterns of sediment connectivity in terraced lands: anthropogenic controls of catchment sensitivity. Land Degrad. Dev. 29, 1198–1210.

Calsamiglia, A., García-Comendador, J., Fortesa, J., López-Tarazón, J., Crema, S., Cavalli, M., Calvo-Cases, A., Estrany, J., 2018b. Effects of agricultural drainage systems on sediment connectivity in a small Mediterranean lowland catchment. Geomorphology 318, 162–171.

Calsamiglia, A., Lucas-Borja, M.E., Fortesa, J., García-Comendador, J., Estrany, J., 2017. Changes in soil quality and hydrological connectivity caused by the abandonment of terraces in a Mediterranean burned catchment. Forests 8, 333.

Cavalli, M., Crema, S., Blok, M., Lucía, A., Comiti, F., Marchi, L., Keesstra, S., 2016. Integrating structural and functional connectivity to characterize sediment dynamics in a small Alpine catchment. In: EGU General Assembly Conference Abstracts, pp. EPSC2016–15034.

Cavalli, M., Trevisani, S., Comiti, F., Marchi, L., 2013. Geomorphometric assessment of spatial sediment connectivity in small Alpine catchments. Geomorphology 188, 31–41.

Cerdà, A., Keesstra, S.D., Rodrigo-Comino, J., Novara, A., Pereira, P., Brevik, E., Prima, S.D., 2017. Runoff initiation, soil detachment and connectivity are enhanced as a consequence of vineyards plantations. J. Environ. Manag. 202, 268–275.

Chorley, R.J., 1971. The Role and Relations of Physical Geography. Edward Arnold.

Cossart, É., Fressard, M., 2017. Assessment of structural sediment connectivity within catchments: insights from graph theory. Earth Surf. Dyn. 5, 253–268.

Crema, S., Cavalli, M., 2018. SedInConnect: a stand-alone, free and open source tool for the assessment of sediment connectivity. Comput. Geosci. 111, 39–45.

Crema, S., Lanni, C., Goldin, B., Marchi, L., Cavalli, M., 2015. Improvement of a free software tool for the assessment of sediment connectivity. In: EGU General Assembly Conference Abstracts, p. 7652.

Di Stefano, C., Ferro, V., 2018. Modelling sediment delivery using connectivity components at the experimental SPA2 basin, Sicily (Italy). J. Mount. Sci. 15 (9), 1868–1880. https://doi.org/10.1007/s11629-018-4956-8.

Díaz, V., Mongil, J., Navarro, J., 2014. Proposal of a new methodology to assess the effectiveness of check-dams. Cuadernos Invest. Geogr. 40, 169–190.

Díaz-Gutiérrez, V., Mongil-Manso, J., Navarro-Hevia, J., Ramos-Díez, I., 2019. Check dams and sediment control: final results of a case study in the upper Corneja River (Central Spain). J. Soils Sediments 19, 451–466.

Dotterweich, M., 2008. The history of soil erosion and fluvial deposits in small catchments of central Europe: deciphering the long-term interaction between humans and the environment—a review. Geomorphology 101 (1–2), 192–208.

Gellis, A., Walling, D., 2011. Sediment source fingerprinting (tracing) and sediment budgets as tools in targeting river and watershed restoration programs. Stream Restoration in Dynamic Fluvial Systems: Scientific Approaches, Analyses, and Tools, vol. 194, pp. 263–291.

Gerrard, A.J., 1981. Soils and Landforms. An Integration of Geomorphology and Pedology. George Allen & Unwin (Publishers) Ltd.

Hamel, P., Chaplin-Kramer, R., Sim, S., Mueller, C., 2015. A new approach to modeling the sediment retention service (InVEST 3.0): case study of the Cape Fear catchment, North Carolina, USA. Sci. Total Environ. 524, 166–177.

Heckmann, T., Cavalli, M., Cerdan, O., Foerster, S., Javaux, M., Lode, E., Smetanová, A., Vericat, D., Brardinoni, F., 2018. Indices of sediment connectivity: opportunities, challenges and limitations. Earth Sci. Rev. 187, 77–108. https://doi.org/10.1016/j.earscirev.2018.08.004.

Hooke, J., Sandercock, P., 2012. Use of vegetation to combat desertification and land degradation: recommendations and guidelines for spatial strategies in Mediterranean lands. Landsc. Urban Plann. 107, 389–400.

Hoonhout, B.M., Vries, S.D., 2016. A process-based model for aeolian sediment transport and spatiotemporal varying sediment availability. J. Geophys. Res. Earth Surf. 121 (8), 1555–1575.

Howell, D.G., Murray, R.W., 1986. A budget for continental growth and denudation. Science 233 (4762), 446–449.

Jing, Y., Zhao, Q., Lu, M., Wang, A., Yu, J., Liu, Y., Ding, S., 2022. Effects of road and river networks on sediment connectivity in mountainous watersheds. Sci. Total Environ. 826, 154189.

Kalantari, Z., Ferreira, C.S.S., Koutsouris, A.J., Ahlmer, A.K., Cerdà, A., Destouni, G., 2019. Assessing flood probability for transportation infrastructure based on catchment characteristics, sediment connectivity and remotely sensed soil moisture. Sci. Total Environ. 661, 393–406.

Keesstra, S., Nunes, J.P., Saco, P., Parsons, T., Poeppl, R., Masselink, R., Cerdà, A., 2018. The way forward: can connectivity be useful to design better measuring and modelling schemes for water and sediment dynamics? Sci. Total Environ. 644, 1557–1572.

Keesstra, S.D., Davis, J., Masselink, R.H., Casalí, J., Peeters, E.T., Dijksma, R., 2019. Coupling hysteresis analysis with sediment and hydrological connectivity in three agricultural catchments in Navarre, Spain. J. Soils Sediments 19, 1598–1612.

Kharismalatri, H.S., Ishikawa, Y., Gomi, T., Sidle, R.C., Shiraki, K., 2018. Evaluating factors for controlling sediment connectivity of landslide materials: a flume experiment. Water 11, 17.

Le Roux, J.J., Sumner, P.D., Lorentz, S.A., Germishuyse, T., 2013. Connectivity aspects in sediment migration modelling using the soil and water assessment tool. Geosciences 3, 1–12.

Liu, Y., Fu, B., 2016. Assessing sedimentological connectivity using WATEM/SEDEM model in a hilly and gully watershed of the Loess Plateau, China. Ecol. Indicat. 66, 259–268.

Lizaga, I., Quijano, L., Palazón, L., Gaspar, L., Navas, A., 2018. Enhancing connectivity index to assess the effects of land use changes in a Mediterranean catchment. Land Degrad. Dev. 29, 663–675.

Llena, M., Vericat, D., Cavalli, M., Crema, S., Smith, M., 2019. The effects of land use and topographic changes on sediment connectivity in mountain catchments. Sci. Total Environ. 660, 899–912.

López-Vicente, M., Kramer, H., Keesstra, S., 2021. Effectiveness of soil erosion barriers to reduce sediment connectivity at small basin scale in a fire-affected forest. J. Environ. Manag. 278, 111510.

López-Vicente, M., Sun, X., Onda, Y., Kato, H., Gomi, T., Hiraoka, M., 2017. Effect of tree thinning and skidding trails on hydrological connectivity in two Japanese forest catchments. Geomorphology 292, 104–114.

Lu, H., Moran, C.J., Prosser, I.P., 2006. Modelling sediment delivery ratio over the Murray Darling basin. Environ. Model. Software 21, 1297–1308.

Marchamalo, M., Hooke, J.M., Sandercock, P.J., 2016. Flow and sediment connectivity in semi-arid landscapes in SE Spain: patterns and controls. Land Degrad. Dev. 27, 1032–1044.

Martini, L., Faes, L., Picco, L., Iroumé, A., Lingua, E., Garbarino, M., Cavalli, M., 2020. Assessing the effect of fire severity on sediment connectivity in central Chile. Sci. Total Environ. 728, 139006.

Masselink, R.J., Heckmann, T., Temme, A.J., Anders, N.S., Gooren, H.P., Keesstra, S.D., 2017. A network theory approach for a better understanding of overland flow connectivity. Hydrol. Process. 31 (1), 207–220.

Masselink, R.J., Keesstra, S.D., Temme, A.J., Seeger, M., Giménez, R., Casalí, J., 2016. Modelling discharge and sediment yield at catchment scale using connectivity components. Land Degrad. Dev. 27, 933–945.

Medeiros, P.H.A., De Araújo, J.C., Mamede, G.L., Creutzfeldt, B., Güntner, A., Bronstert, A., 2014. Connectivity of sediment transport in a semiarid environment: a synthesis for the Upper Jaguaribe Basin, Brazil. J. Soils Sediments 14, 1938–1948.

Mekonnen, M., Keesstra, S.D., Baartman, J.E., Stroosnijder, L., Maroulis, J., 2017. Reducing sediment connectivity through man-made and natural sediment sinks in the Minizr catchment, Northwest Ethiopia. Land Degrad. Dev. 28 (2), 708–717.

Messenzehl, K., Hoffmann, T., Dikau, R., 2014. Sediment connectivity in the high-alpine valley of Val Müschauns, Swiss National Park—linking geomorphic field mapping with geomorphometric modelling. Geomorphology 221, 215–229.

Morgan, R.P.C., 2009. Soil Erosion and Conservation. John Wiley & Sons.

Mukundan, R., Walling, D.E., Gellis, A.C., Slattery, M.C., Radcliffe, D.E., 2012. Sediment source fingerprinting: transforming from a research tool to a management tool 1. JAWRA J. Am. Water Resour. Assoc. 48, 1241–1257.

Naden, P., Murphy, J., Old, G., Newman, J., Scarlett, P., Harman, M., Duerdoth, C., Hawczak, A., Pretty, J., Arnold, A., 2016. Understanding the controls on deposited fine sediment in the streams of agricultural catchments. Sci. Total Environ. 547, 366–381.

Newby, J., Pullar, D., 2009. Using geoprocessing tools to model the potential impact of landcare on the spatial pattern of sediment delivery. In: MODSIM 2009 International Congress on Modelling and Simulation, Modelling and Simulation Society of Australia and New Zealand. Cairns. Citeseer, pp. 148–153.

Novotny, V., Chesters, G., 1989. Delivery of sediment and pollutants from nonpoint sources: a water quality perspective. J. Soil Water Conserv. 44, 568–576.

Parsons, A.J., Bracken, L., Poeppl, R.E., Wainwright, J., Keesstra, S.D., 2015. Introduction to special issue on connectivity in water and sediment dynamics. Earth Surf. Process. Landforms 40 (9), 1275–1277.

Persichillo, M.G., Bordoni, M., Cavalli, M., Crema, S., Meisina, C., 2018. The role of human activities on sediment connectivity of shallow landslides. Catena 160, 261–274.

Peters, D.P., Groffman, P.M., Nadelhoffer, K.J., Grimm, N.B., Collins, S.L., Michener, W.K., Huston, M.A., 2008. Living in an increasingly connected world: a framework for continental-scale environmental science. Front. Ecol. Environ. 6, 229–237.

Phillips, M., 2005. An Assessment of Processes and Strategies for Management of the Penarth Coast. Bath Spa University.

Phillips, S.T., 1999. Lessons from the dust bowl: dryland agriculture and soil erosion in the United States and South Africa, 1900–1950. Environ. Hist. 4 (2), 245–266.

Poeppl, R.E., Keiler, M., Von Elverfeldt, K., Zweimueller, I., Glade, T., 2012. The influence of riparian vegetation cover on diffuse lateral sediment connectivity and biogeomorphic processes in a medium-sized agricultural catchment, Austria. Geogr. Ann. Phys. Geogr. 94, 511–529.

Poeppl, R.E., Keesstra, S.D., Maroulis, J., 2017. A conceptual connectivity framework for understanding geomorphic change in human-impacted fluvial systems. Geomorphology 277, 237–250.

Preston, N., Schmidt, J., 2003. Modelling sediment fluxes at large spatial and temporal scales. In: Long Term Hillslope and Fluvial System Modelling: Concepts and Case Studies from the Rhine River Catchment. Springer.

Ramos-Diez, I., Navarro-Hevia, J., Fernández, R.S.M., Díaz-Gutiérrez, V., Mongil-Manso, J., 2016. Geometric models for measuring sediment wedge volume in retention check dams. Water Environ. J. 30, 119–127.

Rey, F., Burylo, M., 2014. Can bioengineering structures made of willow cuttings trap sediment in eroded marly gullies in a Mediterranean mountainous climate? Geomorphology 204, 564–572.

Rodrigo-Comino, J., Keesstra, S.D., Cerdà, A., 2018. Connectivity assessment in Mediterranean vineyards using improved stock unearthing method, LiDAR and soil erosion field surveys. Earth Surf. Process. Landforms 43 (10), 2193–2206.

Rodrigo-Comino, J., Lucas-Borja, M.E., Bertalan, L., Cerdà, A., 2020. Integrating in situ measurements of an index of connectivity to assess soil erosion processes in vineyards. Hydrol. Sci. J. 65 (4), 671–679.

Romero-Díaz, A., Alonso-Sarriá, F., Martínez-Lloris, M., 2007. Erosion rates obtained from check-dam sedimentation (SE Spain). A multi-method comparison. Catena 71, 172–178.

Saco, P.M., Rodríguez, J.F., Moreno-de las Heras, M., Keesstra, S., Azadi, S., Sandi, S., Rossi, M.J., 2020. Using hydrological connectivity to detect transitions and degradation thresholds: applications to dryland systems. Catena 186, 104354.

Salesa, D., Cerdà, A., 2019. Four-year soil erosion rates in a running-mountain trail in eastern Iberian Peninsula. Cuadernos de investigación geográfica: Geogr. Res. Lett. 309–331.

Salesa, D., Cerdà, A., 2020. Soil erosion on mountain trails as a consequence of recreational activities. A comprehensive review of the scientific literature. J. Environ. Manag. 271, 110990.

Smetanová, A., Paton, E.N., Maynard, C., Tindale, S., Fernández-Getino, A.P., Marqéus Pérez, M.J., Keesstra, S.D., 2018. Stakeholders' perception of the relevance of water and sediment connectivity in water and land management. Land Degrad. Dev. 29 (6), 1833–1844.

Tan, Z., Leung, L.R., Li, H.Y., Cohen, S., 2022. Representing global soil erosion and sediment flux in earth system models. J. Adv. Model. Earth Syst. 14 (1) e2021MS002756.

Thornes, J.B., 1985. The ecology of erosion. Geography 222–235.

Trevisani, S., Cavalli, M., 2016. Topography-based flow-directional roughness: potential and challenges. Earth Surf. Dyn. 4, 343–358.

Turley, M., Hassan, M.A., Slaymaker, O., 2021. Quantifying sediment connectivity: moving toward a holistic assessment through a mixed methods approach. Earth Surf. Process. Landforms 46 (12), 2501–2519.

Turnbull, L., Hütt, M.-T., Ioannides, A.A., Kininmonth, S., Poeppl, R., Tockner, K., Bracken, L.J., Keesstra, S., Liu, L., Masselink, R., 2018. Connectivity and complex systems: learning from a multi-disciplinary perspective. Appl. Netw. Sci. 3, 1–49.

Valentin, C., Poesen, J., Li, Y., 2005. Gully erosion: impacts, factors, and control. Catena 63, 132–153.

Wainwright, J., Turnbull, L., Ibrahim, T.G., Lexartza-Artza, I., Thornton, S.F., Brazier, R.E., 2011. Linking environmental regimes, space and time: interpretations of structural and functional connectivity. Geomorphology 126, 387–404.

Walling, D., Zhang, Y., 2004. Predicting Slope-Channel Connectivity: A National-Scale Approach, vol 288. IAHS PUBLICATION, pp. 107–114.

Walling, D.E., 2013. The evolution of sediment source fingerprinting investigations in fluvial systems. J. Soils Sediments 13, 1658–1675.

Watt, A., Phillips, M.R., Campbell, C.-A., Wells, I., Hole, S., 2019. Wireless sensor networks for monitoring underwater sediment transport. Sci. Total Environ. 667, 160–165.

Wilkes, M.A., Gittins, J.R., Mathers, K.L., Mason, R., Casas-Mulet, R., Vanzo, D., Mckenzie, M., Murray-Bligh, J., England, J., Gurnell, A., 2019. Physical and biological controls on fine sediment transport and storage in rivers. Wiley Interdiscip. Rev. Water 6, e1331.

Wohl, E., Brierley, G., Cadol, D., Coulthard, T.J., Covino, T., Fryirs, K.A., Grant, G., Hilton, R.G., Lane, S.N., Magilligan, F.J., 2019. Connectivity is an emergent property of geomorphic systems. Earth Surf. Process. Landforms 44, 4–26.

Zaimes, G.N., Schultz, R.C., Tufekcioglu, M., 2009. Gully and stream bank erosion in three pastures with different management in southeast Iowa. J. Iowa Acad. Sci.: JIAS 116, 1–8.

CHAPTER

24

Soil properties mapping using the Google Earth Engine platform

Raana Javidan[1], *Omid Rahmati*[2], *Seyed Masoud Soleimanpour*[3] *and Farnoush Mohammadi*[4]

[1]University of Environment, Karaj, Iran; [2]Soil Conservation and Watershed Management Research Institute, Agricultural Research, Education and Extension Organization (AREEO), Kurdistan Agricultural and Natural Resources Research and Education Center, Sanandaj, Iran; [3]Soil Conservation and Watershed Management Research Institute, Fars Agricultural and Natural Resources Research and Education Center, Agricultural Research, Education and Extension Organization (AREEO), Shiraz, Iran; [4]Department of Reclamation of Arid and Mountainous Regions, Faculty of Natural Resources, University of Tehran, Karaj, Iran

24.1 Introduction

Soil, owing to its serious role in food production and also sustainability via the support of vital societal and ecosystem services at the intersection of the atmosphere, lithosphere, hydrosphere, and biosphere, is a complex system (Baveye et al., 2016; Lehmann et al., 2020). At national and regional scales, understanding the processes of soil and its properties has a crucial function to realize the Sustainable Development Goals (SDGs) of the United Nations (Lal et al., 2021). To prepare people to understand the importance of soils in human life, a basic grounding in soil science and effective communication on the SDGs with the general public seems to be an obligation. The formation of soils and the way they are sustained, also their variation across the landscape, and their essential relation to the world's economy—from agriculture, forestry, town and country planning, and gardening, through ecology and conservation—are the main topics that are compulsory to understand among new generations (Bouma, 2018). Expanding our horizons and improving our knowledge about soil and its impressions on our delicate natural environment leads to more sense of responsibility and care. Moreover, learning about soils will prevail among students through their school subjects, which may result in helpful and promising research on soil science at university either as a main subject or as part

Remote Sensing of Soil and Land Surface Processes
https://doi.org/10.1016/B978-0-443-15341-9.00022-8

of geology, botany, agriculture, geography, or biology and human health grade (Brevik et al., 2020).

24.2 Soil properties

24.2.1 Soil health

During the last three decades, some authors have studied the soil health concept. According to the Global Soil Partnership of the Food and Agriculture Organization (FAO), soil health concept focuses on the ability to environmental services including productivity and diversity in terrestrial ecosystems (Costantini and Mocali, 2022). The aim of numerous guidelines at regional, national, or even global scales is to maintain and, of course, recover soil health which results in global health. Determining how much a general method is relevant to the assessment of soil health and its monitoring is aimed to focus, which is concerned about the soil's functional biodiversity. Additionally, it considers the genetic prospects of both the topsoil and the soil profile (Costantini and Mocali, 2022).

Above all, the constant capacity of soil as a functional active ecosystem that assists plants, animals, and humans can describe the word "soil health". Regarding this definition, soil health attaches agricultural and soil science to policy. From a historical point of view, crop production was the chief duty of the soil assessments. Nowadays, it includes more serious subjects such as climate change, water quality, and human health. Due to inadequate purposeful knowledge and the absence of effective methods, despite the pivotal role of soil biodiversity, chemical indicators dominate in quantifying soil health (Lehmann et al., 2020).

24.2.2 Soil carbon

In the terrestrial system, soil represents the largest carbon pool. Compared to the atmosphere, it stores up to five times more carbon. Between the years 2001 and 2020, the number of annual publications relating to both soil and climate change increased by 1800%. Today on average, 18.6% of all publications are related to climate change. Although the soil system has multifaceted impacts on the conditions of the changing climate, the most susceptible population to soil health issues tends to be smallholder farmers. Hence, financial assistance and climate adaptation initiatives must put this in a high priority (Hou, 2022).

Several cycles are connected to the soil and plant systems, soil quality, plant production, and environmental quality, and these all are affected by the climate. Climate conditions, which are altering owing to human activity, have multiple effects on plant nutrition including, decomposition, mineralization, leaching, and losing nutrients in the soil. Soil fertility, plant nutrient accessibility, and climate change reduction are strongly affected by soil carbon sequestration. Accordingly, there is great interest worldwide in the sequestration of soil carbon from atmospheric carbon dioxide and decomposing it in the soil by plants. Soil carbon sequestration (SCseq) and plant nutrition are crucially under the influence of growing soil carbon inputs over outputs and developing efficient management approaches. Thus, the agricultural yield must be boosted for food security, particularly in developing countries (Elbasiouny et al., 2022).

24.2.3 Soil fertility

A lion's share of the human regime is made up of vegetables, and plants provide vitamins and nutrients required to perform countless vital physiological functions in the human body. Nowadays, vegetables contaminated with heavy metals (HMs) and their serious repercussions on human health have become critical matters (Tauqeer et al., 2022). Nevertheless, there is very little public information about the connections between soil and human health, and it is necessary to communicate these connections (Brevik et al., 2019). Environmental processes for the formation and restoration of fertile soils and the key role of plant diversity have been less studied. Needless to mention that productive soils have been a crucial resource for humanity for the last 10 millennia (Furey and Tilman, 2021). Because higher richness of the plant species might be connected to both higher produced plant biomass and more accumulated soil nutrients, the decomposition of the plant tissues will raise soil fertility (Dybzinski et al., 2008; Jenny, 1958; Jobbágy and Jackson, 2004), and plant diversity intensifies this nutritious accumulation (Dijkstra et al., 2007). Hence, the augmented plant biomass of the soil resulting from plant biodiversity leads to even more soil fertility by adding extra nutrients (Ehrenfeld et al., 2005; DeAngelis et al., 2012; Walker and Del Moral, 2003).

24.2.4 Soil biodiversity

One of the factors that affect soil quality is soil biodiversity. Controlling the pathogen populations is under the effect of soil biodiversity. Healthy soils with high cover and high carbon content can lead to better pest control, and lessen disease spread and also allergies can be reduced by exposure to soil microbes. The crop plant resilience can be increased by soil microbes. They also deliver many of the existing antibiotics. Meanwhile, healthy soils can have other benefits as well, including they are less prone to wind and water erosion, they give safe and clean water via filtration, refinement by microbes, and elimination of contaminants, and they also improve good clean air quality. Thus, one of the most significant features of healthy soil is soil biodiversity (Wall et al., 2015). Soil is not only what is above mentioned but also provides necessities for the whole living organisms.

24.2.5 Soil organic matter

Soil organic matter (SOM), refers to the largest terrestrial carbon (C) reservoir, which keeps more carbon in reserve than the combination of vegetation and atmosphere. Hence, the maintenance of SOM is a necessity for human health, climate change mitigation, and soil fertility. Carbon sequestration and soil fertility are influenced by the presence of sufficient amounts of carbon in the soil; therefore, increasing the amount of carbon in agricultural soils leads to higher food security and reduces the susceptibility of the soil to climate change (Soussana et al., 2019; Lal, 2020). Environmental factors, the soil type, and transformation processes, comprising biological interactions, influence the carbon content. Carbon storage is also determined by several natural factors including soil moisture (SM) and water saturation, temperature, topography, texture, acidity, salinity, biomass production, and vegetation (Bot and Benites, 2005; Magdoff and Van Es, 2000). Climate change employing the enlarged occurrence of floods and drought strengthens the cycle of global water. The dynamics of soil carbon are under the influence of the water cycle. Furthermore, moisture disturbances result in inreasing availability of carbon and fluxes in soils (Patel et al., 2021).

24.2.6 Soil moisture

Soil moisture (SM) is one of the most significant hydrologic state factors. The global environment and human society are affected by various Earth and environmental science applications that are affected by SM (Robinson et al., 2008). Some of the possible applications are as follows; forecasting and monitoring of drought conditions (Zhang et al., 2022), prediction of weather and climate changeability (Balehegn et al., 2019), deterrence of natural hazards like landslides, wildfires, dust storms, and floods (Bao et al., 2022), managing and allocating of water resources (Candido et al., 2022), and observing of ecosystem reaction to climate change (Avtar et al., 2022). Because of the importance and widespread application of highly spatially and temporally variable SM data linking energy, water, and carbon cycles, recently, major attempts and studies have been focused on to improve SM measurement and monitoring potentials from the point to global scales (Huang et al., 2022).

Owing to the impacts of SM on various procedures related to corrosion, SM is recognized as a substantial factor in controlling corrosion in soils (Cole and Marney, 2012; Gupta and Gupta, 1979). Erosion of metals from soils is a matter with societal and economic impression. Moreover, substantial economic and environmental damages because of operating expenses related to buried infrastructure, gas and oil pipelines, and tanks can also happen and can cause poisonousness within the biosphere. Therefore, it is vital to understand underground corrosion to boost infrastructure and environmental stability.

The ratio of the water to the voids in the soil defines SM. Unsaturated soils have a degree of saturation of less than one and variables such as pore structure, the soil skeleton, the level of compaction, and the amount of moisture define their properties. The characteristics of water distribution and maintenance in unsaturated soils can be determined based on the degree of saturation and the soil suction. The soil water retention curve (SWRC) demonstrates this relationship (Fredlund et al., 2012).

24.2.7 Soil texture

Most chemical, biological, and physical procedures in soils are in control of soil texture, which is the most critical physical property of soil. It can impact permeability, soil thermal capacity, solute movement, and water holding capacity, which are thoroughly connected with applications of ecological, hydrological modeling, climate, soil pollution control, and smart agricultural management. In both vertical and lateral dimensions, soil texture usually has high spatial heterogeneity over regions and landscapes (Liu et al., 2020). The relative ratio of silt, sand, and clay is the basis of the classification of soil texture. The other chemical, biological, and physical properties of soil are influenced by these individual particle-size fractions (PSFs) that are a type of compositional soil data; moreover, soil fertility is also under the control of PSFs (Kome et al., 2019; Morgan et al., 1998; Wang et al., 2003; Poggio and Gimona, 2017). For instance, fertilizer application rates (Ghiri and Abtahi, 2012), the buffering capacity of soils (Koseva et al., 2010), taxonomic-based diagnostic horizon definitions (Bockheim and Hartemink, 2013), and soil carbon maintenance (Xu et al., 2020) are significantly reliant on textural classes and soil PSFs. Therefore, researchers tend to research to provide high-resolution spatial data connected to soil PSFs.

Forasmuch as soil fertility management is indirectly influenced by texture classification, the classification of soil texture is the main factor for acquiring sustainable agriculture management. A soil texture triangle can be used by soil scientists to specify which of the 12 soil types are detected. Although different texture triangles are available based on the soil condition, the first texture triangle and the most commonly used one is the United States Department of Agriculture (USDA) agro system. M. Whitney of USDA established the textural triangle system in 1911 (Moeys, 2018). The USDA ternary diagram contains 12 different textural classes. The type of classes is determined employing soil fraction percentage. The USDA texture triangle is demonstrated in Fig. 24.1. The proposition of sand, silt, and clay is shown on each side of the soil texture triangle, which is scaled from 0 to 100. The texture class is determined based on the intersection of the three vectors on the triangle.

24.3 Google Earth Engine (GEE)

The GEE as one of the primary web-based programs provides useful data of the Earth to monitor and map geo-spatial issues (Gorelick et al., 2017). It consists of several satellite datasets along with capabilities for image processing (Xiao et al., 2020). The GEE has been commonly utilized for different contexts including rice distribution maps in Northeast Asia (Dong et al., 2016), paddy field fallow (Chong et al., 2020), and wetland distribution (Mahdianpari et al., 2018). Unfortunately, a few application of GEE has been done for soil-related issues (Amani et al., 2020). Simulating the spatial distribution of soil properties using conventional methods in large areas does not reach high prediction accuracy. Also, in

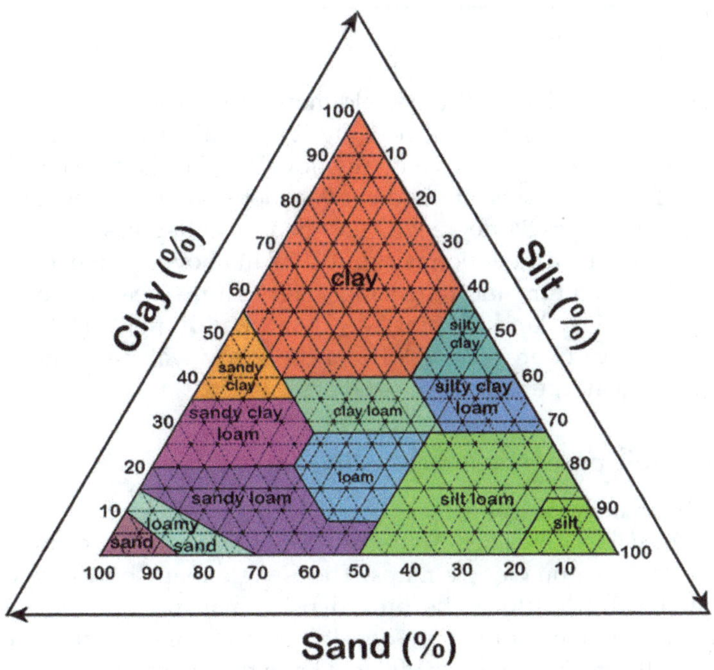

FIGURE 24.1 USDA texture triangle.

addition to the time and cost of using these methods, they need dense and uniform sample points (Robinson and Metternicht, 2006). Digital soil map (DSM) is applied to map the soil functional characteristics, and it increases the efficiency of soil mapping.

24.3.1 Soil properties databases at the GEE platform

The GEE includes several databases in the case of soil properties. These soil databases are characterized by different spatial scales and some of them have been generated for a specific country or region. The development of these soil databases is progressing so that more soil properties can be monitored and the spatial extent of the information can cover larger regions or even all over the world. One of the important soil databases is the OPEN LAND MAP, and its spatial extent covers the whole Earth and different categories of soil characteristics including soil bulk density, clay content, soil organic carbon content, USDA soil taxonomy groups, soil texture class, soil pH, sand content, and soil water content at 33 kPa (field capacity) are freely available. The OPEN LAND MAP datasets are raster layers consisting of information related to 1950–2018. In the description section, the required codes are available to use these datasets or extract the required data related to the study area. There is also a special color pallet in the description section which can be used in illustrating the study area. There are different bands each representing the related data of specified soil depth for instance the b0 band contains soil information related to the topsoil at depth 0 (b200 data related to the depth 200 cm is the maximum depth available here).

24.4 Case study

24.4.1 Description of the study area

To compare the global database with the field monitoring, we analyzed the information on the Talwar river basin in Kurdistan province (Fig. 24.2). This basin is the main source of the Qezalozan and Sefidrud rivers, which eventually flow into the Caspian Sea. The Talwar River basin with an area of 6468.3 km^2 is situated between longitudes 47 degrees 14′ 09″ E and 48 degrees 08′ 32″ E and latitudes 34 degrees 47′ 26″ N and 35 degrees 44′ 20″ N, located in the Kurdistan province. The mean elevation is about 1927 m above the mean sea level. Land use in this basin is mostly agriculture and rangelands. Agriculture areas account for about 85% of the study area, including dry land farming (75%) and paddy fields (10%). Rangeland areas cover about 14% of the study area. About 1% of the study watershed has been assigned to other land use classes (water, urban area, rural area).

24.4.2 Preparing soil properties maps using the GEE

Firstly, the study area should be selected by drawing a rectangle or the shape file of the study area can be used in GEE. Then *Map.centerObject (geometry)* function is used to fit the location of the selected area on Google maps. The Google map shows the code of the study area when we hit the Run button. The bulk density map of the study area is shown in Fig. 24.3, the maximum values of the bulk density were obtained in the middle of the study area, whereas the southern sites in the study area have the lowest bulk density. This map can

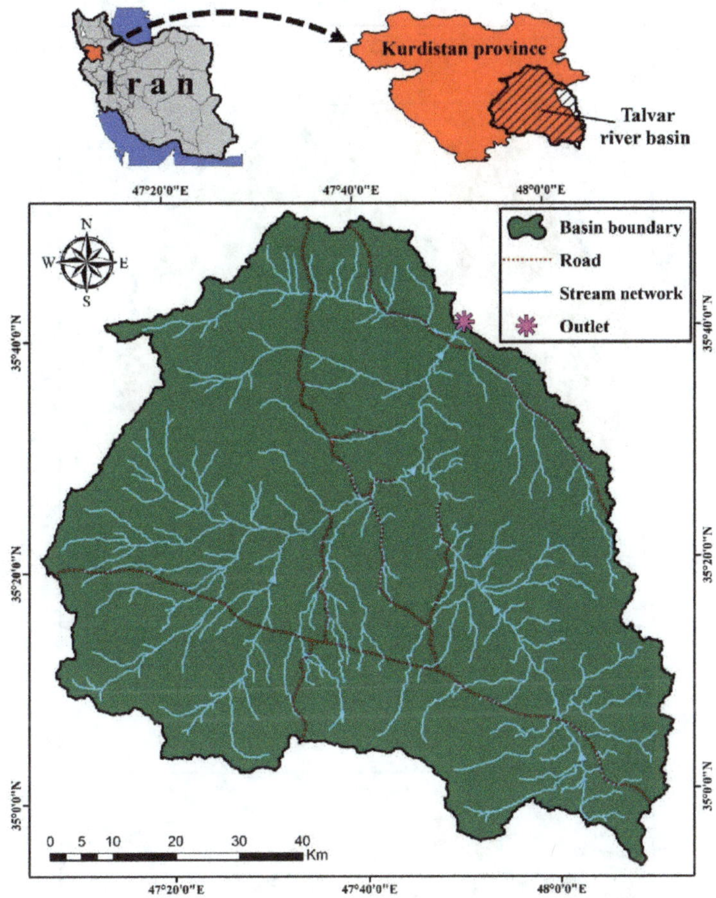

FIGURE 24.2 The geographical location of the Talwar River basin.

provide vital information in terms of spatial patterns and considerable natural variability of the bulk density in the study area.

The clay content map of the Talwar River basin was shown in Fig. 24.4 The clay content varies from 14% to 37% in the study area. Soil with high clay content can be seen in the southern and southwest parts of the Talwar River basin. Overall, the west side of the study area showed the highest clay content, whereas the lowest values belong to the east and southeast sides. More importantly, the spatial variability of the clay content is considerable.

The map of sand content for the Talwar river basin was exhibited in Fig. 24.5. The sand content in the study area ranges from 18% to 61%. A clear increasing spatial gradient was observed from the northwest to the southeast. This map gives important insights to planners, farmers, and experts who work with soil and sediment. For example, soils with high sand contents are suitable for some plants and experts can find appropriate locations for conducting projects. This map is also useful for estimating runoff rates and infiltration potential in hydrological studies.

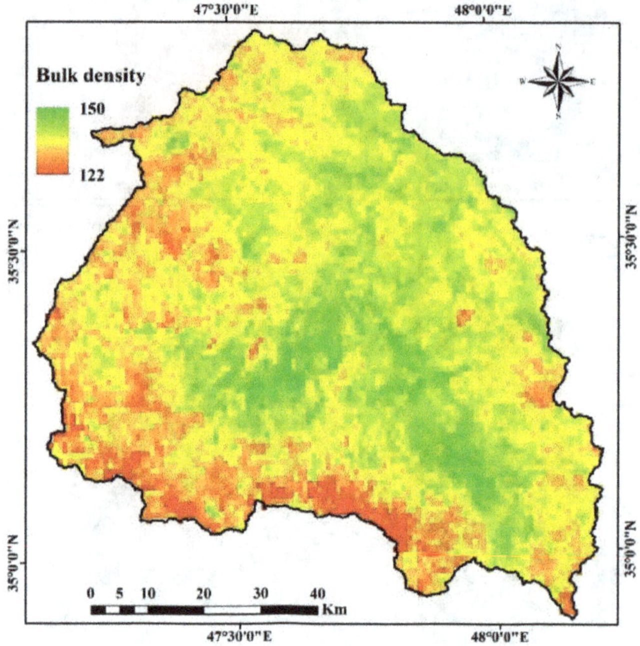

FIGURE 24.3 Bulk density map of the study area.

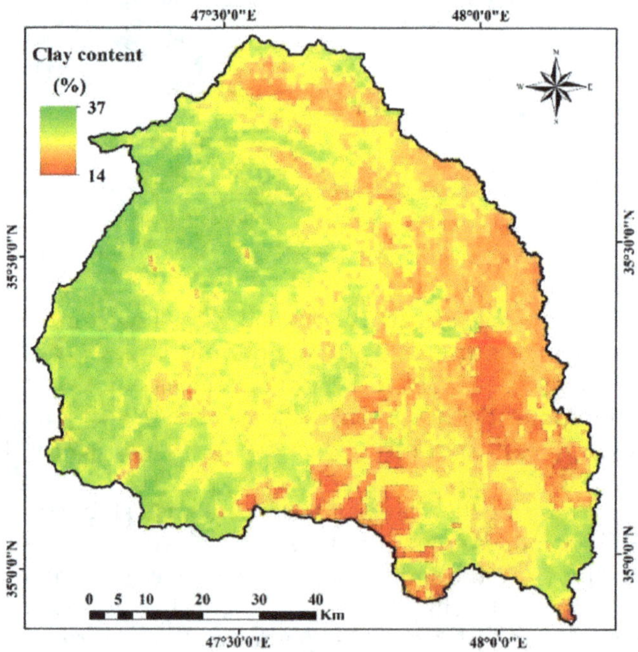

FIGURE 24.4 Clay content map of the study area.

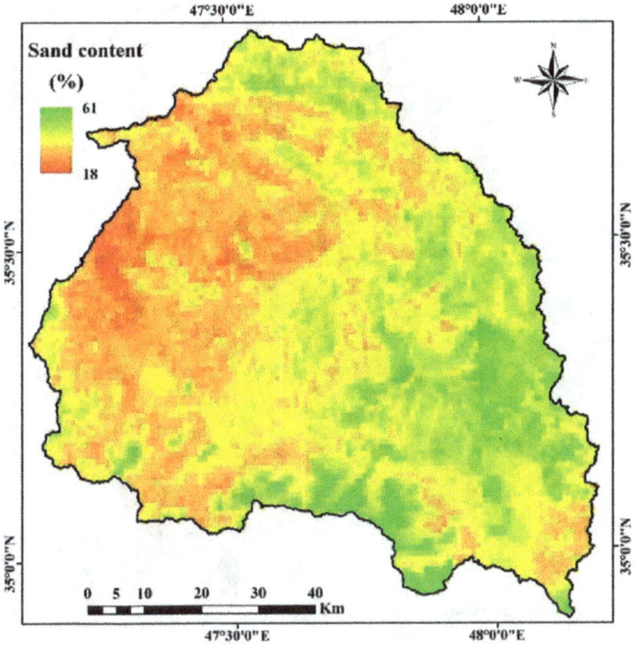

FIGURE 24.5 Sand content map of the study area.

Preparing soil texture maps for large areas has been known as a difficult task. The soil texture map of the study area generated using the GEE is illustrated in Fig. 24.6. The identified soil texture types are silty clay loam, clay loam, sandy loam, sandy clay loam, and loam. Clay loam and loam are the widest soil texture classes in the Talwar River basin. The produced soil texture map supports planners and managers to consider soil texture class in the projects.

The spatial pattern of the soil pH in the study area was displayed in Fig. 24.7. Soil pH is one of the main factors affecting plant ecology and vegetation communities. This parameter ranges from 6.9 to 8.2 in the subject area. The lowest values were in the south of the subject area, but the highest pH values were reported for the middle and northern of the subject area. The spatial variability of the soil pH in the Talwar River basin is clear, and this acidity condition should be taken into account in agricultural and rangeland plans.

The spatial distribution of the soil organic carbon was shown in Fig. 24.8. As can be seen, the soil organic carbon content ranges from 0% to 8%. The high values were reported in the southern and southwest of the study area (green color sites), whereas the lowest organic carbon contents belong to the eastern and middle of the Talwar River basin. In an overview, the soil quality of the study area in terms of the organic carbon contents is poor. The average organic carbon in the Talwar River basin stood at 4.2%. This soil parameter directly affects soil fertility and soil erodibility.

The final output of the soil database of the GEE platform is a water content map. The water content map of the Talwar River basin was shown in Fig. 24.9. The ranges of 9%–31% were

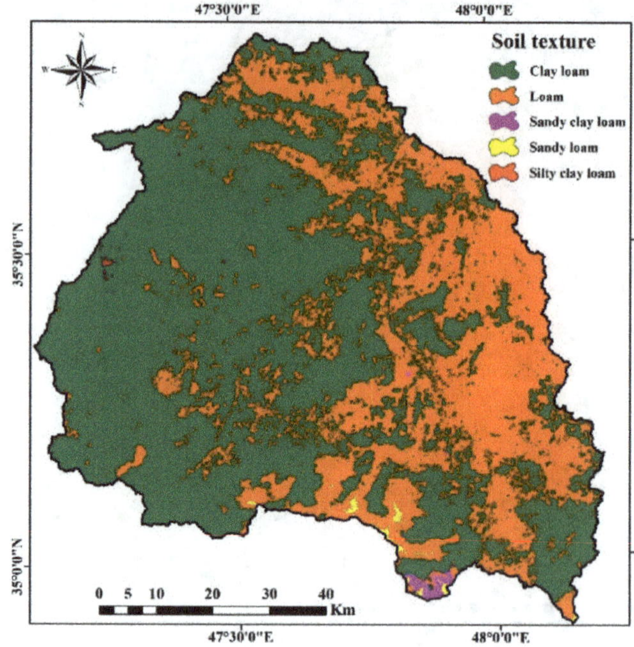

FIGURE 24.6 Soil texture map of the study area.

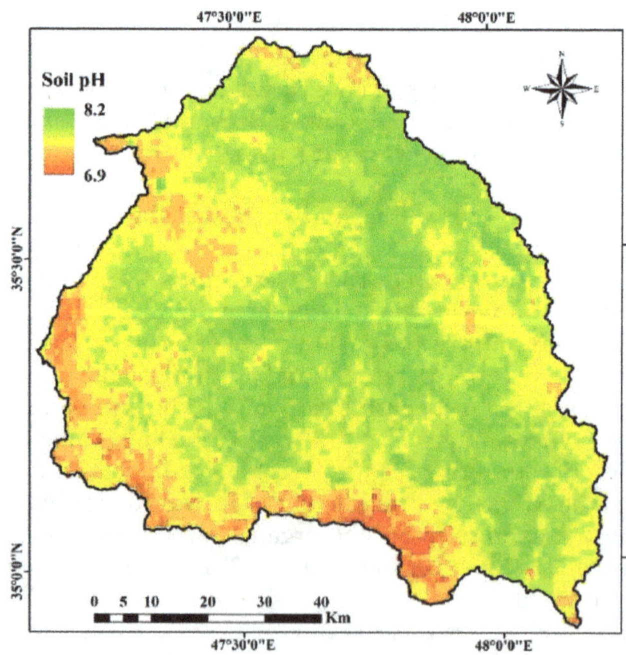

FIGURE 24.7 Soil pH map of the study area.

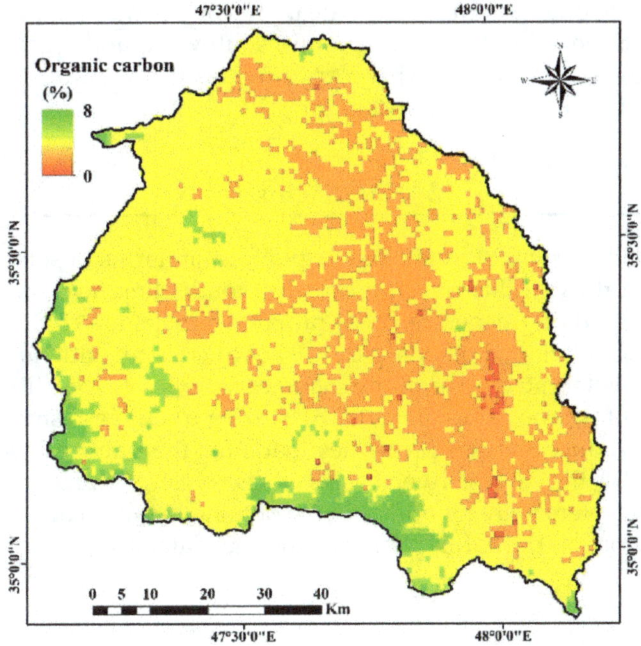

FIGURE 24.8 Soil carbon map of the study area.

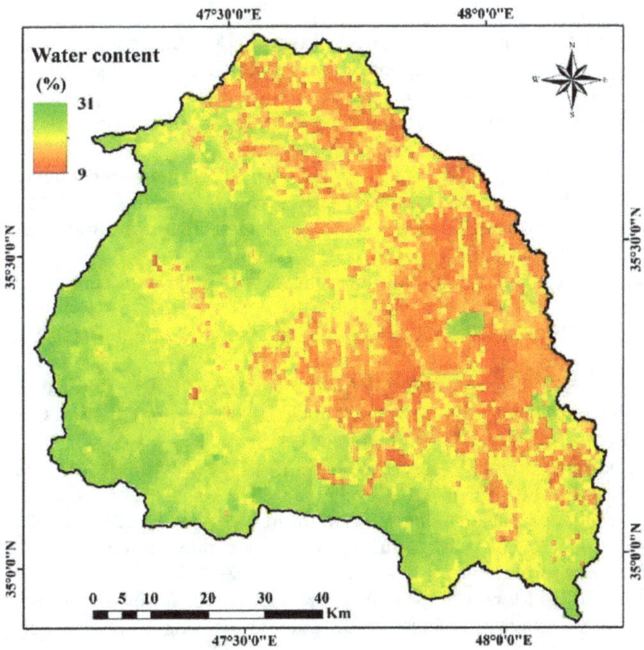

FIGURE 24.9 Water content map of the study area.

obtained for the water content of the subject area. The lowest water contents were in the east, middle, and north of the subject area. The south, southwest, and west of the Talwar River basin had the highest water content. This map provides useful information for agricultural activities, especially in arid and semi-arid regions.

24.5 Conclusions

This chapter explains how GEE is a valuable platform in terms of soil properties mapping. Considering the spatial variability of soil properties, conducting field sampling and applying the interpolation methods for generating soil properties maps are difficult, costly, and time-consuming. The interpolation methods generally work along with uncertainty, especially when a large region is targeted. One of the main advantages of the GEE platform is related to its huge capability for processing big data. According to our case study, the GEE platform could successfully produce the soil properties maps for the Talwar River basin, which has been known as one of the largest river basins in the west of Iran. The spatial patterns of the selected soil parameters were identified in the obtained maps. This study strongly recommends the application of the GEE platform and its databases, especially in developing countries.

References

Amani, M., Ghorbanian, A., Ahmadi, S.A., Kakooei, M., Moghimi, A., Mirmazloumi, S.M., Moghaddam, S.H.A., Mahdavi, S., Ghahremanloo, M., Parsian, S., 2020. Google earth engine cloud computing platform for remote sensing big data applications: a comprehensive review. IEEE J. Sel. Top. Appl. Earth Obs. Rem. Sens. 13, 5326–5350.

Avtar, R., Yunus, A.P., Saito, O., Kharrazi, A., Kumar, P., Takeuchi, K., 2022. Multi-temporal remote sensing data to monitor terrestrial ecosystem responses to climate variations in Ghana. Geocarto Int. 37, 396–412.

Balehegn, M., Balehey, S., Fu, C., Liang, W., 2019. Indigenous weather and climate forecasting knowledge among Afar pastoralists of north eastern Ethiopia: role in adaptation to weather and climate variability. Pastoralism 9, 1–14.

Bao, C., Yong, M., Bueh, C., Bao, Y., Jin, E., Bao, Y., Purevjav, G., 2022. Analyses of the dust storm sources, affected areas, and moving paths in Mongolia and China in early spring. Rem. Sens. 14, 3661.

Baveye, P.C., Baveye, J., Gowdy, J., 2016. Soil "ecosystem" services and natural capital: critical appraisal of research on uncertain ground. Front. Environ. Sci. 4, 41.

Bockheim, J., Hartemink, A., 2013. Distribution and classification of soils with clay-enriched horizons in the USA. Geoderma 209, 153–160.

Bot, A., Benites, J., 2005. The Importance of Soil Organic Matter: Key to Drought-Resistant Soil and Sustained Food Production. Food & Agriculture Org.

Bouma, J., 2018. How Can the Soil Science Discipline Contribute Most Effectively to Achieving the UN Sustainable Development Goals? Soil and Sustainable Development Goals. GeoEcology Essays. Catena Soil Sciences. Schweizerbart Scientific Publishers, Stuttgart, Germany, pp. 15–38.

Brevik, E.C., Steffan, J.J., Rodrigo-Comino, J., Neubert, D., Burgess, L.C., Cerdà, A., 2019. Connecting the public with soil to improve human health. Eur. J. Soil Sci. 70, 898–910.

Brevik, E.C., Dolliver, H., Edinger-Marshall, S., Itkin, D., Johnson-Maynard, J., Liles, G., Mbila, M., Moorberg, C., Sanchez-De Leon, Y., Steffan, J.J., 2020. Undergraduate degrees that train students for soil science careers at universities in the USA and its territories. Soil Sci. Soc. Am. J. 84, 1797–1807.

Candido, L.A., Coêlho, G.A.G., De Moraes, M.M.G.A., Florêncio, L., 2022. Review of decision support systems and allocation models for integrated water resources management focusing on joint water quantity-quality. J. Water Resour. Plann. Manag. 148, 03121001.

Chong, L., Liu, H.-J., Qiang, F., Guan, H.-X., Qiang, Y., Zhang, X.-L., Kong, F.-C., 2020. Mapping the fallowed area of paddy fields on Sanjiang Plain of Northeast China to assist water security assessments. J. Integr. Agric. 19, 1885–1896.

Cole, I.S., Marney, D., 2012. The science of pipe corrosion: a review of the literature on the corrosion of ferrous metals in soils. Corrosion Sci. 56, 5–16.

Costantini, E.A., Mocali, S., 2022. Soil health, soil genetic horizons and biodiversity#. J. Plant Nutr. Soil Sci. 185, 24–34.

Deangelis, D.L., Post, W.M., Travis, C.C., 2012. Positive Feedback in Natural Systems. Springer Science & Business Media.

Dijkstra, F.A., West, J.B., Hobbie, S.E., Reich, P.B., Trost, J., 2007. Plant diversity, CO_2, and N influence inorganic and organic N leaching in grasslands. Ecology 88, 490–500.

Dong, J., Xiao, X., Menarguez, M.A., Zhang, G., Qin, Y., Thau, D., Biradar, C., Moore III, B., 2016. Mapping paddy rice planting area in northeastern Asia with Landsat 8 images, phenology-based algorithm and Google Earth Engine. Rem. Sens. Environ. 185, 142–154.

Dybzinski, R., Fargione, J.E., Zak, D.R., Fornara, D., Tilman, D., 2008. Soil fertility increases with plant species diversity in a long-term biodiversity experiment. Oecologia 158, 85–93.

Ehrenfeld, J.G., Ravit, B., Elgersma, K., 2005. Feedback in the plant-soil system. Annu. Rev. Environ. Resour. 30, 75–115.

Elbasiouny, H., El-Ramady, H., Elbehiry, F., Rajput, V.D., Minkina, T., Mandzhieva, S., 2022. Plant nutrition under climate change and soil carbon sequestration. Sustainability 14, 914.

Fredlund, D., Rahardjo, H., Fredlund, M., 2012. Soil-water Characteristic Curves for Unsaturated Soils. Unsaturated Soil Mechanics in Engineering Practice, pp. 184–272.

Furey, G.N., Tilman, D., 2021. Plant biodiversity and the regeneration of soil fertility. Proc. Natl. Acad. Sci. USA 118 e2111321118.

Ghiri, M.N., Abtahi, A., 2012. Factors affecting potassium fixation in calcareous soils of southern Iran. Arch. Agron Soil Sci. 58, 335–352.

Gorelick, N., Hancher, M., Dixon, M., Ilyushchenko, S., Thau, D., Moore, R., 2017. Google earth engine: planetary-scale geospatial analysis for everyone. Rem. Sens. Environ. 202, 18–27.

Gupta, S., Gupta, B., 1979. The critical soil moisture content in the underground corrosion of mild steel. Corrosion Sci. 19, 171–178.

Hou, D., 2022. Expediting climate-smart soils management. Soil Use Manag. 38, 1–6.

Huang, S., Zhang, X., Chen, N., Ma, H., Fu, P., Dong, J., Gu, X., Nam, W.H., Xu, L., Rab, G., 2022. A novel fusion method for generating surface soil moisture data with high accuracy, high spatial resolution, and high Spatio-temporal continuity. Water Resour. Res. 58 e2021WR030827.

Jenny, H., 1958. Role of the plant factor in the pedogenic functions. Ecology 39, 5–16.

Jobbágy, E.G., Jackson, R.B., 2004. The uplift of soil nutrients by plants: biogeochemical consequences across scales. Ecology 85, 2380–2389.

Kome, G.K., Enang, R.K., Tabi, F.O., Yerima, B.P.K., 2019. Influence of clay minerals on some soil fertility attributes: a review. Open J. Soil Sci. 9, 155–188.

Koseva, I.S., Watmough, S.A., Aherne, J., 2010. Estimating base cation weathering rates in Canadian forest soils using a simple texture-based model. Biogeochemistry 101, 183–196.

Lal, R., Bouma, J., Brevik, E., Dawson, L., Field, D.J., Glaser, B., Hatano, R., Hartemink, A.E., Kosaki, T., Lascelles, B., 2021. Soils and sustainable development goals of the United Nations: an international union of soil sciences perspective. Geoderma Reg. 25, e00398.

Lal, R., 2020. Food security impacts of the "4 per Thousand" initiative. Geoderma 374, 114427.

Lehmann, J., Bossio, D.A., Kögel-Knabner, I., Rillig, M.C., 2020. The concept and future prospects of soil health. Nat. Rev. Earth Environ. 1, 544–553.

Liu, F., Zhang, G.-L., Song, X., Li, D., Zhao, Y., Yang, J., Wu, H., Yang, F., 2020. High-resolution and three-dimensional mapping of soil texture of China. Geoderma 361, 114061.

Magdoff, F., Van ES, H., 2000. Building Soils for Better Crops. Sustainable Agriculture Network Beltsville.

Mahdianpari, M., Salehi, B., Mohammadimanesh, F., Homayouni, S., Gill, E., 2018. The first wetland inventory map of newfoundland at a spatial resolution of 10 m using sentinel-1 and sentinel-2 data on the google earth engine cloud computing platform. Rem. Sens. 11, 43.

Moeys, J., 2018. The Soil Texture Wizard: R Functions for Plotting, Classifying, Transforming and Exploring Soil Texture Data. CRAN. R-Project.

Morgan, R., Quinton, J., Smith, R., Govers, G., Poesen, J., Auerswald, K., Chisci, G., Torri, D., Styczen, M., 1998. The European Soil Erosion Model (EUROSEM): a dynamic approach for predicting sediment transport from fields and small catchments. Earth Surf. Process. Landf. 23, 527–544.

Patel, K.F., Fansler, S.J., Campbell, T.P., Bond-Lamberty, B., Smith, A.P., Roychowdhury, T., Mccue, L.A., Varga, T., Bailey, V.L., 2021. Soil texture and environmental conditions influence the biogeochemical responses of soils to drought and flooding. Commun. Earth Environ. 2, 1–9.

Poggio, L., Gimona, A., 2017. 3D mapping of soil texture in Scotland. Geoderma Reg. 9, 5–16.

Robinson, T., Metternicht, G., 2006. Testing the performance of spatial interpolation techniques for mapping soil properties. Comput. Electron. Agric. 50, 97–108.

Robinson, D.A., Campbell, C.S., Hopmans, J.W., Hornbuckle, B.K., Jones, S.B., Knight, R., Ogden, F., Selker, J., Wendroth, O., 2008. Soil moisture measurement for ecological and hydrological watershed-scale observatories: a review. Vadose Zone J. 7, 358–389.

Soussana, J.-F., Lutfalla, S., Ehrhardt, F., Rosenstock, T., Lamanna, C., Havlík, P., Richards, M., Chotte, J.-L., Torquebiau, E., Ciais, P., 2019. Matching policy and science: rationale for the '4 per 1000-soils for food security and climate'initiative. Soil Tillage Res. 188, 3–15.

Tauqeer, H.M., Turan, V., Iqbal, M., 2022. Production of safer vegetables from heavy metals contaminated soils: the current situation, concerns associated with human health and novel management strategies. Advances in Bioremediation and Phytoremediation for Sustainable Soil Management. Springer.

Walker, L.R., Del Moral, R., 2003. Primary Succession and Ecosystem Rehabilitation. Cambridge University Press.

Wall, D.H., Nielsen, U.N., Six, J., 2015. Soil biodiversity and human health. Nature 528, 69–76.

Wang, W., Dalal, R., Moody, P., Smith, C., 2003. Relationships of soil respiration to microbial biomass, substrate availability and clay content. Soil Biol. Biochem. 35, 273–284.

Xiao, W., Chen, W., He, T., Ruan, L., Guo, J., 2020. Multi-temporal mapping of soil total nitrogen using google earth engine across the shandong province of China. Sustainability 12, 10274.

Xu, H., Liu, K., Zhang, W., Rui, Y., Zhang, J., Wu, L., Colinet, G., Huang, Q., Chen, X., Xu, M., 2020. Long-term fertilization and intensive cropping enhance carbon and nitrogen accumulated in soil clay-sized particles of red soil in South China. J. Soils Sediments 20, 1824–1833.

Zhang, Q., Shi, R., Singh, V.P., Xu, C.-Y., Yu, H., Fan, K., Wu, Z., 2022. Droughts across China: drought factors, prediction and impacts. Sci. Total Environ. 803, 150018.

CHAPTER

25

Supportive role of remote sensing techniques for landslide susceptibility modeling

Abolfazl Jaafari[1] *and Davood Mafi-Gholami*[2]

[1]Research Institute of Forests and Rangelands, Agricultural Research, Education and Extension Organization (AREEO), Tehran, Iran; [2]Department of Forest Sciences, Faculty of Natural Resources and Earth Sciences, Shahrekord University, Shahrekord, Iran

25.1 Introduction

Remote sensing is a term that has been defined in a variety of ways, but no standard definition of it has been established (Shao, 2016). The Cambridge English Dictionary defines "remote" as "far away in distance" and "sensing" as "being aware of something." Therefore, "remote sensing" is literally defined as "being aware of something from a far distance." Based on the radiation wavelengths and energy sources, remote-sensing techniques are grouped into four groups: optical remote sensing (ORS), thermal remote sensing, radar remote sensing, and light detection and ranging (LiDAR) remote sensing (Fig. 25.1). Remote sensing sensors can accurately and quickly acquire spectral information about regional and even global targets, allowing a wide range of geoenvironmental variables to be computed. This has made remote sensing an essential tool for different fields of environmental science.

In the case of landslide susceptibility mapping, remote sensing has been suggested as the most accurate approach compared to field surveys. Guzzetti et al. (2012) argued that field surveys are impractical to identify the boundary of a landslide, specifically in terrains with tall and dense vegetation and hummocky topography. Landslide boundaries are not easily detected in the field since visibility is reduced due to local perspectives, landslide sizes, and the fact that they are indistinct or fuzzy (Santangelo et al., 2010). An accurate and complete landslide mapping can be achieved by observing the slide from a distance. Recent landslide research works have documented the use of remote sensing techniques in three main groups.

https://doi.org/10.1016/B978-0-443-15341-9.00014-9

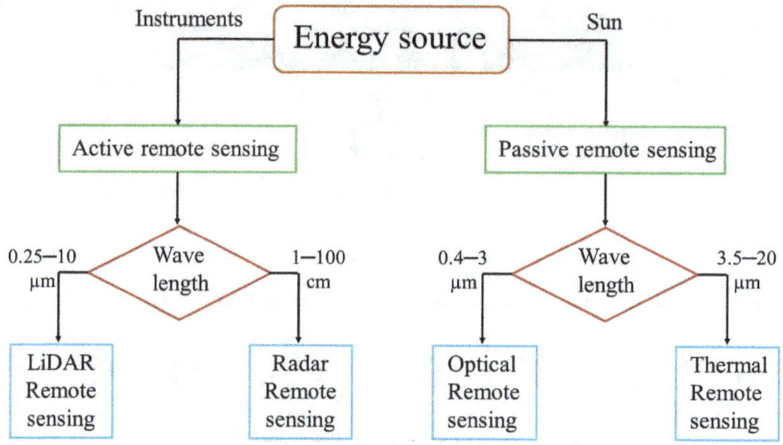

FIGURE 25.1 Classification of remote sensing techniques based on the sources and radiation wavelengths.

1. landslide detection and inventory mapping
2. landslide and surface deformation monitoring
3. landslide susceptibility mapping and risk and hazard analysis

25.2 Data

25.2.1 Synthetic aperture radar

Synthetic aperture radar (SAR) is a type of radar system that uses the motion of the radar antenna to create a large aperture or "virtual" antenna that provides high-resolution images of the ground below. SAR works by transmitting a series of microwave pulses and then measuring the time it takes for the pulses to bounce back from the ground. By processing the data from multiple pulses, SAR can create detailed images of the terrain, even in areas with poor visibility or at night. It is commonly used for remote sensing applications such as mapping, monitoring natural disasters, and military reconnaissance. In contrast to conventional beam-scanning radars that scan a target area from a stationary position, SAR utilizes the movement of the radar antenna to achieve higher spatial resolution. In SAR, a moving platform, such as an aircraft or spacecraft, is configured with a side-looking airborne radar (SLAR). As the SAR devices travel over targets during the period when the target scenes are illuminated, they create a large synthetic antenna aperture. Regardless of whether the aperture is physical (i.e., a large antenna) or synthetic (i.e., a moving antenna), the larger aperture yields higher resolution images. Due to their high spatial and temporal resolution, broad coverage, and ability to avoid weather-caused signal attenuation, SAR data have been broadly applied in landslide research. Many different satellite SAR data, such as ALOS/PALSAR (Qi et al., 2017), COSMO-SkyMed constellation (Schlögel et al., 2017), ERS and Envisat ASAR (Bozzano et al., 2017), TerraSAR-X/TanDEM-X (Du et al., 2017), TerraSAR-X (Yang et al., 2017), GBInSAR (Del Ventisette et al., 2012), and Sentinel-1 (Reyes-Carmona et al., 2020) have been proposed and used in landslide research.

25.2.2 Optical remote sensing

Shao (2016) defines ORS as *"being aware of something from a far distance with light."* ORS relies on sunlight for its energy. ORS sensors record light reflected from things on Earth in a similar way to human eyes that detect sunlight. Given the capability of optical sensors for extracting geomorphic features and the wide coverage, they are suitable for studying shallow soil instabilities (Scaioni et al., 2014) and for generating landslide inventory (Zhao and Lu, 2018). Examples of the optical sensors used in landslide research include ZY-3 satellite high spatial resolution satellite images (Chen et al., 2017), SPOT 1–5, long time-series of Landsat TM/ETM, IRS-1C LISS III, ASTER, and RapidEye (Ghorbanzadeh et al., 2019), and GF-1 (Sun et al., 2017) (Fig. 25.2).

25.2.3 Airborne laser scanning

Airborne laser scanning (ALS) is a remote sensing technique for mapping topography, urban areas, vegetation, and infrastructure. Due to its use as a laser to illuminate the Earth's surface and a photodiode to measure backscatter radiation, ALS is often called LiDAR. As a result of its capabilities to filter overgrown vegetation and model the bare terrain in forested areas, ALS is widely used for mapping deep-seated landslides (Jaboyedoff et al., 2012). Automatic algorithms, including supervised classification approaches as described by Kasai et al. (2009) and conventional signal processing techniques as discussed by Booth et al. (2009), have been applied to this task. At a regional level, terrestrial laser scanning (TLS) has been used for identifying rotational slides that were partial reactivations of old DSGDs (Viero et al., 2010) (Fig. 25.3).

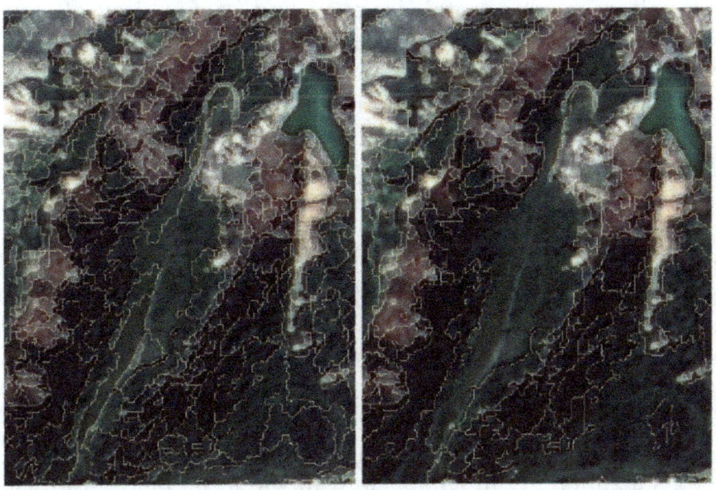

FIGURE 25.2 An example of the GF-1 satellite imagery for landslide detection and classification (Sun et al., 2017).

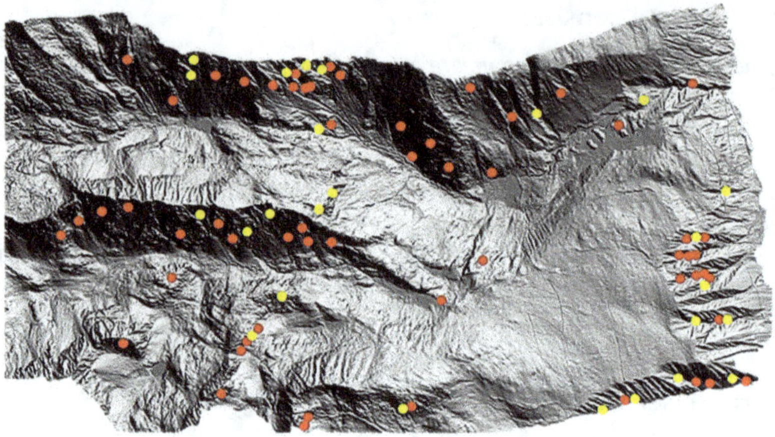

FIGURE 25.3 Light detection and ranging (LiDAR)-supported landslide detection in the Pellizzano, Trentino, northern Italy. The yellow and red dots show the landslide locations used for training and validation of a susceptibility model.

25.3 Method and experiments

25.3.1 Landslide detection

Landslide detection includes all activities aimed at identifying the historical landslides that occurred in a specific landscape to produce inventory maps and databases for further analysis. Malamud et al. (2004) defined "landslide inventory maps as a spatial distribution of landslides represented at a predefined cartographic scale." Landslide detection may involve periodic (i.e., seasonal or multitemporal) investigation of a variety of slope instabilities in large landscapes. Guzzetti et al. (2012) outlined three main assumptions for the preparation of a landslide inventory map (i) since landslides leave morphological signature (i.e., changes in the form, position, appearance, or shape of the topographic surface, and changes in geological, lithological, or land use): on the territory, they can be identified and classified in the field and then spatially mapped using the digital representations of the topographic surface or via the interpretation of aerial photographs and satellite images. For example, Kang et al. (2017) investigated the surface deformation of the landslides that occurred in Guanling, China, using ALOS/PALSAR InSAR. Bozzano et al. (2017) used a procedure that combined field measurements and remote sensing techniques for the investigation of the landslides influencing a river basin in Italy (ii) The same type of landslides (i.e., slide, flow, fall, complex, or compound). produces a similar landslide signature, which provides insights into the age, activity degree, and depth of the surface displacement. Remote sensing provides the opportunity for long-term, regular frequency (days-weeks) measurements to examine and increase our knowledge for the behavior and characteristics of surface displacements (cf. Wasowski and Pisano, 2020). (iii) Knowledge of landslides and information about failures from a given region can be generalized and used to detect and map landslides in other areas. (iv) The recognition of past and recent landslides is key to predict future landslides. Therefore, landslide inventory maps are used as a crucial component for landslide susceptibility modeling and mapping. Mantovani et al. (1996), showed that optical images on a scale ≥1:10,000 were appropriate for the production of such inventory maps. Chen et al. (2017) introduced a framework for mapping landslides using object-oriented techniques, random forests method, and mathematical morphology. The framework utilized high-resolution satellite images from ZY-3 to identify landslides that had occurred in the past. Ghorbanzadeh (2022) evaluated a rule-based object-based image analysis approach based on the ResU-Net deep learning model for landslide detection from multitemporal sentinel-2 images.

25.3.2 Landslide monitoring

Landslide monitoring involves the application of a durable prediction and forewarning system, in conjunction with effective efforts to prevent slope movement. Landslide destruction can be significantly mitigated via monitoring and instrumentation. Integrated geological and geotechnical investigations and monitoring are key to developing reliable plans to correct or prevent landslides.

Satellite imaginary and ground-based radar systems provided opportunities for an increased areal coverage and spatial resolution of surface displacement monitoring data. These tools and geophysical methods effectively complement each other for landslide studies in small to large-scale areas. Pourkhosravani et al. (2022) monitored the displacement caused by landslides mass using persistent scatterer interferometry (PSI) and interferometry synthetic aperture Radar (DInSAR) techniques for the period between 2014 and 2019. Schlögel et al. (2017) validated the satellite observations using differential GPS time-series data and reported on the efficacy of remote sensing techniques for monitoring landslides.

25.3.3 Landslide susceptibility modeling

The probability of a landslide occurring at a specific location due to certain predisposing factors is referred to as landslide susceptibility. Various techniques such as statistics, knowledge-based, and machine learning have been proposed and employed for landslide susceptibility modeling, primarily relying on data obtained through remote sensing methods. Digital elevation models (DEMs) are a crucial component of the data utilized for landslide susceptibility modeling. DEMs are typically produced using data obtained from satellites, drones, and planes. This diverse range of data sources enables the filling of gaps in areas with limited remote data availability. The automatic extraction of DEMs from stereo satellite imagery allows for the utilization of satellite sensors like SPOT-5 (with a resolution of 5–10 m). Various remote sensing techniques can be employed to obtain DEM surfaces.

- SAR interferometry (aka. InSAR): SAR data, which is obtained from antennas captured at nearly the same time, is utilized to create a DEM through SAR interferometry (InSAR). The shuttle radar topography mission (SRTM) is an example of such data collection. Furthermore, some researchers have employed deep learning techniques to rectify urban impact on DEMs generated from InSAR, as demonstrated by the CoastalDEM.
- Stereo photogrammetry: It involves utilizing images from multiple vantage points, typically two or three, in both aerial photography and satellite imagery. This technique

allows for the acquisition of depth and perspective by overlapping the images, similar to how our eyes perceive depth.

- LiDAR (aka. laser altimetry): It measures the elevation of the Earth's surface by analyzing reflected light that has bounced off the ground. This method is similar to DSMs that also use light to determine elevation.
- Digitizing contour lines: Contour lines can be digitized from a map and then interpolated using geospatial software to create DTMs, which are a type of DEM.
- DGPS measurements: They involve field campaigns using specialized devices that rely on satellite information to survey points and determine their location. This method is discrete and requires interpolation to generate a continuous raster.
- Ground surveying: The process of ground surveying involves the use of a theodolite to measure neighboring areas based on established XYZ positions. This task requires highly skilled labor, and, like DGPS, interpolation is necessary to obtain a continuous raster from all the points.

25.4 Conclusions

Landslides are caused via various mechanisms, such as earthquakes, permafrost degradation, intense or long-duration rainstorms, urbanization, and reservoir impoundment. The literature review presented in this chapter revealed that numerous remote sensing techniques, such as SAR, optical, LiDAR, and DEMs derived from airborne, spaceborne, and ground-based platforms, have been used to study landslides worldwide. SAR can detect changes in the ground surface caused by landslides and provide information on the extent and location of landslides. Optical images can be used to identify changes in vegetation cover, soil moisture content, and other factors that may contribute to landslide occurrence. LiDAR and DEMs data can be used to recognize areas with steep slopes or changes in elevation that may indicate landslide-prone areas. By combining various remote sensing data, analytical and numerical methods and taking into account geoenvironmental factors that affect landslides, it is possible to create a model that predicts landslide susceptibility. With the continuous enhancement in temporal and spatial resolution, remote sensing techniques will significantly contribute to landslide monitoring and susceptibility modeling.

References

Booth, A.M., Roering, J.J., Perron, J.T., 2009. Automated landslide mapping using spectral analysis and high-resolution topographic data: Puget Sound lowlands, Washington, and Portland Hills, Oregon. Geomorphology 109 (3–4), 132–147.

Bozzano, F., Mazzanti, P., Perissin, D., Rocca, A., De Pari, P., Discenza, M.E., 2017. Basin scale assessment of landslides geomorphological setting by advanced InSAR analysis. Rem. Sens. 9 (3), 267.

Chen, T., Trinder, J.C., Niu, R., 2017. Object-oriented landslide mapping using ZY-3 satellite imagery, random forest and mathematical morphology, for the Three-Gorges Reservoir, China. Rem. Sens. 9 (4), 333.

Del Ventisette, C., Casagli, N., Fortuny-Guasch, J., Tarchi, D., 2012. Ruinon landslide (Valfurva, Italy) activity in relation to rainfall by means of GBInSAR monitoring. Landslides 9, 497–509.

Du, Y., Xu, Q., Zhang, L., Feng, G., Li, Z., Chen, R.F., Lin, C.W., 2017. Recent landslide movement in Tsaoling, Taiwan tracked by TerraSAR-X/TanDEM-X DEM time series. Rem. Sens. 9 (4), 353.

Ghorbanzadeh, O., Blaschke, T., Gholamnia, K., Meena, S.R., Tiede, D., Aryal, J., 2019. Evaluation of different machine learning methods and deep-learning convolutional neural networks for landslide detection. Rem. Sens. 11 (2), 196.

Ghorbanzadeh, O., Gholamnia, K., Ghamisi, P., 2022. The application of ResU-net and OBIA for landslide detection from multi-temporal sentinel-2 images. Big Earth Data 1–26.

Guzzetti, F., Mondini, A.C., Cardinali, M., Fiorucci, F., Santangelo, M., Chang, K.T., 2012. Landslide inventory maps: new tools for an old problem. Earth Sci. Rev. 112 (1–2), 42–66.

Jaboyedoff, M., Oppikofer, T., Abellán, A., Derron, M.H., Loye, A., Metzger, R., Pedrazzini, A., 2012. Use of LIDAR in landslide investigations: a review. Nat. Hazards 61, 5–28.

Kang, Y., Zhao, C., Zhang, Q., Lu, Z., Li, B., 2017. Application of InSAR techniques to an analysis of the Guanling landslide. Rem. Sens. 9 (10), 1046.

Kasai, M., Ikeda, M., Asahina, T., Fujisawa, K., 2009. LiDAR-derived DEM evaluation of deep-seated landslides in a steep and rocky region of Japan. Geomorphology 113 (1–2), 57–69.

Malamud, B.D., Turcotte, D.L., Guzzetti, F., Reichenbach, P., 2004. Landslide inventories and their statistical properties. Earth Surf. Process. Landforms 29 (6), 687–711.

Mantovani, F., Soeters, R., Van Westen, C.J., 1996. Remote sensing techniques for landslide studies and hazard zonation in Europe. Geomorphology 15 (3–4), 213–225.

Pourkhosravani, M., Mehrabi, A., Pirasteh, S., Derakhshani, R., 2022. Monitoring of Maskun landslide and determining its quantitative relationship to different climatic conditions using D-InSAR and PSI techniques. Geomatics, Nat. Hazards Risk 13 (1), 1134–1153.

Qi, S., Zou, Y., Wu, F., Yan, C., Fan, J., Zang, M., et al., 2017. A recognition and geological model of a deep-seated ancient landslide at a reservoir under construction. Rem. Sens. 9 (4), 383.

Reyes-Carmona, C., Barra, A., Galve, J.P., Monserrat, O., Pérez-Peña, J.V., Mateos, R.M., et al., 2020. Sentinel-1 DInSAR for monitoring active landslides in critical infrastructures: the case of the rules reservoir (Southern Spain. Rem. Sens. 12 (5), 809.

Santangelo, Á., Cardinali, Á., Rossi, Á., Mondini, A.C., Guzzetti, F., 2010. Remote landslide mapping using a laser rangefinder binocular and GPS. Nat. Hazards Earth Syst. Sci. 10 (12), 2539–2546.

Scaioni, M., Longoni, L., Melillo, V., Papini, M., 2014. Remote sensing for landslide investigations: an overview of recent achievements and perspectives. Rem. Sens. 6 (10), 9600–9652.

Schlögel, R., Thiebes, B., Mulas, M., Cuozzo, G., Notarnicola, C., Schneiderbauer, S., et al., 2017. Multi-temporal X-Band radar interferometry using corner reflectors: application and validation at the Corvara Landslide (Dolomites, Italy). Rem. Sens. 9 (7), 739.

Shao, G., 2016. Optical remote sensing. In: International Encyclopedia of Geography: People, the Earth, Environment and Technology, pp. 1–12.

Sun, W., Tian, Y., Mu, X., Zhai, J., Gao, P., Zhao, G., 2017. Loess landslide inventory map based on GF-1 satellite imagery. Rem. Sens. 9 (4), 314.

Viero, A., Teza, G., Massironi, M., Jaboyedoff, M., Galgaro, A., 2010. Laser scanning-based recognition of rotational movements on a deep seated gravitational instability: the Cinque Torri case (North-Eastern Italian Alps. Geomorphology 122 (1–2), 191–204.

Wasowski, J., Pisano, L., 2020. Long-term InSAR, borehole inclinometer, and rainfall records provide insight into the mechanism and activity patterns of an extremely slow urbanized landslide. Landslides 17, 445–457.

Yang, Z., Li, Z., Zhu, J., Preusse, A., Yi, H., Hu, J., et al., 2017. Retrieving 3–D large displacements of mining areas from a single amplitude pair of SAR using offset tracking. Rem. Sens. 9 (4), 338.

Zhao, C., Lu, Z., 2018. Remote sensing of landslides–A review. Rem. Sens. 10 (2), 279.

Further reading

Di Martire, D., Tessitore, S., Brancato, D., Ciminelli, M.G., Costabile, S., Costantini, M., et al., 2016. Landslide detection integrated system (LaDIS) based on in-situ and satellite SAR interferometry measurements. Catena 137, 406–421.

Wasowski, J., Bovenga, F., 2022. Remote sensing of landslide motion with emphasis on satellite multi-temporal interferometry applications: an overview. In: Landslide Hazards, Risks, and Disasters, pp. 365–438.

CHAPTER

26

An overview of remotely sensed fuel variables for the prediction of wildfires

Xingwen Quan[1,2], Binbin He[1], Abolfazl Jaafari[3], Zhanmang Liao[1], Xiangzhuo Liu[4], Chao Yue[5] and Rui Chen[1]

[1]School of Resources and Environment, University of Electronic Science and Technology of China, Chengdu, China; [2]Yangtze Delta Region Institute (Huzhou), University of Electronic Science and Technology of China, Huzhou, China; [3]Research Institute of Forests and Rangelands, Agricultural Research, Education and Extension Organization (AREEO), Tehran, Iran; [4]INRAE, UMR1391 ISPA, Université de Bordeaux, Villenave d'Ornon, France; [5]State Key Laboratory of Soil Erosion and Dryland Farming on the Loess Plateau, Northwest A&F University, Yangling, China

26.1 Introduction

Many different methods/approaches suggested and used for the assessment prediction of wildfires are performed based on a set of explanatory variables that characterize topography, weather/climate, fuel load, and human activities. Among these variables, fuel variables have not been adequately defined and used; some studies simply used the normalized vegetation difference index (NDVI) as an alternative to fuel load (Jaafari et al., 2019; Tien Bui et al., 2017), whereas other variables are more efficient to characterize wildfire fuel property. In this context, this chapter reviews the fuel-related variables, that is, canopy water content (CWC), live fuel moisture content (LFMC), vegetation optical depth (VOD), foliage fuel load (FFL), and aboveground biomass (AGB). LFMC is an essential wildland fire risk and propagate rate driver and is defined as the amount of water in plants as a percentage of the dry mass (Yebra et al., 2013). CWC can assist in determining vegetation's physiological status and is normally expressed as the water content in the canopy (Trombetti et al., 2008). VOD, a metric of the attenuation impacts of electromagnetic (active or passive) radiant energy by the plant canopy, is linked to foliage moisture content (Wigneron et al., 2017). FFL

Remote Sensing of Soil and Land Surface Processes
https://doi.org/10.1016/B978-0-443-15341-9.00008-3

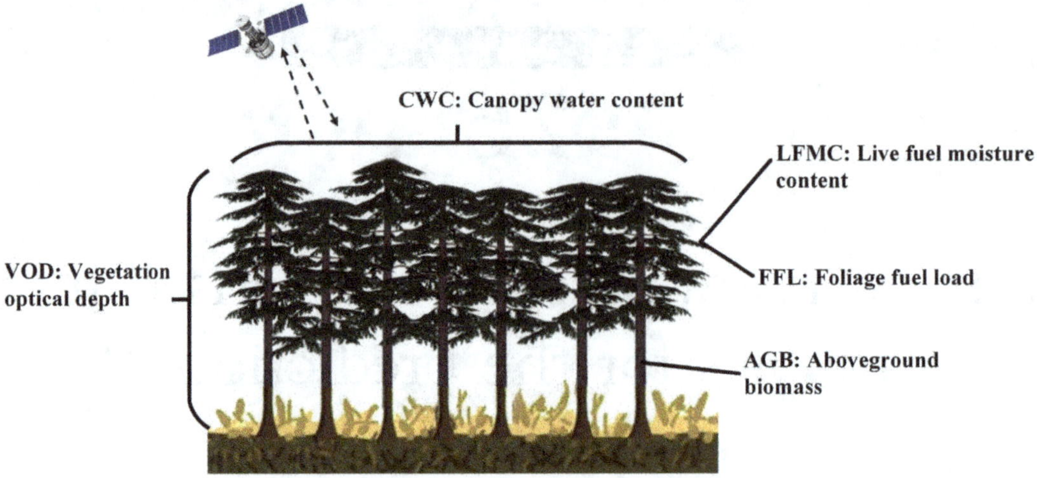

FIGURE 26.1 Sketch picture showing the fuel moisture (live fuel moisture content [LFMC], canopy water content [CWC] and vegetation optical depth [VOD]) and fuel load (foliage fuel load [FFL] and aboveground biomass [AGB]) related variable that can be derived from the satellite observation.

is the critical fuel variable for wildfire occurrence probability assessment and characterized the dry mass of foliage for combustion (Quan et al., 2021a). AGB is the biomass (i.e., dry weight) of the aboveground part of vegetation (Chave et al., 2014). A sketched picture showing these fuel variables is presented in Fig. 26.1, and these variables are entirely derived from the multiple sources of optical and microwave remote sensing data. The following subsection describes these variables in detail.

26.2 Remotely sensed fuel moisture related variables

26.2.1 Live fuel moisture content

In the wildland fire research community, one of the most important variables driving wildfire occurrences and propagation potential is the LFMC (Yebra et al., 2013). Owing to the broad landscape observation image with excellent temporal and geographical detail, LFMC estimated from the distant sensing technology is evident at the local to continental scale nowadays (Quan et al., 2017a,b; Yebra et al., 2013, 2018). The remote sensing–based methods for the LFMC estimation can be broadly classified into the statistical techniques and the reversal of radiative transfer models (RTMs) (Yebra et al., 2013). The former approaches are believed to be simple and have thus been used for sensor-specific and site-dependent LFMC estimation, but they lacking the repeatability to be used in other fields (Yebra et al., 2008). Approaches based on RTMs are more promising as they are based on physical laws that provide an explicit connection between soil-canopy variables and canopy spectra (Houborg et al., 2007). Thus, the RTM-based approaches have the advantage of reproducibility to be applied in different ecoregions (Yebra et al., 2013, 2018).

Typically, in prior research, a single-leaf RTM such as PROSPECT (Jacquemoud and Baret, 1990) or Leaf Incorporating Biochemistry Exhibiting Reflectance and Transmittance Yields (LIBERTY) (Dawson et al., 1998) has been combined with a canopy-level RTM like SAIL, the Markov chain reflectance model (MCRM) (Kuusk, 1995), or GEOSAIL (Huemmrich, 2001) to this end (Quan et al., 2015b; Yebra et al., 2013). The PROSPECT model is a method utilized to compute transmittance and reflection of broad leaves, which considers a leaf as a stack of plates having absorbent and scattering elements. It operates over a spectral range of 400–2500 nm (Jacquemoud and Baret, 1990). The LIBERTY model is another approach utilized to mimic needle transmittance and reflection over the same spectral range. This approach is based on the assumption that a needle structure is composed of spherical cells separated by air voids (Dawson et al., 1998). The SAIL model is a one-dimensional RTM that predicts the homogenized canopy bidirectional reflectivity (Jacquemoud et al., 2006). It addresses the extinction and dispersion of four upward/downward fluxes within the canopy, and it is commonly used to study the interaction of light with vegetation. The SAIL model was originally developed to simulate canopy reflectance, but it was later improved by Kuusk (1991) to incorporate hot spot effects, resulting in the SAILH model. SAILH is simpler to use as it requires only a few input parameters, facilitating the inverse modeling process. The MCRM model is a multispectral RTM (Nilson and Kuusk, 1989) that embraces forest stand geometry's Markov properties, while the GeoSAIL approach integrates a physical system that calculates the quantity of shaded and lit elements in a landscape with the SAIL turbidity media theory to simulate tree crown reflectance and transmission. By utilizing the canopy components' optical characteristics, tree form, solar zenith angle, and canopy cover, the GeoSAIL model can estimate scene reflectance and the fraction of photosynthetically active radiation absorbed or intercepted for forest types (Huemmrich, 2001).

However, this simple model realization is not suitable for resembling the spectra of forest ecosystems, which are distinguished by two-layer canopies (e.g., upper and lower canopy layers of herbaceous vegetation intermixed with some saplings). These developments prompted the creation of more complicated two-layer RTMs, including the two-layer RTM (ACRM) (Kuusk, 2001), which has also been utilized extract biochemical and biophysical data (He et al., 2013; Quan et al., 2014). The ACRM model makes an assumption that the vegetation canopy is comprised of two layers: a major layer and a thin layer on the surface. The identical input parameters affect the optical characteristics of the leaves in both layers (Kuusk, 2001). However, ACRM is not appropriate for estimating LFMC in areas with a two-layered forest where the canopy layer is not uniform but rather varied in composition. To solve this issue, (Quan et al., 2017a) built a coupled RTM to multitemporally retrieve the forest LFMC. This coupled model consisted of a PROGEOSAIL model, simulating the spectra of the upper trees canopy, and a PROSAIL model, simulating herbaceous vegetation intermixed with some saplings in the lower canopy. Their results indicated that the LFMC could be significantly improved when the coupled model was used, which proved the viability of the proposed approach for LFMC retrieval in a two-layer forest. Based on this model, Quan et al. (2021b) produced the first global LFMC product between 2001 and 2020 from moderate resolution imaging spectrometer (MODIS) at a daily and 500 m scale (https://firewatching.cn/world_FMC/). This recent LFMC product provides an opportunity to investigate the relationships between wildfires and LFMC globally.

26.2.2 Canopy water content

The use of remote sensing techniques to retrieve CWC over large areas was developed for years (Houborg et al., 2009; Quan et al., 2015a). CWC can assist in determining the physiological state of vegetation and is found useful for detecting vegetative stress and assessing drought (Houborg et al., 2007). Retrieval is generally dependent on statistical techniques or the reversal of a physical RTM. In the former, statistical approaches are employed to relate the goal variables to the specified spectrum reflectance or vegetation indices adjusted to experimental data, which are straightforward and efficient in terms of computing (Combal et al., 2003; Gitelson et al., 2005). Moreover, the established statistical connections are typically sensor-specific and site-dependent, and the retrieval results vary easily in time and space, which lack generality. The model-based inversion technique, on the other hand, is a promising strategy since the RTM is based on physical rules, which define clear correlations between plant features and waveband reflectance (Houborg et al., 2007).

A RTM must be inverted to retrieve CWC (Kimes et al., 1998). Popular inversion algorithms include numerical optimizations (Combal et al., 2003; Jacquemoud et al., 1996), a look-up table (LUT) (Combal et al., 2002; Houborg et al., 2009; Weiss et al., 2000), and artificial neural network (ANN) (Walthall et al., 2004). The inverted parameters are of great precision in numerical optimization techniques, but the process is time-consuming, especially for a sophisticated RTM. The accuracy level of the LUT and ANN algorithms is heavily reliant on training datasets. Apart from these optimal algorithms, the target parameters might be recovered using the Bayesian network algorithm, which can be utilized as a nonparametric regression model as well as a method of deducing information from multilayered parameters (Marcot et al., 2001). Kalácska et al. (2005) employed a Bayesian network to estimate the LAI of a tropical dry forest, and the results were superior than ANN. Moreover, the Bayesian network could fully utilize ancillary information and derive additional information from the posterior joint probability distribution (PJPD) of target parameters, such as mean value, maximum posterior probability, and information entropy (Qu et al., 2008). Inverse problems are by nature ill-posed (Atzberger, 2004) since the different model parameter combinations can result in nearly identical spectrum reflectance, and errors occur frequently in the retrieval procedure (Combal et al., 2003; Houborg et al., 2009; Weiss et al., 2000). Many researchers have shown how prior information could be used to solve the ill-posed problem efficiently (Combal et al., 2002, 2003), and therefore, the CWC can be retrieved from multiple remote sensing data.

26.2.3 Vegetation optical depth

According to the type of sensor, VOD is classified as active or passive. The retrieval of passive VOD is based on the Tau-Omega (τ-ω) radiative transfer (RT) model, which is used to simulate the brightness temperature (TB) (Mo et al., 1982). Based on this theory, many algorithms were developed to retrieve the VOD from different satellites at different frequencies. The Land Parameter Retrieval Model (LPRM) (Owe et al., 2001) based on the Microwave Polarization Difference Index (MPDI) (Meesters et al., 2005) and the Land Parameter Data Record model (LPDR) (Du et al., 2017) based on the inversion of the land–water emissivity slope index were the most commonly used. The LPRM was implemented in the Advanced

Microwave Scanning Radiometer (AMSR-E, July 2002–11) (Koike et al., 2004) and its successor, Advanced Microwave Scanning Radiometer 2 (AMSR-2, 2012-present) (Imaoka et al., 2012), to produce the VOD at C-band (6.9 or 7.3 GHz) and X-band (10.7 GHz). The LPDR was used to retrieve an alternative X-band VOD from AMSR-E and AMSR-2. What's more, VU University Amsterdam and NASA (VUA-NASA model), a variant of LPRM, was used in the development of the first global long-term (1987–2008) passive VOD dataset at C-band (Liu et al., 2011). The VOD was first retrieved from the multisensors (special sensor microwave imager (SSM/I), TRMM microwave imager (TMI), and AMSR-E) using the VUA-NASA model, and then cumulative distribution frequency (CDF) matching technique was used to adjust the VOD of the other two satellites against VOD of AMSR-E. Recently, Moesinger et al. (2020) develop another new global long-term microwave VOD Climate Archive (VODCA) by using the updated version of LPRM and adding more sensor data (WindSat and AMSR-2). The dataset included the Ku-, X-, and C-band VOD and the time extended to 2018. The soil moisture and ocean salinity (SMOS) (Kerr et al., 2010) and soil moisture active passive (SMAP) (Entekhabi et al., 2010) are two L-band (1.4 GHz) satellites that can also be used to retrieve the VOD. As the SMOS satellite can obtain the TB with multi-angular, the L-band microwave emission of the biosphere (L-MEB) model was developed to simultaneously retrieve soil moisture (SM) and VOD. With the different settings of the parameters in the model, the different visions of SMOS VOD (SMOS level 2, level 3, SMOS-INRA-CESBIO or SMOS-IC V105, SMOS-IC V2) were produced (Fernandez-Moran et al., 2017; Wigneron et al., 2021). SMAP acquired the TB at a single-look incidence angle (40°). Therefore, the single channel algorithm (SCA), LPRM, and multitemporal dual-channel algorithm (MT-DCA) were proposed to estimate SM and VOD (Wigneron et al., 2017). More details about the assessment and intercomparison of passive VOD products over a global scale can be seen in Li et al. (2021).

Compared to passive VOD retrievals, few studies considered about to retrieve VOD as the active backscatter data were usually employed in the retrievals of SM and ocean winds (Konings et al., 2017). To our knowledge, there are only preliminary results acquired from Sentinel-1 in southern France (El Hajj et al., 2019) and two active VOD datasets from ASCAT observations created by Vreugdenhil et al. (2016) and Liu et al. (2021). Those active VOD retrievals are all based on Water Cloud Model (Ulaby et al., 1978) but with different assumptions. Currently, the VOD was used to study the impact of large wildfires on vegetation (Zhang et al., 2021) as all available VOD datasets are at a coarse resolution (~25 km). With the long-term high-resolution active microwave data (Sentinel-1) being free to the public, the high-resolution VOD will be available in the future and used in the study of different size wildfires.

26.3 Remotely sensed fuel load-related variables

26.3.1 Foliage fuel load

Crown fuels, which consist of foliage, branch wood, and arboreal lichens and mosses (Keane et al., 2001), play a critical role in determining the behavior of wildfires, including the rate of crown fire spread and fire intensity (Wagner, 1977). Among these components, foliage is particularly prone to ignition (Gray and Reinhardt, 2003), and thus, the FFL is an essential fuel variable for assessing the likelihood of wildfire occurrence. Various remote sensing—based techniques have been explored to estimate fuel load or biomass at the regional level, utilizing multiple sources of data (García et al., 2011; He et al., 2019; Quan et al., 2017b) through statistical methods. Despite the effectiveness of these techniques, only a small number of the resultant products have, as of yet, been operationally integrated into systems that predict the likelihood of wildfires occurring.

Crown fuel characteristics are of importance to fire management since they affect crown fire initiation, spread, and severity (Mitsopoulos and Dimitrakopoulos, 2007; Wagner, 1977). Crown fuels, consisting of foliage, branch wood, and arboreal lichens and mosses (Keane et al., 2001), are crucial factors in determining the behavior of crown wildfires (Gray and Reinhardt, 2003). To assess the likelihood of a crown wildfire occurring, the FFL is a critical fuel variable, representing the dry weight of all leaves (both living and dead) per unit area of the forest (Mallinis et al., 2013). By accurately estimating and mapping the FFL, fire managers can improve prefire risk warnings and identify safe zones for firefighters to prevent fire spread (Mitsopoulos and Dimitrakopoulos, 2007).

Destructive sampling is used in traditional fuel load (including FFL) collection. Although this is the most precise way, it is time-consuming, labor-intensive, and damaging, with limited applicability to vast areas (Küçük et al., 2008; Mitsopoulos and Dimitrakopoulos, 2007). With the rapid advancement of remote sensing technology over the last two decades, there is a greater application to evaluate fuel load characteristics owing to sufficient temporal-spatial observation and multispectral information in relation to leaf characteristics, which also assesses fuel loads in rugged terrain possibly (Zarco-Tejada et al., 2019).

A widely used technique for estimating fuel load or biomass through remote sensing is the statistical method, which involves establishing correlations between the measurements of fuel load (or biomass) and various sources of remote sensing data. These data sources include multispectral, radar, and light detection and ranging (LiDAR) (García et al., 2011; He et al., 2019; Quan et al., 2017b), and the relationships established can be either linear or nonlinear. These statistical techniques are simple and frequently used; thus, they have been utilized to estimate sensor-specific and site-dependent fuel loads. However, statistical approaches are rarely reproducible enough to be useful in other domains (Yebra et al., 2013).

The inversion of the RTM is an alternative to derive the biophysical and biochemical variables. This approach is a promising method to derive key vegetation variables. Thus, RTM-based approaches have the advantage of reproducibility to be applied to other sites at a local to global scale (Bowyer and Danson, 2004; Quan et al., 2021a; Yebra et al., 2013, 2018).

26.3.2 Aboveground biomass

Accurately quantifying the dynamics of global forest AGB (Le Toan et al., 2011) is crucial for comprehending the worldwide carbon budget and the effects of climate change. Remote sensing presents a promising solution to this task as it allows for large-scale observation with high spatial and temporal resolution (Saatchi and Morel, 2011). Several techniques are used for estimating AGB, including optical, LiDAR, SAR, and passive microwave remote sensing (Koch, 2010), each with their own strengths and limitations.

LiDAR properly assesses vertical forest structure and, as a result, offers the greatest potential for estimating forest AGB (Zhao et al., 2009). The LiDAR-derived parameters, such as forest height, percentage height, and the shape of the echo signal, are widely used (Kimes et al., 2006). Spaceborne LiDAR technology, although promising for forest AGB estimation, currently falls short in providing large-scale continuous coverage (Liu et al., 2015; Qin et al., 2021). Passive microwave instruments, such as AMSR, SMOS, and SMAP, are equipped with a long-term history repository and great temporal resolution, but their limited spatial resolution (over 20 km) restricts comprehensive monitoring and attribution of changes of AGB at smaller scales. Compared to other remote sensing techniques, optical-based AGB estimation using the Landsat series has distinct advantages such as high spatial and temporal resolution and a long-term freely available historical archive. Combining optical data with other sources of information, such as SAR or LiDAR data, has been shown to substantially enhance the accuracy of forest AGB estimation in recent studies (Liao et al., 2020). Although ancillary data can be helpful in estimating changes in forest AGB, their limited availability hinders widespread application. As a result, accurately tracking dynamic changes in AGB at a national or global level may rely heavily on optical data. However, optical data may not be as effective in capturing vertical structure compared to forest density due to its weaker penetration in optical wavelengths. As a result, optical data may become saturated for large AGB (Rahman et al., 2008), which can be a significant limitation in accurately mapping global AGB using optical data alone.

SAR-based AGB estimation mainly includes backscatter-based and InSAR-based (SAR interferometry) methods. Backscatter-based AGB estimate is now the most extensively used approach, in part because backscatter data are very easy to get on a large scale, while the InSAR-based approach is largely influenced by temporal decorrelation for L-band or a shorter wavelength band (Lee et al., 2013). However, the backscatter-based AGB estimation is generally saturated at ca. 100–200 tons/ha for L-band (Bouvet et al., 2018), preventing accurate AGB mapping for high AGB areas. Various SAR techniques can be utilized to extract vertical structure parameters of forests, which can enhance the precision of forest biomass estimation. Although spaceborne SAR offers high temporal and spatial resolution, the current duration of SAR satellite missions is limited, and some data may not be freely accessible. The upcoming P-band BIOMASS mission is expected to bring new sights to global biomass mapping (Le Toan et al., 2011).

26.4 Conclusions

Accurate and timely acquisition of the fuel variables is the key to wildfire risk assessment and fire danger rating systems. Most of the fire danger rating systems obtain their ratings from point data, which performs well at the points where the observations were taken. But it made assumptions about fire hazard levels for large geographic areas through subsequent interpolation, which would introduce uncertainty to the results. Satellite remote sensing offers the potential to address these limitations and the advantages of long-time series, repeatable, large-area coverage, as well as near-real-time free accessibility. This chapter reviewed the fuel moisture (LFMC, CWC, and VOD) and fuel load (FFL and AGB) related variables that can be retrieved from multiple optical and SAR remote sensing data and put an insight to develop an operational wildfire occurrence prediction system. Such a system can provide valuable daily automatically updated and national-scale wildfire occurrence probability information that can assist in wildfire risk early prescription, suppression, and response, as well as improved awareness of wildfire risk to life and property.

References

Atzberger, C., 2004. Object-based retrieval of biophysical canopy variables using artificial neural nets and radiative transfer models. Rem. Sens. Environ. 93, 53–67.

Bouvet, A., Mermoz, S., Le Toan, T., Villard, L., Mathieu, R., Naidoo, L., et al., 2018. An above-ground biomass map of African savannahs and woodlands at 25m resolution derived from ALOS PALSAR. Rem. Sens. Environ. 206, 156–173.

Bowyer, P., Danson, F.M., 2004. Sensitivity of spectral reflectance to variation in live fuel moisture content at leaf and canopy level. Rem. Sens. Environ. 92, 297–308.

Chave, J., Réjou-Méchain, M., Búrquez, A., Chidumayo, E., Colgan, M.S., Delitti, W.B., et al., 2014. Improved allometric models to estimate the aboveground biomass of tropical trees. Global Change Biol. 20, 3177–3190.

Combal, B., Baret, F., Weiss, M., 2002. Improving canopy variables estimation from remote sensing data by exploiting ancillary information. Case study on sugar beet canopies. Agronomie 22, 205–215.

Combal, B., Baret, F., Weiss, M., Trubuil, A., Mace, D., Pragnere, A., et al., 2003. Retrieval of canopy biophysical variables from bidirectional reflectance: using prior information to solve the ill-posed inverse problem. Rem. Sens. Environ. 84, 1–15.

Dawson, T.P., Curran, P.J., Plummer, S.E., 1998. LIBERTY—modeling the effects of leaf biochemical concentration on reflectance spectra. Rem. Sens. Environ. 65, 50–60.

Du, J., Kimball, J.S., Jones, L.A., Kim, Y., Glassy, J., Watts, J.D., 2017. A global satellite environmental data record derived from AMSR-E and AMSR2 microwave Earth observations. Earth Syst. Sci. Data 9, 791–808.

El Hajj, M., Baghdadi, N., Wigneron, J.-P., Zribi, M., Albergel, C., Calvet, J.-C., et al., 2019. First vegetation optical depth mapping from Sentinel-1 C-band SAR data over crop fields. Rem. Sens. 11.

Entekhabi, D., Njoku, E.G., O'Neill, P.E., Kellogg, K.H., Crow, W.T., Edelstein, W.N., et al., 2010. The soil moisture active passive (SMAP) mission. Proc. IEEE 98, 704–716.

Fernandez-Moran, R., Al-Yaari, A., Mialon, A., Mahmoodi, A., Al Bitar, A., De Lannoy, G., et al., 2017. SMOS-IC: an alternative SMOS soil moisture and vegetation optical depth product. Rem. Sens. 9, 457.

García, M., Riaño, D., Chuvieco, E., Salas, J., Danson, F.M., 2011. Multispectral and LiDAR data fusion for fuel type mapping using Support Vector Machine and decision rules. Rem. Sens. Environ. 115, 1369–1379.

Gitelson, A.A., Vina, A., Ciganda, V., Rundquist, D.C., Arkebauer, T.J., 2005. Remote estimation of canopy chlorophyll content in crops. Geophys. Res. Lett. 32, L08403.

Gray, K.L., Reinhardt, E., 2003. Analysis of algorithms for predicting canopy fuel. In: Second International Wildland Fire Ecology and Fire Management Congress and Fifth Symposium on Fire and Forest Meteorology; 2003 November 16-20; Orlando, FL. American Meteorological Society, Boston, MA, p. 11. P5. 8.

He, B., Quan, X., Xing, M., 2013. Retrieval of leaf area index in alpine wetlands using a two-layer canopy reflectance model. Int. J. Appl. Earth Obs. Geoinf. 21, 78–91.

He, J., Loboda, T.V., Jenkins, L., Chen, D., 2019. Mapping fractional cover of major fuel type components across Alaskan tundra. Rem. Sens. Environ. 232, 111324.

Houborg, R., Soegaard, H., Boegh, E., 2007. Combining vegetation index and model inversion methods for the extraction of key vegetation biophysical parameters using Terra and Aqua MODIS reflectance data. Rem. Sens. Environ. 106, 39–58.

Houborg, R., Anderson, M., Daughtry, C., 2009. Utility of an image-based canopy reflectance modeling tool for remote estimation of LAI and leaf chlorophyll content at the field scale. Rem. Sens. Environ. 113, 259–274.

Huemmrich, K.F., 2001. The GeoSail model: a simple addition to the SAIL model to describe discontinuous canopy reflectance. Rem. Sens. Environ. 75, 423–431.

Imaoka, K., Maeda, T., Kachi, M., Kasahara, M., Ito, N., Nakagawa, K., 2012. Status of AMSR2 instrument on GCOM-W1. In: Earth Observing Missions and Sensors: Development, Implementation, and Characterization II. International Society for Optics and Photonics, p. 852815.

Jaafari, A., Zenner, E.K., Panahi, M., Shahabi, H., 2019. Hybrid artificial intelligence models based on a neuro-fuzzy system and metaheuristic optimization algorithms for spatial prediction of wildfire probability. Agric. For. Meteorol. 266, 198–207.

Jacquemoud, S., Baret, F., 1990. PROSPECT: a model of leaf optical properties spectra. Rem. Sens. Environ. 34, 75–91.

Jacquemoud, S., Ustin, S., Verdebout, J., Schmuck, G., Andreoli, G., Hosgood, B., 1996. Estimating leaf biochemistry using the PROSPECT leaf optical properties model. Rem. Sens. Environ. 56, 194–202.

Jacquemoud, S., Verhoef, W., Baret, F., Zarco-Tejada, P.J., Asner, G., Francois, C., et al., 2006. PROSPECT+ SAIL: 15 years of use for land surface characterization. In: Geoscience and Remote Sensing Symposium, 2006. IGARSS 2006. IEEE International Conference on. IEEE, pp. 1992–1995.

Kalácska, M., Sánchez-Azofeifa, G.A., Caelli, T., Rivard, B., Boerlage, B., 2005. Estimating leaf area index from satellite imagery using Bayesian networks. IEEE Trans. Geosci. Remote Sens. 43, 1866–1873.

Keane, R.E., Burgan, R., van Wagtendonk, J., 2001. Mapping wildland fuels for fire management across multiple scales: integrating remote sensing, GIS, and biophysical modeling. Int. J. Wildland Fire 10, 301–319.

Kerr, Y.H., Waldteufel, P., Wigneron, J.-P., Delwart, S., Cabot, F., Boutin, J., et al., 2010. The SMOS mission: new tool for monitoring key elements ofthe global water cycle. Proc. IEEE 98, 666–687.

Kimes, D., Nelson, R., Manry, M., Fung, A., 1998. Review article: attributes of neural networks for extracting continuous vegetation variables from optical and radar measurements. Int. J. Rem. Sens. 19, 2639–2663.

Kimes, D.S., Ranson, K.J., Sun, G., Blair, J.B., 2006. Predicting lidar measured forest vertical structure from multi-angle spectral data. Rem. Sens. Environ. 100, 503–511.

Koch, B., 2010. Status and future of laser scanning, synthetic aperture radar and hyperspectral remote sensing data for forest biomass assessment. ISPRS J. Photogrammetry Remote Sens. 65, 581–590.

Koike, T., Nakamura, Y., Kaihotsu, I., Davaa, G., Matsuura, N., Tamagawa, K., et al., 2004. Development of an advanced microwave scanning radiometer (Amsr-E) algorithm for soil moisture and vegetation water content. Proc. Hydraul. Eng. 48, 217–222.

Konings, A.G., Yu, Y., Xu, L., Yang, Y., Schimel, D.S., Saatchi, S.S., 2017. Active microwave observations of diurnal and seasonal variations of canopy water content across the humid African tropical forests. Geophys. Res. Lett. 44 (5), 2290–2299.

Küçük, Ö., Bilgili, E., Sağlam, B., 2008. Estimating crown fuel loading for calabrian pine and Anatolian black pine. Int. J. Wildland Fire 17.

Kuusk, A., 1991. The hot spot effect in plant canopy reflectance. In: Myneni, R., Ross, J. (Eds.), Photon-Vegetation Interactions. Springer Berlin Heidelberg, pp. 139–159.

Kuusk, A., 1995. A Markov chain model of canopy reflectance. Agric. For. Meteorol. 76, 221–236.

Kuusk, A., 2001. A two-layer canopy reflectance model. J. Quant. Spectrosc. Radiat. Transf. 71, 1–9.

Le Toan, T., Quegan, S., Davidson, M., Balzter, H., Paillou, P., Papathanassiou, K., et al., 2011. The BIOMASS mission: mapping global forest biomass to better understand the terrestrial carbon cycle. Rem. Sens. Environ. 115, 2850–2860.

Lee, S.-K., Kugler, F., Papathanassiou, K.P., Hajnsek, I., 2013. Quantification of temporal decorrelation effects at L-band for polarimetric SAR interferometry applications. IEEE J. Sel. Top. Appl. Earth Obs. Rem. Sens. 6, 1351–1367.

Li, X., Wigneron, J.-P., Frappart, F., Fan, L., Ciais, P., Fensholt, R., et al., 2021. Global-scale assessment and intercomparison of recently developed/reprocessed microwave satellite vegetation optical depth products. Rem. Sens. Environ. 253.

Liao, Z., Van Dijk, A.I., He, B., Larraondo, P.R., Scarth, P.F., 2020. Woody vegetation cover, height and biomass at 25-m resolution across Australia derived from multiple site, airborne and satellite observations. Int. J. Appl. Earth Obs. Geoinf. 93, 102209.

Liu, Y.Y., de Jeu, R.A.M., McCabe, M.F., Evans, J.P., van Dijk, A.I.J.M., 2011. Global long-term passive microwave satellite-based retrievals of vegetation optical depth. Geophys. Res. Lett. 38.

Liu, Y.Y., Dijk, A.I.J.M.V., Jeu, R.A.M.D., Canadell, J.G., Mccabe, M.F., Evans, J.P., et al., 2015. Recent reversal in loss of global terrestrial biomass. Nat. Clim. Change 5, 470–474.

Liu, X., Wigneron, J.-P., Fan, L., Frappart, F., Ciais, P., Baghdadi, N., et al., 2021. ASCAT IB: a radar-based vegetation optical depth retrieved from the ASCAT scatterometer satellite. Rem. Sens. Environ. 264.

Mallinis, G., Mitsopoulos, I., Stournara, P., Patias, P., Dimitrakopoulos, A., 2013. Canopy fuel load mapping of mediterranean pine sites based on individual tree-crown delineation. Rem. Sens. 5, 6461–6480.

Marcot, B.G., Holthausen, R.S., Raphael, M.G., Rowland, M.M., Wisdom, M.J., 2001. Using Bayesian belief networks to evaluate fish and wildlife population viability under land management alternatives from an environmental impact statement. For. Ecol. Manag. 153, 29–42.

Meesters, A.G.C.A., DeJeu, R.A.M., Owe, M., 2005. Analytical derivation of the vegetation optical depth from the microwave polarization difference index. Geosci. Rem. Sens. Lett. IEEE 2, 121–123.

Mitsopoulos, I.D., Dimitrakopoulos, A.P., 2007. Allometric equations for crown fuel biomass of Aleppo pine (*Pinus halepensis* Mill.) in Greece. Int. J. Wildland Fire 16.

Mo, T., Choudhury, B.J., Schmugge, T.J., Wang, J.R., Jackson, T.J., 1982. A model for microwave emission from vegetation-covered fields. J. Geophys. Res. 87.

Moesinger, L., Dorigo, W., de Jeu, R., van der Schalie, R., Scanlon, T., Teubner, I., et al., 2020. The global long-term microwave vegetation optical depth climate archive (VODCA). Earth Syst. Sci. Data 12, 177–196.

Nilson, T., Kuusk, A., 1989. A reflectance model for the homogeneous plant canopy and its inversion. Rem. Sens. Environ. 27, 157–167.

Owe, M., de Jeu, R., Walker, J., 2001. A methodology for surface soil moisture and vegetation optical depth retrieval using the microwave polarization difference index. IEEE Trans. Geosci. Rem. Sens. 39, 1643–1654.

Qin, Y., Xiao, X., Wigneron, J.P., Ciais, P., Moore, B., 2021. Carbon loss from forest degradation exceeds that from deforestation in the Brazilian Amazon. Nat. Clim. Change 11 (5), 442–448.

Qu, Y., Wang, J., Wan, H., Li, X., Zhou, G., 2008. A Bayesian network algorithm for retrieving the characterization of land surface vegetation. Rem. Sens. Environ. 112, 613–622.

Quan, X.W., He, B.B., Wang, Y., Tang, Z., Li, X., 2014. An extended fourier approach to improve the retrieved leaf area index (LAI) in a time series from an alpine wetland. Rem. Sens. 6, 1171–1190.

Quan, X., He, B., Li, X., 2015a. A bayesian network-based method to alleviate the ill-posed inverse problem: a case study on leaf area index and canopy water content retrieval. IEEE Trans. Geosci. Rem. Sens. 53, 6507–6517.

Quan, X.W., He, B.B., Li, X., Tang, Z., 2015b. Estimation of grassland live fuel moisture content from ratio of canopy water content and foliage dry biomass. Geosci. Rem. Sens. Lett. IEEE 12, 1903–1907.

Quan, X., He, B., Yebra, M., Yin, C., Liao, Z., Li, X., 2017a. Retrieval of forest fuel moisture content using a coupled radiative transfer model. Environ. Model. Software 95, 290–302.

Quan, X., He, B., Yebra, M., Yin, C., Liao, Z., Zhang, X., et al., 2017b. A radiative transfer model-based method for the estimation of grassland aboveground biomass. Int. J. Appl. Earth Obs. Geoinf. 54, 159–168.

Quan, X., Li, Y., He, B., Cary, G.J., Lai, G., 2021a. Application of Landsat ETM+ and OLI data for foliage fuel load monitoring using radiative transfer model and machine learning method. IEEE J. Sel. Top. Appl. Earth Obs. Rem. Sens. 14, 5100–5110.

Quan, X., Yebra, M., Riaño, D., He, B., Lai, G., Liu, X., 2021b. Global fuel moisture content mapping from MODIS. Int. J. Appl. Earth Obs. Geoinf. 101.

Rahman, M.M., Csaplovics, E., Koch, B., 2008. Satellite estimation of forest carbon using regression models. Int. J. Rem. Sens. 29, 6917–6936.

Saatchi, S.S., Morel, A., 2011. Benchmark map of forest carbon stocks in tropical regions across three continents. Proc. Natl. Acad. Sci. U.S.A. 108, 9899–9904.

Tien Bui, D., Bui, Q.-T., Nguyen, Q.-P., Pradhan, B., Nampak, H., Trinh, P.T., 2017. A hybrid artificial intelligence approach using GIS-based neural-fuzzy inference system and particle swarm optimization for forest fire susceptibility modeling at a tropical area. Agric. For. Meteorol. 233, 32–44.

Trombetti, M., Riano, D., Rubio, M.A., Cheng, Y.B., Ustin, S.L., 2008. Multi-temporal vegetation canopy water content retrieval and interpretation using artificial neural networks for the continental USA. Rem. Sens. Environ. 112, 203–215.

Ulaby, F., Batlivala, P., Dobson, M., 1978. Microwave backscatter dependence on surface roughness, soil moisture, and soil texture: Part I-bare soil. IEEE Trans. Geosci. Electron. 16, 286–295.

Vreugdenhil, M., Dorigo, W.A., Wagner, W., de Jeu, R.A.M., Hahn, S., van Marle, M.J.E., 2016. Analyzing the vegetation parameterization in the TU-wien ASCAT soil moisture retrieval. IEEE Trans. Geosci. Rem. Sens. 54, 3513–3531.

Wagner, C.V., 1977. Conditions for the start and spread of crown fire. Can. J. For. Res. 7, 23–34.

Walthall, C., Dulaney, W., Anderson, M., Norman, J., Fang, H., Liang, S., 2004. A comparison of empirical and neural network approaches for estimating corn and soybean leaf area index from Landsat ETM+ imagery. Rem. Sens. Environ. 92, 465–474.

Weiss, M., Baret, F., Myneni, R.B., Pragnère, A., Knyazikhin, Y., 2000. Investigation of a model inversion technique to estimate canopy biophysical variables from spectral and directional reflectance data. Agronomie 20, 3–22.

Wigneron, J.P., Jackson, T.J., O'Neill, P., De Lannoy, G., de Rosnay, P., Walker, J.P., et al., 2017. Modelling the passive microwave signature from land surfaces: a review of recent results and application to the L-band SMOS & SMAP soil moisture retrieval algorithms. Rem. Sens. Environ. 192, 238–262.

Wigneron, J.-P., Li, X., Frappart, F., Fan, L., Al-Yaari, A., De Lannoy, G., et al., 2021. SMOS-IC data record of soil moisture and L-VOD: historical development, applications and perspectives. Rem. Sens. Environ. 254.

Yebra, M., Chuvieco, E., Riaño, D., 2008. Estimation of live fuel moisture content from MODIS images for fire risk assessment. Agric. For. Meteorol. 148, 523–536.

Yebra, M., Dennison, P.E., Chuvieco, E., Riaño, D., Zylstra, P., Hunt, E.R., et al., 2013. A global review of remote sensing of live fuel moisture content for fire danger assessment: moving towards operational products. Rem. Sens. Environ. 136, 455–468.

Yebra, M., Quan, X., Riaño, D., Rozas Larraondo, P., van Dijk, A.I.J.M., Cary, G.J., 2018. A fuel moisture content and flammability monitoring methodology for continental Australia based on optical remote sensing. Rem. Sens. Environ. 212, 260–272.

Zarco-Tejada, P.J., Hornero, A., Beck, P.S.A., Kattenborn, T., Kempeneers, P., Hernandez-Clemente, R., 2019. Chlorophyll content estimation in an open-canopy conifer forest with sentinel-2A and hyperspectral imagery in the context of forest decline. Remote Sens. Environ. 223, 320–335.

Zhang, H., Hagan, D.F.T., Dalagnol, R., Liu, Y., 2021. Forest canopy changes in the southern amazon during the 2019 fire season based on passive microwave and optical satellite observations. Rem. Sens. 13.

Zhao, K.G., Popescu, S., Nelson, R., 2009. Lidar remote sensing of forest biomass: a scale-invariant estimation approach using airborne lasers. Rem. Sens. Environ. 113, 182–196.

CHAPTER

27

Improving landslide susceptibility mapping using integration of ResU-Net technique and optimized machine learning algorithms

Fatemeh Rezaie[1,2], *Mahdi Panahi*[3], *Sayed M. Bateni*[4], *Zahra Kalantari*[5,6], *Omid Rahmati*[7], *Saro Lee*[1,2] *and Arip Syaripudin Nur*[3]

[1]Geoscience Data Center, Korea Institute of Geoscience and Mineral Resources (KIGAM), Daejeon, Republic of Korea; [2]Department of Geophysical Exploration, University of Science and Technology, Daejeon, Republic of Korea; [3]Division of Science Education, Kangwon National University, Chuncheon-si, Gangwon-do, Republic of Korea; [4]Department of Civil and Environmental Engineering and Water Resources Research Center, University of Hawaii at Manoa, Honolulu, HI, United States; [5]Department of Physical Geography and Bolin Centre for Climate Research, Stockholm University, Stockholm, Sweden; [6]Department of Sustainable Development, Environmental Science and Engineering (SEED), KTH Royal Institute of Technology, Stockholm, Sweden; [7]Soil Conservation and Watershed Management Research Institute, Agricultural Research, Education and Extension Organization (AREEO), Kurdistan Agricultural and Natural Resources Research and Education Center, Sanandaj, Iran

27.1 Introduction

Landslide is the most destructive geological event involves the movements of rock, soils, and debris down a slope prominently in the mountainous areas due to natural and human-induced factors like earthquakes, heavy rainfall, volcanic eruptions, snowmelt, land-use changes, deforestation, mining, and construction or a combination of these factors

Remote Sensing of Soil and Land Surface Processes
https://doi.org/10.1016/B978-0-443-15341-9.00004-6

(Froude and Petley, 2018). Landslides commonly result in the loss of human life and property; therefore, detecting the probability of landslide occurrences in infrastructures and human settlements is a major concern to the governments and research communities in many parts of the world. A landslide susceptibility (LS) map illustrates the possibility of landslide occurrence in an area based on the local geo-environmental characteristics and terrain condition (Reichenbach et al., 2018). Historical landslide records are used to model any further LS and to determine landslide-prone areas. Currently, satellite imagery is widely used for rapid detection of landslides, generating inventory map and updating field survey observations. The methods used to acquire information from satellite imagery are primarily categorized into two approaches: (1) object-based and (2) pixel-based image analysis. Due to some deficiencies of pixel-based approaches in processing images with very high resolutions, intensive spectral heterogeneity, and intricate textures, object-based image analysis (OBIA) methods have recently received special attention as a knowledge-driven method based on the segmentation process (Ghorbanzadeh et al., 2022). The method critically depends on the expert knowledge to ascertain the properties of the segmentation process as well as classification procedures (Ma et al., 2021; Ghorbanzadeh et al., 2022). OBIA methods enable the user to combine spectral information and spatial features to interpret the images (Rau et al., 2014). However, due to the large variability of the type and size of landslides, a specific set of segmentation parameters cannot be used to accurately detect all possible landslides using the OBIA methods. In the pixel-based method, the radiometric characteristics of each pixel are considered for segmenting the landslide areas; therefore, it is easy to implement.

In the last decade several deep learning algorithms, especially fully convolutional networks (e.g., U-Net and residual U-Net [ResU-Net]), have been utilized to identify landslides and increase the accuracy of semi-automated landslide detection in comparison with manual approaches. Blaschke et al. (2014) combined spectral and spatial data to develop a semi-automated OBIA workflow to detect landslides in north-western Iran. The integration of heterogeneous information increases the accuracy of the delineation of landslides from a satellite imagery with an overall accuracy of 93.07%. Soares et al. (2020) used the U-Net approach to detect landslide segments in southeastern Brazil. Moreover, the influence of various sampling methods and patch sizes on the predictive accuracy of models was scrutinized. The findings showed that models trained with 128×128 tiles achieved the highest F1-score value of 0.55. Qi et al. (2020) assessed the ability of U-Net and ResU-Net to automatically map landslides in Tianshui, China, where 10,000 landslides occurred due to intense rainfall in July 2013. The proposed ResU-Net model performed better in discriminating landslides than the U-Net model where the F1-score value improved by 0.09. Amatya et al. (2021) designed an OBIA-based process to detect landslides triggered by the 2015 Gorkha earthquake in Nepal. The results showed the efficiency and high speed of integration of the mean-shift segmentation method with the random forest (RF) algorithm to map landslide affected points over large areas. The landslide inventory map is the main step of generating LS map and vulnerability assessment.

Various statistical and machine learning (ML) techniques have been utilized to model the LS on nonlinear relationships between landslide effective factors and historical observed landslide points (e.g., frequency ratio, weights-of-evidence, quadratic discriminant analysis, convolutional neural network [CNN], artificial neural networks, support vector machines [SVM], RF, recurrent neural network [RNN], and adaptive neuro-fuzzy inference systems). Moreover, to accurately determine the hyper-parameters of the ML algorithms and enhance the prediction accuracy of the LS map, integrating the ML techniques and metaheuristic optimization algorithms (e.g., genetic algorithm, biogeography-based optimization, water cycle algorithm, artificial bee colony, and imperialist competitive algorithm) have been applied. In bivariate and multivariate statistical methods, the data are analyzed based on assumptions prior to study in terms of the relationship between the effective factors and landslide occurrence (Patra et al., 2018). The ML algorithm as the cost- and time-effective approach has become increasingly important for spatial analysis as large datasets become available. Hakim et al. (2022) assessed the effectiveness of the standard CNN and hybridized with the ICA and gray wolf optimizer (GWO) algorithms to map the LS in Icheon, South Korea. The CNN-GWO showed a superior performance with a higher accuracy to generate the LS map. Additionally, forest-related factors (e.g., diameter, age, and type), lithology, and topographic factors (e.g., slope, terrain ruggedness index and total curvature) significantly influenced the modeling. Youssef et al. (2022) compared the performance of SVM, CNN-1D, and CNN-2D to model the LS at Asir Region, Saudi Arabia. Based on the statistical indices (e.g., kappa coefficient, mean absolute error, and the area under the receiver operating characteristic curve [AUROC]), the CNN-2D was the optimal model with the highest accuracy to identify future landslides. Liu et al. (2021) constructed their model based on the CNN, dense convolutional neural networks (DenseNets), and ResU-Net to map the landslide-susceptible areas in China. In this study, RF algorithm was used to indicate the contribution of effective factors in modeling, where the morphological factors had the greater weights of contribution, followed by hydrological, geological, and spectral factors. Moreover, the DenseNet outperformed the other models in accurately detecting the landslide-susceptible areas. Nhu et al. (2020) evaluated the predictive ability of a deep neural network (DNN) to map the shallow LS. The DNN optimized with the Adam optimization algorithm showed the most accurate prediction performance to map landslide-prone areas in Vietnam. The predictive capability of each model depends entirely on the quantity and quality of the used data as well as the geo-environmental characteristics of the study area. Therefore, selecting the optimized algorithms affects the accuracy and reliability of the results.

This study aims to use a U-net architecture with a residual neural network (ResNet) backbone (also referred to as ResU-Net) to detect the landslide-affected areas and generate a landslide inventory map. Instead of using the U-Net's standard encoder, a ResNet model was used as the encoder to enhance the model's performance. The yield historical data were used to construct the ML models based on the standard support vector regression (SVR) algorithm hybridized by the particle swarm optimization (PSO) and GWO algorithms to construct an LS map. The predictive performance of the models was assessed through AUROC, and the superior model with a higher predictive accuracy was identified. Additionally, the correlation between geo-environmental variables and landslide occurrences was analyzed, and the most influential factor was indicated. The resulting LS map can be utilized to investigate the probability of the occurrence of future landslides and are the first stage of landslide management in areas that suffer from landslides.

27.2 Methodology

Generating a landslide inventory map is the preliminary stage of producing a susceptibility map and a vulnerability assessment. The map can be used to identify the association between the types, distribution, and patterns of landslides with geological and morphological characteristics of the study area. Recently, a freely available image of Sentinel-2 was used to automatically build a landslide inventory map using the ML algorithms. It involved training and testing the ML model (e.g., ResU-Net) based on the postlandslide image to extract all landslides and to generate the map.

Generally, selecting effective factors can affect the predictive accuracy of the LS mapping. A landslide is a complex phenomenon that is stimulated by multiple geo-environmental and anthropogenic factors as well as extreme meteorological events within mountainous regions worldwide. Previous studies have applied various attributes to model the LS based on data availability, geological, topographic, geomorphologic, and hydrological characteristics, and soil properties of the study area (Fonseca et al., 2017; Kadirhodjaev et al., 2020; Arabameri et al., 2021; Fadhillah et al., 2022). Additionally, detecting the linear correlation between the effective factors and removing the dependent variables from the model should be incorporated. The Pearson's correlation coefficient has been widely used to calculate the collinearity between each pair of effective factors. The coefficient values lower than 0.7 showed that the predictor variables were independent of each other (Tehrany et al., 2019). Furthermore, the information gain ratio (IGR) method can be applied to analyze the importance of the factors considered in the LS mapping. Factors with a higher IGR value contribute more to the modeling, and the factors with little or no effect must be eliminated to improve the reliability as well as accuracy of predicting landslide-prone areas (Rezaie et al., 2021).

The final step is to generate SVR model and optimizing the models' hyperparameters using metaheuristic algorithms (e.g., PSO and GWO) to produce a more accurate and reliable LS susceptibility map. The accuracy of classification and predictive ability of the ML can be assessed using statistical approaches. The proposed methodology is explained in the following sections and illustrated in Fig. 27.1.

27.3 Description of ML models

27.3.1 ResU-Net

ResU-Net was developed as a variant of the U-Net algorithm to avoid the vanishing gradient problems and modify the learning performance in semantic segmentation tasks. The U-Net architecture comprises a contracting and expanding path, called encoder and decoder, to capture features with low- and high-resolution through convolution and deconvolution operations, respectively (Liu et al., 2020; Zhang et al., 2022b). The structure of an encoding (downsampling) path is identical to CNN architectures with consecutive convolution blocks and ResNet34 blocks for feature extraction. The decoding (upsampling) path is used to retrieve the vanished information of the feature localization (Khryashchev and Larionov, 2020). The input of U-Net was the stacked nine layers of Sentinel-2 with spatial resolution of 10 and 20, with a total size of $64 \times 64 \times 9$, and the output was the detected landslide

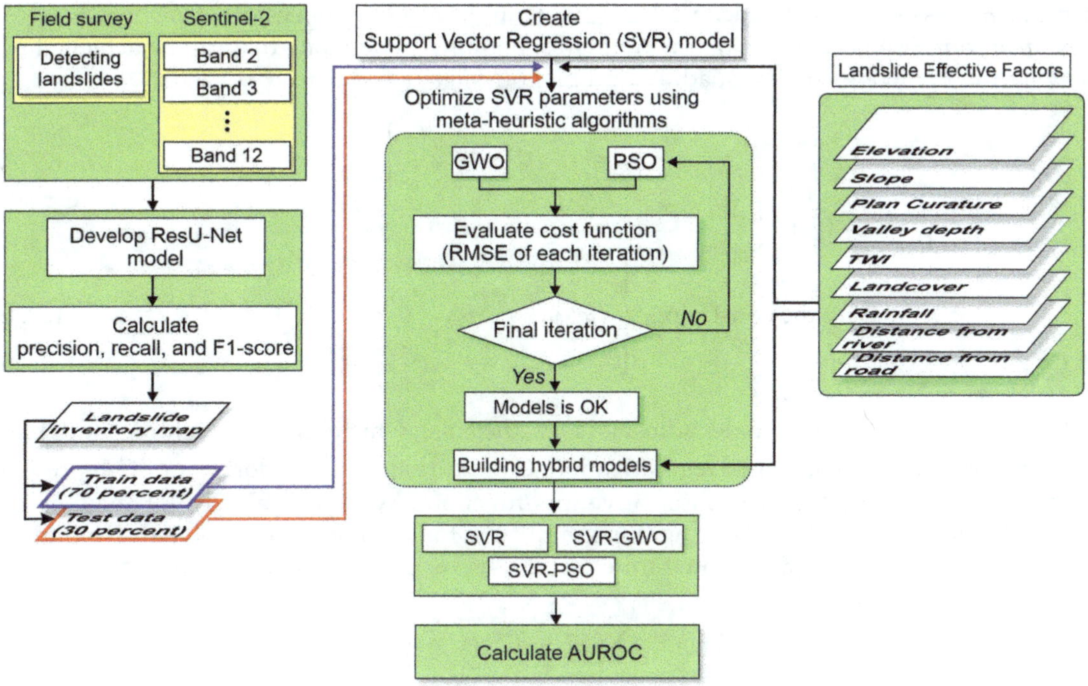

FIGURE 27.1 Procedure to detect landslide-prone areas.

with the size 64 × 64. A single residual block is mathematically expressed as (Ghorbanzadeh et al., 2021):

$$Y_l = H(X_l) + F(X_l, W_l) \tag{27.1}$$

$$X_{l+1} = f(Y_l) \tag{27.2}$$

where X_{l+1} and X_l indicate the output and input of the l-th block, $F(.)$ and $f(Y_l)$ refer to the residual function and activation function, respectively. $H(X_l) = X_l$ is an identity mapping (He et al., 2016).

27.3.2 SVR algorithm

The SVR algorithm proposed by Vapnik (1995) has been widely used to solve forecasting problems (Wei-Mou Hung et al., 2009; Mirarabi et al., 2019; Wang et al., 2022). SVR has the ability to solve nonlinear problems using only a few samples based on the structural risk minimization. Compared with other algorithms, the SVR has a faster convergence rate and can efficiently solve multidimensional problems due to its capability to calculate the relationship between input and output data (Ayodele et al., 2019). A given training dataset with n training data points is defined by $D = \{(x_i, y_i) | x_i \in R^d, y_i \in \{-1, 1\}\}_{i=1}^{n}$, where x_i is an input datum, y_i is associated with each vector x_i, and d indicates the dimension of the input space. The main purpose is determining the optimal separating hyperplane that can be achieved by solving the following convex quadratic problem (Melgani and Bruzzone, 2004):

$$R_{\min}: \left[\frac{1}{2}\|w\|^2 + C\sum_{i=1}^{n} \xi_i + \xi_i^*\right] \tag{27.3}$$

$$\text{With the constraints} \begin{cases} y_i - (w^T\varphi(x_i) + b_i) \leq \varepsilon + \xi_i^* \\ (w^T\varphi(x_i) + b_i) - y_i \leq \varepsilon + \xi_i^* \;\; C > 0 \\ \xi_i, \xi_i^* \geq 0 \; i = 1, 2, \ldots, n \end{cases} \tag{27.4}$$

where $w \in R^n$ and $b \in R$ denote adjustable coefficients, which mean weighting and bias vectors. $\frac{1}{2}\|w\|^2$ and is a factor of regularization, C indicates penalty factor, ξ_i and ξ_i^* represent slack variables, and $\varepsilon > 0$ is an insensitive loss function (Ayodele et al., 2019; Panahi et al., 2020a; Balogun et al., 2021). The Lagrange multiplier function is used to solve the optimization problem as follows (Balasundaram et al., 2014; Awad and Khanna, 2015):

$$f(x) = \sum_{i=1}^{n} (\beta_i - \beta_i^*) k(x, x_i) + b \tag{27.5}$$

where $k(x, x_i) = \langle \varphi(x), \varphi(x_i) \rangle$ is called kernel function, and β_i and β_i^* are obtained by solving a quadratic program. To improve prediction performance, appropriate values must be selected for the kernel function parameters and ε (Wang et al., 2012; Owolabi, 2019; Panahi et al., 2020b). Metaheuristic optimization algorithms can be used to fine-tune the model parameters.

27.3.3 PSO algorithm

The PSO algorithm developed based on the foraging behavior of a swarm of animals (e.g., birds, insects, herds, and fishes) with the advantages of global searching abilities, simple implementation, robust adaptability, and fast convergence speed (Xiong et al., 2022). The swarm works together to find food, and each individual adjusts its search approach according to the learning experiences from their fellow in the group (Umbarkar et al., 2017; Yan et al., 2017). Therefore, each particle can affect the way of searching of other particles. The PSO algorithm starts with the random production of a particle set same as other swarm-based algorithms. Supposing $x_i^t = (x_{i1}^t, x_{i2}^t, x_{i3}^t, \ldots, x_{in}^t)$ and $v_i^t = (v_{i1}^t, v_{i2}^t, v_{i3}^t, \ldots, v_{in}^t)$ are the position and velocity vector of the i-th particle in the t-th iteration, respectively. The particle iteratively updates its current position and velocity using the following formula (Sun and Liu, 2013; Sheng et al., 2022):

$$x_i^{t+1} = x_i^t + v_i^{t+1} - v_{\max} \leq x_i^t \leq v_{\max} \tag{27.6}$$

$$v_i^{t+1} = \omega \cdot v_i^t + c_1 \cdot r_1 \cdot (p_i^t - x_i^t) + c_2 \cdot r_2 \cdot (g_i^t - x_i^t) \tag{27.7}$$

where n denotes the population size, x_i^t is the previous location of the i-th particle, p_i^t is the best location found by the i-th particle, g_i^t is the best location achieved among all the neighbors, and r_1, r_2 are uniformly distributed random numbers from 0 to 1. Additionally, the three parameters ω, c_1, and c_2 are the inertia weight, personal, and social learning factor, respectively (Peng et al., 2022). ω, c_1, and c_2 are controlling local exploitation and global exploration. If c_2 is small and c_1 and ω are large, the search area is expanded by the population to enhance global exploration and increases the convergence speed. Whereas, the larger c_2 and smaller c_1 and ω lead to an improvement in the local search ability and increases the predictive accuracy of the algorithm (Zhang and Zhao, 2021; Cui et al., 2022). The algorithm will continue as long as the fitness function reaches the highest value that shows a more optimum solution (Saini et al., 2014).

27.3.4 GWO algorithm

The GWO algorithm was developed based on mimicking the leadership hierarchy and the hunting mechanism of gray wolves (Mirjalili et al., 2014). Alpha (α), beta (β), delta (δ), and omega (ω) wolves make the hierarchical structure of the wolf pack. Alpha is the leader of the pack, and betas aid the leader in decision-making. Deltas detect the location of the prey, and the omegas hunt and kill the prey (Yin and Sun, 2022). The three main steps of hunting are searching for prey, encircling, and attacking the prey (Balogun et al., 2021), which can be mathematically simulated as the following:

$$\overrightarrow{X_\alpha^*} = \left|\overrightarrow{C_1} . \overrightarrow{X_\alpha}(\mathrm{t}) - \overrightarrow{X}(\mathrm{t})\right|, \overrightarrow{X_\beta^*} = \left|\overrightarrow{C_2} . \overrightarrow{X_\beta}(\mathrm{t}) - \overrightarrow{X}(\mathrm{t})\right|, \overrightarrow{X_\delta^*} = \left|\overrightarrow{C_3} . \overrightarrow{X_\delta}(\mathrm{t}) - \overrightarrow{X}(\mathrm{t})\right| \tag{27.8}$$

$$\overrightarrow{X_1}(\mathrm{t}) = \overrightarrow{X_\alpha}(\mathrm{t}) - \overrightarrow{A_1} . \overrightarrow{X_\alpha^*}, \overrightarrow{X_2}(\mathrm{t}) = \overrightarrow{X_\beta}(\mathrm{t}) - \overrightarrow{A_2} . \overrightarrow{X_\beta^*}, \overrightarrow{X_3}(\mathrm{t}) = \overrightarrow{X_\delta}(\mathrm{t}) - \overrightarrow{A_3} . \overrightarrow{X_\delta^*} \tag{27.9}$$

$$\overrightarrow{X}(t+1) = \frac{\overrightarrow{X_1}(\mathrm{t}) + \overrightarrow{X_2}(\mathrm{t}) + \overrightarrow{X_3}(\mathrm{t})}{3} \tag{27.10}$$

$$\overrightarrow{A} = 2\overrightarrow{a}\overrightarrow{r_1} - \overrightarrow{a} \qquad \overrightarrow{C} = 2\overrightarrow{r_2} \qquad \overrightarrow{a} = 2 - 2\left(\frac{t}{t_{max}}\right) \tag{27.11}$$

where $\overrightarrow{D}$ is the distance vector of the prey from the wolves (α , β, and δ), $\overrightarrow{X}(\mathrm{t})$ denotes the current position of the wolves, and $\overrightarrow{X}(t+1)$ shows the position of gray wolves in the next iteration. t and $t_{\max}$ express the current and maximum number of the iteration, respectively. The first three best wolves' positions in the t-th iteration are shown with X_α, X_β, and X_δ (Nadimi-Shahraki et al., 2022). $\overrightarrow{A}$ and $\overrightarrow{C}$ are coefficient vectors, and $\overrightarrow{a}$ is the convergence factor which is linearly decremented from 2 to 0 with the number of iterations

(Singh and Bansal, 2022). $\vec{r_1}$ and $\vec{r_2}$ are random vectors take values in the range [0,1]. The hunting process is ceased when $C < 1$ and $\vec{A}$ equals the values between −1 and 1. When $C > 1$ and $|A| > 1$, the wolves expand their search scope to find a more suitable prey (Bai and Li, 2022). By contrast, when $-1 < A < 1$, the greater emphasis will be on exploitation (Gao and Zhao, 2019; Li et al., 2021c). The alpha is the best optimized solution followed by β, δ, and ω. The GWO is an appropriate optimization method in solving a nonlinear problem due to the high-speed convergence, low number of adjustable parameters, and proper global exploration ability (Hao and Sobhani, 2021).

27.3.5 Accuracy assessment

The accuracy of detecting landslide affected areas is assessed based on comparison between the resulting map and the inventory data set. F1-score, recall, and precision are commonly used to evaluate the classification accuracy. Recall denotes the proportion of correctly classified landslides to the total pixels representing landslides. Precision represents the ratio of pixels correctly identified as landslides (Lizama et al., 2022). The harmonic average between precision and recall is F1-score that the higher value corresponds to the better classification (Meena et al., 2022). The evaluation parameters are formulated as:

$$\text{Precision} = \frac{\text{TP}}{\text{TP} + \text{FP}} \tag{27.12}$$

$$\text{Recall} = \frac{\text{TP}}{\text{TP} + \text{FN}} \tag{27.13}$$

$$\text{F1} - \text{score} = 2 \times \frac{\text{Precision} \times \text{Recall}}{\text{Precision} + \text{Recall}} \tag{27.14}$$

where TP and FP are defined the number of pixels correctly and incorrectly classified as landslide, respectively. The number of incorrectly classified pixels as nonlandslide is FN.

In addition to assessing the classification accuracy, evaluating the reliability and accuracy of the LS map is essential and commonly indicated based on the AUROC. The goodness-of-fit and predictive performance of the ML model are determined using training and testing datasets, respectively.

27.4 Case study

At 18:07:59.3 (UTC) on September 5, 2018, a powerful earthquake with a moment magnitude of 6.6 hit the Eastern Iburi District of Hokkaido Island in Northern Japan. More than 6000 earthslides and 100 rockslides occurred in a 20 × 20 km^2 area in the vicinity of Atsuma, Hokkaido (Fig. 27.2). Approximately 82% of the total fatalities (42 people) were from shallow earthslides in Atsuma where the maximum seismic intensity of 9.5 MMI was recorded. The majority of shallow earthslides observed within 20 km north of the epicenter (42.690°N by

FIGURE 27.2 Location of study area.

142.006°E). However, some deep occurred within 10 km from the epicenter (Liu et al., 2021). Around 40% of Hokkaido Island is covered with pyroclastic flow deposits and pyroclastic fall deposits. The performance and topography of the Quaternary volcanic soils in Hokkaido are the main cause of the rainfall- and earthquake-induced failures in the region (Kawamura et al., 2019). Moreover, the thickness of the surface geology varies between 4 and 5 m in and adjacent to the epicentral area; therefore, a large number of the landslides were shallow with a planar or spoon-shaped rupture surface (Yamagishi and Yamazaki, 2018). The area experienced heavy rainfall with a total precipitation of 100 mm as a result of Typhoon Jebi on September 3 to 5, just some days before the earthquake occurred (Meteorological Agency of Japan, https://www.jma.go.jp/). Thus, the rain-soaked subsurface deposits of soil were prone to landslides (Aimaiti et al., 2019; Serikawa et al., 2019; Guzzetti et al., 2022; Lee et al., 2022).

27.4.1 Landslide inventory map

In this study, the image of the Sentinel-2 Level-1C (acquisition date: September 10, 2018) was processed to build a landslide inventory map based on the U-Net algorithm with ResNet backbone for image segmentation and landslide detection. In the first stage, the SNAP software developed by the European Space Agency (ESA) was applied for atmospheric corrections. The U-Net model was then trained with the high-resolution bands of Sentinel-2 (10 and 20 m spatial resolution) using a shapefile of premapped landslide polygons to predict the exact location and extent of the all landslides. Zhang et al. (2019) produced the first

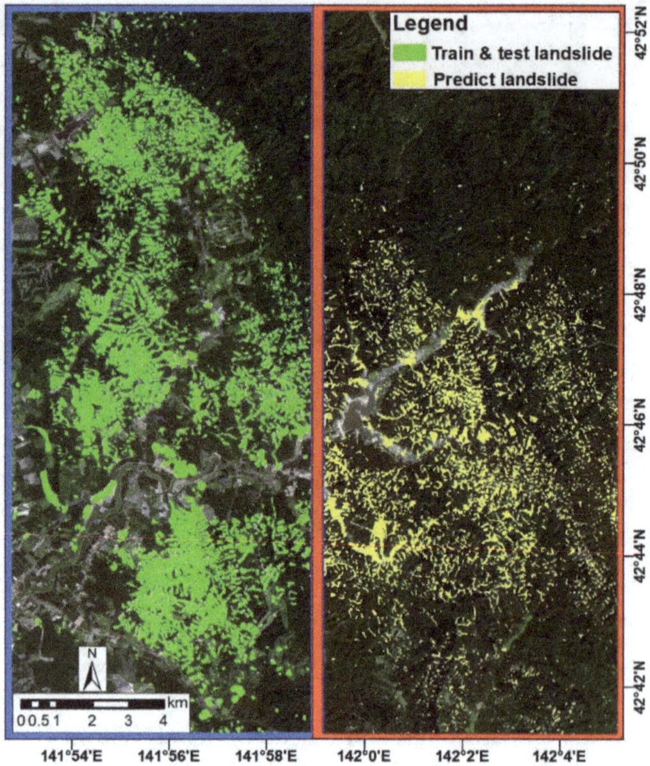

FIGURE 27.3 Training, testing, and predicted landslides presented on Sentinel-2 image.

shapefile of the landslides triggered by the Hokkaido Eastern Iburi earthquake and was used to train and test the U-Net model for landslide detection and prediction of all landslides (Fig. 27.3). In total, 4900 features were obtained from the ResU-Net model. The area of the selected features ranged between 25 m^2 and 1.3 km^2 (Zhang et al., 2019).

In order to build the LS model, the same number of polygons were detected as nonlandslide areas. Then, landslide and nonlandslide samples were coded 1 and 0, respectively. Both groups of data were randomly split into training and testing subsets at a ratio of 70:30 (Jaafari et al., 2019; Huang et al., 2022; Panahi et al., 2022). The final dataset to train the LS model was built through merging both training subsets of landslide and nonlandslide samples (i.e., 70% of the landslide and nonlandslide inventory). Both testing subsets of landslide and nonlandslide samples (i.e., 30% of the landslide and nonlandslide inventory) were combined to generate the testing set.

27.4.2 Landslide effective factors

Nine effective factors were chosen to generate the LS map, namely elevation, slope, plan curvature, valley depth, topographic wetness index (TWI), land cover, rainfall, and distance from river and road to detect landslide-susceptible areas (Fig. 27.4). These factors were

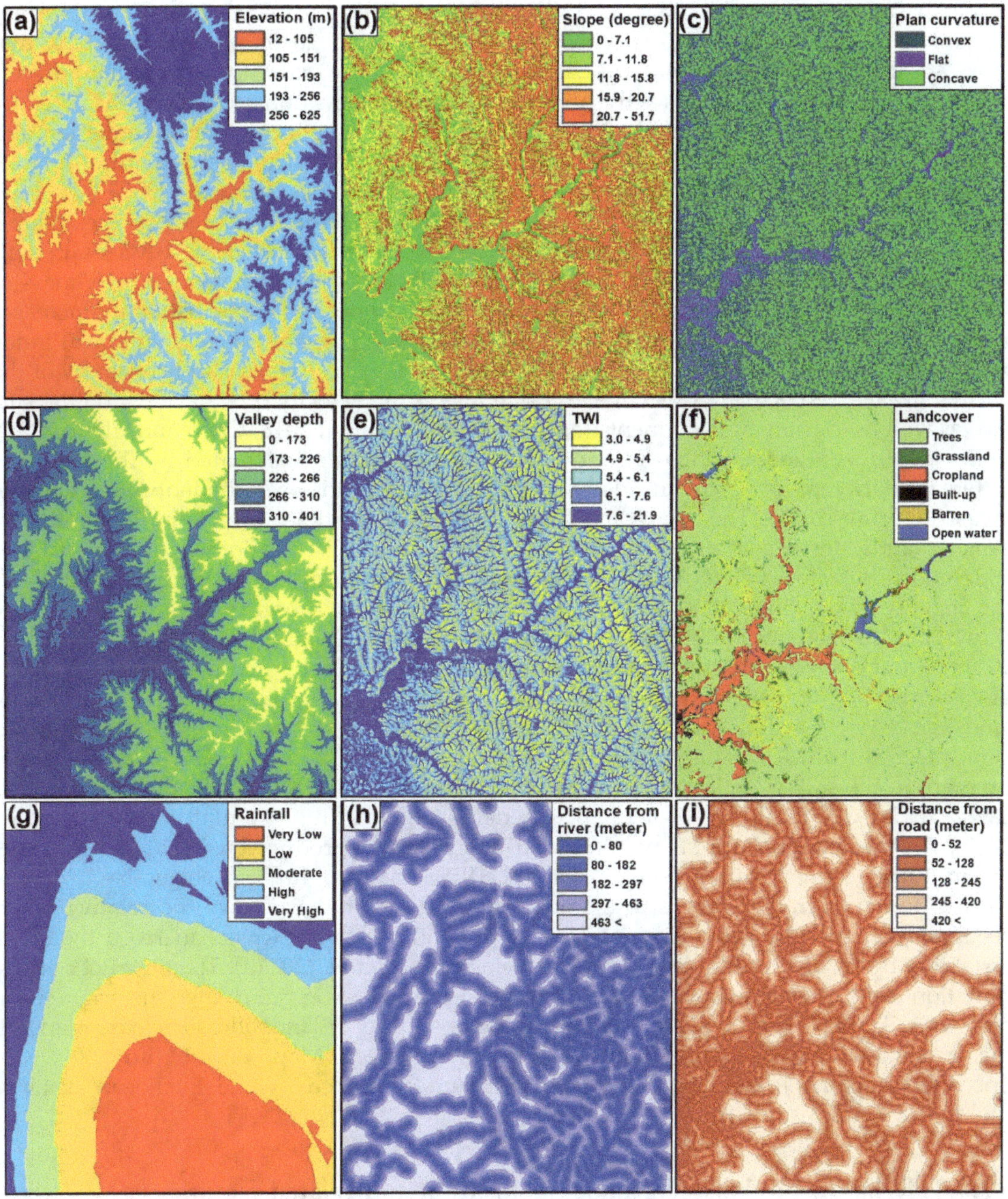

FIGURE 27.4 Landslide effective factors: (A) elevation, (B) slope, (C) plan curvature, (D) valley depth, (E) topographic wetness index (TWI), (F) land cover, (G) rainfall, (H) distance from river, and (I) distance from road.

chosen based on literature review and characteristics of the study area. Elevation has a significant connection with landslide occurrence (Aditian et al., 2018). The studies show that rocks are stronger in lower elevations as well as foothills. The existence of nitrogen and carbon in soil and dehydrogenases activity in areas of higher elevation are the significant cause of landslide occurrences (Ali et al., 2021). The advanced spaceborne thermal emission and reflection radiometer (ASTER) digital elevation model (DEM) with a spatial resolution of 30×30 m was downloaded from the United States Geological Survey (USGS) Website (http://earthexplorer.usgs.gov). Other topographic and hydrologic factors (i.e., slope, plan curvature, valley depth, TWI) were produced based on DEM using SAGA-GIS software (https://saga-gis.sourceforge.io/en/index.html). Downslope movement of a mass of rock, debris, or soil is directly associated with the slope angle, and slope instability increases when the angle is relatively large (Xing et al., 2021). Plan curvature is a geometrical index where positive, zero, and negative value represents convexity, flatness, and concavity of the Earth's surface, respectively (Li et al., 2021a). It affects the velocity of water flow on the surface as well as erosion and sediment transport rates (Oh et al., 2018; Chen and Chen, 2021). Valley depth as a geomorphologic and hydrologic attribute is calculated by the elevation difference between the upstream ridge and the given pixel. It is used to identify the upslope areas (Le et al., 2021). The toe of the slope is more susceptible to erosion by rivers and causing landslides (Kalantar et al., 2020; Ma et al., 2020). TWI is another prominent landslide effective factor that represents the spatial distribution of soil moisture and impacts the stability of the slope. TWI is correlated with infiltration, runoff rates, and lower value shows the decrease in water flowing to a certain point (Naghibi et al., 2020; Li et al., 2021a). Land cover plays a significant role in the increase of slope stability and the decrease of soil erosion. The water in the soil absorbed by the land cover reduces the propensity of landslides (Amir Yazdadi and Ghanavati, 2016). The land cover map for 2018 was obtained from the Sentinel-2A satellite image operated by ESA. Rainfall increases the pore water pressure in the soil or rock and decreases the shear strength of the soil slopes (Miller and Degg, 2012; Anis et al., 2019). The LS is generally high in high rainfall areas (Ali et al., 2021). The mean annual rainfall map from 2015 to 2020 was produced using climate hazards group infraRed precipitation with stations (CHIRPS) dataset (https://chc.ucsb.edu/data/chirps). Distance from rivers is a geomorphological factor that shows the relative water accumulation and the saturation degrees of the soil materials. Rivers increase the slope instability by eroding and hollowing out the slope toe Li et al. (2021a). To construct a transportation route, the manual excavation of slope toe and blasting can change the original stability of soil and rock structures and elevates the possibility of landslides. The base maps for the rivers and roads were taken from Open Street Map portal (https://www.openstreetmap.org/), and the distance maps were generated by the Euclidean distance tool in the ArcGIS software.

27.4.3 Relationship between landslide location and related factors

Frequency ratio (FR) is widely used to quantitatively state the spatial relationships between distribution of landslides occurrence and related factor. The average value of FR is

1. FR > 1 indicates the class of that factor has a strong impact on the LS. For factors that have weak relationships with the LS, the FR is less than 1. The FR of each factor was calculated as (Eq. 27.15):

$$FR = \frac{\%\ \text{pixel of landslide points}}{\%\ \text{pixel class of landslide effective factor}} \tag{27.15}$$

The FR values showed that classes with strong effects on the LS were: (1) elevation = 105−151 m (FR = 1.55), (2) slope = 11.8−15.8 degrees (1.13), (3) plan curvature of convex structure (1.24), (4) valley depth = 266−310 m (1.54), (5) TWI = 7.6 to 21.9 (1.20), (6) land cover of barren areas (5.48), (7) high level of rainfall (1.62), (8) distance from river ranging from = 0 to 80 m (1.10), and (10) distance from road greater than 421 m (1.08). Moreover, the results of the IGR values indicated that elevation (IGR = 0.524) is the most effective factor on predicting landslide-prone areas, followed by valley depth (0.506) as well as land cover (0.226), and the factor of distance from road received the least IGR value (0.005). The results are in line with the studies conducted by Meinhardt et al. (2015), Mahalingam et al. (2016), de Oliveira et al. (2019), Linkha et al. (2019), and Ozioko and Igwe (2020). Elevations also play a controlling role in the stability of the slope, vegetation growth, and climate characteristics. Valley depth is used to detect the distribution of the upslope areas and to quantify the degree of fluvial incision (Pawley et al., 2018). The probability of landslide occurrence is high in regions near the valley due to variation in soil moisture content and stability of slopes (Panahi et al., 2020a, 2022). Land cover affects soil moisture. The high susceptibility of the built up areas is due to the disruption of the natural slope by road construction and the building of houses. Forests and dense vegetation increase the infiltration capacity of soil and dint of tree roots which fixes the soil and reduces the probability of landslide occurring in the area (Anis et al., 2019).

Finally, Pearson's correlation coefficient was considered to detect the relationships of variables in pairs. The highest correlation coefficient (0.646) was calculated between elevation and valley depth, followed by TWI and plan curvature (0.428). Valley depth, TWI, and plan curvature are all DEM-driven variables, and the higher correlation coefficients between the factors were expected.

27.4.4 Landslide susceptibility mapping

The standard and optimized SVR algorithms were conducted using the training dataset (70% of all data) and all factors to obtain the LS maps of the study area. The maps were classified utilizing the quantile method into five categories from very high to very low susceptibility to landslides (Fig. 27.5). All models indicated that the central parts of the study area with potential landslide prone areas were close to rivers and had steep slopes. In terms of the percentage of the high and very high susceptibility classes, the two optimized models exhibited a similar spatial distribution. High and very high susceptible areas account for approximately 40% of the entire study area. However, the standard SVR model classified most of the areas with low and very low LS levels.

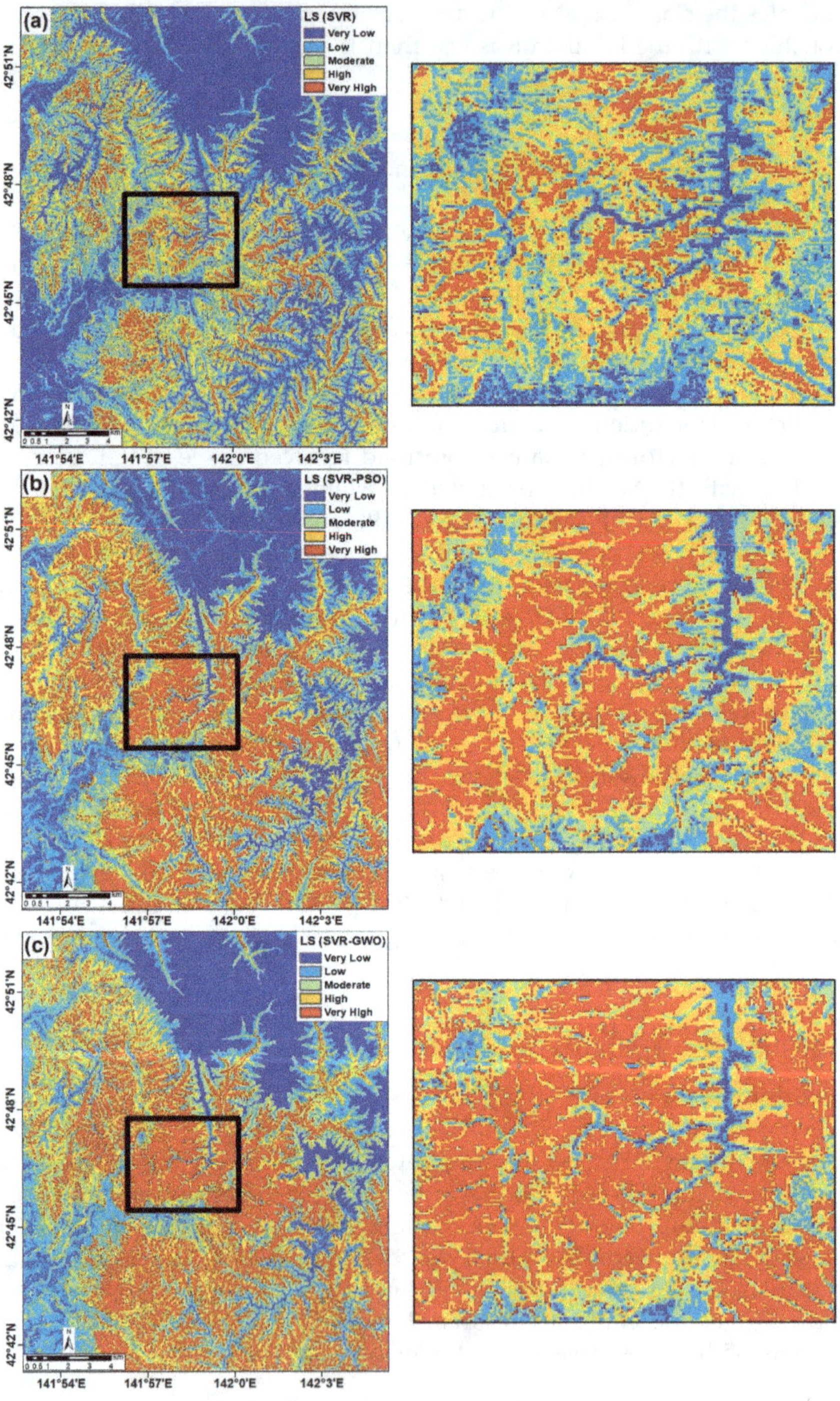

FIGURE 27.5 Landslide susceptibility (LS) maps: (A) support vector regression (SVR), (B) SVR-particle swarm optimization (PSO), and (C) SVR-gray wolf optimizer (GWO).

27.4.5 Models performance

Assessing prediction accuracy and reliability of the ML model are the critical step in producing the LS map. The AUROC analysis is a common way of validating the model prediction. The value of AUROC ranges from 0.5 (random forecasting) to 1 (perfect forecasting) (Fawcett, 2006). The goodness-of-fit and prediction accuracy were evaluated using the training and testing dataset, respectively. Fig. 27.6 shows that SVR-GWO had the highest goodness-of-fit (AUROC = 0.82) and prediction accuracy (AUROC = 0.80). Thus, the model had the superior performance than the others. The GWO could select more optimum values for the SVR hyper-parameters than that of the PSO in both the training and testing phase according to AUROC values. The GWO has more precise performance due to fewer parameters and easy to implement, as well as fast rate of convergence (Zhang et al., 2022a). The advantages of GWO over the other optimization algorithms (e.g., PSO, genetic, bat, and cuckoo) were proven by Balogun et al. (2021), Li et al. (2021b), Panahi et al. (2021), and Zhang and Hong (2021). The LS map revealed the central locations were the most susceptible areas to landslide occurrence. The causes for the high and very high susceptibility in some regions were the steepness of slope, the high level of rainfall and soil moisture as well as proximity to water bodies and roads, which is in accordance with the results achieved by Anis et al. (2019), Kalantar et al. (2020), and Chen and Zhang (2021).

It is noteworthy that the accuracy of the image classification and the precision of the inventory map directly affect the results. In this study, the ability of the ResU-Net model to detect landslide affected areas was quantitatively assessed based on the value of precision (76.6290), recall (72.9173), and F1-score (74.7271). The values suggested that the ResU-Net could correctly detect landslides.

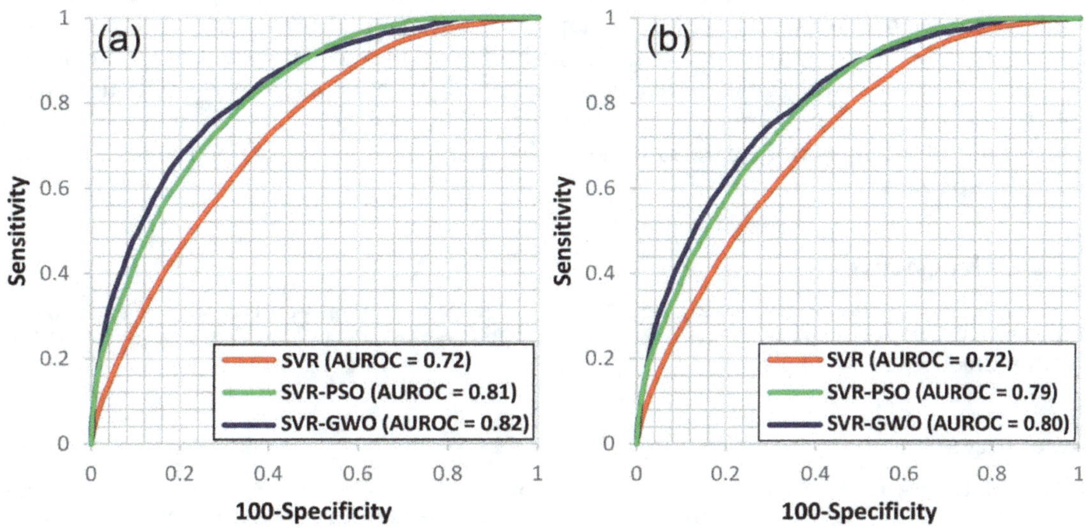

FIGURE 27.6 Evaluation models performance: (A) training dataset and (B) testing dataset.

27.5 Conclusion

LS mapping is an important tool in mitigating the human casualties and socio-economic damage. Identification of areas with very high- and high-susceptibility provides information that can be used to manage risk and develop innovative responses to future landslides. This study describes a systematic comparison and evaluation of the standard and hybridized SVR algorithms for the LS based on the nine factors associated with landslides. The locations of landslides were determined using ResU-Net. In the case of generating an LS map, the SVR-GWO model outperformed in terms of accuracy, with the AUROC values of 0.82 and 0.80, respectively. The result demonstrated the effectiveness and efficiency of the GWO in fine-tuning the SVR hyperparameters. The SVR-GWO model categorized approximately 40% of the total area as high and very high susceptible regions. The results of the IGR indicated that elevation and valley depth are the most important factors driving landslide occurrence.

Similar any study, there are limitations and sources of uncertainties that should be acknowledged. Since, the contextual and geometric information of the features represented in the satellite images were not considered in the pixel-based approaches, the accuracy of landslide inventory map can be increased using OBIA. However, the pixel-based ResU-Net algorithm is able to detect big landslides from dense vegetation, whereas the false-positive detection substantially increases in the neighboring bare land. The problem can be solved by providing appropriate auxiliary data such as topographic and displacement information. Moreover, small landslides can be missed or incorrectly detected using Sentinel-2 data. Therefore, using synthetic aperture radar (SAR) and airborne lidar data that provide high-resolution images of the small ground displacements will be beneficial.

It is important to note that inherent variability and complexity of landslide processes lead to uncertainties in LS mapping. Therefore, performing sensitivity and robustness analysis to assess the impact of different input variables as well as random partitioning of historical data on the accuracy and reliability of the LS map are recommended for future studies. In addition, using deep learning methods (e.g., long short-term neural network, CNN, and RNN) coupled with some novel metaheuristic optimization algorithms (e.g., crow search algorithm, ant colony optimization, chimp optimization algorithm, and germinal center optimization) will be helpful to better capture the nonlinear relationships between input variables and landslide susceptibility.

Acknowledgments

The research was supported by the Basic Research Project of the Korea Institute of Geoscience and Mineral Resources (KIGAM) and the National Research Foundation of Korea (NRF) grant funded by Korea government (MSIT) (No. 2023R1A2C1003095).

References

Aditian, A., Kubota, T., Shinohara, Y., 2018. Comparison of GIS-based landslide susceptibility models using frequency ratio, logistic regression, and artificial neural network in a tertiary region of Ambon, Indonesia. Geomorphology 318, 101–111.

Aimaiti, Y., Liu, W., Yamazaki, F., Maruyama, Y., 2019. Earthquake-induced landslide mapping for the 2018 Hokkaido Eastern Iburi earthquake using PALSAR-2 data. Remote Sens 11.

Ali, S.A., Parvin, F., Vojteková, J., Costache, R., Linh, N.T.T., Pham, Q.B., Vojtek, M., Gigović, L., Ahmad, A., Ghorbani, M.A., 2021. GIS-based landslide susceptibility modeling: a comparison between fuzzy multi-criteria and machine learning algorithms. Geosci. Front. 12, 857–876.

Amatya, P., Kirschbaum, D., Stanley, T., Tanyas, H., 2021. Landslide mapping using object-based image analysis and open source tools. Eng. Geol. 282, 106000.

Amir Yazdadi, E., Ghanavati, E., 2016. Landslide hazard zonation by using AHP (Analytical hierarchy process) model in GIS (geographic information system) environment (case study: Kordan Watershed). Int. J. Progress. Sci. Technol. (IJPSAT) 2, 24–39.

Anis, Z., Wissem, G., Vali, V., Smida, H., Mohamed Essghaier, G., 2019. GIS-based landslide susceptibility mapping using bivariate statistical methods in North-western Tunisia. Open Geosci. 11, 708–726.

Arabameri, A., Chandra Pal, S., Rezaie, F., Chakrabortty, R., Saha, A., Blaschke, T., Di Napoli, M., Ghorbanzadeh, O., Thi Ngo, P.T., 2021. Decision tree based ensemble machine learning approaches for landslide susceptibility mapping. Geocarto Int. 1–35.

Awad, M., Khanna, R., 2015. Support vector regression. In: Efficient Learning Machines. Apress, Berkeley, CA, pp. 67–80.

Ayodele, T.R., Ogunjuyigbe, A.S.O., Amedu, A., Munda, J.L., 2019. Prediction of global solar irradiation using hybridized k-means and support vector regression algorithms. Renew. Energy Focus 29, 78–93.

Bai, Q., Li, H., 2022. The application of hybrid cuckoo search-grey wolf optimization algorithm in optimal parameters identification of solid oxide fuel cell. Int. J. Hydrogen Energy 47, 6200–6216.

Balasundaram, S., Gupta, D., Kapil, 2014. Lagrangian support vector regression via unconstrained convex minimization. Neural Netw. 51, 67–79.

Balogun, A.-L.L., Rezaie, F., Pham, Q.B., Gigović, L., Drobnjak, S., Aina, Y.A., Panahi, M., Yekeen, S.T., Lee, S., 2021. Spatial prediction of landslide susceptibility in western Serbia using hybrid support vector regression (SVR) with GWO, BAT and COA algorithms. Geosci. Front. 12, 101104.

Blaschke, T., Feizizadeh, B., Holbling, D., 2014. Object-based image analysis and digital terrain analysis for locating landslides in the Urmia lake basin, Iran. IEEE J. Sel. Top. Appl. Earth Obs. Remote Sens. 7, 4806–4817.

Chen, X., Chen, W., 2021. GIS-based landslide susceptibility assessment using optimized hybrid machine learning methods. Catena 196.

Chen, W., Zhang, S., 2021. GIS-based comparative study of Bayes network, Hoeffding tree and logistic model tree for landslide susceptibility modeling. Catena 203.

Cui, Y., Meng, X., Qiao, J., 2022. A multi-objective particle swarm optimization algorithm based on two-archive mechanism. Appl. Soft Comput. 119, 108532.

de Oliveira, G.G., Ruiz, L.F.C., Guasselli, L.A., Haetinger, C., 2019. Random forest and artificial neural networks in landslide susceptibility modeling: a case study of the Fão River Basin, Southern Brazil. Nat. Hazards 99, 1049–1073.

Fadhillah, M.F., Hakim, W.L., Panahi, M., Rezaie, F., Lee, C.-W., Lee, S., 2022. Mapping of landslide potential in Pyeongchang-gun, South Korea, using machine learning meta-based optimization algorithms. Egypt. J. Remote Sens. Sp. Sci. 25, 463–472.

Fawcett, T., 2006. An introduction to ROC analysis. Pattern Recogn. Lett. 27, 861–874.

Fonseca, L.D.M., Lani, J.L., Fernandes Filho, E.I., Santos, G.R., dos, Ferreira, W.P.M., Santos, A.M.R.T., 2017. Variability in soil physical properties in landslide-prone areas. Acta Sci. Agron. 39, 109.

Froude, M.J., Petley, D.N., 2018. Global fatal landslide occurrence from 2004 to 2016. Nat. Hazards Earth Syst. Sci. 18, 2161–2181.

Gao, Z.-M., Zhao, J., 2019. An improved grey wolf optimization algorithm with variable weights. Comput. Intell. Neurosci. 1–13.

Ghorbanzadeh, O., Crivellari, A., Ghamisi, P., Shahabi, H., Blaschke, T., 2021. A comprehensive transferability evaluation of U-Net and ResU-Net for landslide detection from Sentinel-2 data (case study areas from Taiwan, China, and Japan). Sci. Rep. 11, 14629.

Ghorbanzadeh, O., Gholamnia, K., Ghamisi, P., 2022. The application of ResU-net and OBIA for landslide detection from multi-temporal sentinel-2 images. Big Earth Data 00, 1–26.

Guzzetti, F., Gariano, S.L., Peruccacci, S., Brunetti, M.T., Melillo, M., 2022. Rainfall and landslide initiation. In: Rainfall. Elsevier, pp. 427–450.

Hakim, W.L., Rezaie, F., Nur, A.S., Panahi, M., Khosravi, K., Lee, C.-W., Lee, S., 2022. Convolutional neural network (CNN) with metaheuristic optimization algorithms for landslide susceptibility mapping in Icheon, South Korea. J. Environ. Manag. 305, 114367.

Hao, P., Sobhani, B., 2021. Application of the improved chaotic grey wolf optimization algorithm as a novel and efficient method for parameter estimation of solid oxide fuel cells model. Int. J. Hydrogen Energy 46, 36454–36465.

He, K., Zhang, X., Ren, S., Sun, J., 2016. Identity Mappings in Deep Residual Networks.

Huang, F., Yan, J., Fan, X., Yao, C., Huang, J., Chen, W., Hong, H., 2022. Uncertainty pattern in landslide susceptibility prediction modelling: effects of different landslide boundaries and spatial shape expressions. Geosci. Front. 13, 101317.

Hung, W.-M., Hong, W.-C., Chen, T.-B., 2009. Application of hybrid genetic algorithm and simulated annealing in a SVR traffic flow forecasting model. In: 2009 IEEE Congress on Evolutionary Computation. IEEE, pp. 728–735.

Jaafari, A., Panahi, M., Pham, B.T.B.T., Shahabi, H., Bui, D.T.D.T., Rezaie, F., Lee, S., 2019. Meta optimization of an adaptive neuro-fuzzy inference system with grey wolf optimizer and biogeography-based optimization algorithms for spatial prediction of landslide susceptibility. Catena 175, 430–445.

Kadirhodjaev, A., Rezaie, F., Lee, M.-J., Lee, S., 2020. Landslide susceptibility assessment using an optimized group method of data handling model. ISPRS Int. J. Geo-Inf. 9, 566.

Kalantar, B., Ueda, N., Saeidi, V., Ahmadi, K., Halin, A.A., Shabani, F., 2020. Landslide susceptibility mapping: machine and ensemble learning based on remote sensing big data. Remote Sens 12, 1737.

Kawamura, S., Kawajiri, S., Hirose, W., Watanabe, T., 2019. Slope failures/landslides over a wide area in the 2018 Hokkaido Eastern Iburi earthquake. Soils Found. 59, 2376–2395.

Khryashchev, V., Larionov, R., 2020. Wildfire segmentation on satellite images using deep learning. In: 2020 Moscow Workshop on Electronic and Networking Technologies (MWENT). IEEE, pp. 1–5.

Le, T.T.T., Tran, T.V., Hoang, V.H., Bui, V.T., Bui, T.K.T., Nguyen, H.P., 2021. Developing a landslide susceptibility map using the analytic hierarchical process in Ta van and Hau Thao Communes, Sapa, Vietnam. J. Disaster Res. 16, 529–538.

Lee, J.-J., Song, M.-S., Yun, H.-S., Yum, S.-G., 2022. Dynamic landslide susceptibility analysis that combines rainfall period, accumulated rainfall, and geospatial information. Sci. Rep. 12, 18429.

Li, B., Wang, N., Chen, J., 2021a. GIS-based landslide susceptibility mapping using information, frequency ratio, and artificial neural network methods in Qinghai Province, Northwestern China. Adv. Civ. Eng. 2021.

Li, E., Yang, F., Ren, M., Zhang, X., Zhou, J., Khandelwal, M., 2021b. Prediction of blasting mean fragment size using support vector regression combined with five optimization algorithms. J. Rock Mech. Geotech. Eng. 13, 1380–1397.

Li, Y., Chen, W., Rezaie, F., Rahmati, O., Davoudi Moghaddam, D., Tiefenbacher, J., Panahi, M., Lee, M.-J., Kulakowski, D., Tien Bui, D., Lee, S., 2021c. Debris flows modeling using geo-environmental factors: developing hybridized deep-learning algorithms. Geocarto Int. 1–25.

Linkha, T.R., Rai, D.K., Lama, F., 2019. Landslide hazard mapping: GIS based susceptibility assessment of Leoutikhola Watershed, Dhankuta, Nepal. Third Pole J. Geogr. Educ. 18, 71–84.

Liu, P., Wei, Y., Wang, Q., Chen, Y., Xie, J., 2020. Research on post-earthquake landslide extraction algorithm based on improved U-net model. Remote Sens 12, 894.

Liu, T., Chen, T., Niu, R., Plaza, A., 2021. Landslide detection mapping employing CNN, ResNet, and DenseNet in the three Gorges Reservoir, China. IEEE J. Sel. Top. Appl. Earth Obs. Remote Sens. 14, 11417–11428.

Lizama, E., Morales, B., Somos-Valenzuela, M., Chen, N., Liu, M., 2022. Understanding landslide susceptibility in northern Chilean Patagonia: a basin-scale study using machine learning and field data. Remote Sens. 14, 907.

Ma, S., Qiu, H., Hu, S., Pei, Y., Yang, W., Yang, D., Cao, M., 2020. Quantitative assessment of landslide susceptibility on the Loess Plateau in China. Phys. Geogr. 41, 489–516.

Ma, L., Zhu, X., Qiu, C., Blaschke, T., Li, M., 2021. Advances of local climate zone mapping and its practice using object-based image analysis. Atmosphere 12, 1146.

Mahalingam, R., Olsen, M.J., O'Banion, M.S., 2016. Evaluation of landslide susceptibility mapping techniques using lidar-derived conditioning factors (Oregon case study). Geomatics Nat. Hazards Risk 7, 1884–1907.

Meena, S.R., Soares, L.P., Grohmann, C.H., van Westen, C., Bhuyan, K., Singh, R.P., Floris, M., Catani, F., 2022. Landslide detection in the Himalayas using machine learning algorithms and U-Net. Landslides 19, 1209–1229.

Meinhardt, M., Fink, M., Tünschel, H., 2015. Landslide susceptibility analysis in central Vietnam based on an incomplete landslide inventory: comparison of a new method to calculate weighting factors by means of bivariate statistics. Geomorphology 234, 80–97.

Melgani, F., Bruzzone, L., 2004. Classification of hyperspectral remote sensing images with support vector machines. IEEE Trans. Geosci. Remote Sens. 42, 1778–1790.

Miller, S., Degg, M., 2012. Landslide susceptibility mapping in north-East Wales. Geomatics Nat. Hazards Risk 3, 133–159.

Mirarabi, A., Nassery, H.R., Nakhaei, M., Adamowski, J., Akbarzadeh, A.H., Alijani, F., 2019. Evaluation of data-driven models (SVR and ANN) for groundwater-level prediction in confined and unconfined systems. Environ. Earth Sci. 78, 489.

Mirjalili, S.M.S.M., Mirjalili, S.M.S.M., Lewis, A., 2014. Grey wolf optimizer. Adv. Eng. Software 69, 46–61.

Nadimi-Shahraki, M.H., Taghian, S., Mirjalili, S., Zamani, H., Bahreininejad, A., 2022. GGWO: Gaze Cues learning-based grey wolf optimizer and its applications for solving engineering problems. J. Comput. Sci. 101636.

Naghibi, S.A., Hashemi, H., Berndtsson, R., Lee, S., 2020. Application of extreme gradient boosting and parallel random forest algorithms for assessing groundwater spring potential using DEM-derived factors. J. Hydrol. 589, 125197.

Nhu, V.H., Hoang, N.D., Nguyen, H., Ngo, P.T.T., Thanh Bui, T., Hoa, P.V., Samui, P., Tien Bui, D., 2020. Effectiveness assessment of Keras based deep learning with different robust optimization algorithms for shallow landslide susceptibility mapping at tropical area. Catena 188, 104458.

Oh, H.-J.H.J., Kadavi, P.R.P.R., Lee, C.-W.C.W., Lee, S., 2018. Evaluation of landslide susceptibility mapping by evidential belief function, logistic regression and support vector machine models. Geomatics Nat. Hazards Risk 9, 1053–1070.

Owolabi, T.O., 2019. Modeling the magnetocaloric effect of manganite using hybrid genetic and support vector regression algorithms. Phys. Lett. Sect. A Gen. At. Solid State Phys. 383, 1782–1790.

Ozioko, O.H., Igwe, O., 2020. GIS-based landslide susceptibility mapping using heuristic and bivariate statistical methods for Iva Valley and environs Southeast Nigeria. Environ. Monit. Assess. 192.

Panahi, M., Gayen, A., Pourghasemi, H.R.H.R., Rezaie, F., Lee, S., 2020a. Spatial prediction of landslide susceptibility using hybrid support vector regression (SVR) and the adaptive neuro-fuzzy inference system (ANFIS) with various metaheuristic algorithms. Sci. Total Environ. 741, 139937.

Panahi, M., Sadhasivam, N., Pourghasemi, H.R.H.R., Rezaie, F., Lee, S., 2020b. Spatial prediction of groundwater potential mapping based on convolutional neural network (CNN) and support vector regression (SVR). J. Hydrol. 588, 125033.

Panahi, M., Khosravi, K., Ahmad, S., Panahi, S., Heddam, S., Melesse, A.M., Omidvar, E., Lee, C.-W., 2021. Cumulative infiltration and infiltration rate prediction using optimized deep learning algorithms: a study in Western Iran. J. Hydrol. Reg. Stud. 35, 100825.

Panahi, M., Rahmati, O., Rezaie, F., Lee, S., Mohammadi, F., Conoscenti, C., 2022. Application of the group method of data handling (GMDH) approach for landslide susceptibility zonation using readily available spatial covariates. Catena 208, 105779.

Patra, S., Mishra, P., Mahapatra, S.C., 2018. Delineation of groundwater potential zone for sustainable development: a case study from Ganga Alluvial Plain covering Hooghly district of India using remote sensing, geographic information system and analytic hierarchy process. J. Clean. Prod. 172, 2485–2502.

Pawley, S.M., Hartman, G., Chao, D., 2018. Modelling the Spatial Distribution of Landslide Susceptible Terrain in the Portions of the Interior Plains and Shield Regions. Canada, GeoEdmonton.

Peng, J., Li, Y., Kang, H., Shen, Y., Sun, X., Chen, Q., 2022. Impact of population topology on particle swarm optimization and its variants: an information propagation perspective. Swarm Evol. Comput. 69, 100990.

Qi, W., Wei, M., Yang, W., Xu, C., Ma, C., 2020. Automatic mapping of landslides by the ResU-Net. Remote Sens. 12, 1–14.

Rau, J.Y., Jhan, J.P., Rau, R.J., 2014. Semiautomatic object-oriented landslide recognition scheme from multisensor optical imagery and DEM. IEEE Trans. Geosci. Remote Sens. 52, 1336–1349.

Reichenbach, P., Rossi, M., Malamud, B.D.B.D., Mihir, M., Guzzetti, F., 2018. A review of statistically-based landslide susceptibility models. Earth Sci. Rev. 180, 60–91.

Rezaie, F., Kim, S.W., Alizadeh, M., Panahi, M., Kim, H., Kim, S., Lee, J., Lee, J., Yoo, J., Lee, S., 2021. Application of machine learning algorithms for geogenic radon potential mapping in Danyang-gun, South Korea. Front. Environ. Sci. 9.

Saini, S., Bt Awang Rambli, D.R., Zakaria, M.N.B., Bt Sulaiman, S., 2014. A review on particle swarm optimization algorithm and its variants to human motion tracking. Math. Probl Eng. 1–16.

Serikawa, Y., Miyajima, M., Yoshida, M., Matsuno, K., 2019. Inclination of houses induced by liquefaction in the 2018 Hokkaido Iburi-tobu earthquake, Japan. Geoenvironmental Disasters 6, 1–9.

Sheng, M., Wang, Z., Liu, W., Wang, X., Chen, S., Liu, X., 2022. A particle swarm optimizer with multi-level population sampling and dynamic p-learning mechanisms for large-scale optimization. Knowl. Based Syst. 242, 108382.

Singh, S., Bansal, J.C., 2022. Mutation-driven grey wolf optimizer with modified search mechanism. Expert Syst. Appl. 194, 116450.

Soares, L.P., Dias, H.C., Grohmann, C.H., 2020. Landslide Segmentation with U-Net: Evaluating Different Sampling Methods and Patch Sizes.

Sun, S., Liu, H., 2013. Particle swarm algorithm. In: Yang, X.-S., Xiao, R., Karamanoglu, M., Cui, Z., Gandomi, A.H. (Eds.), Swarm Intelligence and Bio-Inspired Computation. Elsevier, pp. 137–168.

Tehrany, M.S., Jones, S., Shabani, F., 2019. Identifying the essential flood conditioning factors for flood prone area mapping using machine learning techniques. Catena 175, 174–192.

Umbarkar, A.J., Balande, U.T., Seth, P.D., 2017. Performance Evaluation of Firefly Algorithm with Variation in Sorting for Non-linear Benchmark Problems, p. 020032.

Vapnik, V.N., 1995. The Nature of Statistical Learning Theory, the Nature of Statistical Learning Theory. Springer New York, New York, NY.

Wang, J., Li, L., Niu, D., Tan, Z., 2012. An annual load forecasting model based on support vector regression with differential evolution algorithm. Appl. Energy 94, 65–70. https://doi.org/10.1016/j.apenergy.2012.01.010.

Wang, L., Xia, Y., Lu, Y., 2022. A novel forecasting approach by the GA-SVR-GRNN hybrid deep learning algorithm for oil future prices. Comput. Intell. Neurosci. 1–12.

Xing, Y., Yue, J., Guo, Z., Chen, Y., Hu, J., Travé, A., 2021. Large-scale landslide susceptibility mapping using an integrated machine learning model: a case study in the Lvliang Mountains of China. Front. Earth Sci. 9, 1–15.

Xiong, G., Shuai, M., Hu, X., 2022. Combined heat and power economic emission dispatch using improved bare-bone multi-objective particle swarm optimization. Energy 244, 123108.

Yamagishi, H., Yamazaki, F., 2018. Landslides by the 2018 Hokkaido Iburi-tobu earthquake on September 6. Landslides 15, 2521–2524.

Yan, B., Zhao, Z., Zhou, Y., Yuan, W., Li, J., Wu, J., Cheng, D., 2017. A particle swarm optimization algorithm with random learning mechanism and Levy flight for optimization of atomic clusters. Comput. Phys. Commun. 219, 79–86.

Yin, L., Sun, Z., 2022. Distributed multi-objective grey wolf optimizer for distributed multi-objective economic dispatch of multi-area interconnected power systems. Appl. Soft Comput. 117, 108345.

Youssef, A.M., Pradhan, B., Dikshit, A., Al-Katheri, M.M., Matar, S.S., Mahdi, A.M., 2022. Landslide susceptibility mapping using CNN-1D and 2D deep learning algorithms: comparison of their performance at Asir Region. KSA. Bull. Eng. Geol. Environ. 81, 165.

Zhang, Z., Hong, W.-C., 2021. Application of variational mode decomposition and chaotic grey wolf optimizer with support vector regression for forecasting electric loads. Knowl. Base Syst. 228, 107297.

Zhang, L., Zhao, L., 2021. High-quality face image generation using particle swarm optimization-based generative adversarial networks. Future Generat. Comput. Syst. 122, 98–104.

Zhang, S., Li, R., Wang, F., Iio, A., 2019. Characteristics of landslides triggered by the 2018 Hokkaido Eastern Iburi earthquake, northern Japan. Landslides 16, 1691–1708.

Zhang, B., Li, K., Hu, Y., Ji, K., Han, B., 2022a. Prediction of backfill strength based on support vector regression improved by grey wolf optimization. J. Shanghai Jiaotong Univ.

Zhang, H., Lian, Q., Zhao, J., Wang, Y., Yang, Y., Feng, S., 2022b. RatUNet: residual U-Net based on attention mechanism for image denoising. PeerJ Comput. Sci. 8, e970.

Index

Note: 'Page numbers followed by *f* indicates figures, *t* indicates tables.'

E

F

M

T

www.ingramcontent.com/pod-product-compliance
Lightning Source LLC
LaVergne TN
LVHW080006210926
848481LV00011B/2953